THE ELEMENTS OF

Polymer Science and Engineering

Second Edition

An Introductory Text and Reference
for Engineers and Chemists

THE ELEMENTS OF

Polymer Science and Engineering

Second Edition

An Introductory Text and Reference
for Engineers and Chemists

Alfred Rudin

University of Waterloo

ACADEMIC PRESS

San Diego London Boston
New York Sydney Tokyo Toronto

Copyright © 1999 by Academic Press

ACADEMIC PRESS
a division of Harcourt Brace & Company
525 B Street, Suite 1900, San Diego, CA 92101-4495, USA
http://www.apnet.com

ACADEMIC PRESS
24–28 Oval Road, London NW1 7DX, UK
http://www.hbuk.co.uk/ap/

Library of Congress Cataloging-in-Publication Data
Rudin, Alfred
    The elements of polymer science and engineering /Alfred Rudin.—2nd ed.
        p.   cm.
    Includes bibliographical references and index.
    ISBN 0-12-601685-2 (acid-free paper)
    1. Polymers.   2. Polymerization.
QD381.R8   1998
547′.7—dc21   98-11623
                                                                      CIP

Printed in the United States of America
98 99 00 01 02 MV 9 8 7 6 5 4 3 2 1

To Pearl,
with thanks again
for her wisdom and wit

Contents

Chapter 2   **Basic Principles of Polymer Molecular Weights**

Chapter 3   **Practical Aspects of Molecular Weight Measurements**

Chapter 4   Effects of Polymer Isomerism and Conformational Changes

Chapter 5   Step-Growth Polymerizations

Chapter 6   **Free-Radical Polymerization**

Chapter 7   **Copolymerization**

Chapter 8   **Dispersion and Emulsion Polymerizations**

Chapter 9 Ionic and Coordinated Polymerizations

Chapter 10 Polymer Reaction Engineering

Chapter 11   Mechanical Properties of Polymer Solids and Liquids

Chapter 12   Polymer Mixtures

Preface

*Unprovided with original learning, uninformed in the habits of thinking,
unskilled in the arts of composition, I resolved to write a book.*

—Edward Gibbon

This introductory text is intended as the basis for a two- or three-semester course in synthetic polymers. It can also serve as a self-instruction guide for engineers and scientists without formal training in the subject who find themselves working with polymers. For this reason, the material covered begins with basic concepts and proceeds to current practice, where appropriate.

Space does not permit any attempt to be comprehensive in a volume of reasonable size. I have tried, instead, to focus on those elements of polymer science and technology that are somewhat different from the lines of thought in regular chemistry and chemical engineering and in which a student may need some initial help. Few of the ideas dealt with in this text are difficult, but some involve a mental changing of gears to accommodate the differences between macro- and micromolecules. If this text serves it purpose, it will prepare the reader to learn further from more specialized books and from the research and technological literature.

An active, developing technology flourishes in synthetic polymers because of the great commercial importance of these materials. This technology teaches much that is of value in understanding the basic science and engineering of macromolecules, and the examples in this text are taken from industrial practice.

Polymer molecular weight distributions and averages seem to me to be widely quoted and little understood, even though there is nothing particularly difficult in the topic. This may be because many textbooks present the basic equations for

$\overline{M}_n$, $\overline{M}_w$, and so on with no explanation of their origin or significance. It is regrettable that much good effort is defeated because a worker has an incorrect or imperfect understanding of the meaning and limitations of the molecular weight information at his or her disposal. Chapter 3 focuses on the fundamentals of molecular weight statistics and the measurement of molecular weight averages. It will be more use to a reader who is actively engaged in such experimentation than to a beginning student, and some instructors may wish to treat the material in this chapter very lightly.

I have included an introduction to rubber elasticity in Chapter 4 because it follows logically from considerations of conformational changes in polymers. This material need not be taught in the sequence presented, however, and this topic, or all of Chapter 4, can be introduced at any point that seems best to the instructor. Chapters 5 through 7 are quite orthodox in their plans. I have, however, taken the opportunity to present alkyd calculations as an example of practice in the coatings industry and formulating thermosetting materials.

Chapter 7 deals with free radical copolymerizations. This area has been considerably "worked over" for some years. Recent developments have shown, however, that many of our concepts in this area need reexamination, and I have tried to provide a critical introductory picture of the state of this field. Emulsion polymerization, which is dealt with in Chapter 8, is in an intense state of fermentation, spurred by technological advances and mechanistic insights. I have dispensed with the neat mathematical description of the Smith–Ewart emulsion polymerization model that was given in the first edition of this text, because this theory no longer reflects modern thinking. Newer understanding of this important process can be expressed mathematically, but I believe that would be dauntingly complex for novice students and practitioners. Instead, this chapter is mainly a "how-to" primer, with the object of introducing the reader to the many opportunities that are offered by this versatile technique. The material in this chapter may not be very useful to students who will not be in a position to try the polymerization and some instructors may wish to treat this topic very lightly. I have updated Chapter 9, with particular attention paid to metallocene catalysis, because of the great current importance and potential of this subject.

Chapter 10 is a modest attempt to introduce polymerization reaction engineering. I would hope that this subject will be interesting and useful not just to engineers but to scientists as well, because it is always informative to see how basic concepts are applied in practice. In this connection, the student and practitioner should realize that there are lessons to be learned not only in industrial versions of laboratory-scale reactions but also in how and why some processes are not used. Computer modeling of polymerization processes has not been included in this chapter because of space limitations.

Chapter 11 treats the basic elements of the mechanical properties of polymeric solids and melts. Topics such as fracture mechanics and rheology are touched on lightly, because of their importance to polymer applications. Some instructors may prefer to skip this material, depending on the orientation of their classes. Polymer mixtures, which are of great commercial importance, are treated in Chapter 12. There are two main schools of thought in this area: the scientists who study the statistical thermodynamics of polymer mixtures and the technologists who make and use blends. Neither pays much attention to the other; I have tried to introduce some basic elements of both viewpoints in this chapter to show that each can benefit from the other.

The only references included here are those dealing with particular concepts in greater detail than this text. This omission is not meant to imply that the ideas that are not referenced are my own, any more than the concepts in a general chemistry textbook are those of the author of that book. I lay full claim to the mistakes, however.

The units in this book are not solely in SI terms, although almost all the quantities used are given in both SI and older units. Many active practitioners have developed intuitive understandings of the meanings and magnitudes of certain quantities in non-SI units, and it seems to be a needless annoyance to change these parameters completely and abruptly.

The problems at the end of each chapter are intended to illustrate and expand the text material. A student who understands the material in the chapter should not find these problems time consuming. The problems have been formulated to require numerical rather than essay-type answers, as far as possible, since "hand-waving" does not constitute good engineering or science. The instructor may find an incidental advantage in that answers to such problems can be graded relatively rapidly with the aid of the *Solutions Manual.*

My thanks go to all the students who have endured this course before and after the writing of the first edition, to the scientists and engineers whose ideas and insights form the sum of our understanding of synthetic polymers, and to the users who kindly pointed out errors in the first edition.

Alfred Rudin

Chapter 1

Introductory Concepts and Definitions

Knowledge is a treasure, but practice is the key to it.
—Thomas Fuller, *Gnomologia*

1.1 SOME DEFINITIONS

Some basic concepts and definitions of terms used in the polymer literature are reviewed in this chapter. Much of the terminology in current use in polymer science has technological origins, and some meanings may therefore be understood by convention as well as by definition. Some of these terms are included in this chapter since a full appreciation of the behavior and potential of polymeric materials requires acquaintance with technical developments as well as with the more academic fundamentals of the field. An aim of this book is to provide the reader with the basic understanding and vocabulary for further independent study in both areas.

Polymer technology is quite old compared to polymer science. For example, natural rubber was first masticated to render it suitable for dissolution or spreading on cloth in 1820, and the first patents on vulcanization appeared some twenty years later. About another one hundred years were to elapse, however, before it was generally accepted that natural rubber and other polymers are composed of giant covalently bonded molecules that differ from "ordinary" molecules primarily only in size. (The historical development of modern ideas of polymer constitution is traced by Flory in his classical book on polymer chemistry [1], while Brydson [2] reviews the history of polymer technology.) Since some of the terms we are going to review derive from technology, they are less precisely defined than those the reader may have learned in other branches of science. This should not be cause for alarm, since all the more important definitions that follow are clear in the contexts in which they are normally used.

1.1.1   Polymer

Polymer means "many parts" and designates a large molecule made up of smaller repeating units. Thus the structure of polystyrene can be written

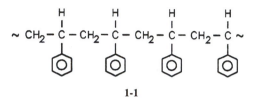

1-1

Polymers generally have molecular weights greater than about 5000 but no firm lower limit need be defined since the meaning of the word is nearly always clear from its use. The word *macromolecule* is a synonym for polymer.

1.1.2   Monomer

A monomer is a molecule that combines with other molecules of the same or different type to form a polymer. Acrylonitrile, $CH_2{=}CHCN$, is the monomer for polyacrylonitrile:

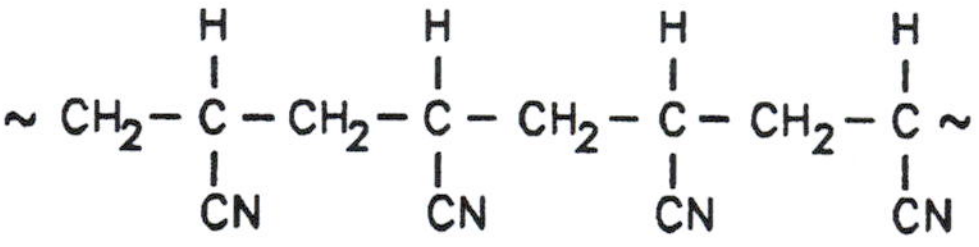

$$\sim CH_2-\underset{\underset{CN}{|}}{\overset{\overset{H}{|}}{C}}-CH_2-\underset{\underset{CN}{|}}{\overset{\overset{H}{|}}{C}}-CH_2-\underset{\underset{CN}{|}}{\overset{\overset{H}{|}}{C}}-CH_2-\underset{\underset{CN}{|}}{\overset{\overset{H}{|}}{C}}\sim$$

1-2

which is the basic constituent of "acrylic" fibers.

1.1.3   Oligomer

An oligomer is a low-molecular-weight polymer. It contains at least two monomer units. Hexatriacontane ($n\text{-}CH_3{-}(CH_2)_{29}{-}CH_3$) is an oligomer of polyethylene

$$\text{\small ⋀⋀ } CH_2CH_2CH_2CH_2CH_2CH_2CH_2CH_2CH_2 \text{ ⋀⋀:}$$

1-3

Generally speaking, a species will be called *polymeric* if articles made from it have significant mechanical strength and oligomeric if such articles are not strong enough to be practically useful. The distinction between the sizes of oligomers and the corresponding polymers is left vague, however, because there is no sharp transition in most properties of interest.

The terms used above stem from Greek roots: *meros* (part), *poly* (many), *oligo* (few), and *mono* (one).

1.1.4   Repeating Unit

The repeating unit of a linear polymer (which is defined below) is a portion of the macromolecule such that the complete polymer (except for the ends) might be produced by linking a sufficiently large number of these units through bonds between specified atoms.

The repeating unit may comprise a single identifiable precursor as in polystyrene (**1-1**), polyacrylonitrile (**1-2**), polyethylene (**1-3**), or poly(vinyl chloride):

$$\sim CH_2-\underset{\underset{Cl}{|}}{\overset{\overset{H}{|}}{C}}-CH_2-\underset{\underset{Cl}{|}}{\overset{\overset{H}{|}}{C}}-CH_2-\underset{\underset{Cl}{|}}{\overset{\overset{H}{|}}{C}}-CH_2-\underset{\underset{Cl}{|}}{\overset{\overset{H}{|}}{C}}\sim$$

1-4

A repeating unit may also be composed of the residues of several smaller molecules, as in poly(ethylene terephthalate):

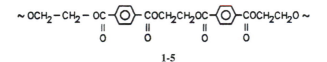

1-5

or poly(hexamethylene adipamide), nylon-6,6:

$$\sim \underset{H}{\overset{H}{N}}-(CH_2)_6-\overset{H}{N}-\underset{O}{\overset{}{C}}-(CH_2)_4-\underset{O}{\overset{}{C}}-\overset{H}{N}-(CH_2)_6-\overset{H}{N}-\underset{O}{\overset{}{C}}-(CH_2)_4-\underset{O}{\overset{}{C}}\sim$$

1-6

The polymers that have been mentioned to this point are actually synthesized from molecules whose structures are essentially those of the repeating units shown. It is not necessary for the definition of the term *repeating unit*, however, that such a

synthesis be possible. For example,

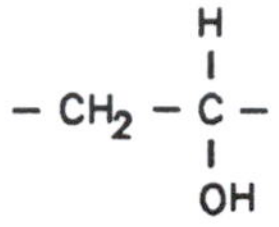

$$-CH_2 - \underset{\underset{OH}{|}}{\overset{\overset{H}{|}}{C}} -$$

1-7

is evidently the repeating unit of poly(vinyl alcohol):

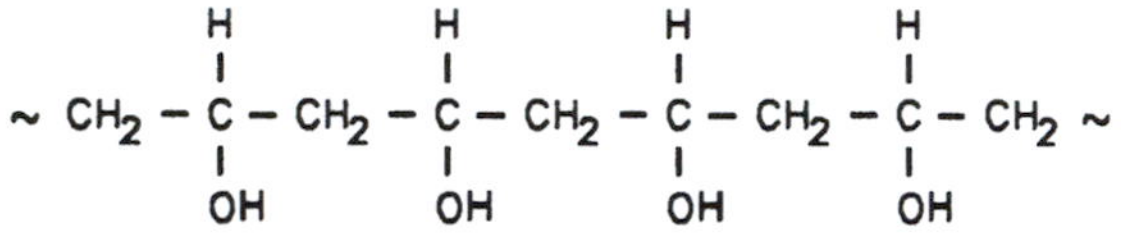

$$\sim CH_2 - \underset{\underset{OH}{|}}{\overset{\overset{H}{|}}{C}} - CH_2 - \underset{\underset{OH}{|}}{\overset{\overset{H}{|}}{C}} - CH_2 - \underset{\underset{OH}{|}}{\overset{\overset{H}{|}}{C}} - CH_2 - \underset{\underset{OH}{|}}{\overset{\overset{H}{|}}{C}} - CH_2 \sim$$

1-8

The ostensible precursor for this polymer is vinyl alcohol,

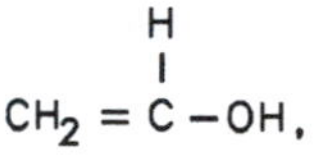

$$CH_2 = \underset{}{\overset{\overset{H}{|}}{C}} - OH,$$

1-9

which does not exist (it is the unstable tautomer of acetaldehyde). Poly(vinyl alcohol), which is widely used as a water-soluble packaging film and suspension agent and as an insolubilized fiber, is instead made by linking units of vinyl acetate,

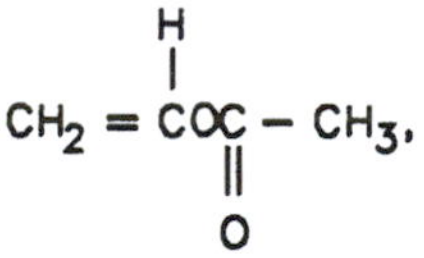

$$CH_2 = \underset{\underset{O}{\|}}{\overset{\overset{H}{|}}{C}}OC - CH_3,$$

1-10

and subsequently subjecting the poly(vinyl acetate) polymer to alcoholysis with ethanol or methanol.

Similarly, protein fibers such as silk are degradable to mixtures of amino acids, but a direct synthesis of silk has not yet been accomplished. Another polymer for which there is no current synthetic method is cellulose, which is composed of β-1,4-linked D-glucopyranose units:

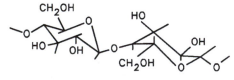

1-11

The concept of an identifiable simple repeating unit loses some of its utility with polymers that are highly branched, with species that consist of interconnected branches, or with those macromolecules that are synthesized from more than a few different smaller precursor monomers. Similar difficulties arise when the final polymeric structure is built up by linking different smaller polymers. This limitation to the definition will become clearer in context when some of these polymer types, like alkyds (Section 1.6), are discussed later in the text. Any deficiency in the general application of the term is not serious, since the concept of a repeating unit is in fact only employed where such groupings of atoms are readily apparent.

1.1.5 Representations of Polymer Structures

Polymer structures are normally drawn as follows by showing only one repeating unit. Each representation below is equivalent to the corresponding structure that has been depicted above:

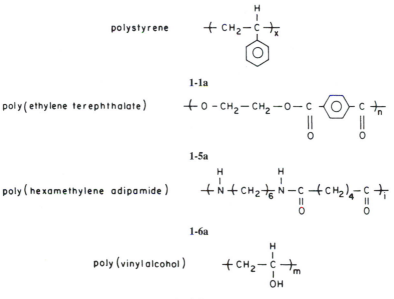

The subscripts x, n, i, m, and so on represent the number of repeating units in the polymer molecule. This number is often not known definitely in commercial synthetic polymer samples, for reasons which are explained later in the text.

This representation of polymer structures implies that the whole molecule is made up of a sequence of such repeating units by linking the left-hand atom shown to the right-hand atom, and so on. Thus, the following structures are all equivalent to **1-5**.

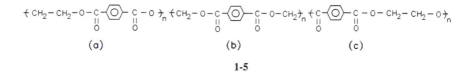

(a)   (b)   (c)

1-5

Formula **(b)** would not normally be written because the nominal break is in the middle of a $-OCH_2CH_2O-$unit, which is the residue of one of the precursors ($HOCH_2CH_2OH$) actually used in syntheses of this polymer by the reaction:

n HOCH$_2$ – CH$_2$ OH + n HOOC —◯— COOH ⟶ $\left(\text{O} - CH_2 - CH_2 - O - \underset{\text{O}}{\overset{\text{O}}{C}} - \text{O} - \underset{\text{O}}{\overset{\text{O}}{C}}\right)_n$ + 2nH$_2$O   (1-1)

1.1.6   End Groups

The exact nature of the end groups is frequently not known and the polymer structure is therefore written only in terms of the repeating unit, as in the foregoing structural representations. End groups usually have negligible effect on polymer properties of major interest. For example, most commercial polystyrenes used to made cups, containers, housings for electrical equipment, and so on have molecular weights of about 100,000. An average polymer molecule will contain 1000 or more styrene residues, compared to two end units.

1.2   DEGREE OF POLYMERIZATION

The term *degree of polymerization* refers to the number of repeating units in the polymer molecule. The degree of polymerization of polyacrylonitrile is y. The definition given here is evidently useful only for polymers which have regular identifiable repeating units.

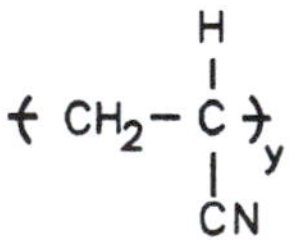

$$\left(CH_2 - \underset{CN}{\overset{H}{\underset{|}{\overset{|}{C}}}} \right)_y$$

1-2a

The term *degree of polymerization* is also used in some contexts in the polymer literature to mean the number of monomer residues in an average polymer molecule. This number will be equal to the one in the definition stated above if the repeating unit is the residue of a single monomer. The difference between the two terms is explained in more detail in Section 5.4.2, where it will be more readily understood. To that point in this book the definition given at the beginning of this section applies without qualification. We use the abbreviation DP for the degree of polymerization defined here and X for the term explained later. (The coining of a new word for one of these concepts could make this book clearer, but it might confuse the reader's understanding of the general literature where the single term *degree of polymerization* is unfortunately used in both connections.)

The relation between degree of polymerization and molecular weight, M, of the same macromolecule is given by

$$M = (DP)M_0 \qquad (1\text{-}2)$$

where M_0 is the formula weight of the repeating unit.

1.3   POLYMERIZATION AND FUNCTIONALITY

1.3.1   Polymerization

Polymerization is a chemical reaction in which the product molecules are able to grow indefinitely in size as long as reactants are supplied. Polymerization can occur if the monomers involved in the reaction have the proper functionalities.

1.3.2   Functionality

The functionality of a molecule is the number of sites available for bonding to other molecules *under the specific conditions* of the polymerization reaction.

A *bifunctional monomer* can be linked to two other molecules under appropriate conditions. Examples are

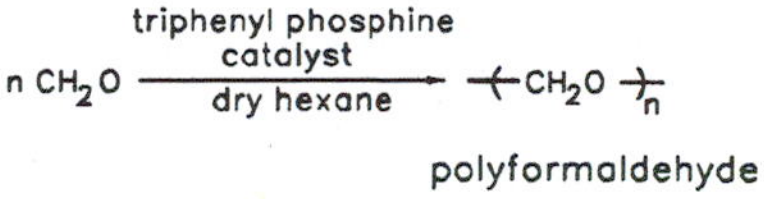

$$n\ CH_2O \xrightarrow[\text{dry hexane}]{\substack{\text{triphenyl phosphine}\\ \text{catalyst}}} \left(CH_2O \right)_n \qquad (1\text{-}3)$$

polyformaldehyde

1-12

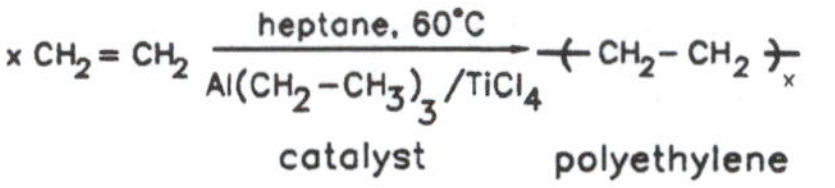

$$x\ CH_2 = CH_2 \xrightarrow[\substack{Al(CH_2-CH_3)_3/TiCl_4 \\ catalyst}]{heptane,\ 60°C} \xrightarrow{} \left(\!CH_2-CH_2\!\right)_x$$

$$\tag{1-4}$$

polyethylene

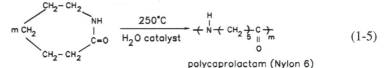

$$\tag{1-5}$$

polycaprolactam (Nylon 6)

1-13

$$\tag{1-6}$$

poly (phenylene oxide)

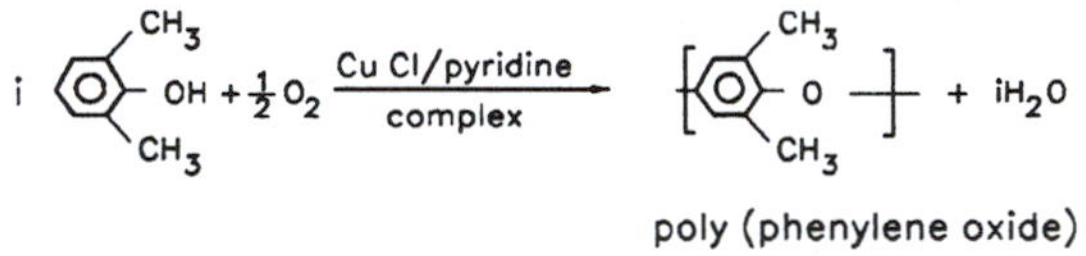

1-14

poly (ethylene terephthalate)

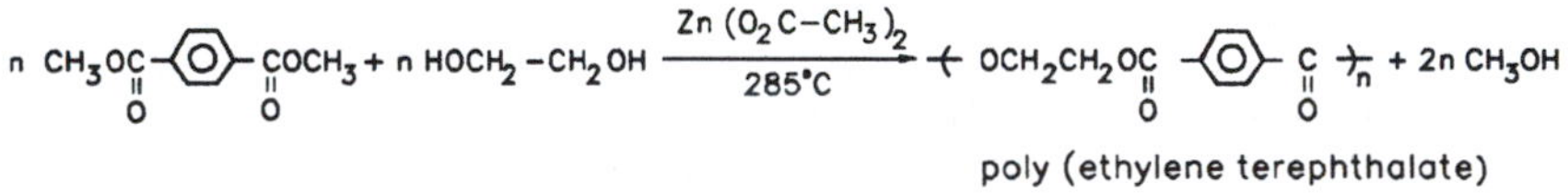

$$\tag{1-7}$$

A *polyfunctional monomer* can react with more than two other molecules to form the corresponding number of new valence bonds during the polymerization reaction. Examples are divinyl benzene,

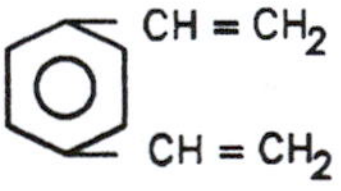

1-15

in reactions involving additions across carbon–carbon double bonds and glycerol or pentaerythritol ($C(CH_2OH)_4$) in esterifications or other reactions of alcohols.

If an a-functional monomer reacts with a b-functional monomer in a nonchain reaction, the functionality of the product molecule is $a + b - 2$. This is because every new linkage consumes two bonding sites. Production of a macromolecule in such reactions can occur only if a and b are both greater than one.

The following points should be noted:

1.   Use of the term *functionality* here is not the same as in organic chemistry where a carbon–carbon double bond, for example, is classified as a single functional group.

2.   Functionality refers in general to the *overall* reaction of monomers to yield products. It is not used in connection with the individual steps in a reaction sequence. The free radical polymerization of styrene, for example, is a chain reaction in which a single step involves attack of a radical with ostensible functionality of 1 on a monomer with functionality 2. The radical is a transient species, however, and the net result of the chain of reactions is linkage of styrene units with each other so that the process is effectively polymerization of these bifunctional monomers.

3.   Functionality is defined only for a given reaction. A glycol, HOROH, has a functionality of 2 in esterification or ether-forming reactions, but its functionality is zero in amide-forming reactions. The same is true of 1,3-butadiene.

$$CH_2 = \overset{\overset{\displaystyle H}{|}}{C} - \overset{\overset{\displaystyle H}{|}}{C} = CH_2$$

1-16

which may have a functionality of 2 or 4, depending on the particular double-bond addition reaction.

4.   The condition that monomers be bi- or polyfunctional is a necessary, but not sufficient, condition for polymerization to occur in practice. Not all reactions between polyfunctional monomers actually yield polymers. The reaction must also proceed cleanly and with good yield to give high-molecular-weight products. For example, propylene has a functionality of 2 in reactions involving the double bond, but free-radical reactions do not produce macromolecules whereas polymerization in heptane at $70°C$ with an $Al(CH_2CH_3)_2Cl/TiCl_3$ catalyst does yield high polymers.

5.   Functionality is a very useful concept in polymer science, and we use it later in this book. There are, however, other definitions than the one given here. All are valuable in their proper contexts.

1.3.3   Latent Functionality

Important commercial use is made of monomers containing functional groups that react under different conditions. This allows chemical reactions on polymers after they have been shaped into desired forms. Some examples follow.

(i)   Vulcanization

Isoprene,

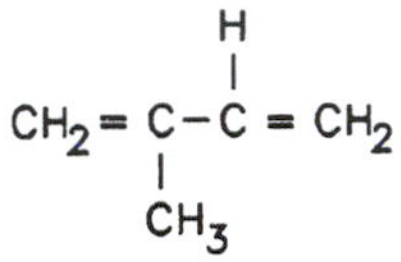

$$CH_2 = \underset{\underset{CH_3}{|}}{C} - \underset{\overset{|}{H}}{C} = CH_2$$

1-17

can be polymerized at about 50°C in *n*-pentane with either butyl lithium or titanium tetrachloride/triisobutyl aluminum catalysts. The product, which is the cis form of 1,4-polyisoprene,

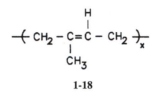

1-18

has a structure close to that of natural rubber. The residual double bond in each re-peating unit is essentially inert toward further polymerization under these reaction conditions, but it activates the repeating unit for subsequent reaction with sulfur during vulcanization after the rubber has been compounded with other ingredients and shaped into articles like tires. The chemistry of the vulcanization reaction is complex and depends partly on ingredients in the mix other than just sulfur and rubber. Initially discrete rubber molecules become chemically linked by structures such as

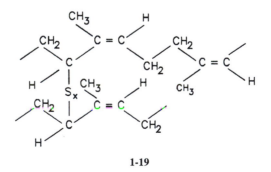

1-19

Here *x* is 1 or 2 in efficient vulcanization systems but may be as high as 8 under other conditions where cyclic and other structures are also formed in the reaction. The rubber article is essentially fixed in shape once it is vulcanized, and it is necessary that this cross-linking reaction not occur before the rubber has been molded into form.

(ii)   Epoxies

Epoxy polymers are used mainly as adhesives, surface coatings, and in combination with glass fibers or cloth, as lightweight, rigid structural materials. Many different epoxy prepolymers are available. (A prepolymer is a low-molecular-weight polymer that is reacted to increase its molecular size. Such reactions are usually carried out during or after a process that shapes the material into desired form. The polyisoprene mentioned in Section 1.3.3.i is not called a prepolymer, since the average molecule contains about 1000 repeating units before vulcanization.) The most widely used epoxy prepolymer is made by condensing epichlorohydrin and bisphenol A (2,2′-bis(4-hydroxyphenyl)propane). The degree of polymerization of prepolymer **1-20** may be varied to produce liquids (as when $x = 0$) or solids which soften at temperatures up to about 140°C (at $x = 14$).

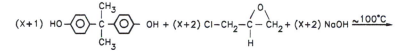

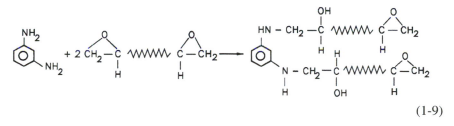

$$+ (X+2)H_2O + (X+2)NaCl$$

1-20

$$(1\text{-}8)$$

The prepolymer can be reacted further with a wide variety of reagents, and its latent functionality depends on the particular reaction. The conversion of an epoxy polymer to an interconnected network structure is formally similar to the vulcanization of rubber, but the process is termed *curing* in the epoxy system. When the epoxy "hardener" is a primary or secondary amine like *m*-phenylene diamine the main reaction is

$$(1\text{-}9)$$

The unreacted terminal epoxide groups can react with other diamine molecules to form a rigid network polymer. In this reaction the functionality of the bisphenol A–epichlorohydrin prepolymer **1-20** will be 2 since hydroxyl groups are not involved and the functionality of each epoxide group is one.

When the hardening reaction involves cationic polymerization induced by a Lewis acid, however, the functionality of each epoxy group is 2 and that of structure **1-20** is 4. The general hardening reaction is illustrated in Eq. (1-10) for initiation by BF_3, which is normally used in this context as a complex with ethylamine, for easier handling.

These examples barely touch the wide variety of epoxy polymer structures and curing reactions. They illustrate the point that the latent functionality of the prepolymers and the chemical and mechanical properties of the final polymeric structures will vary with the choice of ingredients and reaction conditions.

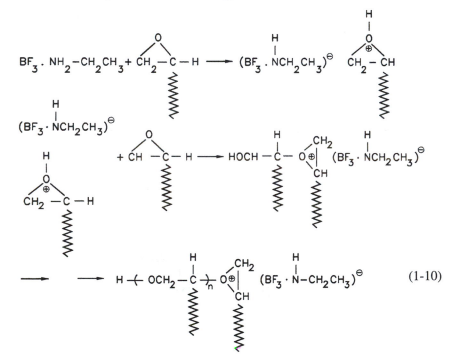

$$(1\text{-}10)$$

1.4 WHY ARE SYNTHETIC POLYMERS USEFUL? [3]

The primary valence bonds and intermolecular forces in polymer samples are exactly the same as those in any other chemical species, but polymers form strong plastics, fibers, rubbers, coatings and adhesives, whereas conventional chemical compounds are useless for the same applications. For example, n-hexatriacontone (molecular weight 437), which was mentioned earlier as an oligomer of polyethylene, forms weak, friable crystals but the chemically identical material polyethylene (**1-3**) can be used to make strong films, pipe, cable jackets, bottles, and so on, provided the polymer molecular weight is at least about 20,000. Polymers are useful materials mainly because their molecules are very large.

The intermolecular forces in hexatriacontane and in polyethylene are essentially van der Waals attractions. When the molecules are very large, as in a polymer, there are so many intermolecular contacts that the sum of the forces holding each molecule to its neighbors is appreciable. It becomes difficult to break a polymeric material since this involves separating the constituent molecules. Deformation of a polymeric structure requires that the molecules move past each other, and this too requires more force as the macromolecules become larger.

Not surprisingly, polymers with higher intermolecular attractions develop more strength at equivalent molecular weight than macromolecules in which intermolecular forces are weaker. Polyamides (nylons) are characterized by the structure

$$\underset{x}{\left(\overset{\overset{\displaystyle H}{|}}{N} - R - \overset{\overset{\displaystyle H}{|}}{N} - \underset{\underset{\displaystyle O}{\|}}{C} - R' - \underset{\underset{\displaystyle O}{\|}}{C} \right)}$$

1-21

in which hydrogen-bonding interactions are important and are mechanically strong at lower degrees of polymerization than hydrocarbon polymers like polyethylene.

A crude, but useful, generalization of this concept can be seen in Fig. 1-1, where a mechanical property is plotted against the average number of repeating units in the polymer. The property could be the force needed to break a standard specimen or any of a number of other convenient characteristics. The intercepts on the abscissa correspond to molecular sizes at which zero strength would be detected by test methods normally used to assess such properties of polymers. (Finite strengths may, of course, exist at lower degrees of polymerization, but they would not be measurable without techniques which are too sensitive to be useful with practical polymeric materials.) "Zero-strength" molecular sizes will be inversely related to the strength of intermolecular attractive forces, as shown in the figure. In general, the more polar and more highly hydrogen-bonded molecules form stronger articles at lower degrees of polymerization.

The epoxy prepolymers mentioned above would lie in the zero-strength region of curve B in Fig. 1-1. They have low molecular weights and are therefore sufficiently fluid to be used conveniently in surface coatings, adhesives, matrices

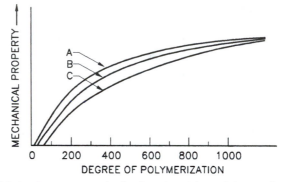

Fig. 1-1.   Relation between strength of polymeric articles and degree of polymerization.   A, aliphatic polyamides (e.g., **1-6**); B, aromatic polyesters (e.g., **1-5**); C, olefin polymers (e.g., **1-3**).

for glass cloth reinforcement, and in casting or encapsulation formulations. Their molecular sizes must, however, eventually be increased by curing reactions such as Eq. (1-9), (1-10), or others to make durable final products. Such reactions move the molecular weights of these polymers very far to the right in Fig. 1-1.

The general reaction shown in Eq. (1-8) can be modified and carried out in stages to produce fairly high-molecular-weight polymers without terminal epoxy groups:

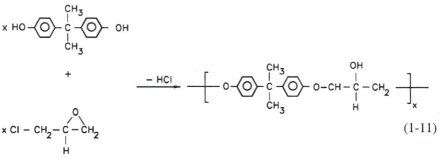

$$(1\text{-}11)$$

1-22

When x in formula **1-22** is about 100, the polymers are known as *phenoxy resins*. Although further molecular weight increase can be accomplished by reaction on the pendant hydroxyls in the molecule, commercial phenoxy polymers already have sufficient strength to be formed directly into articles. They would be in the finite-strength region of curve B in Fig. 1-1. (The major current use for these polymers is in zinc-rich coatings for steel automobile body panels.)

The curves in Fig. 1-1 indicate that all polymer types reach about the same strength at sufficiently high molecular weights. The sum of intermolecular forces on an individual molecule will equal the strength of its covalent bonds if the molecule is large enough. Most synthetic macromolecules have carbon–carbon, carbon–oxygen, or carbon–nitrogen links in their backbones and the strengths of these bonds do not differ very much. The ultimate strengths of polymers with extremely high molecular weights would therefore be expected to be almost equal.

This ideal limiting strength is of more theoretical than practical interest because a suitable balance of characteristics is more important than a single outstanding property. Samples composed of extremely large molecules will be very strong, but they cannot usually be dissolved or caused to flow into desired shapes. The viscosity of a polymer depends strongly on the macromolecular size, and the temperatures needed for molding or extrusion, for example, can exceed those at which the materials degrade chemically. Thus, while the size of natural rubber molecules varies with the source, samples delivered to the factory usually have molecular weights between 500,000 and 1,000,000. The average molecular weight and the viscosity of the rubber are reduced to tractable levels by masticating the polymer with chemicals that promote scission of carbon–carbon bonds and stabilize the fragmented ends. The now less viscous rubber can be mixed with other ingredients of the compound, formed, and finally increased in molecular weight and fixed in shape by vulcanization.

Figure 1-1 shows plateau regions in which further increases in the degree of polymerization have little significant effect on mechanical properties. While the changes in this region are indeed relatively small, they are in fact negligible only against the ordinate scale, which starts at zero. Variations of molecular sizes in the plateau-like regions are often decisive in determining whether a given polymer is used in a particular application.

Polyester fibers, for example, are normally made from poly(ethylene tereph-thalate) (**1-5**). The degree of polymerization of the polymer must be high for use in tire cord, since tires require high resistance to distortion under load and to bruising. The same high molecular weight is not suitable, however, for polymers to be used in injection molding applications because the polymer liquids are very viscous and crystallize slowly, resulting in unacceptably long molding cycle times. Poly(ethylene terephthalate) injection molding grades have degrees of polymer-ization typically less than half those of tire cord fiber grades and about 60% of the DPs of polymers used to make beverage bottles.

Since the strength of an article made of discrete polymer molecules depends on the sum of intermolecular attractions, it is obvious that any process that in-creases the extent to which such macromolecules overlap with each other will result in a stronger product. If the article is oriented, polymer molecules tend to become stretched out and mutually aligned. The number of intermolecular con-tacts is increased at the expense of intramolecular contacts of segments buried in the normal ball-like conformation of macromolecules. The article will be much stronger in the orientation direction. Examples are given below in Section 1.8, on fibers, since orientation is an important part of the process of forming such materials.

If the polymer molecule is stiff, it will have less tendency to coil up on itself, and most segments of a given molecule will contact segments of other macromolecules. A prime example is the aromatic polyamide structure:

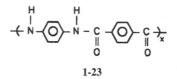

1-23

Dispersions of this polymer in sulfuric acid are spun into fibers which can be stretched to two or three times their original lengths. The products have extremely high strength, even at temperatures where most organic compounds are appreciably decomposed.

The range of molecular sizes in a polymer material is always a key parameter in determining the balance of its processing and performance properties, but these characteristics may also be affected by other structural features of the polymer. This is particularly the case with crystallizable polymers, such as polyethylene (**1-3**), where branching impedes crystallization and affects the stiffness and impact resistance of the final articles. Further mention of branching distributions is made in Section 3.4.6.

1.5   COPOLYMERS

A *homopolymer* is a macromolecule derived from a single monomer, whereas a *copolymer* contains structural units of two or more different precursors.

This distinction is primarily useful when the main chain of the macromolecule consists of carbon–carbon bonds. There is little point in labeling poly(ethylene terephthalate), **1-5**, a copolymer, since this repeating unit obviously contains the residues of two monomers and the polymer is made commercially only by reaction (1-1) or (1-7). Some polyesters are, however, made by substituting about 2 mol % of sodium-2,5-di(carboxymethyl) sulfonate (**1-24**) for the 3,5 isomer for the dimethyl terephthalate in reaction (1-7):

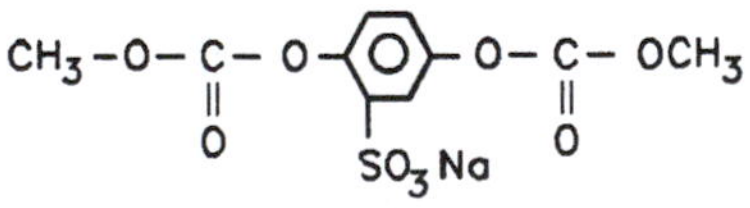

1-24

The eventual product, which is a modified polyester fiber with superior affinity for basic dyes, is often called a *copolymer* to distinguish it from conventional poly(ethylene terephthalate). This is a rather specialized use of the term, however, and we shall confine the following discussion to copolymers of monomers with olefinic functional groups.

The most important classes of copolymers are discussed next.

1.5.1 Random Copolymer

A *random copolymer* is one in which the monomer residues are located randomly in the polymer molecule. An example is the copolymer of vinyl chloride and vinyl acetate, made by free-radical copolymerization (Chapter 7):

$$
\begin{array}{c}
\text{WWW } CH_2 - \overset{\displaystyle H}{\underset{\displaystyle Cl}{C}} - CH_2 - \overset{\displaystyle H}{\underset{\displaystyle Cl}{C}} - CH_2 - \overset{\displaystyle H}{\underset{\displaystyle \underset{\displaystyle C=O}{O}}{C}} - CH_2 - \overset{\displaystyle H}{\underset{\displaystyle Cl}{C}} - CH_2 - \overset{\displaystyle H}{\underset{\displaystyle \underset{\displaystyle C=O}{O}}{C}} - CH_2 - \overset{\displaystyle H}{\underset{\displaystyle Cl}{C}} \text{ WWW}
\end{array}
$$

1-25

The vinyl acetate content of such materials ranges between 3 and 40%, and the copolymers are more soluble and pliable than poly(vinyl chloride) homopolymer. They can be shaped mechanically at lower temperatures than homopolymers with the same degree of polymerization and are used mainly in surface coatings and products where exceptional flow and reproduction of details of a mold surface are needed.

The term *random copolymer* is retained here because it is widely used in polymer technology. A better term in general is *statistical copolymer*. These are primarily copolymers that are produced by simultaneous polymerization of a mixture of two or more comonomers. They include *alternating copolymers*, described below, as well as *random copolymers*, which refer, strictly speaking, to materials in which the probability of finding a given monomer residue at any given site depends only on the relative proportion of that comonomer in the reaction mixture. The reader will find the two terms used interchangeably in the technical literature, but they are distinguished in more academic publications.

1.5.2 Alternating Copolymer

In an *alternating copolymer* each monomer of one type is joined to monomers of a second type. An example is the product made by free-radical polymerization of equimolar quantities of styrene and maleic anhydride:

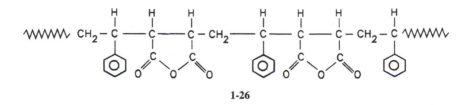

1-26

These low-molecular-weight polymers have a variety of special uses including the improvement of pigment dispersions in paint formulations.

1.5.3   Graft Copolymer

Graft copolymers are formed by growing one polymer as branches on another preformed macromolecule. If the respective monomer residues are coded A and B, the structure of a segment of a graft copolymer would be

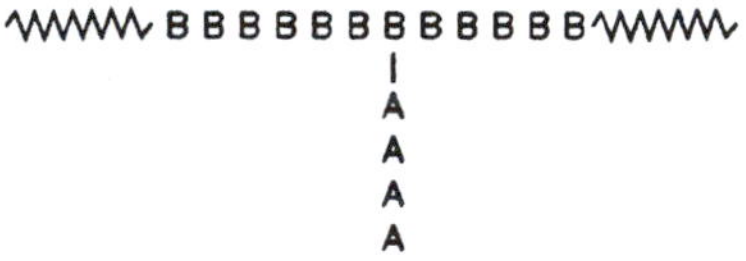

```
ᐟᐠᐟᐠᐟᐠ B B B B B B B B B B B B ᐠᐟᐠᐟᐠᐟ
                    |
                    A
                    A
                    A
                    A
```

1-27

The most important current graft copolymers include impact-resistant polystyrenes, in which a rubber, like polybutadiene (**1-28**) is dissolved in styrene:

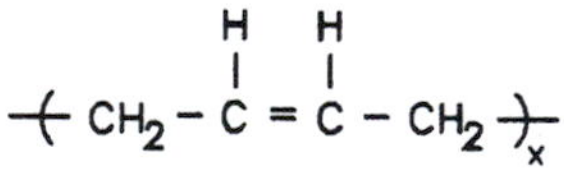

$$
\begin{array}{c}
\quad\quad H \quad\, H \\
\quad\quad | \quad\,\, | \\
\text{—}\!\!\!\left(\, CH_2 - C = C - CH_2 \,\right)\!\!\!\!_x
\end{array}
$$

1-28

When the styrene is polymerized by free-radical initiation, it reacts by adding across the double bonds of other styrene and rubber units, and the resulting product contains polystyrene grafts on the rubber as well as ungrafted rubber and polystyrene molecules. This mixture has better impact resistance than unmodified polystyrene.

A related graft polymerization is one of the preferred processes for manufacture of ABS (acrylonitrile-butadiene-styrene) polymers, which are generally superior to high-impact polystyrene in oil and grease resistance, impact strength, and maximum usage temperature. In this case, the rubber in a polybutadiene-in-water emulsion is swollen with a mixture of styrene and acrylonitrile monomers, which are then copolymerized *in situ* under the influence of a water-soluble free-radical initiator. The dried product is a blend of polybutadiene, styrene-acrylonitrile (usually called SAN) copolymer, and grafts of SAN on the rubber. The graft itself is a random copolymer.

1.5.4 Block Copolymer

Block copolymers have backbones consisting of fairly long sequences of different repeating units.

Elastic, so-called "Spandex" fibers, for example, are composed of long molecules in which alternating stiff and soft segments are joined by urethane

$$-\text{O}-\underset{\underset{\text{O}}{\|}}{\text{C}}-\underset{\text{H}}{\overset{}{\text{N}}}-,$$

1-29

and sometimes also by urea:

$$-\underset{\text{H}}{\overset{}{\text{N}}}-\underset{\underset{\text{O}}{\|}}{\text{C}}-\underset{\text{H}}{\overset{}{\text{N}}}-,$$

1-30

linkages. One variety is based on a hydroxyl-ended polytetrahydrofuran polymer (**1-31**) with a degree of polymerization of about 25, which is reacted with 4,4′-diphenylmethane diisocyanate:

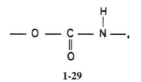

1-31

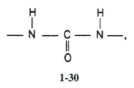   (1-12)

1-32

The isocyanate-ended prepolymer **1-32** is spun into fiber form and is simultaneously treated further with ethylene diamine in aqueous dimethylformamide:

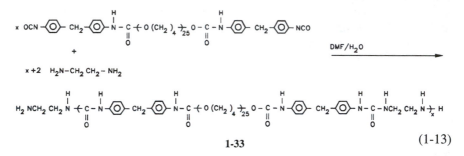

$$(1\text{-}13)$$

The polytetrahydrofuran blocks in the final structure **1-33** are soft segments which permit the molecules to uncoil and extend as the fiber is stretched. The urea linkages produced in reaction (1-13) form intermolecular hydrogen bonds which are strong enough to minimize permanent distortion under stress. The fibers made from this block copolymer snap back to their original dimensions after being elongated to four or five times their relaxed lengths.

1.6   MOLECULAR ARCHITECTURE

A *linear polymer* is one in which each repeating unit is linked only to two others. Polystyrene (**1-1**), poly(methyl methacrylate) (**1-34**), and poly(4-methyl pentene-1) (**1-35**) are called linear polymers although they contain short branches which are part of the monomer structure. By contrast, when vinyl acetate is polymerized by free-radical initiation, the polymer produced contains branches which were not present in the monomers. Some repeating units in these species are linked to three or four other monomer residues, and such polymers would therefore be classified as branched.

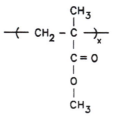

1-34

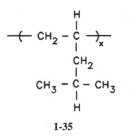

1-35

Branched polymers are those in which the repeating units are not linked solely in a linear array, either because at least one of the monomers has functionality greater than 2 or because the polymerization process itself produces branching points in a polymer that is made from exclusively bifunctional monomers.

An example of the first type is the polymer made, for instance, from glycerol, phthalic anhydride, and linseed oil. A segment of such a macromolecule might look like **1-36**:

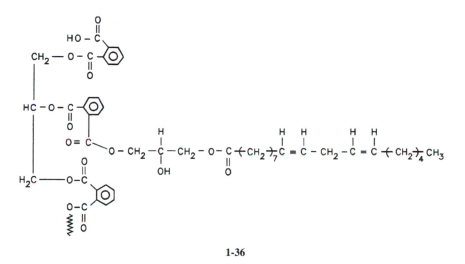

1-36

This is the structure of alkyd polymers, which are the reaction products of acids with di- and polyhydric alcohols. Such polymers are used primarily for surface coatings, which can be caused to react further *in situ* through residual hydroxyl, acid, or olefin groups.

The major example of the second branched polymer type is the polyethylene that is made by free-radical polymerization at temperatures between about 100 and 300°C and pressures of 1000–3000 atm (100–300 MPa). Depending on reaction conditions, these polymers will contain some 20 to 30 ethyl and butyl branches per 1000 carbon atoms and one or a few much longer branches per molecule. They differ sufficiently from linear polyethylene such that the two materials are generally not used for the same applications. Poly(vinyl acetate) polymers resemble polyethylene in that the conventional polymerization process yields branched macromolecules.

By convention, the term *branched* implies that the polymer molecules are discrete, which is to say that their sizes can be measured by at least some of the usual analytical methods described in Chapter 3. A *network polymer* is an interconnected branch polymer. The molecular weight of such polymers is infinite, in the sense that it is too high to be measured by standard techniques. If the average functionality of a mixture of monomers is greater than 2, reaction to sufficiently high conversion yields network structures (p. 174).

Network polymers can also be made by chemically linking linear or branched polymers. The process whereby such a preformed polymer is converted to a network structure is called *cross-linking*. *Vulcanization* is an equivalent term that is used mainly for rubbers. The rubber in a tire is cross-linked to form a network. The molecular weight of the polymer is not really infinite even if all the rubber in the tire is part of a single molecule (this is possible, at least in theory), since the size of the tire is finite. Its molecular weight is infinite, however, on the scale applied in polymer measurements, which require the sample to be soluble in a solvent.

The structure of a *ladder polymer* comprises two parallel strands with regular cross-links, as in polyimidazopyrrolone (**1-37**), which is made from pyromellitic dianhydride (**1-38**) and 1,2,4,5-tetraminobenzene (**1-39**).

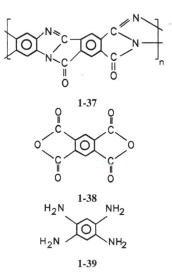

1-37

1-38

1-39

This polymer is practically as resistant as pyrolitic graphite to high temperatures and high-energy radiation.

Ladder polymers are double-strand linear polymers. Their permanence properties are superior even to those of conventional network polymers. The latter are randomly cross-linked, and their molecular weight can be reduced by random scission events. When a chemical bond is broken in a ladder polymer, however, the second strand maintains the overall integrity of the molecule and the fragments of the broken bond are held in such close proximity that the likelihood of their recombination is enhanced.

Space limitations do not permit the description of other varieties of rigid chain macromolecules, such as semiladder and spiro structures, which are of lesser current commercial importance.

1.7 THERMOPLASTICS AND THERMOSETS

A thermoplastic is a polymer which softens and can be made to flow when it is heated. It hardens on cooling and retains the shape imposed at elevated temperature. This heating and cooling cycle can usually by repeated many times if the polymer is properly compounded with stabilizers. Some of the polymers listed earlier which are thermoplastics are polystyrene (**1-1**), polyethylene (**1-3**), poly(vinyl chloride) (**1-4**), poly(ethylene terephthalate) (**1-5**), and so on.

A *thermosetting* plastic is a polymer that can be caused to undergo a chemical change to produce a network polymer, called a *thermoset polymer*. Thermosetting polymers can often be shaped with the application of heat and pressure, but the number of such cycles is severely limited. Epoxies, for which cross-linking reactions are illustrated in Eqs. (1-9) and (1-10), are thermosetting polymers. The structurally similar phenoxies (**1-22**) are usually not cross-linked and are considered to be thermoplastics.

A *thermoset plastic* is a solid polymer that cannot be dissolved or heated to sufficiently high temperatures to permit continuous deformation, because chemical decomposition intervenes at lower temperatures. Vulcanized rubber is an example.

The classification into thermoplastic and thermosetting polymers is widely used although the advances of modern technology tend to blur the distinction between the two. Polyethylene and poly(vinyl chloride) wire coverings and pipe can be converted to thermoset structures by cross-linking their molecules under the influence of high-energy radiation or free radicals released by decomposition of peroxides in the polymer compound. The main advantage of this cross-linking is enhanced dimensional stability under load and elevated temperatures. Polyethylene and poly(vinyl chloride) are classed as thermoplastics, however, since their major uses hinge on their plasticity when heated.

1.8 ELASTOMERS, FIBERS, AND PLASTICS

Polymers can be usefully classified in many ways, such as by source of raw materials, method of synthesis, end use, and fabrication processes. Some classifications have already been considered in this chapter. Polymers are grouped by end use in this section, which brings out an important difference between macromolecules and other common materials of construction. This is that the chemical structure and size of a polymeric species may not completely determine the properties of an article made from such a material. The process whereby the article is made may also exert an important influence.

The distinction between elastomers, fibers, and plastics is most easily made in terms of the characteristics of tensile stress–strain curves of representative samples. The parameters of such curves are nominal stress (force on the specimen divided by the original cross-sectional area), the corresponding nominal strain (increase in length divided by original length), and the modulus (slope of the stress–strain curve). We refer below to the initial modulus, which is this slope near zero strain.

Generalized stress–strain curves look like those shown in Fig. 1-2. For our present purposes we can ignore the yield phenomenon and the fact that such curves are functions of testing temperature, speed of elongation, and characteristics of the particular polymer sample. The nominal stress values in this figure are given in pounds force per square inch (psi) of unstrained area (1 psi $= 6.9 \times 10^3$ N/m^2).

Elastomers recover completely and very quickly from great extensions, which can be up to 1000% or more. Their initial moduli in tension are low, typically up to about a range of 1000 psi (7 MN/m^2) but they generally stiffen on stretching. Within a limited temperature range, the moduli of elastomers increase as the temperature is raised. (This ideal response may not be observed in the case of samples of real vulcanized rubbers, as discussed on page 149). If the temperature is lowered sufficiently, elastomers become stiffer and begin to lose their rapid recovery properties. They will be glassy and brittle under extremely cold conditions.

Figure 1-3a illustrates the response of an elastomer sample to the application and removal of a load at different temperatures. The sample here is assumed to be cross-linked, so that the polymer does not deform permanently under stress.

Fibers have high initial moduli which are usually in the range 0.5×10^6 to 2×10^6 psi (3×10^3 to 14×10^3 MN/m^2). Their extensibilities at break are often lower than 20%. If a fiber is stretched below its breaking strain and then allowed to relax, part of the deformation will be recovered immediately and some, but not all, of the remainder will be permanent (Fig. 1-3b). Mechanical properties of commercial synthetic fibers do not change much in the temperature range between -50 and about 150°C (otherwise they would not be used as fibers).

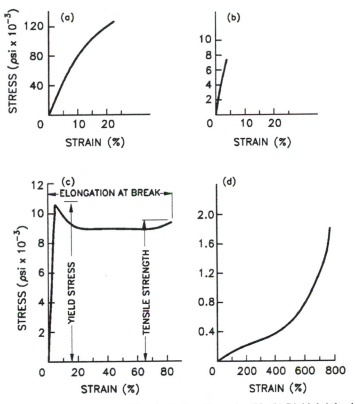

Fig. 1-2. Stress-strain curves. (a) Synthetic fiber, like nylon 66. (b) Rigid, brittle plastic, like polystyrene. (c) Tough plastic, like nylon 66. (d) Elastomer, like vulcanized natural rubber.

[As an aside, we mention that fiber strength (tenacity) and stiffness are usually expressed in units of grams per denier or grams per tex (i.e., grams force to break a one-denier or one-tex fiber). This is because the cross-sectional area of some fibers, like those made from copolymers of acrylonitrile, is not uniform. Denier and tex are the weights of 9000- and 1000-m fiber, respectively.]

Plastics generally have intermediate tensile moduli, usually 0.5×10^5 to 4×10^5 psi (3.5×10^2 to 3×10^3 MN/m^2), and their breaking strain varies from a few percent for brittle materials like polystyrene to about 400% for tough, semicrystalline polyethylene. Their strain recovery behavior is variable, but the elastic component is generally much less significant than in the case of fibers (Fig. 1-3c). Increased temperatures result in lower stiffness and greater elongation at break.

Some chemical species can be used both as fibers and as plastics. The fiber-making process involves alignment of polymer molecules in the fiber direction.

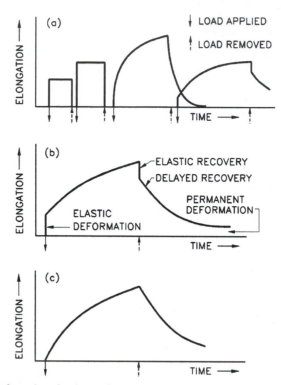

Fig. 1-3. Deformation of various polymer types when stress is applied and unloaded. (a) Cross-linked ideal elastomer. (b) Fiber. (c) Amorphous plastic.

This increases the tensile strength and stiffness and reduces the elongation at break. Thus, typical poly (hexamethylene adipamide) (nylon-66, structure **1-6**) fibers have tensile strengths around 100,000 psi (700 MN/m^2) and elongate about 25% before breaking. The same polymer yields moldings with tensile strengths around 10,000 psi (70 MN/m^2) and breaking elongations near 100%. The macromolecules in such articles are randomly aligned and much less extended.

Synthetic fibers are generally made from polymers whose chemical composition and geometry enhance intermolecular attractive forces and crystallization. A certain degree of moisture affinity is also desirable for wearer comfort in textile applications. The same chemical species can be used as a plastic, without fiber-like axial orientation. Thus most fiber forming polymers can also be used as plastics, with adjustment of molecular size if necessary to optimize properties for particular fabrication conditions and end uses. Not all plastics can form practical fibers, however, because the intermolecular forces or crystallization tendency may be too weak to achieve useful stable fibers. Ordinary polystyrene is an example of such a plastic material, while polyamides, polyesters, and polypropylene are prime examples of polymers that can be used in both areas.

Elastomers are necessarily characterized by weak intermolecular forces. Elastic recovery from high strains requires that polymer molecules be able to assume coiled shapes rapidly when the forces holding them extended are released. This rules out chemical species in which intermolecular forces are strong at the usage temperature or which crystallize readily. The same polymeric types are thus not so readily interchangeable between rubber applications and uses as fibers or plastics.

The intermolecular forces in polyolefins like polyethylene (**1-3**) are quite low, but the polymer structure is so symmetrical and regular that the polymer segments in the melt state are not completely random. The vestiges of solid-state crystallites that persist in the molten state serve as nuclei for the very rapid crystallization that occurs as polyethylene cools from the molten state. As a result, solid polyethylene is not capable of high elastic deformation and recovery because the crystallites prevent easy uncoiling or coiling of the macromolecules. By contrast, random copolymers of ethylene and propylene in mole ratios between about 1/4 and 4/1 have no long sequences with regular geometry. They are therefore noncrystallizing and elastomeric.

1.9 MISCELLANEOUS TERMS

Chain. A linear or branched macromolecule is often called a chain because the repeating units are joined together like links in a chain. Many polymers are polymerized by chain reactions, which are characterized by a series of successive reactions initiated by a single primary event. Here the term *chain* is used to designate a kinetic sequence of reaction events which results in the production of a molecular chain composed of linked repeating units.

Resin. In polymer technology the term *resin* usually means a powdered or granular synthetic polymer suitable for use, possibly with the addition of other nonpolymeric ingredients. Although the meaning is very ill defined, it is listed here because it is widely used.

Condensation and addition polymers. The explanation of these two widely used terms is postponed to Section 5.1 where polymerization processes are considered for the first time in this text.

1.10   POLYMER NOMENCLATURE

Custom then is the great guide of human life.
                    —David Hume, *Concerning Human Understanding*

A systematic IUPAC nomenclature exists for polymers just as it does for organic and inorganic chemicals. This polymer nomenclature is rarely used, however, because a trivial naming system is deeply entrenched through the force of usage. A similar situation prevails with all chemical species which are commercially important commodities. Thus, large-scale users of the compound **1-40**

$$CH_3-\underset{\underset{O}{\|}}{C}-CH_2CH_3$$

1-40

will know it as MEK (methyl ethyl ketone) rather than 2-butanone. The common polymer nomenclature prevails in the scientific as well as the technological literature. [It is not used in *Chemical Abstracts* and reference should be made to Volume 76 (1972) of that journal for the indexing of polymers.]

Reference [4] gives details of the systematic IUPAC nomenclature. The remainder of this section is devoted to a review of the common naming system, a knowledge of which is needed in order to read current literature.

Although the common naming system applies to most important polymers, the system does break down in some cases. When inconsistencies occur resort is made to generally accepted conventions for assignment of names to particular polymers. The nomenclature is thus arbitrary in the final analysis. It usually works quite smoothly because there are probably no more than a few dozen polymers which are of continuing interest to the average worker in the field, and the burden of memorization is thus not excessive. The number of polymeric species which require frequent naming will eventually become too large for convenience in the present system, and a more formal nomenclature will probably be adopted in time.

Although there are no codified rules for the common nomenclature the following practice is quite general.

Polymers are usually named according to their source, and the generic term is "polymonomer" whether or not the monomer is real. Thus we have polystyrene **(1-1)** and poly(vinyl alcohol) **(1-7)**. Similarly, polyethylene is written as **1-3** although the representation $+CH_2+_{2n}$ and the corresponding name "polymethylene" could have been chosen equally well to reflect the nature of the repeating unit.

The monomer name is usually placed in parentheses following the prefix "poly" whenever it includes a substituted parent name like poly(1-butene) **(1-41)** or a multiword name like poly(vinyl chloride) (**1-4**):

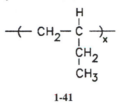

1-41

A particular common name is used even if the polymer could be synthesized from an unusual monomer. Thus structure **1-42** is conventionally called poly(ethylene oxide), since it is derived from this particular monomer.

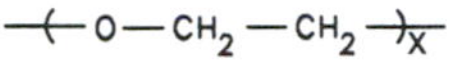

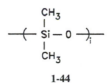

1-42

The same name would ordinarily be used even if the polymer were synthesized from ethylene glycol ($HOCH_2CH_2OH$), ethylene chlorohydrin ($ClCH_2CH_2OH$), or bischloromethyl ether ($ClCH_2OCH_2Cl$). Similarly, structure **1-13** is called polycaprolactam because it is made industrially from the lactam by reaction (1-5), in preference to polymerization of the parent amino acid, $H_2N (CH_2)_5COOH$.

It is useful to digress at this point to review some common names for frequently used vinyl monomers. These are summarized in Fig. 1-4. Alternative names will be apparent, but these are not used by convention. (Thus acrylonitrile could logically be called vinyl cyanide, but this would be an unhappy choice from a marketing point of view.) The polymer name in each case is poly "monomer."

A few polymers have names based on the repeating unit without reference to the parent monomer. The primary examples are silicones, which possess the repeating unit.

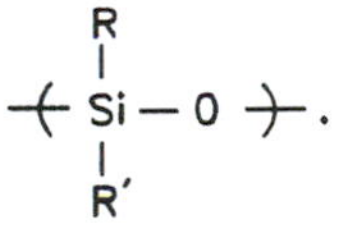

1-43

The most common silicone fluids are based on poly(dimethyl siloxane) with the repeating unit structure:

1-44

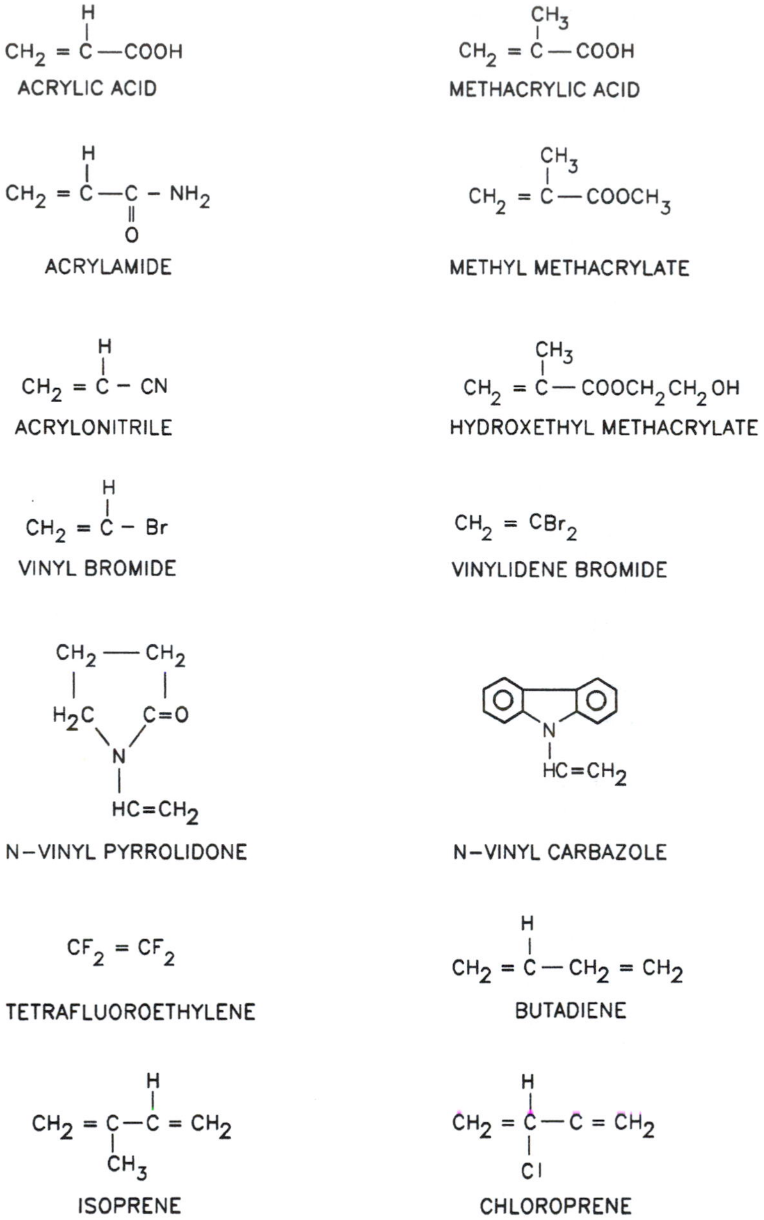

$$CH_2 = \underset{\underset{\displaystyle H}{|}}{C} - COOH$$

ACRYLIC ACID

$$CH_2 = \underset{\underset{\displaystyle CH_3}{|}}{C} - COOH$$

METHACRYLIC ACID

$$CH_2 = \underset{\underset{\displaystyle H}{|}}{C} - \underset{\underset{\displaystyle O}{\parallel}}{C} - NH_2$$

ACRYLAMIDE

$$CH_2 = \underset{\underset{\displaystyle CH_3}{|}}{C} - COOCH_3$$

METHYL METHACRYLATE

$$CH_2 = \underset{\underset{\displaystyle H}{|}}{C} - CN$$

ACRYLONITRILE

$$CH_2 = \underset{\underset{\displaystyle CH_3}{|}}{C} - COOCH_2CH_2OH$$

HYDROXETHYL METHACRYLATE

$$CH_2 = \underset{\underset{\displaystyle H}{|}}{C} - Br$$

VINYL BROMIDE

$$CH_2 = CBr_2$$

VINYLIDENE BROMIDE

N−VINYL PYRROLIDONE

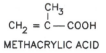

N−VINYL CARBAZOLE

$$CF_2 = CF_2$$

TETRAFLUOROETHYLENE

$$CH_2 = \underset{\underset{\displaystyle H}{|}}{C} - CH_2 = CH_2$$

BUTADIENE

$$CH_2 = \underset{\underset{\displaystyle CH_3}{|}}{C} - C = CH_2$$

ISOPRENE

$$CH_2 = \underset{\underset{\displaystyle Cl}{|}}{C} - C = CH_2$$

CHLOROPRENE

Fig. 1-4.   Some common vinyl monomers.

The nomenclature of copolymers includes the names of the monomers separated by the interfix *co-*. Thus **1-25** would be poly(vinylchloride-co-vinyl acetate). The first monomer name is that of the major component, if there is one. This system applies strictly only to copolymers in which the monomers are arranged more or less randomly. If the comonomers are known to alternate, as in **1-26**, the name would be poly(styrene-*alt*-maleic anhydride). Interfixes may be omitted when the name is frequently used, as in styrene-acrylonitrile copolymers (Section 1.5.3).

When the repeating unit of linear polymers contains other atoms as well as carbon, the polymer can frequently be named from the linking group between hydrocarbon portions. Thus, polymer **1-45**

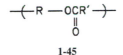

1-45

is evidently a polyester,

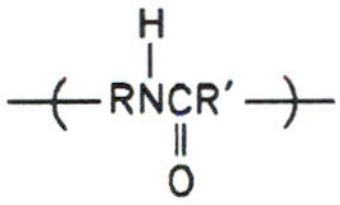

1-46

is a polyamide,

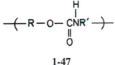

1-47

is a polyurethane,

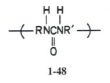

1-48

is a polyurea, and

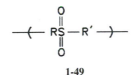

1-49

is a polysulphone. These polymers are generally made by reacting two monomers with the elimination of a smaller molecule [reactions (1-1) and (1-7), for example]. They are thus called condensation polymers (see also Section 5.1).

Condensation polymers are named by analogy with the lower molecular weight esters, amides, and so on. Thus, since the names of all esters end with the suffix *-ate* attached to that of the parent acid (e.g., **1-50**)

$$CH_3 - \underset{\underset{O}{\|}}{C}OCH_3CH_3$$

1-50

is ethyl acetate, polymer **1-5** is named poly(ethylene terephthalate). The parent acid here is terephthalic acid, which is the para isomer. (The ortho diacid is phthalic acid and the meta isomer is isophthalic acid.) The alcohol residue must be a glycol if the polymer is to be linear, and so it is not necessary to use the word *glycol* in the polymer name. The word *ethylene* implies the glycol. Note that the trade name is usually used for the monomers. Thus structure **1-51** would be named poly(tetramethylene terephthalate) or poly(butylene terephthalate) rather than poly(1,4-butane terephthalate).

$$\left(O - CH_2 - CH_2 - CH_2 - CH_2 - O - \underset{\underset{O}{\|}}{C} - \underset{}{\bigcirc} - \underset{\underset{O}{\|}}{C} \right)_n$$

1-51

Polyamides are also known as nylons. They may be named as polyamides. Thus **1-6** is poly(hexamethylene adipamide). This name indicates that the polymeric structure could be made by condensing hexamethylene diamine, $H_2N(CH_2)_6NH_2$, and adipic acid, $HOOC(CH_2)_4COOH$. The dibasic acids are named according to their trivial names: oxalic ($HOOC-COOH$), $HOOC(CH_2)COOH$ malonic, $HOOC(CH_2)_2COOH$ succinic, glutaric, adipic, and so on. (The mnemonic is OMSGAPSAS: oh my, such good apple pie, sweet as sugar. We leave it to the reader to fill in the trivial names after adipic, if and when they are needed.)

There is also an alternative numbering system for synthetic polyamides. Polymers that could be made from amino acids are called nylon-x, where x is the number of carbon atoms in the repeating unit. Thus, polycaprolactam (**1-13**) is nylon-6, while the polymer from ω-aminoundecanoic acid is nylon-11. Nylons from diamines and dibasic acids are designated by two numbers, in which the first represents the number of carbons in the diamine chain and the second the number of carbons in the dibasic acid. Structure **1-6** is thus nylon-6,6. Nylon-6,6 and nylon-6 differ in repeating unit length and symmetry and their physical properties are not identical.

Polymers such as polyamides (**1-13**), polyesters (**1-5**), and so on are not named as copolymers since the chemical structure of the joining linkage in each case shows that the parent monomers must alternate and copolymer nomenclature would therefore be redundant.

There are a few common polymers also in which the accepted name conveys relatively little information about the repeating unit structure. This list includes polycarbonate (**1-52**)

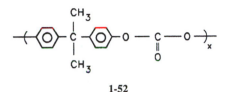

1-52

and poly(phenylene oxide) (**1-14**). ABS polymers (Section 1.5.3) are an important class of thermoplastics which consists of blends and/or graft copolymers. A simple repeating unit and name cannot usually be written for such species.

Graft copolymers like **1-27** are named as poly(A-g-B) with the backbone polymer mentioned before the branch polymer. Examples are poly(ethylene-g-styrene) or starch-g-polystyrene. In block copolymer nomenclature b is used in place of g and the polymers are named from an end of the species. Thus the triblock macromolecule **1-53**

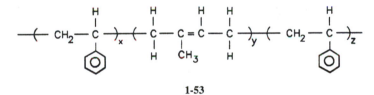

1-53

is called poly(styrene-b-isoprene-b-styrene). When such materials are articles of commerce they are usually designated by the monomer initials, and this structure would be named SIS block copolymer. Reference [5] may be consulted for further details of copolymer nomenclature. Reference [6] lists locations of International Union of Pure and Applied Chemistry recommendations on macromolecular nomenclature.

Note that the common nomenclature generally uses trivial names for monomers as well as the corresponding polymer (**1-1** and **1-2** are examples).

This brief review has emphasized the exceptions more than the regularities of the conventional polymer nomenclature. The reader will find that this jargon is not as formidable as it may appear to be on first encounter. A very little practice is all that is usually needed to recognize repeating units, parent monomer structures, and the common names.

PROBLEMS

An ounce of practice is worth a pound of preaching.

—Proverb

1-1.   Show the repeating unit which would be obtained in the polymerization of the following monomers:

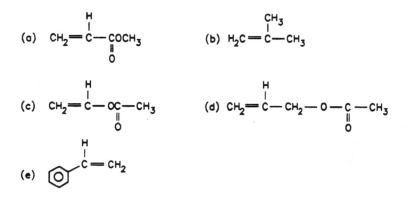

(a) CH_2=$\overset{H}{\underset{}{C}}$—$\overset{}{\underset{O}{C}}OCH_3$

(b) H_2C=$\overset{CH_3}{\underset{}{C}}$—$CH_3$

(c) CH_2=$\overset{H}{\underset{}{C}}$—$\overset{}{\underset{O}{OC}}$—$CH_3$

(d) CH_2=$\overset{H}{\underset{}{C}}$—$CH_2$—$O$—$\overset{}{\underset{O}{C}}$—$CH_3$

(e) $\overset{H}{\underset{}{C}}$=$CH_2$

1-2.   Show the repeating unit which would be obtained by reacting the following:

(a) $HOOC(CH_2)_6 COOH$ and $H_2N(CH_2)_4 NH_2$

(b) $HOCH_2 CH_2 CH_2 CH_2 OH$ and $HOOC$—⬡—$COOH$

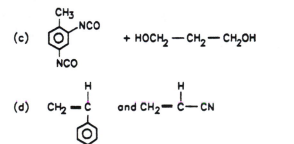

(c)   ⬡ with CH_3, NCO, NCO + $HOCH_2$—CH_2—CH_2OH

(d) CH_2=$\overset{H}{\underset{⬡}{C}}$ and CH_2=$\overset{H}{\underset{}{C}}$—$CN$

(e) $HOCH_2 CH_2 CH_2 CH_2 OH$ and CI—$\overset{}{\underset{O}{C}}$—⬡—$\overset{}{\underset{O}{C}}$—$CI$

1-3. What is the degree of polymerization of each of the following?

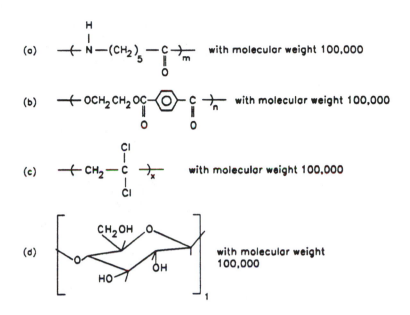

(o) $\xleftarrow{} N - (CH_2)_5 - C \xrightarrow{}_m$ with molecular weight 100,000

(b) $\xleftarrow{} OCH_2CH_2OC - \bigcirc - C \xrightarrow{}_n$ with molecular weight 100,000

(c) $\xleftarrow{} CH_2 - C \xrightarrow{}_x$ with molecular weight 100,000

(d) 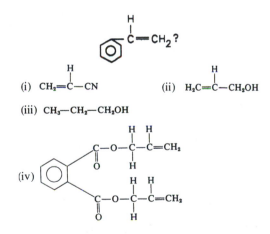 with molecular weight 100,000

1-4. (a) What is the functionality of the following monomers in reactions with styrene?

$$\bigcirc \overset{H}{\underset{}{C}} = CH_2 ?$$

(i) $CH_2 = \overset{H}{\underset{}{C}} - CN$ (ii) $H_2C = \overset{H}{\underset{}{C}} - CH_2OH$

(iii) $CH_3 - CH_2 - CH_2OH$

(iv) $\bigcirc \begin{matrix} C-O-\overset{H}{\underset{H}{C}}-\overset{H}{\underset{}{C}}=CH_2 \\ \overset{\parallel}{O} \\ \\ C-O-\overset{H}{\underset{H}{C}}-\overset{H}{\underset{}{C}}=CH_2 \\ \overset{\parallel}{O} \end{matrix}$

(b) What are their functionalities in reactions with divinyl benzene **(1-15)**?

1-5. Draw structural formulas (one repeating unit) for each of the following polymers:

     (a) poly(styrene-*co*-methyl methacrylate)
     (b) polypropylene
     (c) poly(hexamethylene adipamide)
     (d) polyformaldehyde
     (e) poly(ethylene terephthalate)

1-6. Name the following:

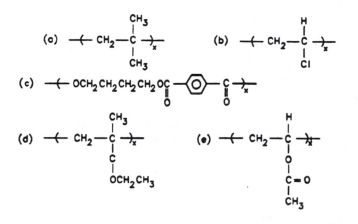

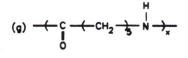

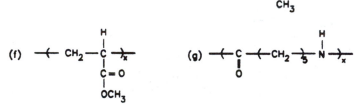

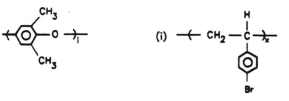

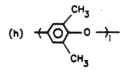

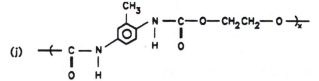

1-7.   Which of the following materials is most suitable for the manufacture of thermoplastic pipe? Briefly tell why.

(a) $CH_3\!\!+\!\!CH_2\!\!\rightarrow_{29}\!\!CH_3$ (b) $CH_3\!\!+\!\!CH_2\!\!\rightarrow_{\overline{14,000}}\!\!CH_3$

(c) $CH_3\!\!+\!\!CH_2\!\!\rightarrow_{\overline{200,000}}\!\!CH_3$

1-8.   What is the functionality of glycerol in the following reactions:
        (a) urethane formation with

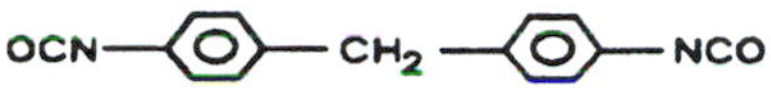

$$OCN-\!\!\langle\bigcirc\rangle\!\!-CH_2-\!\!\langle\bigcirc\rangle\!\!-NCO$$

   (b) esterification with phthalic anhydride
   (c) esterification with acetic acid

$$CH_3\!-\!\underset{\underset{O}{\|}}{C}\!-\!OH$$

   (d) esterification with phosgene?

1-9.   How would you synthesize a block copolymer having segments with the following structures?

$$-\!\!\left(\!CH_2CH_2CH_2CH_2O\!\right)_{\!\!m}\ \text{and}$$

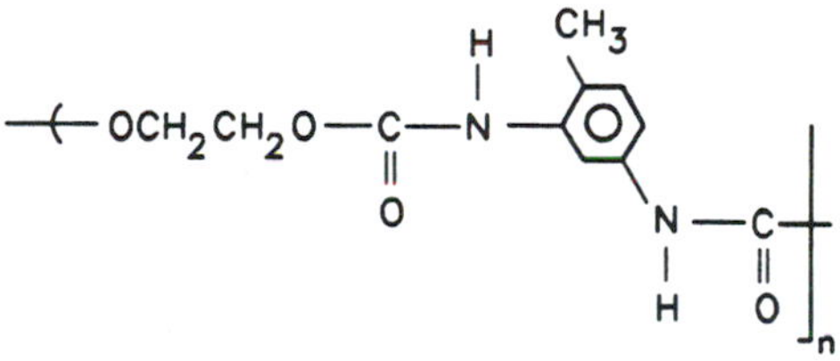

$$-\!\!\left(\!OCH_2CH_2O\!-\!\underset{\underset{O}{\|}}{C}\!-\!\overset{\overset{H}{|}}{N}\!-\!\langle\bigcirc\rangle\!\overset{CH_3}{}\!\underset{\underset{H}{|}}{N}\!-\!\underset{\underset{O}{\|}}{C}\!\right)_{\!\!n}\ \ ?$$

1-10.   Write structural formulas for
        (a) polyethylene
        (b) poly(butylene terephthalate)
        (c) poly(ethyl methacrylate)
        (d) polycarbonate
        (e) poly(1,2-propylene oxalate)

(f) poly(dimethyl siloxane)
(g) polystyrene
(h) polytetrafluoroethylene
(i) poly(methyl acrylate)
(j) poly(vinyl acetate)

1-11. Polyisobutene is used as an elastomer in inner tubes and some cable coatings. It is also used in adhesives and as an additive to adjust the viscosity of motor oils. What is the basic difference in the state of the polymer in these two different applications?

1-12. What is the functionality of the monomer shown

H2N — CH2
CH2
H2C = C — CH2 — C — CH2 — CH2 — C — OH
O

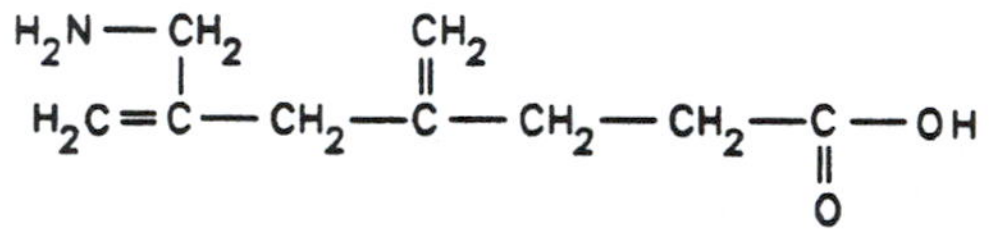

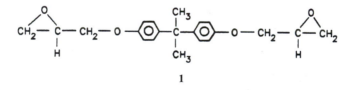

(a) in a free radical or ionic addition reaction through C=C double bonds?
(b) in a reaction which produces amide links?
(c) in a reaction which produces ester links?

1-13. (a) What is the functionality of the diglycidyl ether of bisphenol A (**1**) in a curing reaction with diethylene triamine (**2**)?

O
CH3
O
CH2 — C — CH2 — O
C
O — CH2 — C — CH2
H
CH3
H

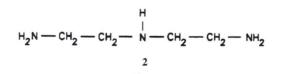

1

(b) What is the functionality of **2** in this epoxide hardening reaction?

H
H2N — CH2 — CH2 — N — CH2 — CH2 — NH2

2

(c) Will this reaction lead to a cross-linked structure?

1-14. Show the repeating unit that would be obtained in the polymerization of the following:

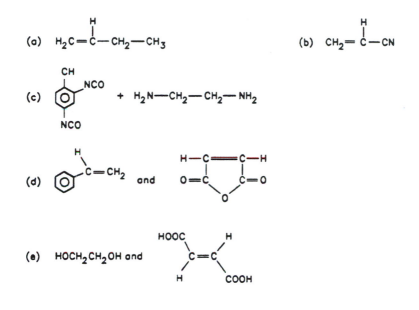

REFERENCES

[1] P. J. Flory, "Principles of Polymer Chemistry." Cornell Univ. Press, Ithaca, NY, 1953.
[2] J. A. Brydson, "Plastic Materials," 3rd ed. Butterworths, London, 1975.
[3] H. F. Mark, *Am. Sci.* **55,** 265 (1967).
[4] W. V. Metanomski, Ed., "IUPAC Compendium of Macromolecular Nomenclature." Blackwel! Scientific publications, Oxford, UK, 1991.
[5] "Source-Based Nomenclature for Copolymers," *Pure Appl. Chem.* **57,** 1427 (1985).
[6] IUPAC Macromol. Chem. Div. (IV), Commission on Macromol. Nomenclature, *Polym. Bull.* **32,** 125 (1994).

Chapter 2

Basic Principles of Polymer Molecular Weights

Die Wahrheit ist das Alle...The truth is the whole.
—G. W. F. Hegel

2.1 IMPORTANCE OF MOLECULAR WEIGHT CONTROL

Both the mechanical properties of solid thermoplastics and their processing behavior at elevated temperatures depend critically on the average size and the distribution of sizes of macromolecules in the sample. This is one reason why the plastics market contains different grades of each polymer. All varieties are often chemically identical, but some of their molecular-weight-dependent properties may differ enough that the polymers cannot be interchanged economically. As a rule of thumb, resistance to deformation increases with increasing average molecular weight. Thus the thermoplastics that are hardest to force into a final shape in the softened state will usually yield the strongest solid articles on subsequent freezing. (Some properties, such as refractive index and hardness at ambient temperatures, are not much dependent on molecular weight, provided this property is in the normal commercial range for the particular polymer type.)

An example of the influence of average molecular weight has been given on page 15, where various grades of thermoplastic polyester were discussed. Plasticized poly(vinyl chloride) sheeting and coated fabric provide a similar illustration in heat sealing applications. If the molecular weight of the polymer is too high, the material will not flow out enough to weld well under normal sealing conditions. If the molecular weight is too low, on the other hand, the plastic may suffer excessive thinning, resulting in a weak weld area or show-through of fabric backing.

The molecular weights of synthetic polymers are much less uniform, within any sample, than those of conventional chemicals. The growth and termination of polymer chains are subject to variations during manufacture that result in the production of a mixture of chemically identical molecules of different sizes, and it is important also to be able to control the distribution of such sizes as well as their average value. A polymer that can crystallize will tend to form brittle articles, for example, if there is much low-molecular-weight crystallizable material in the sample. The presence of appreciable high-molecular-weight material, on the other hand, makes thermoplastic melts more elastic, and this property can be a disadvantage in applications like high-speed wire covering or an advantage in other end uses like extrusion coating of paper.

Molecular weights are not often measured directly for control of production of polymers because other product properties are more convenient experimentally or are thought to be more directly related to various end uses. Solution and melt viscosities are examples of the latter properties. Poly(vinyl chloride) (PVC) production is controlled according to the viscosity of a solution of arbitrary concentration relative to that of the pure solvent. Polyolefin polymers are made to specific values of a melt flow parameter called *melt index*, whereas rubber is characterized by its *Mooney viscosity*, which is a different measure related more or less to melt viscosity. These parameters are obviously of some practical utility, or they would not be used so extensively. They are unfortunately specific to particular polymers and are of little or no use in bringing experience with one polymer to bear on problems associated with another.

Many technical problems that may be encountered, say, with a new thermoplastic, will already have been met and solved with polymers, like rubber, that have been in the marketplace for a comparatively long time. It is not often possible to recognize and use such parallels, however, if the parameters of the molecular weight distributions in the different cases are not measured in the same units. This results in much unnecessary rediscovery of "old" answers, and the engineer or scientist who can interpret both "Mooney" and "melt index" values in terms of statistical parameters of the molecular weight distributions of the respective rubber and thermoplastic may save considerable time and effort.

2.2   PLAN OF THIS CHAPTER

We first review the fundamentals of small particle statistics as these apply to synthetic polymers. This is mainly concerned with the use of statistical moments to characterize molecular weight distributions. One of the characteristics of such a distribution is its central tendency, or average, and the following main topic shows how it is possible to determine various of these averages from measurements of properties of polymer solutions without knowing the parent distribution itself.

Chapter 3 reviews the essentials of practical techniques for measuring average molecular weights and characterizing molecular weight distributions.

2.3   ARITHMETIC MEAN

The distribution of molecular sizes in a polymer sample is usually expressed as the proportions of the sample with particular molecular weights. The mass of data contained in the distribution can be understood more readily by condensing the information into parameters descriptive of various aspects of the distribution. Such parameters evidently must contain less information than the original distribution, but they present a concise picture of the distribution and are indispensable for comparing different distributions.

One such summarizing parameter expresses the central tendency of the distribution. A number of choices are available for this measure, including the median, mode, and various averages, such as the arithmetic, geometric, and harmonic means. Each may be most appropriate for different distributions. The arithmetic mean is usually used with synthetic polymers. This is because it was very much easier, until recently, to measure the arithmetic mean directly than to characterize the whole distribution and then compute its central tendency. The distribution must be known to derive the mode or any simple average except the arithmetic mean. (Some methods like those based on measurement of sedimentation and diffusion coefficients measure more complicated averages directly. They are not used much with synthetic polymers, however, and will not be discussed in this text.)

Various molecular weight averages are current in polymer science. We show here that these are simply arithmetic means of molecular weight distributions. It may be mentioned in passing that the concepts of small particle statistics that are discussed here apply also to other systems, such as soils, emulsions, and carbon black, in which any sample contains a distribution of elements with different sizes.

To define any arithmetic mean A, let us assume unit volume of a sample of N polymer molecules comprising n_1 molecules with molecular weight M_1, n_2 molecules with molecular weight $M_2, \ldots, n_j$ molecules with molecular weight M_j.

$$n_1 + n_2 + \cdots n_j = N \tag{2-1}$$

$$A = \frac{n_1 M_1 + n_2 M_2 + \cdots + n_j M_j}{n_1 + n_2 + \cdots + n_j} = \frac{n_1 M_1 + n_2 M_2 + \cdots n_j M_j}{N} \tag{2-2}$$

$$A = \frac{n_1}{N} M_1 + \frac{n_2}{N} M_2 + \cdots + \frac{n_j}{N} M_j \tag{2-3}$$

The arithmetic mean molecular weight A is given as usual by the total measured quantity (M) divided by the total number of elements. That is, the ratio n_i/N is the proportion of the sample with molecular weight M_i. If we call this proportion f_i, the arithmetic mean molecular weight is given by

$$A = f_1 M_1 + f_2 M_2 + \cdots + f_j M_j = \sum_i f_i M_i \qquad (2\text{-}4)$$

Equation (2-4) defines the arithmetic mean of the distribution of molecular weights. Almost all molecular weight averages can be defined from this equation.

2.3.1 Number Distribution, $\overline{M}_n$

The distribution we have just assumed to define the arithmetic mean is a number distribution, since the record consists of *numbers* of molecules of specified sizes. The sum of these numbers comprises the integral (cumulative) number distribution. Figure 2-1 represents one such distribution. The scale along the abscissa is the molecular weight while that on the ordinate could be the total number of molecules with molecular weights less than or equal to the corresponding value on the abscissa. However, it is easier to compare different distributions if the cumulative figures along the ordinate are expressed as fractions of the total number of molecules in each sample and Fig. 2-1 is drawn in this way. The units of the ordinate are therefore mole fractions and extend from 0 to 1; the integral distribution is now said to be normalized.

In mathematical terms, the cumulative number (or mole) fraction $X(M)$ is defined as

$$X(M) = \sum_i^M x_i \qquad (2\text{-}5)$$

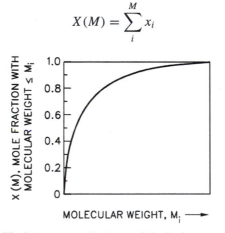

Fig. 2-1. A normalized integral distribution curve.

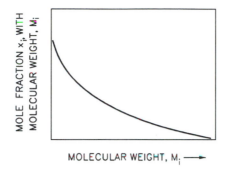

MOLECULAR WEIGHT, M_i ⟶

Fig. 2-2.   A normalized differential number distribution curve.

where x_i is the fraction of molecules with molecular weight M_i. The differential number function is simply the mole fraction x_i, and a plot of these values against corresponding M_i's yields a differential number distribution curve, as in Fig. 2-2. If the distribution is normalized, the area under the x_i–M_i curve in Fig. 2-2 will be unity. (See Section 2.4.2 for units.)

To compile the number distribution we have expressed the proportion of species with molecular weight M_i as the corresponding mole fraction x_i. Substitution of x_i for f_i in Eq. (2-4) shows that the arithmetic mean of the number distribution is

$$A = \sum_i x_i M_i = \overline{M}_n \tag{2-6}$$

This is the definition of number average molecular weight $\overline{M}_n$.

Equivalent definitions follow from simple arithmetic. Since

$$x_i = \frac{n_i}{N} = n_i \Big/ \sum n_i \tag{2-7}$$

$$\overline{M}_n = \sum n_i M_i \Big/ \sum n_i \tag{2-8}$$

where n_i is defined, as above, as the number of polymer molecules per unit volume of sample with molecular weight M_i. Also, if c_i is the total weight of the n_i molecules, each with molecular weight M_i, and w_i is the corresponding weight fraction, then

$$c_i = n_i M_i \tag{2-9}$$

$$w_i = c_i \Big/ \sum c_i = n_i M_i \Big/ \sum n_i M_i \tag{2-10}$$

and

$$\overline{M}_n = \sum c_i \Big/ \sum \frac{c_i}{M_i} = 1 \Big/ \sum \frac{w_i}{M_i} \tag{2-11}$$

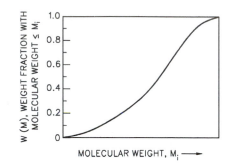

Fig. 2-3.   A normalized integral weight distribution curve.

Since polymer solutions are used for direct determinations of average molecular weights, the symbols n_i and c_i will usually refer respectively to the molar and weight concentrations of macromolecules in such solutions.

2.3.2   Weight Distribution, $\overline{M}_w$

If we had recorded the weight of each species in the sample, rather than the number of molecules of each size, the array of data would be a weight distribution. The situation corresponds to that described for a number distribution. Figure 2-3 depicts a simple integral weight distribution, normalized by recording fractions of the total weight rather than actual weights of the different species.

The integral (cumulative) weight fraction $W(M)$ is given by

$$W(M) = \sum_i w_i \tag{2-12}$$

and is equal to the weight fraction of the sample with molecular weight not greater than M_i. A plot of w_i against M_i yields a differential weight distribution curve, as in Fig. 2-4. As in the case of the number distribution, if $W(M)$ is normalized, the scale of the ordinate in this figure goes from 0 to 1 and the area under the curve equals unity.

The proportion of the sample with size M_i is expressed in the present case as the corresponding weight fraction. Equating w_i and f_i in Eq. (2-4) produces the following expression for the arithmetic mean of the weight distribution:

$$A = \sum_i w_i M_i = \overline{M}_w \tag{2-13}$$

where $\overline{M}_w$ is the weight average molecular weight, which from Eqs. (2-9) and (2-10) may also be expressed as

$$\overline{M}_w \sum M_i c_i \Big/ \sum c_i = \sum M_i^2 n_i \Big/ \sum M_i n_i \tag{2-14}$$

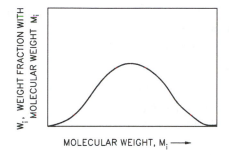

Fig. 2-4.   A normalized differential weight distribution curve.

2.4   MOLECULAR WEIGHT AVERAGES AS
RATIOS OF MOMENTS

2.4.1   Moments in Statistics and Mechanics

We have seen that average molecular weights are arithmetic means of distributions of molecular weights. An alternative and generally more useful definition is in terms of moments of the distribution. This facilitates generalizations beyond the two averages we have considered to this point and clarifies the estimation of parameters related to the breadth and symmetry of the distribution.

The concept of moments was adopted in statistics from the science of mechanics where it was first used in the sense of "importance." The moment of a force about an axis meant the importance of the force in causing rotation about the axis. Similarly, the moment of inertia of a body with respect to an axis expressed the importance of the inertia of the body in resisting a change in the rate of rotation of the body about the axis.

The first moment of a force or weight about an axis is defined as the product of the force and the distance from the axis to the line of action of the force. In this case it is commonly known as the torque. The concept has been extended to more abstract applications such as the moment of an area with respect to a plane and moments of statistical distributions. It is then referred to as the appropriate first moment (the term *torque* is not used).

The second moment of force about the same axis is the product of the force and the square of the distance between its line of action and the axis. This is the moment of inertia. The most direct example of its use is possibly connected with the motion of a rotating body, for which the rotational acceleration caused by an applied torque is calculated by dividing the torque by the moment of inertia of the body. The concept of a second moment has been extended to other less readily pictured applications such as computation of stresses in beams from second moments of cross-sectional areas about particular axes.

By extending the above examples we can say that a moment in mechanics is generally defined as

$$U_j^a = F d^j \qquad (2\text{-}15)$$

where U_j is the jth moment, about a specified line or plane a of a vector or scalar quantity F (for example, force, weight, mass, area), d is the distance from F to the reference line or plane, and j is a number. The moment is named according to the power j to which d is raised. If F is composed of elements F_i each located a distance d_i from the same reference, the moment is the sum of the individual moments of each element

$$U_j^a = \sum_i F_i d_i^j \qquad (2\text{-}16)$$

Mathematically, there is no restriction on the choice of F or j, but use of moments to solve practical mechanics problems usually confines F to the examples listed above and j to values of 1 or 2. The reference line or plane must be specified when the value of the moment is quoted.

In polymer science the mathematical formulation for moments corresponds to that in Eq. (2-16). While the reference line may be located anywhere, the usefulness of choosing the ordinate ($M = 0$) in the graph of the molecular weight distribution (Figs. 2-2 and 2-4) is so great that this reference is usually not mentioned explicitly. The distance d from the reference line is measured along the abscissa in terms of the molecular weight M, and the quantity F is replaced by f_i, the proportion of the polymer with molecular weight M_i. As a matter of utility, j assumes a wider range of values in polymer science than in mechanics. With these differences, which are mainly matters of emphasis, the concepts of moments correspond closely in both disciplines. A general definition of a statistical moment of a molecular weight distribution taken about zero is then

$$U_j' \equiv \sum q_i M_i^j \qquad (2\text{-}17)$$

where q_i is the quantity of polymer in unit volume of the sample with molecular weight M_i and respective values of $q_i = n_i$ (number of molecules, or moles) for an unnormalized number distribution, $= x_i$ (mole fraction) for a normalized number distribution, $= c_i$ (number of grams) for an unnormalized weight distribution, or $= w_i$ (weight fraction) for a normalized weight distribution. In addition, we shall use the notation $_n U$ to refer to a moment of the number distribution and $_w U$ to denote a moment of the weight distribution.

Weight distributions will usually be encountered during analyses of polymer samples. Considerations of polymerization kinetics are often easier in terms of number distributions.

2.4.2 Dimensions

Molecular weight itself is dimensionless. It is the sum of the atomic weights in the formula of the molecule. Atomic weights, in turn, are expressed in terms of dimensionless atomic mass units (amu) which are ratios ($\times 12$) of the masses of the particular atoms to that of the most abundant carbon isotope ^{12}C to which a mass of 12 is assigned. A gram molecular weight, or gram-mole, is the amount of polymer whose weight in grams is numerically equal to the molecular weight (in amu). It is just as correct to use pound-moles or ton-moles if the circumstances so dictate.

The moments of normalized distributions are products of dimensionless frequencies and dimensionless molecular weights or of gram-moles with dimensions of mass. The former moments will be unitless, and the units of the latter will depend on the moment number and on the units of the distribution. Most equations in polymer science imply use of gram-moles, but this is not universal and the dimensions of the particular equation should be checked to determine which units, if any, are being used for molecular weight and concentration quantities.

2.4.3 Arithmetic Mean as a Ratio of Moments

As a general case the ratio of the first moment to the zeroth moment of any distribution defines the arithmetic mean. For an unnormalized number distribution, n_i is the number of moles per unit volume with molecular weight M_i and the zeroth and first moments of the distribution about zero are given respectively by

$$_nU_0' = \sum_i (M_i)^0 n_i = \sum n_i \tag{2-18}$$

$$_nU_1' = \sum_i (M_i)^1 n_i = \sum M_i n_i \tag{2-19}$$

In these symbols the subscript n shows that the moment refers to a number distribution, the numerical subscript is the moment order, and the prime superscript indicates that the moment is taken about the $M = 0$ axis. These equations follow from the definition in Eq. (2-17). The arithmetic mean of the number distribution is the ratio of these moments:

$$A = \frac{_nU_1'}{_nU_0'} = \sum M_i n_i \Big/ \sum n_i = \overline{M}_n \tag{2-20}$$

(Compare Eq. 2-8.)

The arithmetic mean of a weight distribution (the count is in terms of the weight c_i, rather than number of molecules n_i of each species) is likewise given by the ratio of the first to the zeroth moment of the particular distribution about zero. (The notation for moments of weight distributions follows that for number distributions except that the subscript n is replaced by a w).

In these last examples we have chosen unnormalized distributions. If the differential number or weight distribution is normalized, the area under the curve in Figs. 2-2 and 2-4 equals unity. That is,

$$_nU_0' = {}_wU_0' = 1 \text{ (normalized distributions)} \tag{2-21}$$

The arithmetic mean is then numerically equal to the first moment of the normalized distribution, as expressed in Eqs. (2-6) and (2-13).

2.4.4   Extension to Other Molecular Weight Averages

We have seen that $\overline{M}_n$, the arithmetic mean of the number distribution, is equal to the ratio of the first to the zeroth moment of this distribution (Eq. 2-20). If we take ratios of successively higher moments of the number distribution, other average molecular weights are described:

$$\frac{_nU_2'}{_nU_1'} = \sum M_i^2 n_i \bigg/ \sum M_i n_i = \overline{M}_w \tag{2-22}$$

$$\frac{_nU_3'}{_nU_2'} = \sum M_i^3 n_i \bigg/ \sum M_i^2 n_i = \overline{M}_z \tag{2-23}$$

$$\frac{_nU_4'}{_nU_3'} = \sum M_i^4 n_i \bigg/ \sum M_i^3 n_i = \overline{M}_{z+1} \tag{2-24}$$

We may define an average in general as the ratio of successive moments of the distribution. $\overline{M}_n$ and $\overline{M}_w$ are special cases of this definition. The process of taking ratios of successive moments to compute higher averages of the distribution can continue without limit. In fact, the averages usually quoted are limited to $\overline{M}_n$, $\overline{M}_w$, $\overline{M}_z$, and the viscosity average molecular weight $\overline{M}_v$, which is defined later in Section 3.3. We can measure $\overline{M}_n$, $\overline{M}_w$, and $\overline{M}_v$ directly, but it is usually necessary to measure the detailed distribution to estimate $\overline{M}_z$ and higher averages.

Table 2-1 lists averages of the number and weight distributions in terms of these moments.

The reader may notice that any moment about zero of a normalized distribution

$$_nU_j' = \sum x_i(M_i)^j \qquad \text{or} \qquad _wU_j' = \sum w_i(M_i)^j$$

corresponds to the arithmetic mean of the number or weight distribution of $(M_i)^j$, respectively. Respectively, $\overline{M}_n$ and $\overline{M}_w$ are arithmetic means of the number and weight distributions and the source of their names is obvious. The $\overline{M}_z$, $\overline{M}_{z+1}$, and so on, are arithmetic means of the z, $z+1$, etc., distributions. Operational

TABLE 2-1

Moments about Zero and Molecular Weight Averages

(a) Number distribution		
Not normalized	Normalized	Averages
$_nU_0' = \sum_i n_i$	$= \sum_i x_i = 1$	
$_nU_1' = \sum_i M_i n_i$	$= \sum_i M_i x_i$	$\overline{M}_n = {_nU_1'}/{_nU_0'}$
$_nU_2' = \sum_i M_i^2 n_i = \overline{M}_w \overline{M}_n \cdot {_nU_0'}$	$= \sum_i M_i^2 x_i = \overline{M}_w \overline{M}_n$	$\overline{M}_w = {_nU_2'}/{_nU_1'}$
$_nU_3' = \sum_i M_i^3 n_i = \overline{M}_z \overline{M}_w \overline{M}_n \cdot {_nU_0'}$	$= \sum_i M_i^3 x_i = \overline{M}_z \overline{M}_w \overline{M}_n$	$\overline{M}_z = {_nU_3'}/{_nU_2'}$
		$\overline{M}_v{}^* = [{_nU_{a'+1}'}/{_nU_1'}]^{1/a}$
$_nU_j' = \sum M_i^j n_i$	$= \sum M_i^j x_i$	

(b) Weight distribution		
Not normalized	Normalized	Averages
$_wU_{-1}' = \sum_i c_i M^{-1}$	$= \sum_i w_i M_i^{-1}$	
$_wU_0' = \sum_i c_i$	$= \sum_i w_i = 1$	$\overline{M}_n = {_wU_0'}/{_wU_{-1}'}$
$_wU_1' = \sum_i c_i M_i = \overline{M}_w \cdot {_wU_0'}$	$= \sum_i w_i M_i = \overline{M}_w$	$\overline{M}_w = {_wU_1'}/{_wU_0'}$
$_wU_2' = \sum_i c_i M_i^2 = \overline{M}_z \overline{M}_w \cdot {_wU_0'}$	$= \sum_i w_i M_i^2 = \overline{M}_z \overline{M}_w$	$\overline{M}_z = {_wU_2'}/{_wU_1'}$
		$\overline{M}_v^a = ({_wU_a'})^{1/a}$
$_wU_j' = \sum c_i M_i^j$	$= \sum w_i M_i^j$	

[a] $\overline{M}_v$ is derived from solution viscosity measurements through the Mark–Houwink equation $[n] = K\overline{M}_v^a$, where $[n]$ is the limiting viscosity number and K and 1 are constants which depend on the polymer, solvent, and experimental conditions, but not on M (p. 96).

models of these distributions would be too complicated to be useful in polymer science.

Table 2-2 lists various average molecular weights in terms of moments of the number and weight distributions, where the quantity of polymer species with particular sizes are counted in terms of numbers of moles or weights, respectively.

Note that in general a given average is given by

$$\overline{M}_{z+k} = {_nU_{k+3}'}/{_nU_{k+2}'} = {_wU_{k+2}'}/{_wU_{k+1}'} \tag{2-25}$$

The moment orders in the weight distribution are one less than the corresponding orders in the number distributions. (Compare $\overline{M}_n$ formulas.) This symmetry arises

TABLE 2-2

Molecular Weight Averages[a]

	Number distribution			Weight distribution	
	Normalized	Not normalized		Normalized	Not normalized
$\overline{M}_n = \dfrac{_nU'_1}{_nU'_0}$	$= \dfrac{\Sigma x_i M_i}{\Sigma x_i}$	$= \dfrac{\Sigma n_i M_i}{\Sigma n_i}$	$= \dfrac{_wU'_0}{_wU'_{-1}}$	$= \dfrac{\Sigma w_i}{\Sigma(w_i/M_i)}$	$= \dfrac{\Sigma c_i}{\Sigma(c_i/M_i)}$
$\overline{M}_w = \dfrac{_nU'_2}{_nU'_1}$	$= \dfrac{\Sigma x_i M_i^2}{\Sigma x_i M_i}$	$= \dfrac{\Sigma n_i M_i^2}{\Sigma n_i M_i}$	$= \dfrac{_wU'_1}{_wU'_0}$	$= \dfrac{\Sigma w_i M_i}{\Sigma w_i}$	$= \dfrac{\Sigma c_i M_i}{\Sigma c_i}$
$\overline{M}_z = \dfrac{_nU'_3}{_nU'_2}$	$= \dfrac{\Sigma x_i M_i^3}{\Sigma x_i M_i^2}$	$= \dfrac{\Sigma n_i M_i^3}{\Sigma n_i M_i^2}$	$= \dfrac{_wU'_2}{_wU'_1}$	$= \dfrac{\Sigma w_i M_i^2}{\Sigma w_i M_i}$	$= \dfrac{\Sigma c_i M_i^2}{\Sigma c_i M_i}$
$\overline{M}_{z+1} = \dfrac{_nU'_4}{_nU'_3}$	$= \dfrac{\Sigma x_i M_i^4}{\Sigma x_i M_i^3}$	$= \dfrac{\Sigma n_i M_i^4}{\Sigma n_i M_i^3}$	$= \dfrac{_wU'_3}{_wU'_2}$	$= \dfrac{\Sigma w_i M_i^3}{\Sigma w_i M_i^2}$	$= \dfrac{\Sigma c_i M_i^3}{\Sigma c_i M_i^2}$
$\overline{M}_v = \left[\dfrac{_nU'_{a+1}}{_nU'_1}\right]^{1/a}$	$= \left[\dfrac{\Sigma x_i M_i^{a+1}}{\Sigma x_i M_i}\right]^{1/a}$	$= \left[\dfrac{\Sigma n_i M_i^{a+1}}{\Sigma n_i M_i}\right]^{1/a}$	$= [_wU'_a]^{1/a}$	$= [w_i M_i^a]^{1/a}$	$= \left[\dfrac{\Sigma c_i M_i^a}{\Sigma c_i}\right]^{1/a}$

[a] See Appendix 2-A for application of these formulas to mixtures of broad distribution polymers.

because molar and weight concentrations are generally related by Eqs. (2-9) and (2-10). Thus,

$$_nU'_k = \sum n_i M_i^k = \sum (n_i M_i) M_i^{k-1} = \sum c_i M_i^{k-1} = {}_wU'_{k-1} \qquad (2\text{-}26)$$

The viscosity average molecular weight $\overline{M}_v$, which will be discussed later in Section 3.3, is the only average listed in these tables that is not a simple ratio of successive moments of the molecular weight distribution.

2.5 BREADTH OF THE DISTRIBUTION

The distribution of sizes in a polymer sample is not completely defined by its central tendency. The breadth and shape of the distribution curve must also be known, and this is determined most efficiently with parameters derived from the moments of the distribution.

It is always true that $\overline{M}_z > \overline{M}_w > \overline{M}_n$, with the equality occurring only if all species in the sample have the same molecular weight. (This inequality is proven in Section 2.7.) Such monodispersity is unknown in synthetic polymers. The ratio $\overline{M}_w/\overline{M}_n$, or $(\overline{M}_w/\overline{M}_n) - 1$, is commonly taken to be a measure of the polydispersity of the sample. This ratio (the polydispersity index) is not a sound statistical measure of the distribution breadth, and we show later that it is easy to make unjustified inferences from the magnitude of the $\overline{M}_w/\overline{M}_n$ ratio if this parameter is close to unity. However, the use of the polydispersity index is deeply imbedded in polymer science and technology, where it is often called the *breadth of the distribution*. We see later that it is actually related to the variance of the number distribution of the polymer sample. In many cases, when different samples are being compared, any changes in the number distribution will be parallelled by changes in the weight distribution, and so variations in the polydispersity index can substitute for comparisons of the breadth of the weight distributions, which would be more relevant, in general, to the processing and mechanical properties of the materials.

The most widely used statistical measure of distribution breadth is the standard deviation, which can be computed for the number distribution if $\overline{M}_n$ and $\overline{M}_w$ are known. This use of these molecular weight averages provides more information than can be derived from their ratio.

The breadth of a distribution will reflect the dispersion of the measured quantities about their mean. Simple summing of the deviation of each quantity from the mean will yield a total of zero, since the mean is defined such that the sums of negative and positive deviations from its value are balanced. The obvious expedient then is to square the difference between each quantity and the mean of the distribution and add the squared terms. This produces a parameter, called the *variance* of the distribution, which reflects the spread of the observed values about their mean and is independent of the direction of this spread. The positive square root of the variance is called the *standard deviation* of the distribution. Its units are the same as those of the mean.

The standard deviation is calculated from a moment about the mean rather than about zero. The difference between M_i, the molecular weight of any species i and the mean molecular weight A is $M_i - A$, and the jth moment of the normalized distribution about the mean is

$$U_j = \sum f_i (M_i - A)^j \tag{2-27}$$

The absence of a prime superscript on U indicates that the moment is taken with reference to the arithmetic mean.

Since the arithmetic mean is the center of balance of the frequencies in the distribution, the first moment of these frequencies about the mean must be zero:

$$_nU_1 = \sum x_i (M_i - \overline{M}_n) = {}_wU_1 \sum w_i (M_i - \overline{M}_w) = 0 \tag{2-28}$$

The second moment about the mean is the variance of the distribution:

$$_nU_2 = \sum x_i (M_i - \overline{M}_n)^2 = (s_n)^2 \tag{2-29}$$

$$_wU_2 = \sum w_i (M_i - \overline{M}_w)^2 = (s_w)^2 \tag{2-30}$$

where s_n and s_w are the standard deviations of the number and weight distributions, respectively. Thus the standard deviation of the distribution is the square root of the second moment about its arithmetic mean:

$$s = (U_2)^{0.5} \tag{2-31}$$

It remains now to convert U_2 into terms of $\overline{M}_w$ and $\overline{M}_n$. From Eq. (2-29),

$$_nU_2 = \sum x_i (M_i - \overline{M}_n)^2 = \sum x_i \left(M_i^2 - 2M_i \overline{M}_n + \overline{M}_n^2 \right)$$

$$= \sum x_i M_i^2 - 2\overline{M}_n \sum x_i M_i + \overline{M}_n^2 \sum x_i \tag{2-32}$$

$$= {}_nU_2' - 2\overline{M}_n \overline{M}_n + \overline{M}_n^2$$

$$_nU_2 = {}_nU_2' - \overline{M}_n^2 = \overline{M}_w \overline{M}_n - \overline{M}_n^2$$

$$s_n = \left(\overline{M}_w \overline{M}_n - \overline{M}_n^2 \right)^{0.5} \tag{2-33}$$

$$s_n^2 / \overline{M}_n^2 = \overline{M}_w / \overline{M}_n - 1 \tag{2-34}$$

Starting with Eq. (2-30) instead of Eq. (2-29), it is easy to show that

$$s_w^2 / \overline{M}_w^2 = \overline{M}_z / \overline{M}_w - 1 \tag{2-35}$$

If $\overline{M}_w$ and $\overline{M}_n$ of a polymer sample are known, we have information about the standard deviation s_n and the variance of the number distribution. There is no quantitative information about the breadth of the weight distribution of the same sample unless $\overline{M}_z$ and $\overline{M}_w$ are known. As mentioned earlier, it is often assumed that the weight and number distributions will change in a parallel fashion and in this sense the $\overline{M}_w/\overline{M}_n$ ratio is called the breadth of "the" distribution although it actually reflects the ratio of the variance to the square of the mean of the number distribution of the polymer (Eq. 2-34).

Very highly branched polymers, like polyethylene made by free-radical, high-pressure processes, will have $\overline{M}_w/\overline{M}_n$ ratios of 20 and more. Most polymers made by free-radical or coordination polymerization of vinyl monomers have ratios of from 2 to about 10. The $\overline{M}_w/\overline{M}_n$ ratios of condensation polymers like nylons and thermoplastic polyesters tend to be about 2, and this is generally about the narrowest distribution found in commercial thermoplastics.

A truly monodisperse polymer has $\overline{M}_w/\overline{M}_n$ equal to 1.0. Such materials have not been synthesized to date. The sharpest distributions that have actually been made are those of polystyrenes from very careful anionic polymerizations. These have $\overline{M}_w/\overline{M}_n$ ratios as low as 1.04. Since the polydispersity index is only 4% higher than that of a truly monodisperse polymer, these polystyrenes are sometimes assumed to be monodisperse. This assumption is not really justified, despite the small difference from the theoretical value of unity.

For example, let us consider a polymer sample for which $\overline{M}_n = 100,000$, $\overline{M}_w = 104,000$, and $\overline{M}_w\overline{M}_n = 1.04$. In this case s_n is 20,000 from Eq. (2-33). It can be shown, however [1], that a sample with the given values of $\overline{M}_w$ and $\overline{M}_n$ could have as much as 44% of its molecules with molecular weights less than 70,000 or greater than 130,000. Similarly, as much as 10 mol% of the sample could have molecular weights less than 38,000 or greater than 162,000. This polymer actually has a sharp molecular weight distribution compared to ordinary synthetic polymers, but it is obviously not monodisperse.

It should be understood that the foregoing calculations do not assess the symmetry of the distribution. We do not know whether the mole fraction outside the last size limits mentioned is actually 0.1, but we know that it cannot be greater than this value with the quoted simultaneous $\overline{M}_n$ and $\overline{M}_w$ figures. (In fact, the distribution would have to be quite unusual for the proportions to approach this boundary value.) We also do not know how this mole fraction is distributed at the high and low molecular weight ends and whether these two tails of the distribution are equally populated. The $\overline{M}_n$ and $\overline{M}_w$ data available to this point must be supplemented by higher moments to obtain this information.

We should note also that a significant mole fraction may not necessarily comprise a very large proportion of the weight of the polymer. In our last example, if the 10 mol% with molecular weight deviating from $\overline{M}_n$ by at least $\pm 62,000$ were all material with molecular weight 38,000 it would be only 3.8% of the weight of the sample. Conversely, however, if this were all material with molecular weight 162,000, the corresponding weight fraction would be 16.2.

There are various ways of expressing the skewness of statistical distributions. The method most directly applicable to polymers uses the third moment of the distribution about its mean. The extreme molecular weights are emphasized because their deviation from the mean is raised to the third power, and since this power is an odd number, the third moment also reflects the net direction of the deviations.

In mathematical terms,

$$_n U_3 = \sum_i x_i (M_i - \overline{M}_n)^3$$

$$_n U_3 = \sum_i x_i \left(M_i^3 - 3M_i^2 \overline{M}_n + 3M_i \overline{M}_n^2 - \overline{M}_n^3 \right) \tag{2-36}$$

$$_n U_3 = \sum_i x_i M_i^3 - 3\overline{M}_n \sum x_i M_i^2 + 3\overline{M}_n^2 \sum x_i M_i - \overline{M}_n^3 \sum x_i$$

For a normalized distribution,

$$_n U_3 = \overline{M}_z \overline{M}_w \overline{M}_n - 3\overline{M}_n (\overline{M}_w \overline{M}_n) + 3\overline{M}_n^2 (\overline{M}_n) - \overline{M}_n^3 \tag{2-37}$$

$$_n U_3 = \overline{M}_z \overline{M}_w \overline{M}_n - 3\overline{M}_n^2 \overline{M}_w + 2\overline{M}_n^3 \tag{2-38}$$

where $_n U_3$ is positive if the distribution is skewed toward high molecular weights, zero if it is symmetrical about the mean, and negative if it is skewed to low molecular weights.

Asymmetry of different distributions is most readily compared by relating the skewness to the breadth of the distribution. The resulting measure α_3 is obtained by dividing U_3 by the cube of the standard deviation. For the number distribution,

$$_n \alpha_3 = \frac{_n U_3}{s_n^3} = \frac{\overline{M}_z \overline{M}_w \overline{M}_n - 3\overline{M}_n^2 \overline{M}_w + 2\overline{M}_n^3}{\left(\overline{M}_w \overline{M}_n - \overline{M}_n^2 \right)^{3/2}} \tag{2-39}$$

2.6   SUMMARIZING THE MOLECULAR WEIGHT DISTRIBUTION

Complete description of a molecular weight distribution implies a knowledge of all its moments. The central tendency, breadth, and skewness may be summarized by parameters calculated from the moments about zero: U_0', U_1', U_2', and U_3'. These moments also define the molecular weight averages $\overline{M}_n$, $\overline{M}_w$ and $\overline{M}_z$. Note that $\overline{M}_n$ and $\overline{M}_w$ can be measured directly without knowing the distribution but it has not been convenient to obtain $\overline{M}_z$ of synthetic polymers as a direct measurement of a property of the sample. Thus, some information about the breadth of the number distribution can be obtained from $\overline{M}_n$ and $\overline{M}_w$ without analyzing details of the distribution, but the latter information is necessary for estimation of the breadth of the weight distribution and for skewness calculations. This is most conveniently done by means of gel permeation chromatography, which is discussed in Section 3.4.

2.7 $\overline{M}_z \geq \overline{M}_w \geq \overline{M}_n$

Equation (2-34) can be rewritten as

$$\overline{M}_w/\overline{M}_n = s_n^2/\overline{M}_n^2 + 1 \qquad (2\text{-}34a)$$

Since the first term on the right-hand side is the quotient of squared terms, it is always positive or zero. Zero equality is obtained only when the distribution is monodisperse, and s_n then equals zero.
It is obvious then that

$$\overline{M}_w/\overline{M}_n \geq 1$$

with the equality true only for monodisperse polymers.
     Equation (2-35) similarly leads to the conclusion that

$$\overline{M}_z/\overline{M}_w \geq 1$$

and in general,

$$M_{z+j+1} \geq M_{z+j} \geq M_{z+j-1} \geq \cdots \geq \overline{M}_w \geq \overline{M}_n \qquad (2\text{-}40)$$

2.8   INTEGRAL AND SUMMATIVE EXPRESSIONS

The relations presented so far have been in terms of summations for greater clarity. The equations given are valid for a distribution in which the variable (molecular weight) assumes only discrete values. However, differences between successive molecular weights are trivial compared to macromolecular sizes and the accuracy with which these values can be measured. Molecular weight distributions can therefore be regarded as continuous, and integral expressions are also valid.

     In the latter case $q(M)$ is a function of the molecular weight such that the *quantity* of polymer with molecular weight between M and $M + dM$ is given by $q(M)\, dM$. [If the quantity is expressed in moles, then $q(M) = n(M)$; if in mass units, $q(M) = c(M)$.]

The *proportion* of the sample with molecular weight between M and $M + dM$ is given by $f(M)\,dM$ where $f(M)$ is the frequency distribution and $f(M)$ will equal $x(M)$ for a number distribution [or $w(M)$ for a weight distribution].

An arithmetic mean is defined in general as

$$A = \int_0^\infty Mq(M)\,dM \bigg/ \int_0^\infty q(M)\,dM \qquad (2\text{-}41)$$

or in equivalent terms as

$$A = \int_0^\infty Mf(M)\,dM \qquad (2\text{-}42)$$

Equation (2-42) is the integral equivalent of summative equation (2-4). The number fraction of the distribution with molecular weights in the interval M to $M + dM$ is $dx(M) = x(M)\,dM$, and the corresponding weight fraction is $dw(M) = w(M)\,dM$. The following expressions are examples of integral equations that are directly parallel to the summative expressions generally used in this chapter:

$$\overline{M}_n = \int_0^\infty Mx(M)\,dM = 1 \bigg/ \int_0^\infty \frac{w(M)\,dM}{M} \qquad (2\text{-}43)$$

[since $w(M) = (M/\overline{M}_n)x(M)$]

$$\overline{M}_w = \int_0^\infty Mw(M)\,dM \qquad (2\text{-}44)$$

$$\overline{M}_z = \int_0^\infty M^2 w(M)\,dM \bigg/ \int_0^\infty Mw(M)\,dM \qquad (2\text{-}45)$$

Some authors prefer to take the integration limits from $-\infty$ to $+\infty$. The results are equivalent to those shown here for molecular weight distributions because negative values of the variable are physically impossible.

2.9   DIRECT MEASUREMENTS OF AVERAGE MOLECULAR WEIGHTS [2]

It is possible to determine one of the average molecular weights of a polymer sample without knowing the molecular weight distribution. This is accomplished by measuring a chosen property of a solution of the sample. Because this procedure is very widely used it is worthwhile here to identify the assumptions involved:

1.   The value of the property measured is assumed to depend on the nature of the polymer and solvent and on the concentration and temperature of the solution. The functional dependence of the property on the polymer molecular weight is assumed, however, to be the same for all molecular weights in the sample.

2.   It is assumed that all the solute molecules in the sample contribute additively and independently to the measured value of the solution property. This assumption can only be true in extremely dilute solution where the polymer molecules are not affected by each other's existence. Such solutions would have properties very close to those of pure solvent. It is intuitively evident that the conflicting requirements of independence of solute contributions and a magnitude of such contributions sufficient to outweigh experimental uncertainty can only be achieved by making measurements at low, but finite, concentrations and extrapolating these to infinite dilution values.

Suppose that a colligative property of the polymer solution is measured. These are properties that depend on the number of dissolved solute molecules and not on their sizes (see also Section 2.10). Osmotic pressure, vapor pressure lowering, and freezing point depression are some examples of colligative properties. If the value of the property measured is P', then by definition

$$P' = K'N \qquad (2\text{-}46)$$

where K' depends on the solvent and the particular colligative property and N is the molar concentration of polymer.

If assumption 2 just given is true, the polymer sample can be considered without error to be made up of a combination of monodisperse species of various sizes M_i and corresponding molar concentration n_i. Since

$$\sum n_i = N \qquad (2\text{-}1\text{a})$$

and from assumption 1,

$$P_i' = K'n_i \qquad (2\text{-}47)$$

it follows that

$$P' = \sum P_i = K' \sum n_i \qquad (2\text{-}48)$$

Actually, the molar concentration cannot be measured during the experiment because the molecular weight is not yet known. The weight concentration c is the experimental parameter. Obviously

$$c = \sum c_i = \sum n_i M_i \qquad (2\text{-}9\text{a})$$

If the experimental procedure involves measurement of the property per unit weight concentration (P'/c) and subsequent extrapolation of this ratio to infinite dilution, the result is

$$\lim_{c \to 0} \left(\frac{P'}{c} \right) = \sum P_i' \bigg/ \sum c_i = K' \sum n_i M_i = \frac{K'}{\overline{M}_n} \qquad (2\text{-}49)$$

from the definition of $\overline{M}_n$ in Eq. (2-8). (Note that both P' and c tend to zero as the solution is diluted, but their ratio is finite.)

Clearly then, if a colligative property of a polymer solution is measured this provides an estimate of $\overline{M}_n$ of the solute. The choice of the solution property has determined the average molecular weight that the measurement yields.

Alternatively, the property measured could be the intensity of light scattered from the solution. In this case (Section 3.2.2) it turns out that the measured value depends on the product of the weight concentration and molecular weight of the solute. That is

$$P'' = K''cM \tag{2-50}$$

Then, for a heterogeneous polymer comprising concentrations c_i of monodisperse species with molecular weights M_i,

$$P'' = K'' \sum c_i M_i \tag{2-51}$$

and

$$\lim_{c \to 0}[P''/c] = P'' \Big/ \sum c_i = K'' \sum c_i M_i \Big/ \sum c_i = K'' \overline{M}_w \tag{2-52}$$

from Eq. (2-14).

The weight average molecular weight of the sample is determined directly by measurement of the turbidity of the polymer solution.

Ultracentrifugation experiments yield $\overline{M}_z$ values. They are used primarily for biological polymers. The random coil conformation of most synthetic polymers in their solvents makes it difficult to interpret these data, and this method is little used with such materials.

Viscosity average molecular weights are an example of a direct measurement that yields another different average. This technique is discussed in Section 3.3.

2.10   COLLIGATIVE PROPERTIES AND $\overline{M}_n$

Those properties that depend only on the concentration of solute molecules and not on the nature of the solute are called *colligative*. A colligative property is also a measure of the chemical potential (partial molar Gibbs free energy) of the solvent in the solution. We consider ideal solutions first and then show how allowances are made for real solutions.

2.10.1   Ideal Solutions [3]

An ideal solution is one in which the mixing volume and enthalpy effects are zero. In the range of concentrations for which a solution is ideal, the partial molar quantities $\bar{V}$ and $\bar{H}$ of the components are constant. In the case of solvent, these partial molar quantities are equal to the molar quantities for pure solvent, since dilute solutions approach ideality more closely than concentrated ones.

An appropriate definition of an ideal solution is one in which for each component

$$d\mu_i = RT\ d\ln x_i \tag{2-53}$$

where μ_i and x_i are the chemical potential and mole fraction, respectively, of the ith component. (Alternative definitions of ideality can be shown to follow from this expression.) Integrating,

$$\mu_i = \mu_i^0 + RT\ln x_i \tag{2-54}$$

where μ_i^0 is the standard chemical potential. Dilute solutions tend to approach ideality as they approach infinite dilution. That is, Eq. (2-53) becomes valid as the solvent mole fraction approaches unity and all other mole fractions approach zero. Then, if the solvent is labeled component 1: $\mu_1^0 = G_1^0$ the molar Gibbs free energy of solvent. All other μ_i^0 do not concern us here.

It is useful to express Eq. (2-53) in terms of the solute mole fraction x_2. For the arbitrary variable y in general,

$$\ln y = \sum_{K=1}^{\infty}(-1)^{K+1}\frac{(y-1)^K}{K}, \qquad 0 \le y \le 2 \tag{2-55}$$

and since, for a two-component solution,

$$x_1 = 1 - x_2$$

$$\ln x_1 = \ln(1-x_2) = -x_2 - \frac{1}{2}x_2^2 - \frac{1}{3}x_2^3\cdots \tag{2-56}$$

Thus the solvent chemical potential μ_1 follows from Eqs. (2-53) and (2-56) as

$$\mu = \mu_1^0 + RT\ln(1-x_2) = G_1^0 + RT\ln(1-x_2)$$

$$= G_1^0 - RT\left[x_2 + \frac{1}{2}x_2^2 + \frac{1}{3}x_2^3 + \cdots\right] \tag{2-57}$$

In dilute solution, the total number of moles of all species in unit volume will approach n_1, the molar concentration of solvent. Then the mole fraction x_i of any component i can be expressed as

$$x_i = n_i \Big/ \sum n_i \rightarrow \frac{n_i}{n_1} \tag{2-58}$$

as the solution behavior approaches ideality.

If the molar and weight concentrations of solute are n_2 and c_2, respectively, then

$$c_2 = n_2 M \tag{2-59}$$

and

$$x_2 = c_2/Mn_1 \tag{2-60}$$

If the molar volume of pure solvent is V_1^0, with the same volume unit as is used to express the concentrations c_i and n_i (e.g., liters), then $n_1 = 1/V_1^0$ and

$$x_2 = c_2 V_1^0/M \tag{2-61}$$

Substituting in Eq. (2-57),

$$\mu_1 - G_1^0 = -RT V_1^0 \left[c_2/M + \left(V_1^0/2M^2 \right) c_2^2 + \left(\left(V_1^0 \right)^2/3M^3 \right) c_2^3 + \cdots \right] \tag{2-62}$$

Equation (2-62) is the key to the application of colligative properties to polymer molecular weights. We started with Eq. (2-53), which defined an ideal solution in terms of the mole fractions of the components. Equation (2-62), which followed by simple arithmetic, expresses the difference in chemical potential of the solvent in the solution and in the pure state in terms of the mass concentrations of the solute. This difference in chemical potential is seen to be a power series in the solute concentration. Such equations are called *virial equations* and more is said about them on page 65.

It is evident that insertion of corresponding experimental values of c_2, V_1^0, and $(\mu_2 - G_1^0)$ into Eq. (2-62) would provide a measure of the solute molecular weight M. We show in Section 2.10.2 how the difference in the value of a colligative property in pure solvent and solution measures $(\mu_2 - G_1^0)$ and in Section 2.10.3 that the M measured by application of Eq. (2-62) is $\overline{M}_n$ of the polymeric solute.

2.10.2 Osmotic Pressure

Colligative properties reflect the chemical potential of the solvent in solution. Alternatively, a colligative property is a measure of the depression of the activity of the solvent in solution, compared to the pure state. Colligative properties include vapor pressure lowering, boiling point elevation, freezing point depression, and membrane osmometry. The latter property is considered here, since it is the most important of the group as far as synthetic polymers are concerned.

Figure 2-5 is a schematic of apparatus for the measurement of osmotic pressure. A solution is separated from its pure solvent by a semipermeable membrane, which allows solvent molecules to pass but blocks solute. Both components are at the same temperature, and the hydrostatic pressure on each is recorded by means of the heights of the corresponding fluids in capillary columns. The solute cannot distribute itself on both sides of the membrane. The solvent flows initially, however, to dilute the solution, and this flow will continue until sufficient excess hydrostatic

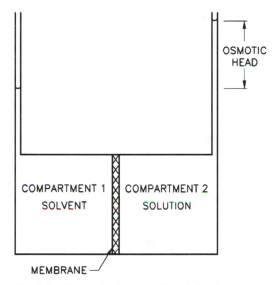

Fig. 2-5.   System for demonstration of osmotic pressure.

pressure is generated on the solution side to block the net flow of solvent. This excess pressure is the osmotic pressure. At thermodynamic equilibrium also, the chemical potential of the solvent will be the same on both sides of the membrane. The relation between this chemical potential, which appears in Eq. (2-62), and the measured osmotic pressure is derived next. Let

μ_1' = chemical potential of the pure solvent in compartment 1 under atmospheric pressure P_1. By definition $\mu_1' = G_1^0$.

μ_1 = chemical potential of the solvent in solution (on side 2) under atmospheric pressure P_1.

μ_1' = chemical potential of the solvent in solution on side 2 under the final pressure $P_1 + \pi$, where π is the osmotic pressure.

The condition for equilibrium is

$$G_1^0 = \mu_1' = \mu_1'' \tag{2-63}$$

The chemical potential of solvent at pressure $P_1 + \pi$ is

$$\mu_1'' = \mu_1 + \int_{P_1}^{P_1+\pi} \left(\frac{\partial \mu_1}{\partial P}\right)_T dP \tag{2-64}$$

In general $(\partial \mu_i / \partial P)_T = \overline{V}_1$. The partial molar volume $\overline{V}_1$ of solvent will be essentially independent of P over the pressure range and will moreover be essentially equal to the molar volume V_1^0 in dilute solutions where osmotic pressure measurements are made. Thus, from Eq. (2-64),

$$\mu_1'' = \mu_1 + \int_{P_1}^{P_{1+\pi}} V_1^0 \, dP \tag{2-65}$$

$$\mu_1'' = \mu_1 + \pi V_1^0 \tag{2-66}$$

From Eq. (2-63), at equilibrium,

$$\pi = -\left(\mu_1 - G_1^0\right)/V_1^0 \tag{2-67}$$

Thus the osmotic pressure π is a direct measure of the chemical potential μ_1 of the solvent in the solution. Equating the terms for $(\mu_1 - G_1^0)$ in Eqs. (2-67) and (2-62),

$$\pi = RTc_2\left[1/M + \left(V_1^0/2M^2\right)c_2 + \cdots\right] \tag{2-68}$$

In the limit of zero concentration

$$\lim_{c_2 \to 0} (\pi/c_2) = RT/M \tag{2-69}$$

which is van't Hoff's law of osmotic pressures.

Other colligative properties can similarly be shown to be related to the left-hand side of Eq. (2-62). Vapor pressure lowering is related, for example, through Raoult's law and Eq. (2-54). Reference should be made to standard introductory physical chemistry textbooks.

The difference $(G_1^0 - \mu_1)$ is measured by the difference $[(P_1 + \pi) - P_1]/V_1^0$ in the osmotic pressure experiment. Other colligative properties are similarly measured in terms of the difference between a property of the pure solvent and that of the solvent in solution, at a particular concentration and common temperature. Specifically, boiling point elevation (ebulliometry) measurements result in

$$\lim_{c_2 \to 0} (\Delta T_b/c_2) = RT_b^2 V_1^0/\Delta H_v M \tag{2-70}$$

where ΔT_b is the difference in temperatures between the boiling point of a solution with concentration c_2 and that of the pure solvent T_b, at the same pressure, and ΔH_v is the latent heat of vaporization of the solvent. Freezing point depression (cryoscopy) measurements yield

$$\lim_{c_2 \to 0} (\Delta T_f/c_2) = RT_f^2 V_1^0/\Delta H_f M \tag{2-71}$$

where the symbols parallel those which apply in ebulliometry and vapor pressure lowering experiments ideally result in

$$\lim_{c_2 \to 0} (\Delta P/c_2) = -P_1^0 V_1^0/M \tag{2-72}$$

where ΔP is the difference between the vapor pressure of the solvent above the solution and P^0, which is the vapor pressure of pure solvent at the same temperature.

2.10.3 Osmotic Pressure Measures, $\overline{M}_n$

Equation (2-68) shows that π is related to the molecular weight of the solute. If the latter is polydisperse in molecular weight then an average value should be inserted into this equation in place of the symbol M.

For a mixture of monodisperse macromolecular species, each with concentration c_i and molecular weight M_i,

$$\pi = RT\frac{c_2}{M} = RT\sum\frac{c_i}{M_i} \tag{2-73}$$

from Eq. (2-69). Since $c_i = n_i M_i$,

$$\frac{\pi}{c_2} = RT\sum n_i \Big/ \sum n_i M_i = \frac{RT}{\overline{M}_n} \tag{2-74}$$

Thus, the reduced osmotic pressure (π/c) measures $\overline{M}_n$.

2.11 VIRIAL EQUATIONS

The real solutions used to study the characteristics of macromolecular solutes are rarely ideal even at the highest dilutions that can be used in practice. The expressions derived earlier for ideal solutions are therefore invalid in the experimental range. It is useful, however, to retain the form of the ideal equations and express the deviation of real solutions in terms of empirical parameters. Thus the usual practice in micromolecular thermodynamics is to retain Eq. (2-62) but substitute fictitious concentrations, called *activities*, for the experimental solute concentrations. In polymer science, on the other hand, the measured concentrations are taken as accurate and deviations from ideality are expressed in the coefficients of the concentration terms.

For example, the osmotic pressure of an ideal solution is given by Eq. (2-68) as

$$-\frac{\left(\mu_1 - G_1^0\right)}{c_2 V_1^0} = \frac{\pi}{c_2} = \frac{RT}{M} + \frac{RT V_1^0}{2M^2}c_2 + \frac{RT\left(V_1^0\right)^2}{3M^3}c_2^2 + \cdots \tag{2-75}$$

The osmotic data of a real solution are then expressed in a parallel form as

$$\pi/c = RT[1/\overline{M}_n + A_2 c + A_3 c^2 + \cdots] \tag{2-76}$$

where A_2 and A_3, the second and third virial coefficients, would be determined in the final analysis by the fitting of corresponding π and c_2 data to Eq. (2-75). (In a bicomponent solution the subscript 2, which refers to solute, is often deleted.)

Unfortunately, there is no uniformity in the exact form of the virial equations used in polymer science. Alternatives to Eq. (2-76) include

$$\pi/c = (\pi/c)_{c=0}[1 + \Gamma_2 c + \Gamma_3 C^2 + \cdots] \qquad (2\text{-}77)$$

and

$$\pi/c = (RT/\overline{M}_n) + Bc + Cc^2 + \cdots \qquad (2\text{-}78)$$

The three forms are equivalent if

$$B = RTA_2 = (RT/M)\Gamma_2 \qquad (2\text{-}79)$$

Authors may report virial coefficients without specifying the equation to which they apply, and this can usually be deduced only by inspecting the units of the virial coefficient. Thus A_2 has units of mol cm^3/g^2 if M is a gram-mole with units of g/mol. If M is in g, however, then A_2 is in cm^3/g. The units of Γ_2 and B may depend on the particular units chosen for R, c_2, and M.

2.12   VIRIAL COEFFICIENTS

In polymer science, the ideal form of the thermodynamic equations is preserved and the nonideality of polymer solutions is incorporated in the virial co-efficients. At low concentrations, the effects of the c_2^2 terms in any of the equations will be very small, and the data are expected to be linear with intercepts which yield values of $\overline{M}_n^{-1}$ and slopes which are measures of the second virial coefficient of the polymer solution. Theories of polymer solutions can be judged by their success in predicting nonideality. This means predictions of second virial coefficients in practice, because this is the coefficient that can be measured most accurately. Note in this connection that the intercept of a straight line can usually be determined with more accuracy than the slope. Thus many experiments which are accurate enough for reasonable average molecular weights do not yield reliable virial coefficients. Many more data points and much more care is needed if the experiment is intended to produce a reliable slope and consequent measure of the second virial coefficient.

A number of factors influence the magnitude of the second virial coefficient. These include the nature of the polymer and solvent, the molecular weight distribution of the polymer and its mean molecular weight, concentration and temperature of the solution, and the presence or absence of branching in the polymer chain.

The second virial coefficient decreases with increasing molecular weight of the solute and with increased branching. Both factors tend to result in more compact structures which are less swollen by solvent, and it is generally true that better solvents result in more highly swollen macromolecules and higher virial coefficients.

The virial coefficients reflect interactions between polymer solute molecules because such a solute excludes other molecules from the space that it pervades. The *excluded volume* of a hypothetical rigid spherical solute is easily calculated, since the closest distance that the center of one sphere can approach the center of another is twice the radius of the sphere. Estimation of the excluded volume of flexible polymeric coils is a much more formidable task, but it has been shown that it is directly proportional to the second virial coefficient, at given solute molecular weight.

Most polymers are more soluble in their solvents the higher the solution temperature. This is reflected in a reduction of the virial coefficient as the temperature is reduced. At a sufficiently low temperature, the second virial coefficient may actually be zero. This is the Flory theta temperature, which is defined as that temperature at which a given polymer species of infinite molecular weight would be insoluble at great dilution in a particular solvent. A solvent, or mixture of solvents, for which such a temperature is experimentally attainable is a theta solvent for the particular polymer.

Theta conditions are of great theoretical interest because the diameter of the polymer chain random coil in solution is then equal to the diameter it would have in the amorphous bulk polymer at the same temperature. The solvent neither expands nor contracts the macromolecule, which is said to be in its "unperturbed" state. The theta solution allows the experimenter to obtain polymer molecules which are unperturbed by solvent but separated from each other far enough not to be entangled. Theta solutions are not normally used for molecular weight measurements, because they are on the verge of precipitation. The excluded volume vanishes under theta conditions, along with the second virial coefficient.

APPENDIX 2A: MOLECULAR WEIGHT AVERAGES OF BLENDS OF BROAD DISTRIBUTION POLYMERS

When broad distribution polymers are blended, $\overline{M}_n$, $\overline{M}_w$, $\overline{M}_z$, etc., of the blend are given by the corresponding expressions listed in Table 2-2 but the M_i's in this case are the appropriate average molecular weights of the broad distribution components of the mixture. Thus, for such a mixture,

$$(\overline{M}_n)_{\text{mixture}} = 1 \bigg/ \sum \frac{w_i}{(\overline{M}_n)_i} \qquad (2\text{-}11a)$$

$$(\overline{M}_w)_{\text{mixture}} = \sum w_i (\overline{M}_w)_i \qquad (2\text{-}13a)$$

and so on.

As a "proof," consider a mixture formed of a grams of a monodisperse polymer A (with molecular weight M_A) and b grams of a monodisperse polymer B (with molecular weight M_B). This blend, which we call mixture 1, contains weight fraction $(w_A)_1$ of polymer A and weight fraction $(w_B)_1$ of polymer B.

$$(w_A)_1 = a/(a+b) \text{ and } (w_B)_1 = b/(a+b)$$

The number average molecular weight $(\overline{M}_n)_1$ of this mixture is

$$(\overline{M}_n)_1 = \frac{a+b}{a/M_A + b/M_B} \tag{2-11b}$$

$$(\overline{M}_n)_1 = \left[\frac{a}{(a+b)M_A} + \frac{b}{(a+b)M_B} \right]^{-1} \tag{2A-1}$$

If mixture 2 is produced by blending c grams of A and d grams of B, then similarly

$$(w_A)_2 = c/(c+d) \text{ and } (w_B)_2 = d/(c+d)$$

while

$$(\overline{M}_n)_2 = \left[\frac{c}{(c+d)M_A} + \frac{d}{(c+d)M_B} \right]^{-1} \tag{2A-2}$$

Now we blend e grams of mixture 1 with f grams of mixture 2. The weight fraction w_1 of mixture 1 is $e/(e+f)$ and that of mixture 2 is $w_2 = f/(e+f)$. The weight fraction of polymer A in the final mixture is $w_1(w_A)_1 + w_2(w_A)_2 = w_A$ and that of polymer B is $w_1(w_B)_1 + w_2(w_B)_1 = w_B$.

$$w_1(w_A)_1 + w_2(w_A)_2 = \frac{e}{e+f}\left(\frac{a}{a+b}\right) + \frac{f}{e+f}\left(\frac{c}{c+d}\right) = w_A \tag{2A-3}$$

$$w_1(w_B)_1 + w_2(w_B)_2 = \frac{e}{e+f}\left(\frac{b}{a+b}\right) + \frac{f}{e+f}\left(\frac{d}{c+d}\right) = w_B \tag{2A-4}$$

The number of average molecular weight of the final blend is

$$(\overline{M}_n)_{\text{blend}} = \left(\frac{w_A}{M_A} + \frac{w_B}{M} \right)^{-1} \text{ by definition} \tag{2-11c}$$

Substituting

$$(\overline{M}_n)_{\text{blend}} = \left[\left(\frac{e}{e+f}\right)\left(\frac{a}{a+b}\right) \Big/ M_A + \left(\frac{f}{e+f}\right)\left(\frac{c}{c+d}\right) \Big/ M_A \right.$$
$$\left. + \left(\frac{e}{e+f}\right)\left(\frac{b}{a+b}\right) \Big/ M_B + \left(\frac{f}{e+f}\right)\left(\frac{d}{c+d}\right) \Big/ M_B \right]^{-1}$$

$$= \left[w_1 \left(\frac{a}{a+b} \right) \middle/ M_A + w_2 \left(\frac{c}{c+d} \right) \middle/ M_A \right.$$

$$\left. + w_1 \left(\frac{b}{a+b} \right) \middle/ M_B + w_2 \left(\frac{d}{c+d} \right) \middle/ M_B \right]^{-1}$$

$$= \left[w_1 \left(\left[\frac{M_A(a+b)}{a} \right]^{-1} + \left[\frac{M_B(a+b)}{b} \right]^{-1} \right) \right.$$

$$\left. + w_2 \left(\left[\frac{M_A(c+d)}{c} \right]^{-1} + \left[\frac{M_B(c+d)}{d} \right]^{-1} \right) \right]^{-1} \qquad (2A\text{-}5)$$

$$(\overline{M}_n)_{\text{blend}} = \left[\frac{w_1}{(\overline{M}_n)_1} + \frac{w_2}{(\overline{M}_n)_2} \right]^{-1} \qquad (2A\text{-}6)$$

This is equivalent to Eq. (2-11) with $(\overline{M}_n)_1$ substituted for M_i.

Similar expressions can be developed in a straightforward manner for $\overline{M}_w$, $\overline{M}_z$ and so on.

PROBLEMS

2-1. If equal weights of polymer A and polymer B are mixed, calculate $\overline{M}_w$ and $\overline{M}_n$ of the mixture

$$\text{Polymer A: } \overline{M}_n = 35,000, \quad \overline{M}_w = 90,000$$

$$\text{Polymer B: } \overline{M}_n = 150,000, \quad \overline{M}_w = 300,000$$

2-2. Calcium stearate $(Ca(OOC(CH_2)_{16}CH_3)_2)$ is sometimes used as a lubricant in the processing of poly(vinyl chloride). A sample of PVC compound containing 2 wt% calcium stearate was found to have $\overline{M}_n = 25,000$. What is $\overline{M}_n$ of the balance of the PVC compound?

2-3. If equal weights of "monodisperse" polymers with molecular weights of 5000 and 50,000 are mixed, what is $\overline{M}_z$ of the mixture?

2-4. Calculate $\overline{M}_n$ and $\overline{M}_w$ for a sample of polystyrene with the following composition (i = degree of polymerization; % is by weight). Calculate the variance of the number distribution of molecular weights

i	20	25	30	35	40	45	50	60	80	>80
wt.%	30	20	15	11	8	6	4	3	3	0

2-5. What value would $\overline{M}_z$ have for a polymer sample for which $\overline{M}_v = \overline{M}_n$?

2-6.   The measured diameters of a series of spheres follow:

Number of spheres	Diameter (cm)
2	1
3	2
4	3
2	4

(a)   Calculate the number average diameter $(\overline{D}_n)$.

(b)   Calculate the weight average diameter $(\overline{D}_w)$. [Weight of a sphere $\propto$ volume $\propto$ (diameter)^3.]

2-7.   A chemist dissolved a 50-g sample of a polymer in a solvent. He added nonsolvent gradually and precipitated out successive polymer-rich phases, which he separated and freed of solvent. Each such specimen (which is called a *fraction*) was weighted, and its number average molecular weight $\overline{M}_n$ was determined by suitable methods. His results follow:

Fraction no.	Weight (g)	$\overline{M}_n$
1	1.5	2,000
2	5.5	50,000
3	22.0	100,000
4	12.0	200,000
5	4.5	500,000
6	1.5	1,000,000

Assume that each fraction is monodisperse and calculate $\overline{M}_n$, $\overline{M}_w$, and a measure of the breadth of the number distribution for the recovered polymer. (*Note*: This is not a recommended procedure for measuring molecular weight distributions. The fractions obtained by the method described will not be monodisperse and the molecular weight distributions of successive fractions will overlap. The assignment of a single average molecular weight to each fraction is an approximation that may or may not be useful in particular cases.)

2-8.   The degree of polymerization of a certain oligomer sample is described by the distribution function

$$w_i = K(i^3 - i^2 + 1)$$

where w_i is the weight fraction of polymer with degree of polymerization i and i can take any value between 1 and 10, inclusively.

(a)   Calculate the number average degree of polymerization.

(b)   What is the standard deviation of the weight distribution?

(c)   Calculate the z average degree of polymerization.

(d)   If the formula weight of the repeating unit in this oligomer is 100 g mol^{-1}, what is $\overline{M}_z$ of the polymer?

REFERENCES

[1]   G. Herdan, "Small Particle Statistics." Butterworths, London, 1960, p. 281.

[2]   M. Gordon, "High Polymers," Addison-Wesley, Reading, MA, 1963.

[3]   C. Tanford, "Physical Chemistry of Macromolecules." Wiley, New York, 1961.

Chapter 3

Practical Aspects of Molecular Weight Measurements

Now what I want is, Facts. Facts alone are wanted in life.
—Charles Dickens, *Hard Times*

3.1 $\overline{M}_n$ METHODS

Most of the procedures for measuring $\overline{M}_n$ rely on colligative solution properties (Section 2.10). These properties include osmotic pressure, boiling point elevation, freezing point depression, and vapor pressure lowering. They are all
described thermodynamically by the ideal solution of Eq. (2-62) or its analogs,
such as Eq. (2-76), for real solutions. At given solute concentration, these relations show that the effect of the dissolved species on the chemical potential of
the solvent decreases with increasing solute molecular weight. Therefore, any
colligative property measurement must be very sensitive if it is to be useful with
high-molecular-weight solutes like synthetic polymers.

Membrane osmometry is the most sensitive and accurate colligative property technique. Consider, for example, a polystyrene with $\overline{M}_n$ around 200,000.
Trial-and-error experience has shown that molecular weight measurements with
similar samples are best made by starting with solutions in good solvents (like
toluene in this case) at concentrations around 10 g/liter and making successive
dilutions from this value. The initial polymer concentration is then [(10 g/liter)
(mol/200,000 g)] $= 5 \times 10^{-5} M$. At room temperature RT is of the order of 23
L-atm and Eq. (2-69) indicates that the osmotic pressure would be about 10^{-3} atm.
This corresponds to the pressure exerted by a column of organic solvent about
14 mm high. Modern membrane osmometers measure pressures with precisions

of the order of ±0.2 mm so that the measurement uncertainty here is of the order of $\pm2\%$. From Eq. (2-70) the boiling point of the solution would be about 1×10^{-4}°C higher than that of the solvent at the same pressure. This is approximately at the limit of conventient temperature measuring devices and thus boiling point elevation is not a suitable method for measuring $\overline{M}_n$ of polymers of this size. Similarly, the vapor pressure lowering would be of the order of 2×10^{-4} mm Hg (2.7×10^{-3} Pa) at 25°C (Eq. 2-72) and could not be measured reliably. Toluene would not be used as a solvent for freezing point depression measurements, because its freezing point is inconveniently low. If our sample were dissolved instead in a material like naphthalene, the difference between the freezing points of the solvent and that of a 1% (10 g/liter) solution would only be about 4×10^{-4}°C. It is obvious then that membrane osmometry is the only colligative property measurement that is practical for direct measurements of $\overline{M}_n$ of high polymers.

Two other techniques that are also used to measure $\overline{M}_n$ are not colligative properties in the strict sense. These are based on end-group analysis and on vapor phase osmometry. Both methods, which are limited to lower molecular weight polymers, are described later in this chapter. Some general details of the various procedures for measuring $\overline{M}_n$ directly are reviewed in this section.

3.1.1   Membrane Osmometry

The practical range of molecular weights that can be measured with this method is approximately 30,000 to one million. The upper limit is set by the smallest osmotic pressure which can be measured at the concentrations which can be used with polymer solutions. The lower limit depends on the permeability of the membrane toward low-molecular-weight polymers. The rule that "like dissolves like" is generally true for macromolecular solutes, and so the structure of solvents can be similar to that of oligomeric species of the polymer solute. Thus low-molecular-weight polystyrenes will permeate through a membrane which passes a solvent like toluene. The net result of this less than ideal semipermeability of real membranes is a tendency for the observed osmotic pressure to be lower than that which would be read if all the solute were held back. From Eq. (2-69), the molecular weight calculated from the zero concentration intercept will then be too high. Membrane osmometry is normally not used with lower molecular weight polymers for which vapor phase osmometry (Section 3.1.2) is more suitable for $\overline{M}_n$ measurements. The membrane leakage error is usually not serious with synthetic polymers with $\overline{M}_n \gtrsim 30,000$.

Osmometers consist basically of a solvent compartment separated from a solution compartment by a semipermeable membrane and a method for measuring the equilibrium hydrostatic pressures on the two compartments. In static osmometers this involves measurements of the heights of liquid in capillary tubes attached to the solvent and solution cells (Fig. 2-5).

Osmotic equilibrium is not reached quickly after the solvent and solution first contact the membrane. Periods of a few hours or more may be required for the pressure difference to stabilize, and this equilibration process must be repeated for each concentration of the polymer in the solvent. Various ingenious procedures have been suggested to shorten the experimental time. Much of the interest in this problem has waned, however, with the advent of high-speed automatic osmometers.

Modern osmometers reach equilibrium pressure in 10–30 min and indicate the osmotic pressure automatically. Several types are available. Some commonly used models employ sensors to measure solvent flow through the membrane and adjust a counteracting pressure to maintain zero net flow. Other devices use strain gauges on flexible diaphragms to measure the osmotic pressure directly.

The membrane must not be attacked by the solvent and must permit the solvent to permeate fast enough to achieve osmotic equilibrium in a reasonable time. If the membrane is too permeable, however, large leakage errors will result. Cellulose and cellulose acetate membranes are the most widely used types with synthetic polymer solutions. Measurements at the relatively elevated temperatures needed to dissolve semicrystalline polymers are hampered by a general lack of membranes that are durable under these conditions.

Membrane osmometry provides absolute values of number average molecular weights without the need for calibration. The results are independent of chemical heterogeneity of the polymer, unlike light scattering data (Section 3.2). Membrane osmometry measures the number average molecular weight of the whole sample, including contaminants, although very low molecular weight materials will equilibrate on both sides of the membrane and may not interfere with the analysis. Water-soluble polyelectrolyte polymers are best analyzed in aqueous salt solutions, to minimize extraneous ionic effects.

Careful experimentation will usually yield a precision of about ±5% on replicate measurements of $\overline{M}_n$ of the same sample in the same laboratory. Interlaboratory reproducibility is not as good as the precision within a single location and the variation in second virial coefficient results is greater than in $\overline{M}_n$ determinations.

The raw data in osmotic pressure experiments are pressures in terms of heights of solvent columns at various polymer concentrations. The pressure values are usually in centimeters of solvent (h) and the concentrations, c, may be in grams per cubic centimeter, per deciliter (100 cm^3), or per liter, and so on. The most direct application of these numbers involves plotting (h/c) against c and extrapolating to $(h/c)_0$ at zero concentration. The column height h is then converted to osmotic pressure π by

$$\pi = \rho h g \qquad\qquad (3\text{-}1)$$

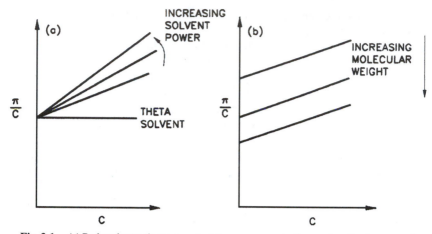

Fig. 3-1. (a) Reduced osmotic pressure (π/c) versus concentration, c, plots for the same polymer sample in different solvents. (b) (π/c) versus c plots for different molecular weight samples of the same polymer type in a common solvent.

where ρ is the density of the solvent and g is the gravitational acceleration constant. The value of $\overline{M}_n$ follows from

$$(\pi/c)_0 = RT/\overline{M}_n \tag{3-2}$$

(cf. Eq. 2-78). It is necessary to remember that the units of R must correspond to those of $(\pi/c)_0$. Thus with h (cm), ρ (g cm^{-3}), and g (cm s^{-2}), R should be in ergs mol^{-1}K^{-1}. For R in J mol^{-1}K^{-1}, h, ρ and g should be in SI units.

The second virial coefficient follows from the slope of the straight-line portion of the (π/c)–c plot essentially by dropping the c^2 terms in Eqs. (2-76)-(2-78).

It is to be expected that measurements of the osmotic pressures of the same polymer in different solvent should yield a common intercept. The slopes will differ (Fig. 3-1a), however, since the second virial coefficient reflects polymer–solvent interactions and can be related, for example, to the Flory–Huggins interaction parameter χ (Chapter 12) by

$$A_2 = \left(\frac{1}{2} - \chi\right) \Big/ L V_1^0 v_2^2 \tag{3-3}$$

Here v_2 is the specific volume of the polymer, L is Avogadro's number, V_1^0 is the molar volume of the solute, and χ is an interaction energy per mole of solvent divided by RT. When $\chi = 0.5$, $A_2 = 0$ and the solvent is a theta solvent for the particular polymer. Better solvents have lower χ values and higher second virial coefficients.

It may be expected also that different molecular weight samples of the same polymer should yield the same slopes and different intercepts when the osmotic pressures of their solutions are measured in a common solvent. This situation, which is shown in Fig. 3-1b, is not realized exactly in practice because the second virial coefficient is a weakly decreasing function of increasing polymer molecular weight.

3.1.2   Vapor Phase Osmometry

Regular membrane osmometry is not suitable for measurements of $\overline{M}_n$ below about 30,000 because of permeation of the solute through the membrane. Other colligative methods must be employed in this range, and vapor pressure lowering can be considered in this connection. The expanded virial form of Eq. (2-72) for this property is

$$\frac{\Delta P}{P_1^0} = \frac{\mu_1 - G_1^0}{RT} = -V_1^0 c_2 \left[\frac{1}{M} + B c_2 + C c_2^2 + \cdots \right] \qquad (3\text{-}4)$$

(Recall Eqs. 2-68 and 2-78). Direct measurement of ΔP is difficult because of the small magnitude of the effect. (At 10 g/liter concentration in benzene, a polymer with $\overline{M}_n$ equal to 20,000 produces a vapor pressure lowering of about 2×10^{-3} mm Hg at room temperature. The limits of accuracy of pressure measurements are about half this value.) It is more accurate and convenient to convert this vapor pressure difference into a temperature difference. This is accomplished in the method called *vapor phase osmometry*. The procedure is also known as vapor pressure osmometry or more accurately as thermoelectric differential vapor pressure lowering.

In the vapor phase osmometer, two matched thermistors are located in a thermostatted chamber which is saturated with solvent vapor. A drop of solvent is placed on one thermistor and a drop of polymer solution of equal size on the other thermistor. The solution has a lower vapor pressure at the test temperature (Eq. 2-72), and so the solvent condenses on the solution thermistor until the latent heat of vaporization released by this process raises the temperature of the solution sufficiently to compensate for the lower solvent activity. At equilibrium, the solvent has the same vapor pressure on the two temperature sensors but is at different temperatures.

Ideally the vapor pressure difference ΔP in Eq. (3-4) corresponds to a temperature difference ΔT, which can be deduced from the Clausius–Clapeyron equation

$$\Delta T = \Delta P R T^2 / \Delta H_v P_1^0 \qquad (3\text{-}5)$$

where ΔH_v is the latent heat of vaporization of the solvent at temperature T. With the previous equation

$$\frac{\Delta T}{c_2} = \frac{RT^2}{\Delta H_v} V_1^0 \left[\frac{1}{M} + B c_2 + C c_2^2 + \cdots \right] \qquad (3\text{-}6)$$

Thus the molecular weight of the solute can be determined in theory by measuring $\Delta T/c_2$ and extrapolating this ratio to zero c_2. (Since ΔT is small in practice, T may be taken without serious error as the average temperature of the two thermistors or as the temperature of the vapor in the apparatus.)

In fact, thermal equilibrium is not attained in the vapor phase osmometer, and the foregoing equations do not apply as written since they are predicated on the existence of thermodynamic equilibrium. Perturbations are experienced from heat conduction from the drops to the vapor and along the electrical connections. Diffusion controlled processes may also occur within the drops, and the magnitude of these effects may depend on drop sizes, solute diffusivity, and the presence of volatile impurities in the solvent or solute. The vapor phase osmometer is not a closed system and equilibrium cannot therefore be reached. The system can be operated in the steady state, however, and under those circumstances an analog of expression (3-6) is

$$\frac{\Delta T}{c_2} = k_s \left[\frac{1}{\overline{M}_n} + Bc_2 + Cc_2^2 + \cdots \right] \tag{3-7}$$

where k_s is an instrument constant. Attempts to calculate this constant *a priori* have not been notably successful and the apparatus is calibrated in use for a given solvent, temperature, and thermistor pair, by using solutes of known molecular weight. The operating equation is

$$\overline{M}_n = \frac{k}{(\Delta\Omega/c)_{c=0}} \tag{3-8}$$

where k is the measured calibration constant and $\Delta\Omega$ is the imbalance in the bridge (usually a resistance) which contains the two thermistors.

There is some question as to whether the calibration is independent of the molecular weight of the calibration standards, in some VPO instruments. It is convenient to use low-molecular-weight compounds, like benzil and hydrazo-benzene, as standards since these materials can be obtained in high purity and their molecular weights are accurately known. However, molecular weights of polymeric species which are based on the calibrations of some vapor phase instruments may be erroneously low. The safest procedure involves use of calibration standards which are in the same molecular weight range, more or less, as the unknown materials which are to be determined. Fortunately, the low-molecular-weight anionic polystyrenes which are usually used as gel permeation chromatography standards (Section 3.4.3) are also suitable for vapor phase osmometry standards. Since these products have relatively narrow molecular weight distributions all measured average molecular weights should be equal to each other to within experimental uncertainty. The $\overline{M}_v$ average (Section 3.3.1) of the polystyrene should be considered as the standard value if there is any uncertainty as to which average is most suited for calibration in vapor phase osmometry.

Vapor phase osmometers differ in design details. The most reliable instruments appear to be those which incorporate platinum gauzes on the thermistors in order to ensure reproducible solvent and solution drop sizes. In any case, the highest purity solvents should be used with this technique, to ensure a reasonably fast approach to steady state conditions.

The upper limit of molecular weights to which the vapor phase osmometer can be applied is usually considered to be 20,000 g mol^{-1}. Newer, more sensitive machines have extended this limit to 50,000 g mol^{-1} or higher. The measurements are convenient and relatively rapid and this is an attractive method to use, with the proper precautions.

3.1.3   Ebulliometry

In this method, the boiling points of solutions of known concentration are compared to that of the solvent, at the same temperature. The apparatus tends to be complicated, and errors are possible from ambient pressure changes and the tendency of polymer solutions to foam. Present-day commercial ebulliometers are not designed for molecular weight measurements in the range of major interest with synthetic polymers. The method is therefore only used in laboratories which have designed and built their own equipment.

3.1.4   Cryoscopy (Freezing Point Depression)

This is a classical method for measurement of molecular weights of micromolecular species. The equipment is relatively simple. Problems include the elimination of supercooling and selection of solvents which do not form solid solutions with solutes and do not have solid phase transitions near their freezing temperatures.

Cryoscopy is widely used in clinical chemistry (where it is often called "osmometry") but is seldom used for synthetic polymers.

3.1.5   End-Group Determinations

End-group analysis is not a colligative property measurement in the strict sense of the concept. It can be used to determine $\bar{M}_n$ of polymer samples if the substance contains detectable end groups, and the number of such end groups per molecule is known beforehand. (Recall that a branched molecule can have many ends.)

Since the concentration of end groups varies inversely with molecular weight, end group methods tend to become unreliable at higher molecular weights. They can be used, where they are applicable at all, up to $\overline{M}_n$ near 50,000.

End group analysis has been applied mainly to condensation polymers, since these materials must have relatively reactive end groups if they are to polymerize. If such polymers are prepared from two different bifunctional monomers the products can contain either or both end group types, and the concentrations of both are preferably measured for the most reliable molecular weight determinations.

The most important measurement techniques of this type rely on chemical analysis, with some use of radioisotope and spectroscopic analyses as well.

The value of $\overline{M}_n$ is derived from the experimental data according to

$$\overline{M}_n = rea/p \tag{3-9}$$

where r is the number of reactive groups per macromolecule, e is the equivalent weight of reagent, a is the weight of polymer, and p is the amount of reagent used. Thus, if 2.7 g of a polyester that is known to be linear and to contain acid groups at both ends requires titration with 15 mL 0.1 N alcoholic KOH to reach a phenolphthalein endpoint

$$\overline{M}_n = 2 \times 56 \times 2.7[1000/(15 \times 0.1 \times 56)] = 3600$$

Here $r = 2$, $e = 56$ (mol wt of KOH), $a = 2.7$, and the term in brackets is p.

The value of $\overline{M}_n$ cannot be calculated in many cases of practical interest because r in Eq. (3-9) is not known. This is particularly true when the branching character or composition of the polymer is uncertain. End group analysis is very useful, up to a point, in calculating how to react the polymer further. Thus, it is common practice to use parameters like the saponification number (number of milligrams of KOH that react with the free acid groups and the ester groups in 1 g of polymer), acid number (number of milligrams of KOH required to neutralize 1 g of polymer), acetyl number (number of grams of KOH that react with the acetic anhydride required to acetylate the hydroxyl groups in 1 g of polymer), and so on.

3.2   LIGHT SCATTERING

Data obtained from light scattering measurements can give information about the weight average molecular weight $\overline{M}_w$, about the size and shape of macromolecules in solution, and about parameters which characterize the interaction between the solvent and polymer molecules.

Light may be regarded as a periodically fluctuating electric field associated with a periodic magnetic field. The electric and magnetic field vectors are in phase with

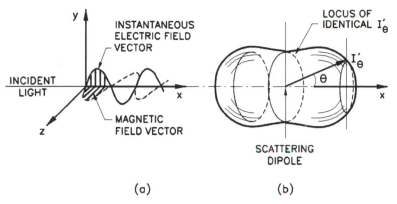

Fig. 3-2.   (a) Light depicted as a transverse wave. (b) Scattering envelope for point scatterer with unpolarized incident light.

each other and are perpendicular to each other and to the direction of propagation of the light. If the light wave travels in the x direction, then the electric and magnetic vectors would vibrate in the z and y directions, respectively (see Fig. 3-2a, for example).

The rate of energy flow per unit area (flux) is proportional to the vector product of the electric and magnetic field vectors. Since the latter two are at right angles to each other and are in phase and proportional to each other, the flux of light energy depends on the square of the scalar magnitude of the electric vector. Experimentally, we are primarily concerned with the time-average flux, which is called the intensity, I, and which is proportional to the square of the amplitude of the electric vector of the radiation.

When a light wave strikes a particle of matter which does not absorb any radiation the only effect of the incident field is a polarization of the particle. (Quantum, Raman, and Doppler effects can be ignored in this application.) If the scattering center moves relative to the light source, the frequency of the scattered light is shifted from the incident frequency by an amount proportional to the velocity component of the scatterer perpendicular to the direction of the light beam. Such very small, time-dependent frequency changes are undetectable with conventional light scattering equipment and can be neglected in the present context. They form the basis of quasi-elastic light scattering, which has applications to polymer science that are outside the scope of this text.

According to classical theory, the electrons and nuclei in the particle oscillate about their equilibrium positions in synchrony with the electric vector of the incident radiation. If the incident light wave is being transmitted along the x direction and the electric field vector is in the y direction, then a fluctuating dipole will be induced in the particle along the y direction. Each oscillating dipole is itself a source of electromagnetic radiation. If the electron in the dipole were moving at constant velocity, its motion would constitute an electric current and generate a steady magnetic field. The fluctuating dipole is equivalent, however, to an accelerating charge and behaves like a miniature dipole transmission antenna. Since the dipoles oscillate with the same frequency as the incident light, the "scattered" light radiated by these dipoles also has this same frequency.

The net result of this interaction of light and a scattering particle is that some of the energy which was associated with the incident ray will been radiated in directions away from the initial line of propagation. Thus the intensity of light transmitted through the particle along the incident beam direction is diminished by the amount radiated in all other directions by the dipoles in the particle.

Classical electromagnetic theory shows that the intensity of light radiated by a small isotropic scatterer is

$$\frac{I'_\theta}{I_0} = \frac{8\pi^4}{\lambda^4 r^2}\alpha^2(1 + \cos^2\theta) \tag{3-10}$$

where I'_θ is the light intensity a distance r from the scattering entity and θ is the angle between the direction of the incident beam and the line between the scattering center and viewer. In this equation I_0 is the incident light intensity, λ (lambda) is the wavelength of the incident light, and α (alpha), which is discussed below, is the excess polarizability of the particle over its surroundings. Figure 3-2b represents the scattering envelope for an isotropic scatterer with unpolarized incident light. The envelope is symmetrical about the plane corresponding to $\theta = \pi/2$. The scattered intensity in the forward direction equals that in the reverse direction, and both are twice that scattered transversely because of the term in $(1 + \cos^2\theta)$.

The scattered intensity is proportional to λ^{-4}. Thus the shorter wavelengths are scattered more intensely than longer wavelength light. (This is why the sky is blue.) In light scattering experiments, more intense signals can be obtained by using light of shorter wavelength. If there are N independent scatters in volume V, the combined intensity of scattering at distance r from the center of the (small) volume and angle θ to the incident beam will be simply (N/V) as great as that recorded above for a single scatterer. This is because interference and enhancement effects will cancel each other on the average if the scatterers are independent.

If c is the weight of particles per unit volume, then the mass per particle is Vc/N and

$$Vc/N = M/L \tag{3-11}$$

where M is the molecular weight of the scattering material and L is Avogadro's constant. Then, for N/V scatterers per unit volume,

$$\frac{I'_\theta}{I_0} = \left(\frac{Lc}{M}\right)\frac{8\pi^4}{\lambda^4 r^2}\alpha^2(1 + \cos^2\theta) \tag{3-12}$$

No average M is implied here, because we assume for the moment that all scatterers are monodisperse in molecular weight.

The excess polarizability cannot be determined experimentally, but it can be shown to be given by

$$\alpha = \frac{n_0 M}{2\pi L}\left(\frac{dn}{dc}\right) \tag{3-13}$$

where n_0 and n are the refractive indices of the suspending medium and suspension of scatterers, respectively. Substituting,

$$\frac{I'_\theta}{I_0} = \frac{2\pi^2}{r^2\lambda^4}\frac{(1+\cos^2\theta)n_0^2}{L}\left(\frac{dn}{dc}\right)^2 Mc \tag{3-14}$$

The equations outlined so far are similar to those which apply to solutions of monodisperse macromolecules, although the reasoning in the latter case is different from the classical Rayleigh treatment which led to the preceding results. We can nevertheless extend Eq. (3-14) to polymer solutions by the reasoning given below. This makes for clarity of presentation but is not a rigorous development of the final expressions which are used in light scattering experiments.

It must be recognized that solutions are subject to fluctuations of solvent density and solute concentration. The liquid is considered to be made up of volume elements. Each element is smaller than the wavelength of light so that it can be treated as a single scattering source. The elements are large enough, however, to contain very many solvent molecules and a few solute molecules. There will be time fluctuations of solute concentration in a volume element. Because of local variations in temperature and pressure, there will also be fluctuations in solvent density and refractive index. These latter contribute to the scattering from the solvent as well as from the solution, and solvent scattering is therefore subtracted from the experimental solution scattering at any given angle. The work necessary to establish a certain fluctuation in concentration is connected with the dependence of osmotic pressure (π) on concentration, such that the M term in Eq. (3-14) is effectively replaced by $RT/(d\pi/dc)$.

For a monodisperse solute, Eq. (2-76) is

$$\frac{\pi}{c} = RT\left[\frac{1}{M} + A_2c + A_3c^2 + \cdots\right] \tag{3-15}$$

(Recall that the A_i are virial coefficients.) Then

$$\frac{d\pi}{dc} = RT\left[\frac{1}{M} + 2A_2c + 2A_3c^2 + \cdots\right] \tag{3-16}$$

and the equivalent of Eq. (3-14) for a solution of monodisperse polymer is

$$\frac{I'_\theta}{I_0} = \frac{2\pi n_0^2}{\lambda^4 r^2 L}\frac{(dn/dc)^2(1+\cos^2\theta)c}{(M^{-1}+2A_2c+3A_3c^2+\cdots)} \tag{3-17}$$

3.2.1   Terminology

Some of the factors in the foregoing equation are instrument constants and are determined independently of the actual light scattering measurement. These include n_0 (refractive index of pure solvent at the experimental temperature and wavelength); L (Avogadro's constant); λ, which is set by the experimenter; and r, an instrument constant.

It is convenient to lump a number of these parameters into the reduced scattering intensity R_θ, which is defined for unit volume of a scattering solution as

$$R_\theta = \frac{I_\theta' r^2}{I_0(1 + \cos^2 \theta)} = \frac{2\pi^2 n_0^2 (dn/dc)^2 c}{\lambda^4 L(M^{-1} + 2A_2 c + 3A_3 c^2 + \cdots)} \tag{3-18}$$

and to define the optical constant K, such that

$$K = \frac{2\pi^2 n_0^2 (dn/dc)^2}{L\lambda^4} \tag{3-19}$$

Thus,

$$Kc/R_\theta = 1/M + 2A_2 c + 3A_3 c^2 + \cdots \tag{3-20}$$

and K contains only quantities which are directly measurable, while R_θ has dimensions of length^{-1} because it is defined per unit volume. When the viewing angle is $\pi/2$, R_θ becomes equal to the Rayleigh ratio:

$$R_{90} = (I_\theta'/I_0)r^2 \tag{3-21}$$

Note that R_θ is apparently independent of I_0, θ, and r. If the scattering solute molecules are small compared to the wavelength of light, it is only necessary to measure I_θ' as a function of enough values of θ to show that R_θ is indeed independent of θ. Then the data at a single scattering angle ($\pi/2$ is convenient) can be used in the form of Eq. (3-20) to yield a plot of Kc/R_θ against c with intercept $1/M$ and limiting slope at low c equal to $2A_2$. (This simple technique cannot be used with polymeric solutes which have dimensions comparable to the wavelength of light. Effects of large scatterers are summarized in Section 3.2.3.)

An alternative treatment of the experimental data involves consideration of the fraction of light scattered from the primary beam in all directions per unit length of path in the solution. A beam of initial intensity I_0 decreases in intensity by an amount $\tau I_0 dx$ while traversing a path of length dx in a solution with turbidity τ. The resulting beam has intensity I, and thus

$$I = I_0 e^{-\tau x} \tag{3-22}$$

or

$$\tau = -\ln(I_0/I)/x \tag{3-23}$$

The total scattering can be obtained by integrating I'_θ (Eq. 3-17) over a sphere of any radius r and it can then be shown that

$$\tau = \frac{16}{3}\pi R_\theta \qquad (3\text{-}24)$$

It is customary also to define another optical constant H such that

$$H = \frac{16\pi K}{3} = \frac{32\pi^3}{3}\frac{n_0^2(dn/dc)^2}{\lambda^4 L} \qquad (3\text{-}25)$$

and thus

$$\frac{Hc}{\tau} = \frac{Kc}{R_\theta} = \frac{1}{M} + 2A_2 c + 3A_3 c^2 + \cdots \qquad (3\text{-}26)$$

3.2.2   Effect of Polydispersity

For a solution in the limit of infinite dilution, Eq. (3-26) becomes

$$\tau = HcM \qquad (3\text{-}27)$$

If the solute molecules are independent agents and contribute additively to the observed turbidity, one can write

$$\tau = \sum \tau_i = H \sum c_i M_i \qquad (3\text{-}28)$$

where τ_i, c_i, and M_i refer to the turbidity, weight concentration, and molecular weight of monodisperse species i which is one of the components of the mixture that makes up the real polymer sample. Then

$$\lim_{c \to 0} \frac{\tau}{c} = \tau \bigg/ \sum c_i = H \sum c_i M_i \bigg/ \sum c_i = H\overline{M}_w \qquad (3\text{-}29)$$

Thus the light scattering method measures the weight average molecular weight of the solute. (This was pointed out in Section 2.9.)

3.2.3   Scattering from Large Particles

The equations to this point assume that each solute molecule is small enough compared to the wavelength of incident light to act as a point source of secondary radiation, so that the intensity of scattered light is symmetrically distributed as shown in Fig. 3-2b. If any linear dimension of the scatterer is as great as about $\lambda/20$, however, then the secondary radiations from dipoles in various regions of

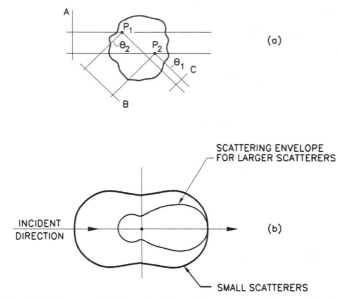

Fig. 3-3.   (a) Interference of light scattered from different regions of a scatterer with dimensions comparable to the wavelength of the scattered light. (b) Scattering envelopes for small and large scatterers. These scattering envelopes are cylindrically symmetrical about the direction of the incident light.

the scatterer may vary in phase at a given viewing point. The resulting interference will depend on the size and shape of the scatterer and on the observation angle. The general effect can be illustrated with reference to Fig. 3-3a, in which a scattering particle with dimensions near λ is shown. Two scattering points, P_1 and P_2, are shown. At plane A all the incident light is in phase. Plane B is drawn perpendicular to the light which is scattered at angle θ_2 from the incident beam. The distance $AP_1B < AP_2B$ so that light which was in phase at A and was then scattered at the two dipoles P_1 and P_2 will be out of phase at B.

Any phase difference at B will persist along the same viewing angle until the scattered ray reaches the observer. The phase difference causes an interference and reduction of intensity at the observation point. A beam is also shown scattered at a smaller angle θ_1, with a corresponding normal plane C. The length difference $OP_1C - OP_2C$ is less than $OP_2B - OP_1B$. (At the smaller angle θ_1, $AP_2 > AP_1$ while $P_2C < P_1C$, so the differences in two legs of the paths between planes A and C tend to compensate each other to some extent. At the larger angle θ_2, however, $AP_2 > AP_1$ and $P_2B > P_1B$.) The interference effect will therefore be greater the larger the observation angle, and the radiation envelope will not be symmetrical. The scattering envelopes for large and small scatterers are compared schematically in Fig. 3-3b. Both envelopes are cylindrically symmetrical about the incident ray, but that for the large scatterers is no longer symmetrical about a plane through the scatterer and normal to the incident direction. This effect is called *disymmetry*.

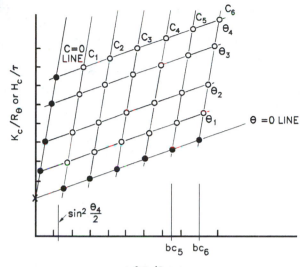

Fig. 3-4.   Zimm plot for simultaneous extrapolation of light scattering data to zero angle (θ) and zero concentration (c). The symbols are defined in the text. o, experimental points; • extrapolated points; x, double extrapolation.

Interference effects diminish as the viewing angle approaches zero degrees to the incident light. Laser light scattering photometers are now commercially available in which scattering can be measured accurately at angles at least as low as 3°. The optics of older commercial instruments which are in wide use are restricted to angles greater than about 30° to the incident beam. Zero angle intensities are estimated by extrapolation. It is always necessary to extrapolate the data to zero concentration, for reasons which are evident from Eq. (3-17). Conventional treatment of ligh scattering data will also involve an extrapolation to zero viewing angle.

The double extrapolation to zero θ and zero c is effectively done on the same plot by the Zimm method. The rationale for this method follows from calculations for random coil polymers which show that the ratio of the observed scattering intensity at an angle θ to the intensity which would be observed if there were no destructive inference is a function of the parameter $\sin^2(\theta/2)$. Zimm plots consist of graphs in which Kc/R_θ is plotted against $\sin^2(\theta/2) + bc$, where b is an arbitrary scale factor chosen to given an open set of data points. (It is often convenient to take $b = 100$.) In practice, intensities of scattered light are measured at a series of concentrations, with several viewing angles at each concentration. The Kc/R_θ (or Hc/τ) values are plotted as shown in Fig. 3-4. Extrapolated points at zero angle, for example, are the intersections of the lines through the Kc/R_θ values for a fixed c and various θ values with the ordinates at the corresponding bc values. Similarly, the zero c line traverses the intersections of fixed θ, variable c experimental points with the corresponding $\sin(\theta/2)$ ordinates. The zero angle and zero concentration lines intercept at the ordinate and the intercept equals $\overline{M}_w^{-1}$.

In many instances, Zimm plots will curve sharply downward at lower values of $\sin(\theta/2)$. This is usually caused by the presence of either (or both) very large polymer entities or large foreign particles like dust. The large polymers may be aggregates of smaller molecules or very large single molecules. If large molecules or aggregates are fairly numerous, the plot may become banana shaped. Double extrapolation of the "zero" lines is facilitated in this case by using a negative value of b to spread the network of points.

It is obviously necessary to clarify the solvent and solutions carefully in order to avoid spurious scattering from dust particles. This is normally done by filtration through cellulose membranes with 0.2 to 0.5 μm-diameter pores.

If a laser light scattering photometer is used, the scattered light can be observed at angles only a few degrees off the incident beam path. In that case extrapolation to zero angle is not needed and the Zimm plot can be dispensed with. The turbidities at several concentrations are then plotted according to Eq. (3-26). A single concentration observation is all that is needed if the concentration is low (the $A_3 c^2$ term is Eq. 3-26 becomes negligible) and if the second virial coefficient A_2 is known. However, A_2 is weakly dependent on molecular weight and better accuracy is generally realized if the scattered light intensities are measured at several concentrations.

3.2.4   Light Scattering Instrumentation

Light scattering photometers include a light source and means for providing a collimated light beam incident on the sample, as well as for detecting the intensity of scattered light as a function of a (usually) limited series of angles. A cell assembly to hold the solution and various components for control and readout of signals are also included. Scattered light is detected with a photomultiplier. At any angle, solvent scattering is subtracted from solution scattering to ensure that the scattering which is taken into account in the molecular weight calculation is due to solute alone.

Since the square of the specific refractive index increment (dn/dc) appears in the light scattering equations, this value must be accurately known in order to measure $\overline{M}_w$. (An error of $x\%$ in dn/dc will result in a corresponding error of about $2x\%$ in $\overline{M}_w$.) The value of dn/dc is needed at infinite dilution, but there is very little concentration dependence for polymer concentrations in the normal range used for light scattering. The required value can therefore be obtained from

$$\frac{n - n_0}{c} = \frac{\Delta n}{c} = \frac{dn}{dc} \tag{3-30}$$

where n and n_0 are the refractive indices of solution with concentration c and of solvent, respectively. In practice, dn/dc values may be positive or negative. Their absolute values are rarely >0.2 cm^3/g. Thus, if a solution of 10 g/liter concentration is being used, the $n - n_0$ value would be 2×10^{-3} and this would have to be measured within $\pm 2 \times 10^{-5}$ to obtain $dn/dc \pm 1\%$. Conventional refractometers are not suitable for measurements of this accuracy, and a direct measurement of Δn is obviously preferable to individual measurements of n and n_0. The preferred method for measuring dn/dc is differential refractometry, which measures the refraction of a light beam passing through a divided cell composed of solvent and solution compartments that are separated by a transparent partition.

3.2.5   Light Scattering from Copolymers [1]

The foregoing analysis of the scattering of light from polymer solutions relied on the implicit assumption that all polymer molecules had the same refractive index. This rule does not hold for copolymers since the intensity of light scattered at a particular angle from a solution with given concentration depends not only on the mean molecular weight of the solute but also on the heterogeneity of the chemical composition of the polymer. The true weight average molecular weight of a binary copolymer can be determined, in principle, by measuring the scattering of light from its solutions in at least three solvents with different refractive indices. These measurements also yield estimates of parameters which characterize the heterogeneity of the chemical composition of the solute.

The basic method yields good measurements of $\overline{M}_w$ but the heterogeneity parameters are generally found not to be credible for statistical copolymers. This may be due to a dependence of dn/dc on polymer molecular weight, at low molecular weights [2].

3.2.6   Radius of Gyration from Light Scattering Data

A radius of gyration in general is the distance from the center of mass of a body at which the whole mass could be concentrated without changing its moment of rotational inertia about an axis through the center of mass. For a polymer chain, this is also the root-mean-square distance of the segments of the molecule from its center of mass. The radius of gyration is one measure of the size of the random coil shape which many synthetic polymers adopt in solution or in the amorphous bulk state. (The radius of gyration and other measures of macromolecular size and shape are considered in more detail in Chapter 4.)

The radius of gyration, r_g, of a polymer in solution will depend on the molecular weight of the macromolecule, on its constitution (whether or not and how it is branched), and on the extent to which it is swollen by the solvent. An average radius of gyration can be determined from the angular dependence of the intensities of scattered light.

We saw in Section 3.2.3 that the light scattered from large particles is less intense than that from small scatterers except at zero degrees to the incident beam. This reduction in scattered light intensity depends on the viewing angle (cf. Fig. 3-3b), on the size of the solvated polymer, and on its general shape (whether it is rodlike, a coil, and so on). A general relation between these parameters can be derived [3], and it is found that the effects of molecular shape are negligible at low viewing angles. The relevant equation (for zero polymer concentration) is

$$\lim_{c \to 0} \frac{Kc}{R_\theta} = \frac{1}{\overline{M}_w} \left(1 + \frac{16\pi^2}{3\lambda^2} r_g^2 \sin^2 \frac{\theta}{2} + \cdots \right) \tag{3-31}$$

The limiting slope of the zero concentration line of the plot of Kc/R_θ against $\sin^2 \theta/2$ (Fig. 3-4) gives $(16\pi^2/3\lambda^2 \overline{M}_w)r_g$. The mutual intercept of the zero concentration and zero angle lines gives $\overline{M}_w^{-1}$, and the limiting slope of the zero angle line can be used to obtain the second virial coefficient as indicated by Eq. (3-20).

For a polydisperse polymer, the average molecular weight from light scattering is $\overline{M}_w$, but the radius of gyration which is estimated is the z average.

3.3   DILUTE SOLUTION VISCOMETRY

The viscosity of dilute polymer solutions is considerably higher than that of the pure solvent. The viscosity increase depends on the temperature, the nature of the solvent and polymer, the polymer concentration, and the sizes of the polymer molecules. This last dependence permits estimation of an average molecular weight from solution viscosity. The average molecular weight which is measured is the viscosity average $\overline{M}_v$, which differs from those described so far in this text. Before viscosity increase data are used to calculate $\overline{M}_v$ of the solute it is necessary, however, to eliminate the effects of solvent viscosity and polymer concentration. The methods whereby this is achieved are described in this section.

The procedures outlined below do not remove the effects of polymer–solvent interactions, and so $\overline{M}_v$ of a particular polymer sample will depend to some extent on the solvent used in the solution viscosity measurements (Section 3.3.1).

Solution viscosity measurements require very little investment in apparatus and can be carried out quite rapidly with certain shortcuts described in Section 3.3.4. As a result, this is the most widely used method for measuring a polymer molecular

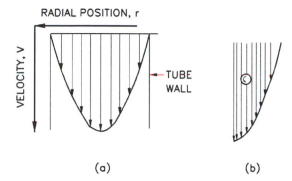

Fig. 3-5. (a) Variation of the velocity of laminar flow with respect to the distance r from the center of a tube. (b) Sphere suspended in a flowing liquid.

weight average. Solution viscosities are also used, without explicit estimation of molecular weights, for quality control of some commercial polymers, including poly(vinyl chloride) and poly(ethylene terephthalate).

We first consider briefly why a polymer solution would be expected to have a higher viscosity than the liquid in which it is dissolved. We think initially of a suspension of solid particles in a liquid. The particles are wetted by the fluid, and the suspension is so dilute that the disturbance of the flow pattern of the suspending medium by one particle does not overlap with that caused by another. Consider now the flow of the fluid alone through a tube which is very large compared to the dimensions of a suspended particle. If the fluid wets the tube wall its velocity profile will be that shown in Fig. 3-5a. Since the walls are wetted, liquid on the walls is stationary while the flow rate is greatest at the center of the tube. The flow velocity v increases from the wall to the center of the tube. The difference in velocities of adjacent layers of liquid (velocity gradient $= dv/dr$) is greatest at the wall and zero in the center of the tube.

When one layer of fluid moves faster than the neighboring layer, it experiences a retarding force F due to intermolecular attractions between the materials in the two regions. (If there were no such forces the liquid would be a gas.) It seems intuitively plausible that the magnitude of this force should be proportional to the local velocity gradient and to the interlayer area A. That is,

$$F = \eta(dv/dr)A \qquad (3\text{-}32)$$

where the proportionality constant η (eta) is the coefficient of viscosity or just the viscosity. During steady flow the driving force causing the fluid to exit from the tube will just balance the retarding force F. A liquid whose flow fits Eq. (3-32) is called a Newtonian fluid; η is independent of the velocity gradient. Polymer solutions which are used for molecular weight measurements are usually Newtonian. More concentrated solutions or polymer melts are generally not Newtonian in the sense that η may be a function of the velocity gradient and sometimes also of the history of the material.

Now consider a particle suspended in such a flowing fluid, as in Fig. 3-5b. Impingement on the particle of fluid flowing at different rates causes the suspended entity to move down the tube and also to rotate as shown. Since the particle surface is wetted by the liquid, its rotation brings adhering liquid from a region with one velocity into a volume element which is flowing at a different speed. The resulting readjustments of momenta cause an expenditure of energy which is greater than that which would be required to keep the same volume of fluid moving with the particular velocity gradient, and the suspension has a higher viscosity than the suspending medium.

Einstein showed that the viscosity increase is given by

$$\eta = \eta_0(1 + \omega\phi) \tag{3-33}$$

where η and η_0 are the viscosities of the suspension and suspending liquid, respectively, ϕ is the volume fraction of suspended material, and ω (omega) is a factor which depends on the general shape of the suspended species. In general, rigid macromolecules, having globular or rodlike shapes, behave differently from flexible polymers, which adopt random coil shapes in solution. Most synthetic polymers are of the latter type, and the following discussion focuses on their behavior in solution.

The effects of a dissolved polymer are similar in some respects to those of the suspended particles described earlier. A polymer solution has a higher viscosity than the solvent, because solvent which is trapped inside the macromolecular coils cannot attain the velocities which the liquid in that region would have in the absence of the polymeric solute. (Appendix 3A provides an example of an industrial application of this concept.) Thus the polymer coil and its enmeshed solvent have the same effect on the viscosity of the mixture as an impenetrable sphere, but this hypothetical equivalent sphere may have a smaller volume than the real solvated polymer coil because some of the solvent inside the coil can drain through the macromolecule.

The radius of the equivalent sphere is considered to be a constant, while the volume and shape of the real polymer coil will be changing continuously as a result of rotations about single bonds in the polymer chain and motions of the segments of the polymer. Nevertheless, the time-averaged effects of the real, solvent-swollen polymer can be taken to be equal to that of equivalent smaller, impenetrable spherical particles.

For spheres and random coil molecules, the shape factor ω in Eq. (3-33) is 2.5 and this equation becomes

$$\eta/\eta_0 - 1 = 2.5\phi \tag{3-34}$$

If all polymer molecules exist in solution as discrete entities, without overlap, and each solvated molecule has an equivalent volume V and molecular weight M (the polymer is monodisperse), then the volume fraction ϕ (phi) of solvent-swollen polymer coils at a concentration c (g cm^{-3}) is

$$\phi = LcV/M \qquad (3\text{-}35)$$

where L is Avogadro's number. The two preceding equations yield

$$\frac{1}{c}\left(\frac{\eta - \eta_0}{\eta_0}\right) = \frac{2.5LV}{M} \qquad (3\text{-}36)$$

In the entity on the left-hand side of Eq. (3-36), the contribution of the polymer solute to the solution viscosity is adjusted for solvent viscosity since the term in parentheses is the viscosity increase divided by the solvent viscosity. The term is also divided by c to compensate for the effects of polymer concentration, but this expedient is not effective at finite concentrations where the disturbance of flow caused by one suspended macromolecule can interact with that from another solute molecule. The contributions of the individual macromolecules to the viscosity increase will be independent and additive only when the polymer molecules are infinitely far from each other. In other words, the effects of polymer concentration can only be eliminated experimentally when the solution is very dilute. Of course, if the system is too dilute, $\eta - \eta_0$ will be indistinguishable from zero. Therefore, solution viscosities are measured at low but manageable concentrations and these data are used to extrapolate the left-hand side of Eq. (3-36) to zero concentration conditions. Then

$$[\eta] \equiv \lim_{c \to 0} \frac{1}{c}\left(\frac{\eta - \eta_0}{\eta_0}\right) = \lim_{c \to 0} \frac{2.5LV}{M} \qquad (3\text{-}37)$$

The term in brackets on the left-hand side of Eq. (3-37) is called the *intrinsic viscosity* or *limiting viscosity number*. It reflects the contribution of the polymeric solute to the difference between the viscosity of the mixture and that of the solvent. The effects of solvent viscosity and polymer concentration have been removed, as outlined earlier. It now remains to be seen how the term on the right-hand side of Eq. (3-37) can be related to an average molecular weight of a real polymer molecule. To do this we first have to express the volume V of the equivalent hydrodynamic sphere as a function of the molecular weight M of a monodisperse solute. Later we substitute an average molecular weight of a polydisperse polymer for M in the monodisperse case.

If radius of gyration (Section 3.2.6) of a solvated polymer coil is r_g, then the radius of the equivalent sphere r_e will be Hr_g, where H is a fraction which allows for the likelihood that some of the solvent inside the macromolecular volume can drain through the polymer chain. Intuitively, we can see that the solvent deep inside the polymer will move with about the same velocity as its neighboring polymer chain segments while that in the outer regions of the macromolecule will be able to flow more in pace with the local solvent flow lines. Values of H have been calculated theoretically [4]. Since r_g can be measured directly from light scattering experiments (Section 3.2.6) it is possible to determine H by measuring $[\eta]$ and r_g in the same solvent. Data from a number of different investigators show that H is 0.77 [5].

Under theta conditions the polymer coil is not expanded (or contracted) by the solvent and is said to be in its unperturbed state. The radius of gyration of such a macromolecule is shown in Section 4.4.1 to be proportional to the square root of the number of bonds in the main polymer chain. That is to say, if M is the polymer molecular weight and M_0 is the formula weight of its repeating unit, then

$$r_e = Hr_g \propto H(M/M_0)^{1/2} \tag{3-38}$$

Since the volume of the equivalent sphere equals $\frac{4}{3}\pi r_e^3$, then Eqs. (3-37) and (3-38) show that the intrinsic viscosity of solutions of unsolvated (unperturbed) macromolecules should be related to M by

$$[\eta] = \frac{10\pi}{3} LH^3 r_g^3 \propto \frac{10\pi LH^3}{3(M_0)^{3/2}} M^{1/2} \tag{3-39}$$

The intrinsic viscosity in a theta solution is labeled $[\eta]_\theta$. Equation (3-39) can thus be expressed as follows for theta conditions:

$$[\eta]_\theta = K_\theta M^{0.5} \tag{3-40}$$

Flory and Fox [4] have provided a theoretical expression for K_θ which is in reasonable agreement with experimental values.

In a better solution than that provided by a theta solvent the polymer coil will be more expanded. The radius of gyration will exceed the r_g which is characteristic of the bulk amorphous state or a theta solution. If the polymer radius in a good solvent is α_η times its unperturbed r_g, then the ratio of hydrodynamic volumes will be equal to a α_η^3 and its intrinsic viscosity will be related to $[\eta]_\theta$ by

$$[\eta]/[\eta]_\theta = \alpha_\eta^3 \tag{3-41}$$

or

$$[\eta] = K_\theta \alpha_\eta^3 M^{0.5} \tag{3-42}$$

The lower limit of α_η is obviously 1, since the polymer is not soluble in media which are less hospitable than theta solvents. In a good solvent $\alpha_\eta > 1$ and increases with M according to $\alpha_\eta = \lambda M^\Delta$, where λ and Δ are positive and $\Delta = 0$ under theta conditions [6, 7]. Then

$$[\eta] = \lambda^3 K_\theta M^{(0.5+3\Delta)} = K M^a \tag{3-43}$$

where K and a are constants for fixed temperature, polymer type, and solvent.

Equation (3-43) is the Mark–Houwink–Sakurada (MHS) relation. It appeared empirically before the underlying theory which has just been summarized.

To this point we have considered the solution properties of a monodisperse polymer. The MHS relation will also apply to a polydisperse sample, but M in this equation is now an average value where we denote $\overline{M}_v$ the viscosity average molecular weight. Thus, in general,

$$[\eta] = K\overline{M}_v^a \tag{3-44}$$

The constants K and a depend on the polymer type, solvent, and solution temperature. They are determined empirically by methods described in Sections 3.3.2 and 3.4.3. It is useful first, however, to establish a definition of $\overline{M}_v$ analogous to those which were developed for $\overline{M}_w$, $\overline{M}_n$, and so on in Chapter 2.

3.3.1   Viscosity Average Molecular Weight $\overline{M}_v$

We take the Mark–Houwink–Sakurada equation (Eq. 3-44) as given. We assume also that the same values of K and a will apply to all species in a polymer mixture dissolved in a given solvent. Consider a whole polymer to be made up of a series of i monodisperse macromolecules each with concentration (weight/volume) c_i and molecular weight M_i. From the definition of $[\eta]$ in Eq. (3-37),

$$\eta_i/\eta_0 - 1 = c_i[\eta_i] \tag{3-45}$$

where η_i is the viscosity of a solution of species i at the specified concentration, and $[\eta_i]$ is the intrinsic viscosity of this species in the particular solvent. Recall that

$$c_i = n_i M_i \tag{3-46}$$

where n_i is the concentration in terms of moles/volume. Also,

$$[\eta_i] = KM_i^a \tag{3-44a}$$

and so

$$\eta_i/\eta_0 - 1 = n_i KM_i^{a+1} \tag{3-47}$$

If the solute molecules in a solution of a whole polymer are independent agents, we may regard the viscosity of the solution as the sum of the contributions of the i monodisperse species that make up the whole polymer. That is,

$$\left[\frac{\eta}{\eta_0} - 1\right]_{\text{whole}} = \sum_i \left(\frac{\eta_i}{\eta_0} - 1\right) = K\sum_i n_i M_i^{a+1} \tag{3-48}$$

From Eq. (3-37),

$$[\eta] = \lim_{c \to 0} \frac{1}{c} \left(\frac{\eta}{\eta_0} - 1 \right)_{whole} = \lim_{c \to 0} \frac{K}{c} \sum n_i M_i^{a+1} \qquad (3\text{-}49)$$

However,

$$c = \sum c_i = \sum n_i M_i \qquad (3\text{-}50)$$

and so

$$[\eta] = \lim_{c \to 0} \left(K \sum n_i M_i^{a+1} \Big/ \sum n_i M_i \right) \qquad (3\text{-}51)$$

with Eq. (3-44),

$$[\eta] = K \overline{M}_v^a = \lim_{c \to 0} \left(K \sum_i n_i M_i^{a+1} \Big/ \sum_i n_i M_i \right) \qquad (3\text{-}52)$$

Then, in the limit of infinite dilution,

$$\overline{M}_v = \left[K \sum n_i M_i^{a+1} \Big/ \sum n_i M_i \right]^{1/a} \qquad (3\text{-}53)$$

Alternative definitions follow from simple arithmetic:

$$\overline{M}_v = \left[\sum w_i M_i^a \right]^{1/a} \qquad (3\text{-}54)$$

In terms of moments,

$$\overline{M}_v = \left[\eta U'_{a+1} \right]^{1/a} = \left[w U'_a \right]^{1/a} \qquad (3\text{-}55)$$

Note that $\overline{M}_v$ is a function of the solvent (through the exponent a) as well as of the molecular weight distribution of the polymer. Thus a given polymer sample can be characterized only by a single value of $\overline{M}_n$ or $\overline{M}_w$, but it may have different $\overline{M}_v$'s depending on the solvent in which $[\eta]$ is measured. Of course, if the sample were monodisperse, $\overline{M}_v = \overline{M}_w = \overline{M}_n = \ldots$. In general, the broader the molecular weight distribution, the more $\overline{M}_v$ may vary in different solvents.

Note that Eq. (3-53) defines $\overline{M}_n$ with $a = -1$ and $\overline{M}_w$ with $a = 1$. For polymers that assume random coil shapes in solution, $0.5 \leq a \leq 0.8$, and $\overline{M}_v$ will be much closer to $\overline{M}_w$ than to $\overline{M}_n$ because a is closer to 1 than to -1. Also, $\overline{M}_v$ is much easier to measure than $\overline{M}_w$ once K and a are known, and it is often convenient to assume that $\overline{M}_v \simeq \overline{M}_w$. This approximation is useful but not always very reliable for broad distribution polymers.

Another interesting result is available from consideration of Eqs. (3-51) and (3-44) which yield

$$[\eta] = \sum n_i M_i [\eta]_i \Big/ \sum n_i M_i \qquad (3\text{-}56)$$

With Eq. (3-46),

$$[\eta] = \sum c_i [\eta]_i \Big/ \sum c_i = \sum w_i [\eta]_i \qquad (3\text{-}57)$$

This last relation shows that the intrinsic viscosity of a mixture of polymers is the weight average value of the intrinsic viscosities of the components of the mixture in the given solvent. (Compare Eq. 2-13 for the weight average of a molecular size.)

3.3.2 Calibration of the Mark–Houwink–Sakurada Equation

Since $\overline{M}_v$ depends on the exponent a as well as the molecular weight distribution this average molecular weight is not independent of the solvent unless the molecular weight distribution of the polymer sample is very narrow. In the limit of monodispersity w_i in Eq. (3-54) approaches 1 and $\overline{M}_v = \overline{M}_i = $ any average molecular weight of the sample.

The classical method for determining K and a relies on fractionation (p. 455) to divide a whole polymer sample into subspecies with relatively narrow molecular weight distributions. An average molecular weight can be measured on each such subspecies, which is called a *fraction*, by osmometry ($\overline{M}_n$) or light scattering ($\overline{M}_w$), and the measured average can be equated to a solvent-independent $\overline{M}_v$ if the distribution of the sample is narrow enough. The intrinsic viscosities of a number of such characterized fractions are fitted to the equation

$$\ln[\eta] = \ln K + a \ln(\overline{M}_v) \qquad (3\text{-}44b)$$

to yield the MHS constants for the particular polymer–solvent system.

Since actual fractions are not really monodisperse it is considered better practice to characterize them by light scattering than by osmometry because $\overline{M}_v$ is closer to $\overline{M}_w$ than to $\overline{M}_n$.

Although the initial calibration is actually in terms of the relation between $[\eta]$ and $\overline{M}_w$ or $\overline{M}_n$, as described, Eq. (3-44) can only be used to estimate $\overline{M}_v$ for unknown polymers. It cannot be employed to estimate $\overline{M}_w$ (or $\overline{M}_n$ as the case may be) for such samples unless the unknown is also a fraction with a molecular weight distribution very similar to those of the calibration samples. An important class of polymers which constitutes an exception to this restriction consists of linear polyamides and polyesters polymerized under equilibrium conditions (Chapter 5). In these cases the molecular weight distributions are always random (Section 5.4.3) and the relation

$$[\eta] = K\overline{M}_n^a \qquad (3\text{-}58)$$

can be applied. For this group (in which $\overline{M}_w = 2\overline{M}_n$), whole polymers can be used for calibration, and $\overline{M}_w$ and $\overline{M}_n$ can be obtained from solution viscosities. The most important examples of this exceptional class of polymers are commercial fiber-forming nylons and poly(ethylene terephthalates).

The procedure described for calibration of K and a is laborious because of the required fractionation process. The two constants are derived as described from the intercept and slope of a linear least squares fit to $[\eta]$–M values for a series of fractionated polymers. Experimentally, K and a are found to be inversely correlated. If different laboratories determine these MHS constants for the same polymer–solvent combination, the data set which yields the higher K value will produce the lower a. Thus, $\overline{M}_v$ from Eq. (3-44) is often essentially the same for different K and a values provided the molecular weight ranges of the samples used in the two calibration processes overlap.

We have assumed so far that K and a are fixed for a given polymer type and solvent and do not vary with polymer molecular weight. This is not strictly true. Oligomers (less than about 100 repeating units in most vinyl polymers) often conform to

$$[\eta] = K\overline{M}_v^{0.5} \tag{3-59}$$

with K and the exponent a independent of the solvent. The MHS constants determined for higher molecular weight species may depend on the molecular weight range, however. Tabulations of such constants therefore usually list the molecular weights of the fractions for which the particular K and a values were determined. Table 3-1 presents such a list for some common systems of more general interest.

TABLE 3-1

Mark–Houwink–Sakurada Constants[a]

Polymer	Solvent	Temperature (°C)	Molecular weight range of calibration samples ($M \times 10^{-4}$)	$K \times 10^3$ ($cm^3 g^{-1}$)	a
Polystyrene	Toulene	25	1–160	17	0.09
Polystyrene	Benzene	25	0.4–1	100	0.50
Polyisobutene	Decalin	25	>500	22	0.70
Poly(methyl acrylate)	Acetone	25	28–160	5.5	0.77
Poly(methyl methacrylate)	Acetone	25	8–137	7.5	0.70
Poly(vinyl alcohol)	Water	30	1–80	43	0.64
Nylon-6	Trifluoroethanol	25	1.3–10	53.6	0.74
Cellulose triacetate	Acetone	20	2–14	2.38	1.0

[a] From M. Bandrup and E. Immergut, Eds., "Polymer Handbook," 3rd ed. Wiley, New York, 1989.

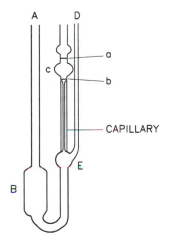

UBBELOHDE VISCOMETER

Fig. 3-6.   Ubbelohde suspended level viscometer.

An alternative procedure for determining the MHS constants from gel permeation chromatography is given on page 109, after the latter technique has been described.

3.3.3   Measurement of Intrinsic Viscosity

Laboratory devices are available to measure intrinsic viscosities without human intervention. These are useful where many measurements must be made. The basic principles involved are the same as those in the glass viscometers which have long been used for this determination. An example of the latter is the Ubbelohde suspended level viscometer shown in Fig. 3-6. In this viscometer a given volume of polymer solution with known concentration is delivered into bulb B through stem A. This solution is transferred into bulb C by applying a pressure on A with column D closed off. When the pressure is released, any excess solution drains back into bulb B and the end of the capillary remains free of liquid. The solution flows from C through the capillary and drains around the sides of the bulb E. The volume of fluid in B exerts no effect on the rate of flow through the capillary because there is no back pressure on the liquid emerging from the capillary. The flow time t is taken as the time for the solution meniscus to pass from mark a to mark b in bulb C above the capillary. The solution in D can be diluted by adding more solvent through A. It is then raised up into C, as before, and a new flow is obtained.

The flow time is related to the viscosity η of the liquid by the Hagen–Poiseuille equation:

$$\eta = \pi P r^4 t / 8 Q l \tag{3-60}$$

where P is the pressure drop along the capillary which has length l and radius r from which a volume Q of liquid exits in time t. It is necessary to compare the flow behavior of pure solvent with that of solution of concentration c. We will subscript the terms related to solvent behavior with zeros. The average hydrostatic heads, h and h_0, are the same during solvent and solution flow in this apparatus, because t is the time taken for the meniscus to pass between the same fiducial marks a and b. Then the mean pressures driving the solvent and solution are $h\rho_0 g$ and $h\rho g$, respectively, where g is the gravitational acceleration constant and ρ is a density (compare Eq. 3-1). For dilute solutions ρ is very close to ρ_0 and it follows from Eq. (3-60) that

$$\eta/\eta_0 = t/t_0 \tag{3-61}$$

where t_0 is flow time for the solvent and t that for the solution. Thus the ratio of viscosities needed in Eq. (3-37) can be obtained from flow times without measuring absolute viscosities.

The intrinsic viscosity $[\eta]$ is defined in the above equation as a limit at zero concentration. The η/η_0 ratios which are actually measured are at finite concentrations, and there are a variety of ways to estimate $[\eta]$ from these data. The variation in solution viscosity (η) with increasing concentration can be expressed as a power series in c. The equations usually used are the Huggins equation [8]:

$$\frac{1}{c}\left(\frac{\eta}{\eta_0} - 1\right) = \frac{1}{c}\left(\frac{t}{t_0} - 1\right) = [\eta] + k_{\mathrm{H}}[\eta]^2 c \tag{3-62}$$

and the Kraemer equation [9]:

$$c^{-1}\ln(\eta/\eta_0) = [\eta] - k_1[\eta]^2 c \tag{3-63}$$

It is easily shown that both equations should extrapolate to a common intercept equal to $[\eta]$ and that $k_{\mathrm{H}} + k_1$ should equal 0.5. The usual calculation procedure involves a double extrapolation of Eqs. (3-62) and (3-63) on the same plot, as shown in Fig. 3-7. This data handling method is generally satisfactory. Sometimes experimental results do not conform to the above expectations. This is because the real relationships are actually of the form

$$c^{-1}(\eta/\eta_0 - 1) = [\eta] + k_{\mathrm{H}}[\eta]^2 c + k'_{\mathrm{H}}[\eta]^3 c^2 + \cdots \tag{3-64}$$

and

$$c^{-1}\ln(\eta/\eta_0) = [\eta] - k_1[\eta]^2 c - k'_1[\eta]^3 c^2 - \cdots \tag{3-65}$$

and the preceding equations are truncated versions of these latter virial expressions in concentration. No two-parameter solution such as Eq. (3-62) or (3-63) is universally valid, because it forces a real curvilinear relation into a rectilinear form. The

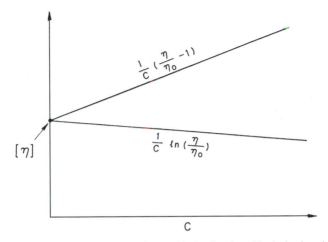

Fig. 3-7. Double extrapolation for graphical estimation of intrinsic viscosity.

power series expressions may be solved directly by nonlinear regression analysis [10], but this is seldom necessary unless it is desired to obtain very accurate values of $[\eta]$ and the slope constants k_{H} and k_1.

The term k_{H} in Eq. (3-62) is called "Huggins constant." Its magnitude can be related to the breadth of the molecular weight distribution or branching of the solute. Unfortunately, the range of k_{H} is not large (a typical value is 0.33) and it is not determined very accurately because Eq. (3-62) fits a chord to the curve of Eq. (3-64), and the slope of this chord is affected by the concentration range in which the curve is used.

A useful initial concentration for solution viscometry of most synthetic polymers is about 1 g/100 cm^3 solvent. High-molecular-weight species may require lower concentrations to produce a linear plot of $c^{-1}(\eta/\eta_0 - 1)$ against c (Fig. 3-7), which does not curve away from the c axis at the high concentrations. At very low concentrations, such plots may also curve upward. This effect is thought to be due to absorption of polymer on the capillary walls and can be eliminated by avoiding such high dilutions.

3.3.4 Single-Point Intrinsic Viscosities

It is interesting to note that the intrinsic viscosity can often be determined to within a few percent from a relative flow time measurement at a single concentration only. A number of such mathematical techniques have been proposed. Several of these are very useful but all should be verified with standard

TABLE 3-2

Solution Viscosity Nomenclature

Name	Symbol	Definition
Solution viscosity	η	
Solvent viscosity	η_0	
Relative viscosity	η_r	η/η_0
Specific viscosity	η_{sp}	$\eta_r - 1 = \eta/\eta_0 - 1$
Reduced specific viscosity	η_{sp}/c	$(\eta_r - 1)/c$
Inherent viscosity	η_{inh}	$c^{-1} \ln \eta_r = c^{-1} \ln(\eta/\eta_0)$
Intrinsic viscosity	$[\eta]$	$\lim\limits_{c \to 0} c^{-1}(\eta/\eta_0 - 1)$

multipoint determinations for new polymer–solvent combinations or new ranges of polymer molecular weights.

The equation of Solomon and coworkers [11],

$$[\eta] = \frac{\sqrt{2}}{c}\left(\frac{\eta}{\eta_0} - 1 - \ln\left(\frac{\eta}{\eta_0}\right)\right)^{\frac{1}{2}} \tag{3-66}$$

is easy to calculate and applies to many common polymer solution. Reference [12] cites other manual calculation procedures for single-point $[\eta]$'s.

3.3.5 Solution Viscosity Terminology

There are two terminologies in this field. The trivial nomenclature is vague and misleading but it is widely used. The IUPAC nomenclature [13] is clear, concise, and rarely used. Table 3-2 lists the common names and usual symbols.

The intrinsic viscosity is also called the limiting viscosity number. Many of these so-called viscosity terms are not viscosities at all. Thus η_r and η_{sp} are actually unitless ratios of viscosities; $[\eta]$ is a ratio of viscosities divided by a concentration. The units of $[\eta]$ are reciprocal concentration and are commonly quoted in cubic centimeters per gram (cm^3/g) or deciliters per gram (dl/g). [A deciliter (dL) equals 100 cm^3.] The units of K in Eq. (3-44) must correspond to those of $[\eta]$.

The viscosities we have been using in this section are conventional dynamic viscosities with units of poises or Pas. They are to be distinguished from kinematic viscosities which have units in stokes and equal the dynamic viscosity divided by the density of the liquid. The latter are widely used in technology where the mass flow rate is measured in preference to the volumetric flow rate of Eq. (3-60).

3.3.6   Solution Viscosities in Polymer Quality Control

Solution viscosities are involved in quality control of a number of commercial polymers. Production of poly(vinyl chloride) polymers is usually monitored in terms of relative viscosity (η/η_0) while that of some fiber forming species is related to IV [inherent viscosity, $c^{-1}\ln(\eta/\eta^0)$]. The magnitudes of these parameters depends primarily on the choices of concentration and solvent and to some extent on the solution temperature. There is no general agreement on these experimental conditions and comparison of such data from different manufacturers is not always straightforward.

3.3.7   Copolymers and Branched Polymers

The relationships used to this point assume implicity that the hydrodynamic radius of a polymer molecule is a single-valued function of the size of the macromolecule. This is not true if copolymer composition or molecular shape is also changing. Branched polymers, for example, are more compact than their linear analogs at given molecular weight. They will therefore exhibit lower intrinsic viscosities. The change in solution viscosity depends on the frequency and nature of the branching. The extent of solvation may similarly vary with the chemical composition of a copolymer. For these reasons, MHS relations are not readily established for polymers like low-density polyethylene, where branching varies with polymerization conditions or for styrene-butadiene copolymers in which the copolymer composition can be varied widely.

3.4   SIZE EXCLUSION CHROMATOGRAPHY

Size exclusion chromatography (SEC) is a column fractionation method in which solvated polymer molecules are separated according to their sizes. The technique is also known as gel permeation chromatography (GPC). The separation occurs as the solute molecules in a flowing liquid move through a stationary bed of porous particles. Solute molecules of a given size are sterically excluded from some of the pores of the column packing, which itself has a distribution of pore sizes. Larger solute molecules can permeate a smaller proportion of the pores and thus elute from the column earlier than smaller molecules.

Size exclusion chromatography provides the distribution of molecular sizes from which average molecular weights can be calculated with the formulas summarized in Chapter 2. SEC is not a primary method as usually practiced; it requires calibration in order to convert raw experimental data into molecular weight distribution.

SEC can be used for analytical purposes or as a method to produce fractions with narrower molecular weight distributions than those of starting polymer. We confine ourselves here to the former application.

3.4.1   Experimental Arrangement

In SEC analyses a liquid is pumped continually through columns packed with porous gels which are wetted by the fluid. A variety of such packings can be used. The most common type and the first to be employed widely with synthetic polymers consists of polystyrene gels. These porous beads are highly cross-linked so that they can be packed firmly without clogging the columns when the solvent flows through them under pressure. They are also highly porous and are made with controlled pore sizes. This combination of properties is achieved by copolymerization of styrene and divinylbenzene in mixed solvents, which are good solvents for the monomers but have marginal affinity for polystyrene [14]. The most common packings for GPC in aqueous systems (also called gel filtration and gel chromatography) are cross-linked dextran or acrylamide polymers and porous glass.

A dilute solution of polymer in the GPC solvent is injected into the flowing eluant. In the column, the molecules with smaller hydrodynamic volumes can diffuse into and out of pores in the packing, while larger solute molecules are excluded from many pores and travel more in the interstitial volume between the porous beads. As a result, smaller molecules have longer effective flow paths than larger molecules and their exit from the GPC column set is relatively delayed.

The initial pulse of polymer solution which was injected into the column entry becomes diluted and attenuated as the different species are separated on the gel packing. The column effluent is monitored by detectors which respond to the weight concentration of polymer in the flowing eluant. The most common detector is a differential refractometer. Spectrophotometers, which operate at fixed frequencies, are also used as alternative or auxiliary detectors. Some special detectors which are needed particularly for branched polymers or copolymers are mentioned in Section 3.4.4.

It is also necessary to monitor the volume of solvent which has passed through the GPC column set from the time of injection of the sample (this is called the elution volume or the retention volume). Solvent flow is conveniently measured by means of elapsed time since sample injection, relying implicitly on a constant solvent pumping rate. As an added check on this assumption, flow times may be ratioed to those of a low-molecular-weight marker that provides a sharp elution peak at long flow times.

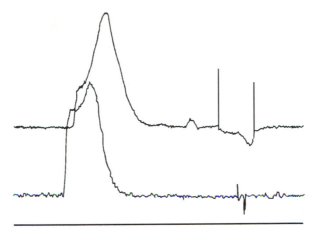

Fig. 3-8.   Typical GPC raw data. The units of the vertical axis depend on the detectors used, while those on the horizontal axis are elapsed time. In this case the lower curve is that of the differential refractometer, while the upper curve is the trace produced by a continuous viscometer (which is described briefly in Section 3.4.4). The curve proceeds from left to right.

The raw data in gel permeation chromatography consist of a trace of detector response, proportional to the amount of polymer in solution, and the corresponding elution volumes. A typical SEC record is depicted in Fig. 3-8. It is normal practice to use a set of several columns, each packed with porous gel with a different porosity, depending on the range of molecular sizes to be analyzed.

3.4.2   Data Interpretation

The differential refractive index detector response on the ordinate of the SEC chromatogram in Fig. 3-8 can be transformed into a weight fraction of total polymer while suitable calibration permits the translation of the elution volume axis into a logarithmic molecular weight scale.

To normalize the chromatogram, a baseline is drawn through the recorder trace, and chromatogram heights are taken for equal small increments of elution volume. (Accurate operation requires that the baseline be straight through the whole chromatogram.) An ordinate corresponding to a particular elution volume is converted to a weight fraction by dividing by the sum of the heights of all the ordinates under the trace. (Recall the mention of normalization in Section 2.3.)

Corrections for instrumental broadening (also called axial dispersion) are also sometimes applied [15]. This phenomenon arises because of eddy diffusion and molecular diffusion at the leading and trailing edges of the pulse of polymer solution [16]. The result is a symmetrical, Gaussian spreading of the GPC chromatogram in which the observed range of elution volumes exceeds that which corresponds to the real range of solute sizes. The calculated $\overline{M}_n$ is lowered and the calculated $\overline{M}_w$ is raised to a lesser extent as a consequence. A related phenomenon involves a skewing of the GPC trace toward higher elution volumes and lower molecular weights. This results from the radial distribution of velocities in fluid flow (Fig. 3-5a). Its importance varies with the viscosity of the solution and depends therefore on the high-molecular-weight tail of the polymer molecular weight distribution. A number of procedures have been proposed to correct the raw GPC trace for instrumental broadening [17]. Such adjustments can be neglected for most synthetic polymers with $\overline{M}_w/\overline{M}_n \gtrsim 2$. Skewing corrections require independent measurements of $\overline{M}_n$ by osmometry or $\overline{M}_w$ by light scattering.

When a differential refractometer is used as a detector, instrumental broadening of the GPC chromatogram is compensated to some extent by another effect due to the tendency for specific refractive indexes of polymer solutions to decrease with decreasing molecular weight in the low-molecular-weight range.

3.4.3   Universal Calibration for Linear Homopolymers

We consider now how the elution volume axis of the raw chromatogram can be translated into a molecular weight scale. A series of commercially available anionically polymerized polystyrenes is particularly well suited for calibration of GPC columns. These polymers are available with a range of molecular weights and have relatively narrow molecular weight distributions (Section 2.5). When such a sample is injected into the GPC column set, the resulting chromatogram is narrower than that of a whole polymer, but it is not a simple spike because of the band broadening effects described earlier and because the polymer standard itself is not actually monodisperse. Since the distribution is so narrow, however, no serious error is committed by assigning the elution volume corresponding to the peak of the chromatogram to the molecular weight of the particular polystyrene. (The molecular weight distributions of most of these samples are sharp enough that all experimental average molecular weights are essentially equivalent to within experimental error.) Thus a series of narrow distribution polystyrene samples yields a set of GPC chromatograms as shown in Fig. 3-9. The peak elution volumes and corresponding sample molecular weights produce a calibration curve (see Fig. 3-11 later) for *polystyrene* in the particular GPC solvent and column set. It turns out that combinations of packing pore sizes which are generally used result in more or less linear calibration curves when the logarithms of the polystyrene molecular weights are plotted against the corresponding elution volumes. It remains now to translate this polystyrene calibration curve to one which will be effective in the same apparatus and solvent for other linear polymers. (Branched polymers and copolymers present

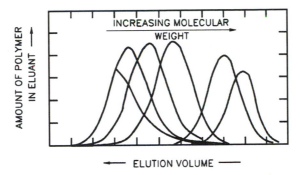

Fig. 3-9.   Gel permeation chromatography elution curves for anionic polystyrene standards used for calibration. The polystyrene standard samples were measured separately; use of a mixture of polymers may cause elution volumes of very high molecular weight standards to be erroneously low [18].

complications and are discussed separately later.) This technique is called a *universal calibration*, although we shall see that it is actually not universally applicable.

Studies of GPC separations have shown that polymers appear in the eluate in inverse order of their hydrodynamic volumes in the particular solvent. This forms the basis of a universal calibration method since Eq. (3-37) is equivalent to

$$\ln([\eta]M) = \ln(2.5L) + \ln\left(\lim_{c \to 0} V\right) \tag{3-67}$$

The product $[\eta]M$ is a direct function of the hydrodynamic volume of the solute at infinite dilution. Two different polymers which appear at the same elution volume in a given solvent and particular GPC column set therefore have the same hydrodynamic volumes and the same $[\eta]M$ characteristics.

The conversion of a calibration curve for one polymer (say, polystyrene, as in Fig. 3-10) to that for another polymer can be accomplished directly if the Mark–Houwink–Sakurada equations are known for both species in the GPC solvent. From Eq. (3-43), one can write

$$[\eta]_i M_i = K_i M_i^{a_i+1} \tag{3-68}$$

where the subscript refers to the polymer type. Thus, at equal elution volumes

$$[\eta]_1 M_1 = K_1 M_1^{a_1+1} = [\eta]_2 M_2 = K_2 M_2^{a_2+1} \tag{3-69}$$

and the molecular weight of polymer 2, which appears at the same elution volume as polymer 1 with molecular weight M_1, is given by

$$\ln M_2 = \frac{1+a_1}{1+a_2} \ln M_1 + \frac{1}{1+a_2} \ln\left(\frac{K_1}{K_2}\right) \tag{3-70}$$

The polystyrene calibration curve of Fig. 3-10 can be translated into that of any other polymer for which the MHS constants are known [19].

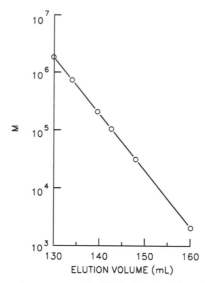

Fig. 3-10.   Polystyrene calibration curve for GPC, where M is the molecular weight of the anionic polystyrene standard samples.

Note that the universal calibration relations apply to polymeric solutes in very dilute solutions. The component species of whole polymers do indeed elute effectively at zero concentration but sharp distribution fractions will be diluted much less as they move through the GPC columns. Hydrodynamic volumes of solvated polymers are inversely related to concentration and thus elution volumes may depend on the concentration as well as on the molecular weights of the calibration samples. To avoid this problem, the calibration curve can be set up in terms of hydrodynamic volumes rather than molecular weights. A general relation [20] is

$$V = \frac{4\pi [\eta] M}{\mu + 4\pi L c([\eta] - [\eta]_\theta)} \tag{3-71}$$

Here c is set equal to the concentration of the solution of a sharp distribution calibration standard used to establish the calibration curve, and $[\eta]_\theta$ (Eq. 3-40) can be calculated *a priori* [4]. For whole polymers, all species elute effectively at zero c and

$$V = 4\pi [\eta] M / \mu \tag{3-72}$$

The constant μ in Eqs. (3-71) and (3-72) must equal $10\pi L$ (instead of the numerical value given in [20]) in order to coincide with Eq. (3-37). This procedure is necessary for high-molecular-weight polymer standards in good solvents. In other cases, the hydrodynamic volume calibration is equivalent to the infinite dilute

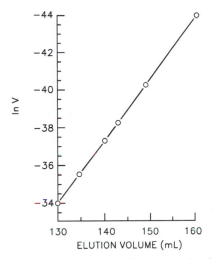

Fig. 3-11.   Universal calibration relation in terms of hydrodynamic volume V and elution volume.

$[\eta]M$ method. With this modification, the calibration curve for narrow distribution standards is converted to the form shown in Fig. 3-11, using Eq. (3-71) to translate M to hydrodynamic volume V. The curve is then applied to analysis of whole polymers through use of Eq. (3-72).

When Mark–Houwink constants for a particular polymer are not known, they can be estimated from GPC chromatograms and other data on whole polymers of the particular type [21]. It is not necessary to use fractionated samples in this method of determining K and a.

A parameter J is defined as the product of intrinsic viscosity and molecular weight of a monodisperse species i. That is,

$$J_i \equiv [\eta]_i M_i \tag{3-73}$$

With Eq. (3-68),

$$[\eta]_i = J_i^{a(a+1)} K^{1/(a+1)} \tag{3-74}$$

and from Eq. (3-57),

$$[\eta] = K^{1/(a+1)} \sum w_i J_i^{a/a+1} \tag{3-75}$$

If two samples of the unknown polymer are available with different intrinsic viscosities, then

$$\frac{[\eta]_1}{[\eta]_2} = \sum_i w_{1i} J_{1i}^{a/a+1} \Big/ \sum w_{2i} J_{2i}^{a/(a+1)} \tag{3-76}$$

Here the w_i are available from the ordinates of the gel permeation chromatogram and the J_i from the universal calibration curve of elution volume against hydrodynamic volume through Eq. (3-67) or (3-71). The intrinsic viscosities must be in the GPC solvent in this instance, of course. A simple computer calculation produces the best fit a to Eq. (3-76) and this value is inserted into Eq. (3-74) to calculate K. These MHS constants can be used with Eq. (3-70) to translate the polystyrene calibration curve to that for the new polymer.

Note that this procedure need not be restricted to determination of MHS constants in the GPC solvent alone [22]. The ratio of intrinsic viscosities in Eq. (3-76) can be measured in any solvent of choice as long as the w_i and J_i values for the two polymer samples of interest are available from GPC in a common, other solvent. The first step in the procedure is the calculation of K and a in the GPC solvent as outlined in the preceding paragraph. The intrinsic viscosities of the same two polymers are also measured in a common other solvent. The data pertaining to this second solvent will be designated with prime superscripts to distinguish them from values in the GPC solvent. In the second solvent,

$$J_i' \equiv [\eta']_i M_i \tag{3-73a}$$

For a species of given molecular weight M_i, Eqs. (3-73a) and (3-73) yield

$$J_i' \equiv [\eta']_i J_i / [\eta]_i \tag{3-77}$$

With Eq. (3-43)

$$J_i' = J_i (K'/K) M_i^{a'-a} \tag{3-78}$$

Then for the non-GPC solvent, Eq. (3-76) becomes

$$\frac{[\eta']_1}{[\eta']_2} = \sum w_{1i} J^{a'/(a+1)} \Big/ \sum w_{2i} j^{a'/(a+1)} \tag{3-79}$$

As above, the w_i and J_i values are available from the GPC experiment and intrinsic viscosities of the two polymer samples in the GPC solvent. The exponent a can be calculated as described in connection with Eq. (3-76).

The computed best fit value of a' in Eq. (3-79) can be now used to calculate K' from:

$$K' = [\eta'] K^{a'/(a+1)} \Big/ \sum w_i J_i^{a'(a+1)} \tag{3-80}$$

Here the MHS constants K and a for the GPC solvent are used with the exponent a' and the measured intrinsic viscosity $[\eta']$ of a single polymer sample in the non-GPC solvent.

This procedure is much less tedious than the method described in Section 3.3.2 for measurement of MHS constants. It may not necessarily produce the same K and a values as the standard fractionation method described earlier. This is because K and a are inversely correlated, as mentioned, and are also not entirely independent of the molecular weight range of the samples used. Essentially the same K and

a should be obtained if the two samples used with Eq. (3-76) or (3-79) and the fractions used with Eq. (3-44b) have similar molecular weights.

Because Eq. (3-53) defines $\overline{M}_w$ when $a = 1$, it is possible to estimate $\overline{M}_w$ of a sample by measuring $\overline{M}_v$ for the polymer in two or three solvents with different values of the exponent a. A plot of $\overline{M}_v$ against a is linear and extrapolates to $\overline{M}_w$ at $a = 1$ [23]. This procedure is fairly rapid if single-point intrinsic viscosities (Section 3.3.4) are used. It can be employed as an alternative to light scattering although the latter technique is more reliable and gives other information in addition to the weight average molecular weight. The GPC method outlined here is a convenient procedure to generate the MHS constants for this approximation of $\overline{M}_w$ from solution viscosity measurements.

It is possible in principle to derive K and a from a single whole polymer sample for which $[\eta]$ in the GPC solvent and $\overline{M}_n$ are known [21]. This method is less reliable than the preceding procedure which involved intrinsic viscosities of two samples because the computations of $\overline{M}_n$ can be adversely affected by skewing and instrumental broadening of the GPC chromatogram.

3.4.4   Branched Polymers

Equation (3-39) links the intrinsic viscosity of a polymer sample to the radii of gyration r_g of its molecules while Eq. (3-72) relates the hydrodynamic volume V of a solvated molecule to the product of its molecular weight and intrinsic viscosity. The separation process in GPC is on the basis of hydrodynamic volume, and the universal calibration described in Section 3.4.3 is valid only if the relation between V and r_g is the same for the calibration standards and the unknown samples.

Branched molecules of any polymer are more compact than linear molecules with the same molecular weight. They will have lower intrinsic viscosities (Section 3.3.7) and smaller hydrodynamic volumes, in a given solvent, and will exit from the GPC columns at higher elution volumes. Universal calibration (preceding section) cannot be used to analyze polymers whose branching or composition is not uniform through the whole sample. Generally useful techniques that apply to such materials, as well as to the linear homopolymers that are amenable to universal calibration, involve augmenting the concentration detector (which is often a differential refractometer, as mentioned) with detectors that measure the molecular weights to the polymers in the SEC eluant. These are continuous viscometers and light scattering detectors. The former are used to measure the intrinsic viscosity of the eluting polymer at each GPC retention time. The universal calibration relation of Eq. (3-67) or (3-72) is equivalent to

$$[\eta]M = [\eta]_{lin}M_{lin} \tag{3-81}$$

where the unsubscripted values refer to the branched polymer and the subscript lin refers to its linear counterpart, which appears at the same elution volume (or to the narrow distribution polystyrene or other polymer used as a standard for universal calibration). When $[\eta]$ of each fraction is measured, the molecular weight of the branched polymer which elutes at any given retention volume is available from the relation of Eq. (3-81). This procedure is also applicable to copolymers, if the variable copolymer composition does not affect the response of the concentration detector that is used along with the viscometer.

With a light scattering photometer and a concentration detector like a differential refractometer, the molecular weight distribution of the unknown polymer is obtained directly without need for the universal calibration procedure of the preceding section. This is by application of Eq. (3-26) to each successive "slice" of the GPC chromatogram. The virial coefficient terms in this equation are best set equal to zero, since their molecular weight dependence (Section 2.11.1) is not known *a priori*. Various designs of light scattering detectors are now available, differing primarily in the number and magnitude of viewing angles used. Low-angle light scattering (using laser light) eliminates the need for angular correction of the observed turbidity (Section 3.2.3), whereas photometers operating at right angles to the incident light beam are less sensitive to adventitious dust.

The three SEC detector types in common use at the time of writing—differential refractometer, continuous viscometer, and light scattering photometer—differ in sensitivity. The differential refractometer signal scales as the concentration, c, of the polymer solute. The viscometer signal is proportional to cM^a (p. 98), with the exponent a about equal to 0.7 for most polymer solutions used in this analysis. The light scattering signal scales as cM (Eq. 3-29). When all three detectors are employed simultaneously, the light scattering device is most sensitive to large species and relatively insensitive to low-molecular-weight polymer, while the reverse selectivity applies to the differential refractometer. Current continuous viscometers are intermediate in performance and are the most generally useful detectors. The analytical technique should, however, be tailored to the specific characteristics of the polymer of interest.

In a multidetector SEC apparatus it is necessary to match the output of the detector that senses eluant concentration with the signals of the detectors that sense molecular weight directly. To do this, the analyst should match the different signals at equal hydrodynamic volumes in the different detectors [24].

3.4.5 Aqueous SEC

Most synthetic polymers are analyzed in organic solvents, using appropriate SEC column packings in which the only interaction between the macromolecular solute and the packing is steric. Separation of the polymeric species is inversely related to their hydrodynamic volumes because the flow paths of the larger species are shortened by their inability to sample all the pores as they move with the flowing solvent. The same basic SEC technique is used to characterize polymers that are soluble primarily in water. Here, however, the procedure is more likely to be complicated by polymer-packing interactions. The packings consist of derivatized silica or cross-linked hydrophillic gels, in contrast to the cross-linked polystyrene or similar substrates used in organic phase SEC. Both the packing and solute contain polar groups and interactions may prevent purely steric separation. Efficient analyses of different water-soluble polymers are often quite specific to the particular material and more specialized references should be consulted for information.

3.4.6 Inhomogeneous Polymers

A polymer sample may consist of a mixture of species whose compositions differ enough to affect the responses of both the concentration-dependent detector and the molecular-weight-sensitive detector in a multidetector system. Examples are mixtures of different polymers or copolymers (Chapter 7) whose composition is not independent of molecular size. Conventional GPC cannot be used reliably to characterize such mixtures, but an on-line viscometer can be employed to measure molecular weight averages independent of any compositional variations [25]. Remember, of course, that such data characterize the mixture as a whole, and not just the major component.

Some polymers are homogeneous with respect to overall chemical composition but vary enough in branch frequency or comonomer spacing that important physical properties may be affected. A prime example is copolymers of ethylene and alpha-olefins (so-called linear low density polyethylene, LLDPE). Here the relative frequency of comonomer placements is reflected in changes in branch frequency, which influence the processing behavior of the polymer. In such cases, even apparent identity of overall chemical composition and molecular weight distributions does not guarantee the same physical properties. A more complete analysis of the polymer structure then requires characterization of branch frequency as well as SEC molecular weight data. A useful technique to assess branching of such polymers is temperature rising elution fractionation [26].

3.4.7 MALDI-MS [27]

MALDI-MS refers to matrix-assisted laser desorption–ionization mass spectroscopy. It is also called MALFI-TOF, because the mass spectrometer is a time-of-flight version. In this technique, the polymer is mixed with a molar excess of a salt, for cationization, and deposited on a probe surface. A UV laser is pulsed at the mixture, vaporizing a layer of the target area. Collisions between cations and polymer in the cloud of debris form charged polymer molecules. These are extracted and accelerated to a fixed kinetic energy by application of a high potential. They are diverted into a field-free chamber, where they separate during flight into groups of ions according to their mass/charge ratios. The output of a detector at the end of the drift region is converted to a mass spectrum on the basis of the time elapsed between the initiation of the laser pulse and the arrival of the charged species at the detector. Lighter ions travel faster and reach the detector earlier. Responses from a multiplicity of laser shots are combined to improve the signal/noise character of the mass spectrum.

MALDI-TOF is a useful technique at present for low-molecular-weight polymers. Application to most commercially important polymers is problematic at the time of writing, however, because these materials have high mean molecular weights and broad molecular weight distributions [28].

APPENDIX 3A: MULTIGRADE MOTOR OILS [29]

Certain polymers act to improve the *viscosity index* (VI) in crankcase lubricants. The principles involved are those described in Section 3.3. Most internal combustion engines are designed to function most efficiently by maintaining approximately constant engine torque over the wide temperature range which the lubricating oil may experience. If the crankcase oil is too viscous at low temperatures, the starting motor will have difficulty in cranking the engine and access of the lubricant may be impeded to all the components it is designed to protect. On the other hand, if the oil is too fluid at the engine's operating temperature, which can exceed 200°C, it may fail to prevent wear of metal parts and may be consumed too fast during running of the vehicle. Oil-soluble polymers are used as VI improvers to counteract the tendency for the lubricant's viscosity to drop with increasing crankcase temperature.

VI improvers require oxidation resistance without generation of corrosive by-products, thermal stability, compatibility with the other additives in the lubricant package, shear stability and solubility, or, rather, absence of separation over the operating range of the engine. This balance of properties is achieved by use of certain polymers at 0.5–3.0% level. The commercially important VI improvers are polymethacrylates, ethylene-propylene copolymers, and hydrogenated styrene-diene copolymers. At low temperatures the oil is a relatively poor solvent for the polymer and the macromolecules tend to shrink into small coils which add very little to the viscosity of the mixture. At high temperatures, however, the oil is a better solvent and swells the polymer coils. The result is a greater hydrodynamic volume of the macromolecular solute, higher viscosity of the solution (cf. Eq. 3-36), and compensation for the increased fluidity of the base oil.

PROBLEMS

3-1. Two "monodisperse" polystyrenes are mixed in equal quantities by weight. One polymer has molecular weight of 39,000 and the other molecular weight of 292,000. What is the intrinsic viscosity of the blend in benzene at 25°C? The Mark–Houwink constants for polystyrene/benzene are $k = 9.18 \times 10^{-5}$ dl/g and $a = 0.74$.

3-2. The following are data from osmotic pressure measurements on a solution of a polyester in chloroform at 20°C. The results are in terms of centimeters of solvent. The density of $HCCl_3$ is 1.48 g cm^{-3}. Find $\overline{M}_n$.

Concentration (g/dl)	h(cm $HCCl_3$)
0.57	2.829
0.28	1.008
0.17	0.521
0.10	0.275

3-3. The relative flow times (t/t_0) of a poly(methyl methacrylate) polymer in chloroform are given below.
(a) Determine $[\eta]$ by plotting η_{sp}/c and η_{inh} against c.
(b) Find $\overline{M}_v$ for this polymer. $[\eta] = 3.4 \times 10^{-5} \overline{M}_v^{0.80}$ (dl/g).

Concentration (g/dl)	t/t_0
0.20	1.290
0.40	1.632
0.60	2.026

3-4. The Mark–Houwink constants for polystyrene in tetrahydrofuran at 25°C are $K = 6.82 \times 10^{-3}$ cm^3/g and $a = 0.766$. The intrinsic viscosity of poly(methyl methacrylate) in the same solvent is given by

$$[\eta] = 1.28 \times 10^{-2} \overline{M}_v^{0.69} \text{ cm}^3/\text{g}$$

Show how this information can be used to construct a calibration curve for poly(methyl methacrylate) in gel permeation chromatography, using anionic polystyrenes as calibration standards.

3-5. A polymer with true molecular weight averages $\overline{M}_n = 430{,}000$ and $\overline{M}_w = 1{,}000{,}000$ is contaminated with 3% by weight of an impurity with molecular weight 30,000. What effects does this contamination have on the average molecular weights determined by light scattering and by membrane osmometry?

3-6. Polyisobutene A has a molecular weight around 3000 and polyisobutene B has a molecular weight around 700,000. Which techniques would be best for direct measurement of $\overline{M}_n$ and $\overline{M}_w$ of each sample?

3-7. The Mark–Houwink relation for polypropylene in o-dichlorobenzene at 130°C was calibrated as follows. A series of sharp fractions of the polymer was obtained by fractionation, and the molecular weight of each fraction was determined by membrane osmometry in toluene at 90°C. The samples were then dissolved in o-dichlorobenzene at 130°C and their intrinsic viscosities ($[\eta]$) were measured. The resulting data fitted an expression of the form

$$\ln[\eta] = \ln K + a \ln \overline{M}_n$$

where K and a are the desired Mark–Houwink constants. It was concluded from this that intrinsic viscosities of all polypropylene polymers in the appropriate molecular weight range could be represented by

$$[\eta] = K\overline{M}_n^a$$

Is this conclusion correct? Justify your answer very briefly.

3-8. What molecular weight measurement methods could be used practically to determine the following?

(a) $\overline{M}_n$ of a soluble polymer from reaction of glycol, phthalic anhydride, and acetic acid. The approximate molecular weight of this sample is known to be about 1200.

(b) $\overline{M}_w$ of a polystyrene with molecular weight about 500,000.

(c) $\overline{M}_n$ of high-molecular-weight styrene-methyl methacrylate copolymer with uncertain styrene content.

3-9. In an ideal membrane osmometry experiment a plot of π/cRT against c is a straight line with intercept $1/M$. Similarly, an ideal light scattering experiment at zero viewing angle yields a straight line plot of Hc/τ against c with intercept $1/M$. For a given polymer sample, solvent, and temperature,

(a) Are the M values the same from osmometry and light scattering?

(b) Are the slopes of the straight-line plots the same?

Explain your answers briefly.

3-10. A sample of poly(hexamethylene adipamide) weighs 4.26 g and is found to contain 4×10^{-3} mol COOH groups by titration with alcoholic KOH. From this information $\overline{M}_n$ of the polymer is calculated to be 2100. What assumption(s) is (are) made in this calculation?

3-11. A dilute polymer solution has a turbidity of 0.0100 cm^{-1}. Assuming that the solute molecules are small compared to the wavelength of the incident light, calculate the ratio of the scattered to incident light intensities at 90° angle to the incident beam and 20 cm from 2 ml of solution. Assume that all the solution is irradiated.

3-12. Solution viscosities for a particular polymer and solvent are plotted in the form $(\eta - \eta_0)/(c\eta_0)$ against c where η is the viscosity of a solution of polymer with concentration c g cm^{-3} and η_0 is the solvent viscosity. The plot is a straight line with an intercept of 1.50 cm^3 g^{-1} and a slope of 0.9 cm^6 g^{-2}. Give the magnitude and units of Huggins's constant for this polymer–solvent pair.

3-13. The following average molecular weights were measured by gel permeation chromatography of a poly(methyl methacrylate) sample:

$$\overline{M}_n\ 2.15 \times 10^5, \qquad \overline{M}_v\ 4.64 \times 10^5, \qquad \overline{M}_w\ 4.97 \times 10^5$$
$$\overline{M}_z\ 9.39 \times 10^5, \qquad \overline{M}_{z+1}\ 1.55 \times 10^6, \qquad \overline{M}_{z+2}\ 2.22 \times 10^6$$

Provide quantiative estimates of the breadth and skewness of the weight distribution of molecular weights.

3-14. Einstein's equation for the viscosity of a dilute suspension of spherical particles is

$$\eta/\eta_0 = 1 + 2.5\phi \tag{3-34}$$

where ϕ is the volume fraction of suspended material. Express the intrinsic viscosity (in deciliters per gram) as a function of the apparent specific volume (reciprocal density) of the solute.

3-15. This multipart question question illustrates material balance calculations used, for example, in formulating polyurethanes. Refer to Section 1.5.4 for some of the reactions of isocyanate groups. This problem is an extension of the concepts mentioned in Section 3.1.5 on end-group determinations. Some useful definitions follow:

Equivalent weight, E = weight of compound per active group for a given reaction. (total weight) / (equivalent weight) = no. of equivalents.

$$E = \frac{\overline{M}_n}{f} \tag{3-15-i}$$

where f is the functionality, i.e., the number of chemically effective groups per molecule for the reaction of interest (Section 1.3.2). In hydroxyl-terminated polymers, which are often called polyols, E follows from the definition of the term hydroxyl number, OH, in Section 3.1.5, as:

$$E = \frac{56.1(1000)}{\text{OH}} \tag{3-15-ii}$$

Then:

$$\overline{M}_n = \frac{56.1(1000)f}{OH} = Ef \qquad (3\text{-}15\text{-iii})$$

For a mixture of polyols A and B,

$$(OH)_{mix} = (OH)_A w_A + (OH)_B w_B \qquad (3\text{-}15\text{-iv})$$

where w is the weight fraction.

$$(\overline{M}_n)_{mix} = 56100 \frac{f_A f_B}{(OH)_A f_B w_A + (OH)_B f_A(1 - w_A)} \qquad (3\text{-}15\text{-v})$$

Percent isocyanate = the percent by weight of isocyanate (NCO) groups present. The amine equivalent, AE, is the weight of sample which reacts with 1 gram-equivalent weight of dibutyl amine (reaction 1–13).

$$AE = \overline{M}_n/(\text{no. of reactive groups}) \qquad (3\text{-}15\text{-vi})$$

$$AE = \frac{4200}{\%NCO} = \frac{(\text{formula wt. of NCO})(100\,g\,\text{compound})}{g\,NCO} \qquad (3\text{-}15\text{-vii})$$

$$\text{for isocyanates: } E = \frac{4200}{\%NCO} = AE \qquad (3\text{-}15\text{-viii})$$

For toluene diisocyanate (TDI, often 80 parts 2,4 isomer and 20 parts 2,6 isomer): $\%NCO = (42)2(100)/174 = 48$. Since $42 = $ the formula weight of the isocyanate NCO group.

Isocyanate index (index number) = 100(actual amount of isocyanate used)/ (equivalent amount of isocyanate required). An excess of isocyanate groups is used in some applications like flexible foam. The analytical values required for isocyanate formulas are the isocyanate value, hydroxyl number, residual acid value (acid number), and water content. The last two parameters reflect the following reactions:

$$\sim NCO + -COOH \rightarrow \sim \overset{H}{\underset{\underset{O}{\|}}{N-C-O-}} \qquad (3\text{-}15\text{-ix})$$

   (a) TDI is used to make a foam expanded with the carbon dioxide produced by reaction with 3 parts of water per 100 parts of a polyester having hydroxyl and acid functionalities (a polyester polyol) with an OH number = 62 mg KOH/g and an acid number (acid value) = 2.1 mg KOH/g. Calculate the amount of TDI required to provide isocyanate indexes of 100 and 105.

$$2 -NCO + H_2O \rightarrow \overset{\overset{\displaystyle H \quad\quad H}{\displaystyle |\quad\quad\quad |}}{-N-\underset{\underset{\displaystyle O}{\displaystyle \|}}{C}-N-} + CO_2 \qquad (3\text{-}15\text{-}x)$$

(b)   Design an isocyanate-ended prepolymer (low-molecular-weight polymer intended for subsequent reaction, as in Eq. 1-13, for example), consisting of equal weights of a triol with molecular weight 3200 and a diol with molecular weight 1750. Use TDI as the diisocyanate monomer to provide 3% free isocyanate, by weight, in the final prepolymer.

(c)   MDI is the acronym for 4,4'-diisocyanato-diphenylmethane. Its structure is shown in Eq. (1-12). A prepolymer is made from an MDI (572 parts) and a polyol (512 parts). The equivalent weight of the MDI is 143 and that of the polyol is 512. Calculate the available NCO, in %, in the prepolymer.

(d)   How much MDI, at 98 isocyanate index, is required to react with 100 parts of a polyether polyol with hydroxyl number of 28 mg KOH/g, an acid value of 0.01 mg KOH/g, and a water content of 0.01% (by weight), blended with 4.0 parts of ethylene glycol and 2 parts of m-phenylene diamine? [30]

REFERENCES

[1]   H. Benoit and D. Froelich, *in* "Light Scattering from Polymer Solutions" (M. B. Huglin, ed.). Academic Press, New York, 1972.
[2]   T. C. Chau and A. Rudin, *Polymer (London)* **15**, 593 (1974).
[3]   C. Tanford, "Physical Chemistry of Macromolecules." Wiley, New York, 1961.
[4]   P. J. Flory and T. G. Fox, *J. Am. Chem. Soc.* **73**, 1904 (1951).
[5]   C. M. Kok and A. Rudin, *Makromol. Chem. Rapid Commun.* **2**, 655 (1981).
[6]   P. J. Flory and W. R. Krigbaum, *J. Chem. Phys.* **18**, 1806 (1956).
[7]   D. K. Carpenter and L. Westerman, *in* "Polymer Molecular Weights" (P. E. Slade, Jr., ed.), Part II. Dekker, New York, 1975.
[8]   M. L. Huggins, *J. Am. Chem. Soc.* **64**, 2716 (1942).
[9]   E. O. Kraemer, *Ind. Eng. Chem.* **30**, 1200 (1938).
[10]   A. Rudin, G. B. Strathdee, and W. B. Edey, *J. Appl. Polym. Sci.* **17**, 3085 (1973).
[11]   O. F. Solomon and I. Z. Ciuta, *J. Appl. Polym. Sci.* **6**, 683 (1962); O. F. Solomon and B. S. Gotesman, *Makromol. Chem.* **104**, 177 (1967).
[12]   A. Rudin and R. A. Wagner, *J. Appl. Polym. Sci.* **19**, 3361 (1975).
[13]   M. L. Higgins and J. J. Hermans, *J. Polym. Sci.* **8**, 257 (1952).
[14]   J. C. Moore, *J. Polym. Sci. Part A* **2**, 835 (1964).
[15]   L. H. Tung, *J. Appl. Polym. Sci.* **10**, 375 (1966).
[16]   J. H. Duerksen, *Sep. Sci.* **5**, 317 (1970).
[17]   N. Friis and A. E. Hamielec, *Adv. Chromatogr.* **13**, 41 (1975).
[18]   C. M. Kok and A. Rudin, *Makromol. Chem.* **182**, 280 (1981).
[19]   Z. Grubisic, P. Rempp, and H. Benoit, *J. Polym. Sci. Part B* **5**, 753 (1967).

[20]   H. K. Mahabadi and A. Rudin, *Polym. J.* **11**, 123 (1979); A. Rudin and H. L. W. Hoegy, *J. Polym. Sci. Part A-1* **10**, 217 (1972).

[21]   A. R. Weiss and E. Cohn-Ginsberg, *J. Polym. Sci. Part B* **7**, 379 (1969).

[22]   C. J. B. Dobbin, A. Rudin, and M. F. Tchir, *J. Appl. Polym. Sci.* **25**, 2985 (1980); *ibid.***27**, 1081 (1982).

[23]   A. Rudin, G. W. Bennett, and J. R. McLaren, *J. Appl. Polym. Sci.* **13**, 2371 (1969).

[24]   M. G. Pigeon and A. Rudin, *J. Appl. Polym. Sci.* **46**, 763 (1992).

[25]   R. Amin Senayei, K. Suddaby, and A. Rudin, *Makromol. Chem.* (1993).

[26]   M. G. Pigeon and A. Rudin, *J. Appl. Polym. Sci.* **51**, 303 (1994).

[27]   H. S. Creel, *Trends Polym. Sci.* **1**, 336 (1993).

[28]   B. Thomson, K. Suddaby, A. Rudin, and G. Lajoie, *Eur. Polym. J.,* **32**, 239 (1996).

[29]   M. K. Mishra and R. G. Saxton, *Chemtech*, 35 (April 1995).

[30]   G. Woods, "The ICI Polyurethanes Book," 2nd ed. Wiley, New York, 1990.

Chapter 4

Effects of Polymer Isomerism and Conformational Changes

With a name like yours, you might be any shape, almost.
—Lewis Carroll, *Through the Looking Glass*

Three types of isomerism are important in macromolecular species. These involve constitutional, configurational, and conformational variations. These terms are defined and illustrated. Their usage in macromolecular science is very much the same as in micromolecular chemistry.

4.1   CONSTITUTIONAL ISOMERISM

The constitution of a molecule specifies which atoms in the molecule are linked together and with what types of bonds.

Isobutane (**4-1**) and *n*-butane (**4-2**) are familiar examples of constitutional isomers. Each has the molecular formula C_4H_{10} but the C and H atoms are joined differently in these two molecules. In polymers the major types of constitutional differences involve positional isomerism and branching.

$$
\begin{array}{c}
\text{CH}_3 \\
| \\
\text{CH}_3 - \text{C} - \text{H} \\
| \\
\text{CH}_3
\end{array}
\qquad\qquad
\text{CH}_3 - \text{CH}_2 - \text{CH}_2 - \text{CH}_3
$$

4-1   4-2

4.1.1   Positional Isomerism

Vinyl and vinylidene monomers are basically unsymmetrical because the two
ends of the double bond are distinguishable (ethylene and tetrafluorethylene are
exceptions). One C of the double bond can be arbitrarily labeled the head and the
other the tail of the monomer, as shown in the formula for vinyl fluoride (**4-3**).

$$
\text{head} \quad CH_2 = \underset{\underset{F}{|}}{\overset{\overset{H}{|}}{C}} \quad \text{tail}
$$

4-3

In principle, the monomer can be enchained by head-to-tail linkages or head-
to-head, tail-to-tail enchainments (**4-4**). Poly(vinyl fluoride) actually has about
15% of its monomers in the head-to-head, tail-to-tail mode. This is exceptional,
however. Head-to-tail enchainment appears to be the predominant or exclusive
constitution of most vinyl polymers because of the influence of resonance and
steric effects.

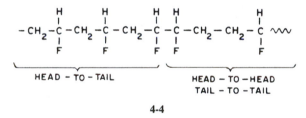

4-4

Vinyl monomers polymerize by attack of an active center (**4-5**) on the double
bond. Equation (4-1) represents head-to-tail enchainment:

$$
\text{WW } CH_2 - \underset{\underset{X}{|}}{\overset{\overset{Y}{|}}{C}} \bullet + CH_2 = \underset{\underset{X}{|}}{\overset{\overset{Y}{|}}{C}} \quad \longrightarrow \quad \text{WW } CH_2 - \underset{\underset{X}{|}}{\overset{\overset{Y}{|}}{C}} - CH_2 - \underset{\underset{X}{|}}{\overset{\overset{Y}{|}}{C}} \bullet \qquad (4\text{-}1)
$$

$$
\qquad\qquad\qquad\qquad \textbf{4-5} \qquad\qquad\qquad\qquad\qquad \textbf{4-6}
$$

while Eq. (4-2) shows the sequence of events in head-to-head, tail-to-tail polymerization:

$$
\text{WW } CH_2 - \underset{\underset{X}{|}}{\overset{\overset{Y}{|}}{C}} \ast + CH_2 = \underset{\underset{X}{|}}{\overset{\overset{Y}{|}}{C}} \quad \longrightarrow \quad \text{WW } CH_2 - \underset{\underset{X}{|}}{\overset{\overset{Y}{|}}{C}} - \underset{\underset{X}{|}}{\overset{\overset{Y}{|}}{C}} - CH_2 \ast \qquad (4\text{-}2)
$$

4-7

The active center may be a free-radical, ion, or metal–carbon bond (Chapter 6). In any event the propagating species **4-6** will be more stable than its counterpart **4-7** if the unpaired electron or ionic charge can be delocalized across either or both substituents X and Y. When X and/or Y is bulky there will be more steric hindrance to approach of the two substituted C atoms than in attack of the active center on the methylene C as in reaction (4-1). Poly(vinyl fluoride) contains some head-to-head linkages because the F atoms are relatively small and do not contribute significantly to the resonance stabilization of the growing macroradical.

Positional isomerism is not generally an important issue in syntheses of polymers with backbones which do not consist exclusively of enchained carbons. This is because the monomers which form macromolecules such as poly(ethylene terephthalate) (**1-5**) or nylon-6,6 (**1-6**) are chosen so as to produce symmetrical polymeric structures which facilitate the crystallization needed for many applications of these particular polymers. Positional isomerism can be introduced into such macromolecules by using unsymmetrical monomers like 1,2-propylene glycol (**4-8**), for example. This is what is done in the synthesis of some film-forming polymers like alkyds (Section 5.4.2) in which crystallization is undesirable.

$$CH_3 - \overset{\displaystyle H}{\underset{\displaystyle OH}{C}} - CH_2OH$$

4-8

It has been suggested that tail-to-tail linkages in vinyl polymers may constitute weak points at which thermal degradation may be initiated more readily than in the predominant head-to-tail structures.

Polymers of dienes (hydrocarbons containing two C—C double bonds) have the potential for head-to-tail and head-to-head isomerism and variations in double-bond position as well. The conjugated diene butadiene can polymerize to produce 1,4 and 1,2 products:

$$
CH_2 = \overset{\displaystyle H}{\underset{\displaystyle 2}{C}} - \overset{\displaystyle H}{\underset{\displaystyle 3}{C}} = CH_2 \tag{4-3}
$$

1,2–polybutadiene

1,4 –polybutadiene

The C atoms in the monomer are numbered in reaction (4-3) and the polymers are named according to the particular atoms involved in the enchainment. There is no 3,4-polybutadiene because carbons 1 and 4 are not distinguishable in the monomer structure. This is not the case with 2-substituted conjugated butadienes like isoprene:

$$
CH_2 = C \underset{\underset{CH_3}{|}}{} C = CH_2 \quad \longrightarrow
$$

with H above the second C

1,2 – polyisoprene

$$
\begin{array}{c} CH_3 \\ | \\ +CH_2 - C \, \overset{}{\underset{}{\big)}}_x \\ | \\ H - C = CH_2 \end{array}
$$

1,2 – polyisoprene

$$
\begin{array}{c} H \\ | \\ +CH_2 - C \, \overset{}{\underset{}{\big)}}_x \\ | \\ C - CH_3 \\ \| \\ CH_2 \end{array}
$$

3,4 – polyisoprene

$$
\begin{array}{c} CH_3 \quad H \\ | \qquad | \\ +CH_2 - C = C - CH_2 \, \overset{}{\underset{}{\big)}}_x \end{array}
$$

1,4 – polyisoprene

(4-4)

Each isomer shown in reaction (4-4) can conceivably also exist in head-to-tail or head-to-head, tail-to-tail forms and thus there are six possible constitutional isomers of isoprene or chloroprene (structure of chloroprene is given in Fig. 1-4), to say nothing of the potential for mixed structures.

The constitution of natural rubber is head-to-tail 1,4-polyisoprene. Some methods for synthesis of such polymers are reviewed in Chapter 9.

Unconjugated dienes can produce an even more complicated range of macromolecular structures. Homopolymers of such monomers are not of current commercial importance but small proportions of monomers like 1,5-cyclooctadiene are copolymerized with ethylene and propylene to produce so-called EPDM rubbers. Only one of the diene double bonds is enchained when this terpolymerization is carried out with Ziegler–Natta catalysts (Section 9.5). The resulting small amount of unsaturation permits the use of sulfur vulcanization, as described in Section 1.3.3.

4.1.2 Branching

Linear and branched polymer structures were defined in Section 1.6. Branched polymers differ from their linear counterparts in several important aspects. Branches in crystallizable polymers limit the size of ordered domains because branch points cannot usually fit into the crystal lattice. Thus branched polyethylene is generally less rigid, dense, brittle, and crystalline than linear polyethylene, because the former polymer contains a significant number of relatively short branches. The branched, low-density polyethylenes are preferred for packaging at present because the smaller crystallized regions which they produce provide transparent, tough films. By contrast, the high-density, linear polyethylenes yield plastic bottles and containers more economically because their greater rigidity enables production of the required wall strengths with less polymer.

A branched macromolecule forms a more compact coil than a linear polymer with the same molecular weight, and the flow properties of the two types can differ significantly in the melt as well as in solution. Controlled introduction of relatively long branches into diene rubbers increases the resistance of such materials to flow under low loads without impairing processability at commercial rates in calenders or extruders. The high-speed extrusion of linear polyethylene is similarly improved by the presence of a few long branches per average molecule.

Branching may be produced deliberately by copolymerizing the principal monomer with a suitable comonomer. Ethylene and 1-butene can be copolymerized with a diethylaluminum chloride/titanium chloride (Section 9.5) and other catalysts to produce a polyethylene with ethyl branches:

$$CH_2 = CH_2 + CH_2 = \overset{\overset{\displaystyle H}{|}}{\underset{\underset{\displaystyle CH_3}{|}}{\underset{CH_2}{|}}}{C} \longrightarrow \text{\small\Lambda\Lambda\Lambda} CH_2 - CH_2 - CH_2 - \overset{\overset{\displaystyle H}{|}}{\underset{\underset{\displaystyle CH_3}{|}}{\underset{CH_2}{|}}}{C} - CH_2 - CH_2 \text{\small\Lambda\Lambda\Lambda} \qquad (4\text{-}5)$$

The extent to which this polymer can crystallize under given conditions is controlled by the butene concentration.

Copolymerization of a bifunctional monomer with a polyfunctional comonomer produces branches which can continue to grow by addition of more monomer. An example is the use of divinylbenzene (**4-9**) in the butyl lithium initiated polymerization of butadiene (Section 9.2). The diene has a functionality of 2 under these conditions whereas the functionality of **4-9** is 4. The resulting

$$H - C \equiv CH_2$$

$$H - C \equiv CH_2$$

4-9

elastomeric macromolecule contains segments with structure **4-10**. Long branches such as these can interconnect and form cross-linked, network structures depending on the concentration of polyfunctional comonomer and the fractions of total monomers which have been polymerized. The reaction conditions under which this undesirable occurrence can be prevented are outlined in Section 7.9.

$$\text{\small{WW}CH}_2 - \overset{\overset{\text{H}}{|}}{\text{C}} = \overset{\overset{\text{H}}{|}}{\text{C}} - \text{CH}_2 - \text{CH}_2 - \overset{\overset{\text{H}}{|}}{\text{C}} - \text{CH}_2 - \overset{\overset{\text{H}}{|}}{\text{C}} = \overset{\overset{\text{H}}{|}}{\text{C}} - \text{CH}_2 \text{\small{WW}}$$

$$\text{\small{WW}CH}_2 - \overset{\overset{\text{H}}{|}}{\underset{\underset{\text{H}}{|}}{\text{C}}} - \text{CH}_2 - \overset{\overset{\text{H}}{|}}{\text{C}} = \overset{\overset{\text{H}}{|}}{\text{C}} - \text{CH}_2 \text{\small{WW}}$$

4-10

Another type of branching occurs in some free-radical polymerizations of monomers like ethylene, vinyl chloride, and vinyl acetate in which the macro-radicals are very reactive. So-called "self-branching" can occur in such polymer-izations because of atom transfer reactions between such radicals and polymer molecules. These reactions, which are inherent in the particular polymerization process, are described in Chapter 6.

Although the occurrence of constitutive isomerism can have a profound effect on polymer properties, the quantitiative characterization of such structural varia-tions has been difficult. Recent research has shown that the ^{13}C chemical shifts of polymers are sensitive to the type, length, and distribution of branches as well as to positional isomerism and stereochemical isomerism (Section 4.2.2). This tech-nique has great potential when the bands in the polymer spectra can be assigned unequivocally.

4.2 CONFIGURATIONAL ISOMERISM

Configuration specifies the relative spatial arrangement of bonds in a molecule (of given constitution) without regard to the changes in molecular shape which can arise because of rotations about single bonds. A change in configuration requires the breaking and reforming of chemical bonds. There are two types of configura-tional isomerism in polymers and these are analogous to geometrical and optical isomerism in micromolecular compounds.

4.2.1 Geometrical Isomerism

When conjugated dienes polymerize by 1,4-enchainment, the polymer backbone contains a carbon–carbon double bond. The two carbon atoms in the double bond cannot rotate about this linkage and two nonsuperimposable configurations are therefore possible if the substituents on each carbon differ from each other. For example, the two monomers maleic acid (**4-11**, *cis*) and fumaric acid (**4-12**, *trans*)

$$H-C-COOH \qquad\qquad H-C-COOH$$
$$\| \qquad\qquad\qquad\qquad \|$$
$$H-C-COOH \qquad HOOC-C-H$$

$$\textbf{4-11} \qquad\qquad\qquad \textbf{4-12}$$

are geometrical isomers. Natural rubber is the all-*cis* isomer of 1,4-polyisoprene and has the structure shown in **1-18**.

The molecules in solid *trans* isomers pack more tightly and crystallize more readily than *cis* isomers. (The melting point of fumaric acid is 160°C higher than that of maleic acid.) These corresponding differences in polymers are also major. The 1,4-*cis*-polydienes are rubbers, whereas the *trans* isomers are relatively low melting thermoplastics.

Isomerism in diene polymers can be measured by infrared and nuclear magnetic resonance spectroscopy. Some of the polymerization methods described in Chapter 9 allow the production of polydienes with known controlled constitutions and geometrical configurations.

Cellulose (**1-11**) and amylose starch do not contain carbon–carbon double bonds but they are also geometrical isomers. Both consist of 1,4-linked D-glucopyranose rings, and the difference between them is in configuration at carbon 1. As a result, cellulose is highly crystalline and is widely applied as a structural material while the more easily hydrolyzed starch is used primarily as food.

4.2.2   Stereoisomerism

Stereoisomerism occurs in vinyl polymers when one of the carbon atoms of the monomer double bond carries two different substituents. It is formally similar to the optical isomerism of organic chemistry in which the presence of an asymmetric carbon atom produces two isomers which are not superimposable. Thus glyceraldehyde exists as two stereoisomers with configurations shown in **4-13**. (The dotted lines denote bonds below and the wedge signifies bonds above the plane of the page.) Similarly, polymerization of a monomer with structure

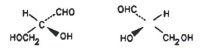

$$\textbf{4-13}$$

4-14 (where X and Y are any substituents that are not identical) yields polymers in which every other carbon atom in the chain is a site of steric isomerism.

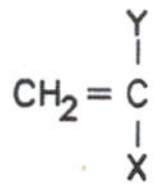

4-14

Such a site, labeled C^x in **4-15**, is termed a *pseudoasymmetric* or *chiral* carbon atom.

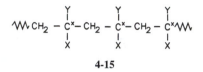

4-15

The two glyceraldehyde isomers of **4-13** are identical in all physical properties except that they rotate the plane of polarized light in opposite directions and form enantiomorphous crystals. When more than one asymmetric center is present in a low-molecular-weight species, however, stereoisomers are formed which are not mirror images of each other and which may differ in many physical properties. An example of a compound with two asymmetric carbons (a diastereomer) is tartaric acid, **4-16**, which can exist in two optically active forms (D and L, mp 170°C), an optically inactive form (meso, mp 140°C), and as an optically inactive mixture (DL racemic, mp 206°C).

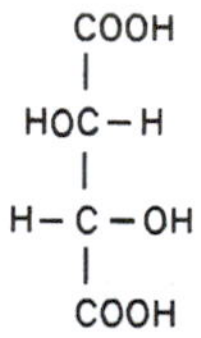

$$
\begin{array}{c}
COOH \\
| \\
HOC-H \\
| \\
H-C-OH \\
| \\
COOH
\end{array}
$$

4-16

Vinyl polymers contain many pseudoasymmetric sites, and their properties are related to those of micromolecular compounds which contain more than one asymmetric carbon. Most polymers of this type are not optically active. The reason for this can be seen from structure **4-15**. Any C^x has four different substituents: X, Y, and two sections of the main polymer chain that differ in length. Optical activity is influenced, however, only by the first few atoms about such a center, and these will be identical regardless of the length of the whole polymer chain. This is why the carbons marked C^x are not true asymmetric centers. Only those C^x centers near the ends of macromolecules will be truly asymmetric, and there are too few chain ends in a high polymer to confer any significant optical activity on the molecule as a whole.

Each pseudoasymmetric carbon can exist in two distinguishable configurations. To understand this, visualize Maxwell's demon walking along the polymer backbone. When the demon comes to a particular carbon C^x she will see three substituents: the polymer chain, X, and Y. If these occur in a given clockwise order (say, chain, X, and Y), C^x has a particular configuration. The substituents could also lie in the clockwise order: chain, Y, and X, however, and this is a different configuration. Thus every C^x may have one or another configuration. This configuration is fixed when the polymer molecule is formed and is independent of any rotations of the main chain carbons about the single bonds which connect them.

The configurational nature of a vinyl polymer has profound effects on its physical properties when the configurations of the pseudoasymmetric carbons are regular and the polymer is crystallizable. The usual way to picture this phenomenon involves consideration of the polymer backbone stretched out so that the bonds between the main chain carbons form a planar zigzag pattern. In this case the X and Y substituents must lie above and below the plane of the backbone, as shown in Fig. 4-1. If the configurations of successive pseudoasymmetric carbons are regular, the polymer is said to be *stereoregular* or *tactic*. If all the configurations are the same, the substituents X (and Y) will all lie either above or below the plane when the polymer backbone is in a planar zigzag shape. Such a polymer is termed *isotactic*. This configuration is depicted in Fig. 4-1a. Note that it is not possible to distinguish between all-D and all-L configurations in polymers because the two ends of the polymer chain cannot be identified. The structure in Fig. 4-1a is thus identical to its mirror image in which all the Y substituents are above the plane.

If the configurations of successive pseudoasymmetric carbons differ, a given substituent will appear alternatively above and below the reference plane in this planar zigzag conformation (Fig. 4-1b). Such polymers are called *syndiotactic*.

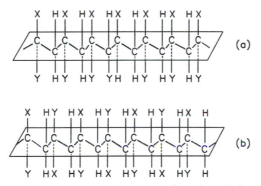

Fig. 4-1.   (a) Isotactic polymer in a planar zigzag conformation. (b) Syndiotactic polymer in a planar zigzag conformation.

When the configurations at the C^x centers are more or less random, the polymer is not stereoregular and is said to be *atactic*. Polymerizations which yield tactic polymers are called *stereospecific*. Some of the more important stereospecific polymerizations of vinyl polymers are described briefly in Chapter 9.

The reader should note that stereoisomerism does not exist if the substituents X and Y in the monomer **4-14** are identical. Thus there are no configurational isomers of polyethylene, polyisobutene, or poly(vinylidene chloride). It should also be clear that 1,2-poly-butadiene (reaction 4-3) and the 1,2- and 3,4-isomers of polyisoprene can exist as isotactic, syndiotactic, and atactic configurational isomers. The number of possible structures of polymers of conjugated dienes can be seen to be quite large when the possibility of head-to-head and head-to-tail isomerism is also taken into account.

It may also be useful at this point to reiterate that the stereoisomerism which is the topic of this section is confined to polymers of substituted ethylenic monomers. Polymers with structures like **1-5** or **1-6** do not have pseudoasymmetric carbons in their backbones.

The importance of stereoregularity in vinyl polymers lies in its effects on the crystallizability of the material. The polymer chains must be able to pack together in a regular array if they are to crystallize. The macromolecules must have fairly regular structures for this to occur. Irregularities like inversions in monomer placements (head-to-head instead of head-to-tail), branches, and changes in configuration generally inhibit crystallization. Crystalline polymers will be high melting, rigid, and difficultly soluble compared to amorphous species with the same constitution. A spectacular difference is observed between isotactic polypropylene, which has a crystal melting point of 176°C, and the atactic polymer which is a rubbery amorphous material. Isotactic polypropylene is widely used in fiber, cordage, and automotive and appliance applications and is one of the world's major plastics. Atactic polypropylene is used mainly to improve the low-temperature properties of asphalt.

Isotactic and syndiotactic polymers will not have the same mechanical properties, because the different configurations affect the crystal structures of the polymers. Most highly stereoregular polymers of current importance are isotactic.

[There are a few exceptions to the general rule that atactic polymers do not crystallize. Poly(vinyl alcohol) (**1-8**) and poly(vinyl fluoride) are examples. Some monomers with identical 1,1-substituents like ethylene, vinylidene fluoride, and vinylidene chloride crystallize quite readily, and others like polyisobutene do not. The concepts of configurational isomerism do not apply in these cases for reasons given above.]

Stereoregularity has relatively little effect on the mechanical properties of amorphous vinyl polymers in which the chiral carbons are trisubstituted. Some differences are noted, however, with polymers in which X and Y in **4-14** differ and neither is hydrogen. Poly(methyl methacrylate) (Fig. 1-4) is an example of the latter polymer type. The atactic form, which is the commercially available product, remains rigid at higher temperatures than the amorphous isotactic polymer.

Completely tactic and completely atactic polymers represent extremes of stereoisomerism which are rarely encountered in practice. Many polymers exhibit intermediate degrees of tacticity, and their characterization requires specification of the overall type and extent of stereoregularity as well as the lengths of the tactic chain sections. The most powerful method for analyzing the stereochemical nature of polymers employs nuclear magnetic resonance (NMR) spectroscopy for which reference should be made to a specialized text [1]. Readers who delve into the NMR literature will be aided by the following brief summary of some of the terminology that is used [2]. It is useful to refer to sequences of two, three, four, or five monomer residues along a polymer chain as a dyad, triad, tetrad, or pentad, respectively. A dyad is said to be *racemic* (r) if the two neighboring monomer units have opposite configurations and *meso* if the configurations are the same. To illustrate, consider a methylene group in a vinyl polymer. In an isotactic molecule the methylene lies in a plane of symmetry. This is a meso structure.

In a syndiotactic region, the methylene group is in a racemic structure

In a triad, the focus is on the central methine between two neighboring monomer residues. An isotactic triad (mm) is produced by two successive meso placements:

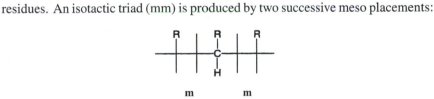

A syndiotactic triad (rr) results from two successive racemic additions:

Similarly, an atactic triad is produced by opposite monomer placements, i.e., (mr) or (rm). The two atactic triads are indistinguishable in an NMR analysis.

The dyads in commercial poly(vinyl chloride) (PVC) are about 0.55% racemic, indicating short runs of syndiotactic monomer placements. The absence of a completely atactic configuration is reflected in the low levels of crystallinity in this polymer, which have a particular influence on the processes used to shape it into useful articles.

4.3   POLYMER CONFORMATION

The conformation of a macromolecule of given constitution and configuration specifies the spatial arrangements of the various atoms in the molecule that may occur because of rotations about single bonds. Molecules with different conformations are called conformational isomers, rotamers, or conformers.

Macromolecules in solution, melt, or amorphous solid states do not have regular conformations, except for certain very rigid polymers described in Section 4.6 and certain polyolefin melts mentioned on page 139. The rate and ease of change of conformation in amorphous zones are important in determining solution and melt viscosities, mechanical properties, rates of crystallization, and the effect of temperature on mechanical properties.

Polymers in crystalline regions have preferred conformations which represent the lowest free-energy balance resulting from the interplay of intramolecular and intermolecular space requirements. The configuration of a macromolecular species affects the intramolecular steric requirements. A regular configuration is required if the polymer is to crystallize at all, and the nature of the configuration determines the lowest energy conformation and hence the structure of the crystal unit cell.

Considerations of minimum overlap of radii of nonbonded substituents on the polymer chain are useful in understanding the preferred conformations of macromolecules in crystallites. The simplest example for our purposes is the polyethylene (**1-3**) chain in which the energy barriers to rotation can be expected to be similar to those in n-butane. Figure 4-2 shows sawhorse projections of the conformational isomers of two adjacent carbon atoms in the polyethylene chain and the corresponding rotational energy barriers (not to scale). The angle of rotation is that between the polymer chain substitutents and is taken here to be zero when the two chain segments are as far as possible from each other.

When the two chain segments would be visually one behind the other if viewed along the polymer backbone, the conformations are said to be *eclipsed.*

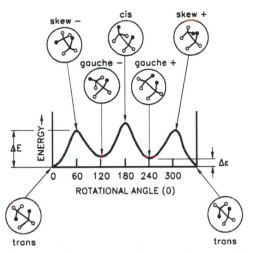

Fig. 4-2. Torsional potentials about adjacent carbon atoms in the polyethylene chain. The white circles represent H atoms and the black circles represent segments of the polymer chain.

The other extreme conformations shown are ones in which the chain substituents are staggered. The latter are lower energy conformations than eclipsed forms because the substituents on adjacent main chain carbons are further removed from each other. The lowest energy form in polyethylene is the staggered *trans* conformation. This corresponds to the planar zigzag form shown in another projection in Fig. 4-1. It is also called an all-*trans* conformation. This is the shape of the macromolecule in crystalline regions of polyethylene.

The conformation of a polymer in its crystals will generally be that with the lowest energy consistent with a regular placement of structural units in the unit cell. It can be predicted from a knowledge of the polymer configuration and the van der Waals radii of the chain substituents. (These radii are deduced from the distances observed between different molecules in crystal lattices.) Thus, the radius of fluorine atoms is slightly greater than that of hydrogen, and the all-*trans* crystal conformation of polyethylene is too crowded for poly(tetrafluoroethylene) which crystallizes instead in a very extended 13_1 helix form. Helices are characterized by a number of f_j, where f is the number of monomer units per j complete turns of the helix. Polyethylene could be characterized as a 1_1 helix in its unit cell.

Helical conformations occur frequently in macromolecular crystals. Isotactic polypropylene crystallizes as a 3_1 helix because the bulky methyl substituents on every second carbon atom in the polymer backbone force the molecule from a *trans/trans/trans*... conformation into a *trans/gauche/trans/gauche*... sequence with angles of rotation of $0°$ (*trans*) followed by a $120°$ (*gauche*) twist.

In syndiotactic polymers, the substituents are further apart because the configurations of successive chiral carbons alternate (cf. Fig. 4-1). The *trans/trans/trans...* planar zigzag conformation is generally the lowest energy form and is observed in crystals of syndiotactic 1,2-poly(butadiene) and poly(vinyl chloride). Syndiotactic polypropylene can also crystallize in this conformation but a *trans/trans/gauche/gauche...* sequence is slightly favored energetically.

Polyamides are an important example of polymers which do not contain pseudoasymmetric atoms in their main chains. The chain conformation and crystal structure of such polymers is influenced by the hydrogen bonds between the carbonyls and NH groups of neighboring chains. Polyamides crystallize in the form of sheets, with the macromolecules themselves packed in planar zigzag conformations.

The difference between the energy minima in the *trans* and *gauche* staggered conformations is labeled $\Delta\epsilon$ in Fig. 4-2. When this energy is less than the thermal energy RT/L provided by collisions of segments, none of the three possible staggered forms will be preferred. If this occurs, the overall conformation of an isolated macromolecule will be a random coil. When $\Delta\epsilon > RT/L$, there will be a preference for the *trans* state. We have seen that this is the only form in the polyethylene crystallite.

The time required for the transition between *trans* and *gauche* states will depend on the height of the energy barrier ΔE in Fig. 4-2. If $\Delta E < RT/L$, the barrier height is not significant and *trans/gauche* isomerizations will take place in times of the order of 10^{-11} sec. When a macromolecule with small ΔE is stretched into an extended form, the majority of successive carbon–carbon links will be *trans*, but *gauche* conformations will be formed rapidly when the molecule is permitted to relax again. As a result, the overall molecular shape will change rapidly from an extended form to a coiled, ball shape. This is the basis of the ideal elastic behavior outlined in more detail in Section 4.5. Note that a stretched polymer molecule will recoil rapidly to a random coil shape only if (1) there is no strong preference for any staggered conformation over another ($\Delta\epsilon$ is small; there is little difference between the energy minima) and (2) if the rotation about carbon–carbon bonds in the main chain is rapid (ΔE is small; the energy barriers between staggered forms are small). If condition (1) holds but (2) does not, the polymer sample will respond sluggishly when the force holding it in an extended conformation is removed.

The *trans* staggered conformation is a lower energy form than either of the *gauche* staggered forms of polyethylene. The difference is much less for polyisobutene, however, as illustrated in Fig. 4-3. Here the chain substituent on the rear carbon shown is either between a methyl and polymer chain or between two methyl groups on the other chain carbon atom. Since no conformation is favored, this polymer tends to assume a random coil conformation. The polymer is elastomeric and can be caused to crystallize only by stretching. However, rotations between

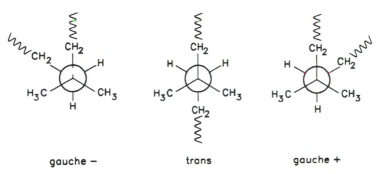

Fig. 4-3.   Newman projections of staggered conformations of adjacent carbons in the main chain of polyisobutene.

staggered conformations require sufficient energy for the chain to overcome the high barrier represented by crowded eclipsed forms, and polyisobutene does not retain its elastic character at temperatures as low as those at which more resilient rubbers can be used.

The preference for *trans* conformations in hydrocarbon polymer chains may be affected by the polymer constitution. *Gauche* conformations become more energetically attractive when atoms with lone electron pairs (like O) are present in the polymer backbone, and polyformaldehyde (or polyoxymethylene), **1-12**, crystallizes in the all-*gauche* form.

4.4   MOLECULAR DIMENSIONS IN THE AMORPHOUS STATE

Polymers differ from small molecules in that the space-filling dimensions of macro-molecules are not fixed. This has some important consequences, one of which is that certain polymers behave elastically when they are deformed. The nature of this rubber elasticity and its connection with changes in the dimensions of elastomeric polymers are explored in this and the subsequent section of this chapter.

Portions of polymer molecules which are in crystalline regions have overall dimensions and space-filling characteristics that are determined by the particular crystal habit which the macromolecule adopts. Here, however, we are concerned with the sizes and shapes of flexible polymers in the amorphous (uncrystallized) condition. It will be seen that the computation of such quantities provides valuable insights into the molecular nature of rubber elasticity.

4.4.1   Radius of Gyration and End-to-End Distance of Flexible Macromolecules

The extension in space of a polymer molecule is usually characterized by two average dimensions. These are the end-to-end distance d and the radius of gyration r_g. The end-to-end distance is the straight-line distance between the ends of a linear molecule in a given conformation. It can obviously change with the overall molecular shape and will vary with time if the macromolecule is dynamically flexible. The radius of gyration was defined on page 89 as the square root of the average squared distance of all the repeating units of the molecule from its center of mass. This definition follows from the fact that r_g of a body with moment of inertia I and mass m is defined in mechanics as

$$r_g \equiv (I/m)^{1/2} \tag{4-6}$$

If we consider the polymer chain to be an aggregate of repeating units each with identical mass m_i and variable distance r_i from the center of mass of the macro-molecule, then I by definition is

$$I \equiv \sum m_i r_i^2 \tag{4-7}$$

(I is the second moment of mass about the center of mass, cf. Section 2.4.1) Then from the above equations,

$$r_g^2 \equiv \left(\sum m_i r_i^2 \right) \Big/ \sum m_i \tag{4-8}$$

and the average value of r_g will be

$$\langle r_g \rangle = \langle r_g^2 \rangle^{1/2} = \left\langle \left(\sum m_i r_i^2 \right) \right\rangle \Big/ \sum m_i \right\rangle^{1/2} \tag{4-9}$$

where the $\langle \rangle$ means an average.

The radius of gyration is directly measurable by light scattering (p. 90), neutron scattering, and small angle X-ray scattering experiments. The end-to-end distance is not directly observable and has no significance for branched species which have more than two ends. A unique relationship exists between r_g and d for high-molecular-weight linear macromolecules that have random coil shapes:

$$r_g = d/\sqrt{6} \tag{4-10}$$

The end-to-end distance is more readily visualized than the radius of gyration and is more directly applicable in the following molecular explanation of rubber elasticity. The derivations in the following section therefore focus on d rather than r_g.

4.4.2   Root Mean Square End-to-End Distance of Flexible Macromolecules

(i)   Freely Oriented Chain

The simplest calculation is based on the assumption that a macromolecule comprises $\sigma + 1$ (sigma + 1) elements of equivalent size which are joined by σ bonds of fixed length l. (If the bonds differ in length, an average value can be used in this calculation. Here we assume that all bonds are equivalent.) All angles between successive bonds are equally probable. Such a chain is illustrated in Fig. 4-4 where each bond is represented by a vector $\mathbf{l}_i$. The end-to-end vector in a given conformation is

$$\mathbf{d} = \sum_{i=1}^{\sigma} \mathbf{l}_i \qquad (4\text{-}11)$$

Now it is convenient to recall the meaning of the dot product of two vectors. For vectors $\mathbf{a}$ and $\mathbf{b}$, the dot (or scalar) product is equal to the product of their lengths and the cosine of the angle between them. That is,

$$\mathbf{a} \cdot \mathbf{b} = ab \cos \theta \qquad (4\text{-}12)$$

where θ is the bond angle (which is the supplement of the valence angle, for C—C

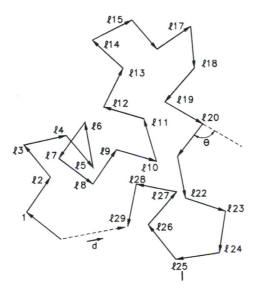

Fig. 4-4.   An unrestricted macromolecule. The bond lengths l_i are fixed and equal and there are no preferred bond angles.

bonds) and a and b are the respective bond lengths. The dot product is a scalar quantity and the dot product of a vector with itself is just the square of the length of the vector. To obtain the end-to-end scalar distance, we take the dot product of $\mathbf{d}$. Because the single chain can take any of an infinite number of conformations, we will compute the average scalar magnitude of $\mathbf{d}$ over all possible conformations. That is,

$$\langle d^2 \rangle = \langle \mathbf{d} \cdot \mathbf{d} \rangle = \left\langle \left(\sum_{i=1}^{\sigma} \mathbf{l}_i \right) \cdot \left(\sum_{j=1}^{\sigma} \mathbf{l}_j \right) \right\rangle \tag{4-13}$$

The dummy subscripts i and j indicate that each term in the second sum is multiplied by each term in the first sum.

If θ is the angle between the positive directions of any two successive bonds, then

$$\mathbf{l}_i \cdot \mathbf{l}_{i+1} = l_i l_{i+1} \cos \theta \tag{4-12a}$$

All values of θ are equally probable for a chain with unrestricted rotation, and since $\cos \theta = -\cos(-\theta)$

$$\mathbf{l}_i \cdot \mathbf{l}_{i+1} = l_i l_{i+1} \langle \cos \theta \rangle = 0 \tag{4-14}$$

Thus all dot products in Eq. (4-13) vanish except when each vector is multiplied with itself. We are left with

$$\langle d^2 \rangle = \left\langle \sum_{i=1}^{\sigma} l_i^2 \right\rangle = \sigma l^2 \tag{4-15}$$

The extended or contour length of the freely oriented macromolecule will be σl. Its rms end-to-end distance will be $l\sigma^{1/2}$ from the above equation. It can be seen that the ratio of the average end-to-end separation to the extended length is $\sigma^{-1/2}$. Since σ will be of the order of a few hundred even for moderately sized macromolecules, d will be on the average much smaller than the chain end separation in the fully extended conformation.

The average chain end separation which has been calculated gives little information about the magnitudes of this distance for a number of macromolecules at any instant. When this distribution of end-to-end distances is calculated, it is found, not surprisingly, that it is very improbable that the two ends of a linear molecule will be very close or very far from each other. It can also be shown that the density of chain segments is greatest near the center of a macromolecule and decreases toward the outside of the random coil.

(ii)   Real Polymer Chains

The random-flight model used in the previous section underestimates the true dimensions of polymer molecules, because it ignores restrictions to completely free orientation resulting from fixed valence bond angles and steric effects. It also fails to allow for the long-range effects which result from the inability of two segments of the chain to occupy the same space at the same time.

The effects of fixed rather than unrestricted bond angles can be readily computed, and it is found that the random flight relation of Eq. (4-15) is modified to

$$\langle d^2 \rangle = \sigma l^2 (1 + \cos\theta)/(1 - \cos\theta) \tag{4-16}$$

For a polyethylene chain, l is the C—C bond distance (1.54×10^{-10} m) and θ is $180°$ minus the tetrahedral bond angle $= 180° - 109° \, 28' = 70° \, 32'$. $\cos\theta$ then is 0.33 and $\langle d^2 \rangle = 2\sigma l^2$. The actual chain end-to-end distance is thus expanded by a factor of $\sqrt{2}$ for this reason.

When some conformations are preferred over others (e.g., in Fig. 4-2), the chain dimensions are further expanded over those calculated, and Eq. (4-16) becomes

$$\langle d^2 \rangle = \sigma l^2 \left[\frac{1 + \cos\theta}{1 - \cos\theta} \right] \left[\frac{1 + \langle\cos\phi\rangle}{1 - \langle\cos\phi\rangle} \right] \tag{4-17}$$

where $\langle\cos\phi\rangle$ is the average value of the cosine of the rotation angle ϕ. For free rotation, all values of ϕ are equally probable, $\cos\phi$ is zero, and Eq. (4-17) reduces to Eq. (4-16). In a completely planar zigzag conformation, all rotamers are trans and $\phi = 0$. Then $\langle\cos\phi\rangle = 1$, and the model breaks down in this limit. Nevertheless, it does show that the chain becomes more and more extended the closer the rotation angle is to zero. The values of $\langle\cos\phi\rangle$ can be calculated if the functional dependence of potential energy of a sequence of bonds on the bond angle is known. For small molecules, this can be deduced from infrared spectra. Figure 4-2 showed this relation approximately for a normal paraffin.

The value of $\langle\cos\phi\rangle$ will depend on temperature, of course, since the molecule will have sufficient torsional motion to overcome the energy barriers hindering rotation when the temperature is sufficiently high.

The barriers to rotational motion which have been observed in micromolecular species are incorporated into computer programs along with interatomic potentials, bond angles and so on to model the lowest energy conformations of macromolecules in specified environments. Such molecular dynamics studies have shown that the lowest energy state of polymers in their "melt" condition is not necessarily that with the highest entropy. In particular, molten polyethylene molecules do not resemble a bowl of spaghetti. Rather, the overall conformation with the lowest energy is one which comprises a significant fraction of shapes in which the chains are folded back on themselves in an expanded version of the polyethylene crystal structure described in Chapter 11.

When the average end-to-end distance of a macromolecular coil is given by Eq. (4-17), the polymer is said to be in its "unperturbed" state. Its dimensions then are determined only by the characteristics of the molecule itself. In general, the end-to-end distance of a dissolved macromolecule is greater than that in its unperturbed state because the polymer coil is swollen by solvent. If the actual average end-to-end distance in solution is $\langle d^2 \rangle^{1/2}$, then

$$\langle d^2 \rangle^{1/2} = \langle d_0^2 \rangle^{1/2} \alpha \tag{4-18}$$

where the subscript zero refers to the unperturbed dimensions. The expansion coefficient α can be considered to be practically equal to the coefficient α_η, which was introduced on page 94 in connection with the ratio of intrinsic viscosities of a particular polymer in a good solvent and under theta conditions. If a random coil polymer is strongly solvated in a particular solvent, the molecular dimensions will be relatively expanded and α will be large. Conversely, in a very poor solvent α can be reduced to a value of 1. This corresponds to theta conditions under which the end-to-end distance is the same as it would be in bulk polymer at the same temperature (p. 67).

(iii)   The Equivalent Random Chain [3]

The real polymer chain may be usefully approximated for some purposes by an equivalent freely jointed chain. It is obviously possible to find a randomly jointed model which will have the same end-to-end distance as a real macromolecule with given molecular weight. In fact, there will be an infinite number of such equivalent chains. There is, however, only one equivalent random chain which will fit this requirement and the additional stipulation that the real and phantom chains also have the same contour length.

   If both chains have the same end-to-end distance, then

$$\langle d^2 \rangle = \langle d_e^2 \rangle = \sigma_e l_e^2 \tag{4-19}$$

where the unsubscripted term refers to the real chain and the subscript e designates the equivalent random chain. Also if both have the same contour length D, then

$$D = D_e = \sigma_e l_e \tag{4-20}$$

From Eqs. (4-19) and (4-20):

$$l_e = \langle d^2 \rangle / D \tag{4-21}$$

and

$$\sigma_e = D^2 / \langle d^2 \rangle \tag{4-22}$$

   To illustrate the use of this model, consider a *cis*-polyisoprene chain (**1-18**). The length of the isoprene unit is found to be 4.60×10^{-10} m, and the fully extended (contour) length is D is therefore $4.60 \times 10^{-10}\, n$m, where n is the number of repeating units in the macromolecule. It has also been calculated that $\langle d^2 \rangle^{1/2}$ for high-molecular-weight polyisoprene is $= (2.01\sqrt{\sigma})10^{-10}$ m. Since there are four bonds per isoprene unit we expect that $\langle d^2 \rangle = (2.01^2 4n)10^{-10}$ m $= (16.2n)10^{-10}$ m. Hence $l_e = (3.52)10^{-10}$ m and $\sigma_e = 1.31n$. That is to say, there are 0.77 isoprene repeating units (3.1 C—C bonds) per equivalent random link.

4.5   RUBBER ELASTICITY

4.5.1   Qualitative Description of Elastomer Behavior

Unvulcanized rubber consists of a large number of flexible long molecules with a structure that permits free rotation about single bonds in the primary chain. On deformation the molecules are straightened, with a decrease in entropy. This results in a retractive force on the ends of the polymer molecules. The molecular structure of the flexible rubber molecules makes it relatively easy for them to take up statistically random conformations under thermal motion. This property is a result of the weak intermolecular attractive forces in elastomers and distinguishes them chemically from other polymers which are more suitable for use as plastics or fibers.

It is important to understand that flow and deformation in high polymers results from local motion of small segments of the polymer chain and not from concerted, instantaneous movements of the whole molecule. High elasticity results from the ability of extended polymer chains to regain a coiled shape rapidly. Flexibility of segments of the molecule is essential for this property, and this flexibility results from relative ease of rotation about the axis of the polymer chain. Figure 4-5 illustrates the mechanism of a segmental jump by rotation about two carbon–carbon bonds in a schematic chain molecule. The hole in the solid structure is displaced to the right, in this scheme, as the three-carbon segment jumps to the left. Clearly, such holes (which are present in wastefully packed, i.e., noncrystalline polymers) can move through the structure.

If molecules are restrained by entanglement with other chains or by actual chemical bonds (cross-links) between chains, deformation is still possible because of cooperative motions of local segments. This presupposes that the number of chain atoms between such restraints is very much larger than the average size of segments involved in local motions. Ordinary vulcanized natural rubber contains 0.5–5 parts (by weight) of combined sulfur vulcanizing agent per 100 parts of rubber. Approximately one of every few hundred monomer residues is cross-linked in a typical rubber with good properties (the molecular weight of the chain

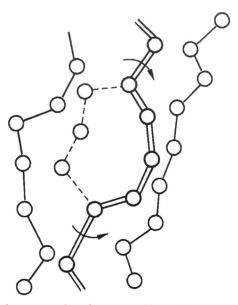

Fig. 4-5. Schematic representation of a segmental jump by rotation about two carbon–carbon bonds in a macromolecular chain [4].

regions between cross-links is 20,000–25,000 in such a hydrocarbon rubber). If the cross-link density is increased, for example, by combining 30–50 parts of sulfur per 100 parts rubber, segmental motion is severely restricted. The product is a hard, rigid nonelastomeric product known as "ebonite" or "hard rubber."

High elasticity is attributed to a shortening of the distance between the ends of chain molecules undergoing sufficient thermal agitation to produce rotations about single bonds along the main chains of the molecules. The rapid response to application and removal of stress which is characteristic of rubbery substances requires that these rotations take place with high frequency at the usage temperatures.

Rotations about single bonds are never completely free, and energy barriers are encountered as substituents on adjacent chain atoms are turned away from staggered conformations (Fig. 4-2). These energy barriers are smallest for molecules without bulky or highly polar side groups. Unbranched and relatively symmetrical chains are apt to crystallize on orientation or cooling, however, and this is undesirable for high elasticity because the crystallites hold their constituent chains fixed in the lattice. Some degree of chain irregularity caused by copolymerization can be used to reduce the tendency to crystallize. If there are double bonds in the polymer chain as in 1,4-polydienes like natural rubber, the *cis* configuration produces a lower packing density; there is more free space available for segmental jumps and the more irregular arrangement reduces the ease of crystallization. Thus *cis*-polyisoprene (natural rubber) is a useful elastomer while *trans*-polyisoprene is not.

The molecular requirements of elastomers can be summarized as follows:

1.   The material must be a high polymer.
2.   Its molecules must remain flexible at all usage temperatures.
3.   It must be amorphous in its unstressed state. (Polyethylene is not an elastomer, but copolymerization of ethylene with sufficient propylene reduces chain regularity sufficiently to eliminate crystallinity and produce a useful elastomer.)
4.   For a polymer to be useful as an elastomer, it must be possible to introduce cross-links in such a way as to bond a macroscopic sample into a continuous network. Generally, this requires the presence of double bonds or chemically functional groups along the chain.

Polymers which are not cross-linked to form infinite networks can behave elastically under transient stressing conditions. They cannot sustain prolonged loads, however, because the molecules can flow past each other to relieve the stress, and the shape of the article will be deformed by this creep process. [Alternatives to cross-linking are mentioned on pages 20 and 315.]

Polybutadiene with no substituent groups larger than hydrogen has greater resilience than natural rubber, in which a methyl group is contained in each isoprene repeating unit. Polychloroprenes (neoprenes) have superior oil resistance but lose their elasticity more readily at low temperatures since the substituent is a bulky, polar chlorine atom. (The structures of these monomers are given in Fig. 1-4.)

4.5.2   Rubber as an Entropy Spring

Disorder makes nothing at all, but unmakes everything.
—John Stuart Blackie

Bond rotations and segmental jumps occur in a piece of rubber at high speed at room temperature. A segmental movement changes the overall conformation of the molecule. There will be a very great number of equi-energetic conformations available to a long chain molecule. Most of these will involve compact rather than extended contours. There are billions of compact conformations but only one fully extended one. Thus, when the ends of the molecule are far apart because of uncoiling in response to an applied force, bond rotations after release of the force will turn the molecule into a compact, more shortened state just by chance. About 1000 individual C—C bonds in a typical hydrocarbon elastomer must change conformation when a sample of fully extended material retracts to its shortest state at room temperature [5]. There need not be any energy changes involved in this change. It arises simply because of the very high probability of compact compared to extended conformations.

An elastomer is essentially an *entropy spring*. This is in contrast to a steel wire, which is an *energy spring*. When the steel spring is distorted, its constituent atoms are displaced from their equilibrium lowest energy positions. Release of the applied force causes a retraction because of the net gain in energy on recovering the original shape. An energy spring warms on retraction. An ideal energy spring is a crystalline solid with Young's modulus about 10^{11}–10^{12} dyn/cm (10^{10}–10^{11}N/m^2). It has a very small ultimate elongation. The force required to hold the energy spring at constant length is inversely proportional to temperature. In thermodynamic terms $(\partial U/\partial l)_T$ is large and positive, where U is the internal energy thermodynamic state function.

An ideal elastomer has Young's modulus about 10^6–10^7 dyn/cm^2 (10^5–10^6 N/m^2) and reversible elasticity of hundreds of percent elongation. The force required to hold this entropy spring at fixed length falls as the temperature is lowered. This implies that $(\partial U/\partial l)_T = 0$.

(i)   Ideal Elastomer and Ideal Gas

An ideal gas and an ideal elastomer are both entropy springs.

The molecules of an ideal gas are independent agents. By definition, there is no intermolecular attraction. The pressure of the gas on the walls of its container is due to random thermal bombardment of the molecules on the walls. The tension of rubber against restraining clamps is due to random coiling and uncoiling of chain molecules. The molecules of an ideal elastomer are independent agents. There is no intermolecular attraction, by definition. (If there is appreciable intermolecular attraction, the material will not exhibit high elasticity, as we saw earlier.)

Gas molecules tend to their most likely distribution in space. The molecules of an ideal elastomer tend to their most probable conformation, which is that of a random coil. The most probable state in either case is that in which the entropy is a maximum.

If the temperature of an ideal gas is increased at constant volume, its pressure rises in direct proportion to the temperature. Similarly, the tension of a rubber specimen at constant elongation is directly proportional to temperature. An ideal gas undergoes no temperature change on expanding into a vacuum. An ideal rubber retracting without load at constant volume undergoes no temperature change. Under adiabatic conditions, an ideal gas cools during expansion against an opposing piston, and a stretched rubber cools during retraction against a load.

Table 4-1 lists the thermodynamic relations between pressure, volume, and temperature of an ideal gas and its internal energy U and entropy S. We see that the definition of an ideal gas leads to the conclusion that the pressure exerted by

TABLE 4-1

Ideal Gas as an Entropy Spring

First and second laws of thermodynamics applied to compression of a gas:

$$dU = dq + dw \qquad \text{(i)}$$

where U = internal energy function, dq = heat absorbed by substance, and dw = work done on substance by its surroundings.

$$dU = T\,dS - P\,dV \qquad \text{(ii)}$$

Equation (ii) yields

$$P = T(\partial S/\partial V)_T - (\partial U/\partial V)_T = -(\partial A/\partial V)_T \qquad \text{(iii)}$$

where S = entropy and A = Helmholtz free energy $\equiv U - TS$.
From (iii), the pressure consists of two terms:

entropy contribution: $\qquad T(\partial S/\partial V)_T$ (called kinetic pressure)
internal energy contribution: $\quad -(\partial U/\partial V)_T$ (called internal pressure)

To evaluate the terms in Eq. (iii) experimentally, substitute

entropy contribution: $\qquad T(\partial S/\partial V)_T = T(\partial P/\partial T)_V$

Thus,

$$P = T(\partial P/\partial T)_V - (\partial U/\partial V)_T \qquad \text{(iiia)}$$

internal energy contribution to the total pressure: $\qquad P - T(\partial P/\partial T)_V \qquad \text{(iv)}$
Definition of *ideal gas* is gas which obeys the equation of state:

$$PV = nRT$$

and for which the internal energy U is a function of temperature only, i.e.,

$$(\partial U/\partial V)_T = (\partial U/\partial P)_T = 0 \qquad \text{(ideal gas)} \qquad \text{(vi)}$$

It follows that

$$P = T(\partial P/\partial T)_V, \qquad \text{(ideal gas)} \qquad \text{(vii)}$$

and the pressure is due only to the entropy contribution.

---

such a material is entirely due to an entropy contribution. If an ideal gas confined at a certain pressure were allowed to expand against a lower pressure, the increase in volume would result in the gas going to a state of greater entropy. The internal energy of the ideal gas is not changed in expanding at constant temperature.

(ii) Thermodynamics of Rubber Elasticity

In an ideal gas we considered the relations between the thermodynamic properties S and U, on the other hand, and the state variables P, V, and T of the substance. With an ideal elastomer we shall be concerned with the relation between U and S and the state variables force, length, and temperature.

The first law of thermodynamics defines the internal energy from

$$dU \equiv dq + dw \qquad \text{(4-23)}$$

(The increase dU in any change taking place in a system equals the sum of the energy added to the system by the heat process, dq, and the work performed on it, dw.)

The second law of thermodynamics defines the entropy change dS in any reversible process:

$$T\,dS = dq_{\text{rev}} \tag{4-24}$$

(A reversible process is the thermodynamic analog of frictionless motion in mechanics. When a process has been conducted reversibly, we can, by performing the inverse process in reverse, set the system back in precisely its initial state, with zero net expenditure of work in the overall process. The system and its surroundings are once again exactly as they were at the beginning. A reversible process is an idealization which constitutes a limit that may be approached but not attained in real processes.)

For a reversible process, Eqs. (4-23) and (4-24) yield

$$dU = T\,dS + dw \tag{4-25}$$

We define the Helmholtz free energy A as

$$A \equiv U - TS \tag{4-26}$$

(This is a useful thermodynamic quantity to characterize changes at constant volume of the working substance.)   For a change at constant temperature, from Eq. (4-26),

$$dA = dU - T\,dS \tag{4-27}$$

Combining Eqs. (4-26) and (4-25)

$$dA = dw \tag{4-28}$$

That is, the change in A in an isothermal process equals the work done on the system by its surroundings. Conventionally, when gases and liquids are of major interest the work done on the system is written $dw = -P\,dV$. When we consider elastic solids, the work done by the stress is important. If tensile force is f and l is the initial length of the elastic specimen in the direction of the force, the work done in creating an elongation dl is

$$dw = f\,dl \tag{4-29}$$

If a hydrostatic pressure P is acting in addition to the tensile force f, the total work on the system is

$$dw = f\,dl - P\,dV \tag{4-30}$$

In the case of rubbers, dV is very small and if $P = 1$ atm, $P\,dV$ is less than $10^{-3}f\,dl$. Thus we can neglect $P\,dV$ and use Eq. (4-29).

From Eqs. (4-28) and (4-29),

$$(\partial A/\partial l)_T = (\partial w/\partial l)_T = f \tag{4-31}$$

That is, the tension is equal to the change in Helmoholtz free energy per unit extension. From Eqs. (4-27) and (4-29),

$$(\partial A/\partial l)_T = (\partial U/\partial l)_T - (\partial S/\partial l)_T = f \tag{4-32}$$

Thus the force consists of an internal energy component and an entropy component [compare (iii) of Table 4-1 for the pressure of a gas].

To evaluate Eq. (4-32) experimentally, we proceed in an analogous fashion to the method used to estimate the entropy component of the pressure of a gas (Table 4-1). From Eq. (4-26), for any change,

$$dA = dU - T\,dS - S\,dT \tag{4-33}$$

For a reversible change, from Eqs. (4-25) and (4-29),

$$dU = f\,dl + T\,dS \tag{4-34}$$

Combining the last two equations,

$$dA = f\,dl - S\,dT \tag{4-35}$$

Thus, by partial differentiation,

$$(\partial A/\partial l)_T = f \qquad \text{(this is Eq. 4-31)}$$

$$(\partial A/\partial T)_l = -S \tag{4-36}$$

Since

$$\frac{\partial}{\partial l}\left(\frac{\partial A}{\partial T}\right)_l = \frac{\partial}{\partial T}\left(\frac{\partial A}{\partial l}\right)_T \tag{4-37}$$

we can substitute Eqs. (4-31) and (4-36) into Eq. (4-37) to obtain

$$(\partial S/\partial l)_T = -(\partial f/\partial T)_l \tag{4-38}$$

This gives the entropy change per unit extension, $(\partial S/\partial T)_T$, which occurs in Eq. (4-32), in terms of the temperature coefficient of tension at constant length $(\partial f/\partial T)_l$, which can be measured. With Eq. (4-38), Eq. (4-32) becomes

$$(\partial U/\partial l)_T = f - T(\partial f/\partial T)_l \tag{4-39}$$

where $(\partial U/\partial l)_T$ is the internal energy contribution to the total force. [Compare Eq. (iv) in Table 4-1 for a gas.]

Figure 4-6 shows how experimental data can be used with Eqs. (4-38) and (4-39) to determine the internal energy and entropy changes accompanying deformation of an elastomer. Such experiments are simple in principle but difficult in practice because it is hard to obtain equilibrium values of stress.

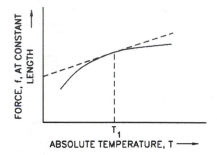

Fig. 4-6.   Experimental measurement of $(\partial f/\partial l)_T$ and $(\partial U/\partial l)_T$. The slope of the tangent to the curve at temperature $T_1 = (\partial f/\partial T)_l$ at T_1. This equals $(\partial S/\partial l)_T$, which equals entropy change per unit extension when the elastomer is extended isothermally at T_1. The intercept on the force axis equals $(\partial U/\partial l)_T$ since this corresponds to $T = 0$ in Eq. (4-39). The intercept is the internal energy change per unit extension.

For an ideal elastomer $(\partial U/\partial l)_T$ is zero and Eq. (4-39) reduces to

$$f = T(\partial f/\partial T)_l \qquad (4\text{-}40)$$

in complete analogy to Eq. (vii) of Table 4-1 for an ideal gas. In real elastomers, chain uncoiling must involve the surmounting of bond rotational energy barriers and this means that the internal energy term $(\partial U/\partial l)_T$ cannot be identically zero, however.

If the internal energy contribution to the force at constant length and sample volume is f_e, its relative contribution is

$$\frac{f_e}{f} = 1 - \frac{T}{f}\left(\frac{\partial f}{\partial T}\right)_{V,l} \qquad (4\text{-}41)$$

Various measurements have shown that f_e/f is about 0.1–0.2 for polybutadiene and *cis*-polyisoprene elastomers. These polymers are essentially but not entirely entropy springs.

(iii)   Stress–Strain Properties of Cross-Linked Elastomers

Consider a cube of cross-linked elastomer with unit dimensions. This specimen is subjected to a tensile force f. The ratio of the increase in length to the unstretched length is the nominal strain ϵ (epsilon), but the deformation is sometimes also expressed as the extension ratio Λ (lambda):

$$\Lambda = \lambda/\lambda_0 = 1 + \epsilon \qquad (4\text{-}42)$$

where λ and λ_0 are the stretched and unstretched specimen lengths, respectively. With a cube of unit initial dimensions, the stress τ is equal to f. (Recall the definitions of stress, normal strain, and modulus on page 24.) Also, in this special case $d\lambda = \lambda_0\, d\Lambda = d\Lambda$, and so Eqs. (4-38) and (4-40) are equivalent to

$$\tau = -T\,(\partial S/\partial \Lambda)_{T,V} \tag{4-43}$$

Statistical mechanical calculations [3] have shown that the entropy change is given by

$$\Delta S = -\frac{1}{2}N\kappa(\Lambda^2 + (2/\Lambda) - 3) \tag{4-44}$$

where N is the number of chain segments between cross-links per unit volume and κ (kappa) is Boltzmann's constant (R/L). Then, from Eq. (4-43),

$$\tau = N\kappa T(\Lambda - 1/\Lambda^2) \tag{4-45}$$

Equation (4-45) is equivalent to

$$\tau = (\rho RT/M_c)(\Lambda - 1/\Lambda^2) \tag{4-46}$$

where ρ is the elastomer density (gram per unit volume), M_c is the average molecular weight between cross-links $(M_c = \rho L/N)$, and L is Avogadro's constant.

Equation (4-46) predicts that the stress–strain properties of an elastomer that behaves like an entropy spring will depend only on the temperature, the density of the material, and the average molecular weight between cross-links. In terms of nominal strain this equation is approximately

$$\tau = (\rho RT/M_c)(3\epsilon + 3\epsilon^2 + \cdots \tag{4-47}$$

and at low strains, Young's modulus, Y, is

$$Y \equiv d\tau/d\epsilon = 3\rho RT/M_c \tag{4-48}$$

The more tightly cross-linked the elastomer, the lower will be M_c and the higher will be its modulus. That is, it will take more force to extend the polymer a given amount at fixed temperature. Also, because the elastomer is an entropy spring, the modulus will increase with temperature.

Equation (4-46) is valid for small extensions only. The actual behavior of real cross-linked elastomers in uniaxial extension is described by the Mooney–Rivlin equation which is similar in form to Eq. (4-46):

$$\tau = (C_1 + C_2/\Lambda)(\Lambda - 1/\Lambda^2) \tag{4-49}$$

Here C_1 and C_2 are empirical constants, and C_1 is often assumed to be equal to $\rho RT/M_c$.

(iv)   Real and Ideal Rubbers

To this point, we have emphasized that the retractive force in a stretched ideal elastomer is directly proportional to its temperature. In a cross-linked, real elastomer that has been reinforced with carbon black, as is the usual practice, the force to produce a given elongation may actually be seen to *decrease* with increased temperature. This is because the anchor regions that hold the elastomer chains together are not only chemical cross-links, as assumed in the ideal theory. They also comprise physical entanglements of polymer molecules and rubber-carbon black adsorption sites. Entanglements will be more labile at higher temperatures, where the molecular chains are more flexible, with a net decrease in the number of effective intermolecular anchor points, an increase in M_c, and a decrease in the retractive force, according to Eq. (4-48).

4.6   RODLIKE MACROMOLECULES

Very rigid macromolecules are at the opposite end of the spectrum of properties from elastomers, which are characterized by weak intermolecular forces, a high degree of molecular flexibility, and an absence of regular intermolecular order.

Aromatic polyamides and polyesters are examples of stiff chain polymers. Poly(p-phenylene terephthalamide) (Kevlar^{TM}, **1-23**) can be made by reaction (4-50) in a mixture of hexamethylphosphoramide and N-methylpyrrolidone:

$$(4\text{-}50)$$

Polyamide-hydrazides (**4-17**)

$$(4\text{-}51)$$

and aromatic polyesters like poly(p-hydroxybenzoic acid) (**4-18**)

$$(4\text{-}52)$$

also provide rodlike species. These polymers behave in solution like logs on a pond rather than like random coils. They exhibit liquid crystalline properties where they have the short-range order of nematic mesophases. The liquid crystals are readily oriented under shear and can be used to produce very highly oriented ultrastrong fibers. On a specific weight basis they are stronger and stiffer than steel or glass and are used to reinforce flexible and rigid composites like tires, conveyor belts, and body armor as well as to make industrial and military protective clothing.

PROBLEMS

4-1. Which of the following monomers can conceivably form isotactic polymers?

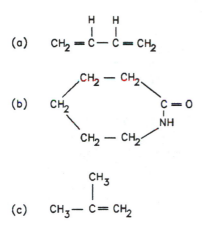

(e) $HOCH_2 CH_2 OH + HO - \overset{\displaystyle O}{\underset{\displaystyle \|}{C}} - \langle O \rangle - \overset{\displaystyle O}{\underset{\displaystyle \|}{C}} - OH$?

4-2. Draw projection diagrams (planar zigzag) for
    (a)   the syndiotactic polymer produced by 1,2 enchainment of isoprene.
    (b)   isotactic poly(3,4-isoprene).

4-3. Polyethylene and polyisobutene are both hydrocarbon polymers and have intermolecular forces of similar magnitude. Yet, one polymer is a plastic and fiber-former and the other is an elastomer. Comment briefly on the reason for this difference in properties.

4-4. Poly(vinyl alcohol) is made by free-radical polymerization of vinyl acetate and subsequent base-catalyzed transesterification with methanol to yield the alcohol polymer.
    (a)   Write an equation showing the transesterification of one repeating unit of poly(vinyl acetate) to one of poly(vinyl alcohol).
    (b)   Before the advent of nuclear magnetic resonance spectroscopy, one way of determining head-to-head structures in poly(vinyl alcohol) was by means of the following difference in diol reactions:

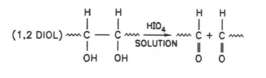

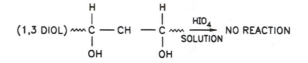

With a particular poly(vinyl alcohol) sample no periodic acid is consumed, within the limits of analytical accuracy. This indicates no apparent (1,2-diol) cleavage. However, the viscosity average molecular weight of the sample decreased from 250,000 to 100,000. Explain these results in terms of the structures of poly(vinyl acetate) and poly(vinyl alcohol). [The analytical technique is described by P. J. Flory and F. S. Leutner, *J. Polym. Sci.* **3**, 880 (1948); **5**, 267 (1950).]

4-5. Polyethylene terephthalate (**1-5**) is a crystallizable polymer which is melted and shaped at temperatures greater than 270°C because of its high crystal melting point. It was thought that substitution of propylene glycol for the ethylene glycol in reaction (1-1) would produce a polyester which would still be crystallizable and hence rigid, but would also be processable at lower temperatures because of the increased hydrocarbon character of the polymer backbone. Should the glycol used here be 1,3-propane diol or the 1,2-isomer? Justify your answer briefly.

4-6. About one out of every 150 chain carbon atoms is cross-linked in a typical natural rubber (*cis*-polyisoprene) compound with good properties. The density of such a vulcanizate is $0.97\,\mathrm{g\,cm^{-3}}$ at 25°C. The gas constant $R = 8.3 \times 10^7$ ergs mol $K^{-1} = 1.987$ cal mol K^{-1}. Estimate the modulus of the sample at low extensions.

4-7. Given an SBR rubber that has $\overline{M}_n = 100,000$ before cross-linking, calculate the stress at 100% elongation of the cross-linked elastomer. The density of the vulcanizate (without fillers) with 23.5 mol% styrene in the polymer is 0.98 gcm^{-3}. Take the temperature to be 25°C. Express your answers in units of MN/m². Assume

(a)   average molecular weight between cross-links is 10,000.

(b)   average molecular weight between cross-links is 5000.

(c)   Calculate the modulus at very low extension for part (a).

4-8. (a) Calculate the rms-end-to-end distance for a macromolecule in molten polypropylene. Take the molecular weight to be 10^5, tetrahedral carbon angle $= 109.5°$, and the C—C bond length $= 1.54 \times 10^{-8}$ cm. Assume free rotation.

(a)   How extensible is this molecule? (That is, what ratio does its extended length bear to the average chain end separation?)

(b)   Would the real macromolecule be more or less extensible than the model used for this calculation? Explain briefly.

4-9.   Polymer A contains x freely oriented segments each of length l_a, and polymer B contains y freely oriented segments with length l_b. One end of A is attached to an end of B. What is the average end-to-end distance of the new molecule?

REFERENCES

[1]   J. C. Randall, "Polymer Sequence Determination." Academic Press, New York, 1977.
[2]   T. Radiotis and G. R. Brown, *J. Chem. Ed.*, **72**, 133 (1995).
[3]   L. R. G. Treloar, "The Physics of Rubber Elasticity," 3rd ed. Oxford (Clarendon) University Press, London and New York, 1975.
[4]   M. Gordon, "High Polymers." Addison-Wesley, Reading, MA, 1963.
[5]   H. Mark, *ChemTech*, 220 (April 1984).

Chapter 5

Step-Growth Polymerizations

Lest men suspect your tale untrue, keep probability in view.
—John Gay, *The Painter Who Pleased Nobody and Everybody*

5.1 CONDENSATION AND ADDITION POLYMERS

There are many possible ways to classify polymers. Each may be useful depending on the interests of the classifier. Examples include typing according to the source of the product (e.g., naturally occurring polymers, entirely synthetic macromolecules, or those derived by chemical modification of naturally occurring polymers), chemical structure (e.g., polyolefin, polyamide, etc.), polymer texture during use (rubbery, glassy, partially crystalline), area of application (adhesive, fiber, etc.), and so on. An important classification divides macromolecules into *addition* and *condensation polymers*. This distinction was made by W.H. Carothers [1], who invented nylon-6,6 and made many fundamental contributions to our knowledge and control of polymerizations.

Carother's classification into condensation and addition polymers is no longer valid for its intended purpose because of the many advances in technology since the idea was proposed. It is nevertheless still deeply entrenched in current thinking, and it will be necessary for the reader to understand its current meaning in order to read the polymer literature with ease. In this section we review the original classification, show why it is no longer generally applicable, summarize the current accepted meaning of the terms *addition* and *condensation* polymers, and then turn our attention to a useful, alternative classification which focuses on polymerization processes rather than the products of such processes. This line of reasoning takes us into a more detailed consideration of polymerizations in this and succeeding chapters.

155

A *condensation polymer* is one in which the repeating unit lacks certain atoms which were present in the monomer(s) from which the polymer was formed or to which it can be degraded by chemical means. Condensation polymers are formed from bi- or polyfunctional monomers by reactions which involve elimination of some smaller molecule. Polyesters (e.g., **1-5**) and polyamides like **1-6** are examples of such thermoplastic polymers. Phenol-formaldehyde resins (Fig. 5-1) are thermosetting condensation polymers. All these polymers are directly synthesized by condensation reactions. Other condensation polymers like cellulose (**1-11**) or starches can be hydrolyzed to glucose units. Their chemical structure indicates that their repeating units consist of linked glucose entities which lack the elements of water. They are also considered to be condensation polymers although they have not been synthesized yet in the laboratory.

In *addition polymers*, by contrast, the recurring units have the same structures as the monomer(s) from which the polymer was formed. Examples are polystyrene (**1-1**), polyethylene (**1-3**), styrene-maleic anhydride copolymers (**1-26**), and so on.

The difficulty with these definitions is that the same macromolecular structure can be made by different reaction pathways. This situation occurs particularly when cyclic and linear monomers can produce the same polymer. Thus nylon-6 can be made by either of two reactions:

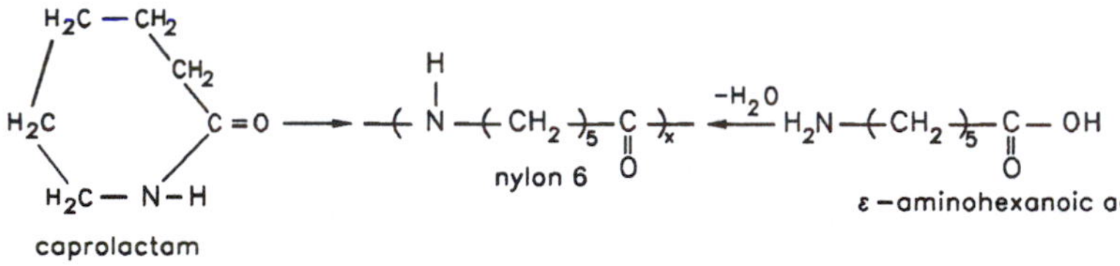

The polyamide made from caprolactam is technically an addition polymer by the above definition, while the product made from the amino acid would be a condensation polymer. Actually, only the caprolactam synthesis is used commercially, and the product (polycaprolactam = nylon-6) is called a condensation polymer because all polyamides are so classified.

Similarly, polyethylene can be made from ethylene, dihaloalkanes, or diazomethane as follows:

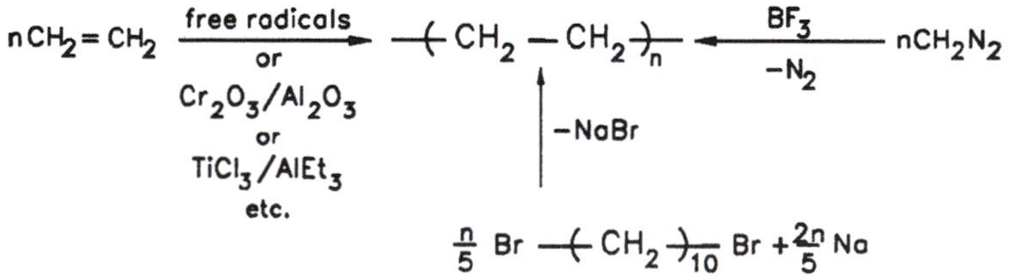

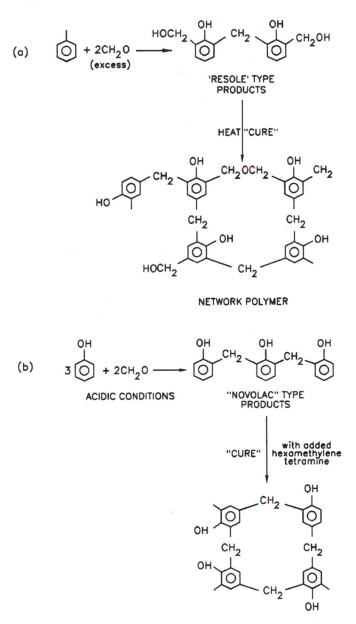

Fig. 5-1.   Phenol-formaldehyde polymers.   (a) Character of "resole"-type resins normally produced with excess formaldehyde under alkaline conditions. (b) "Novolac"-type resins normally made with excess phenol under acidic conditions.

This polymer (which is only made from ethylene in practice) is labeled an addition polymer.

By convention, polymers whose main chains consist entirely of C—C bonds are generally classified as addition polymers while those in which hetero atoms (O, N, S, Si) are present in the polymer backbone are considered to be condensation polymers. [An exception is polyformaldehyde (**1-12**), which is an addition polymer.]

It will be clear from these examples that this classification is now somewhat arbitrary. Many authors still refer to addition and condensation polymers in their current, conventional sense, however, and it is therefore necessary for the informed reader to understand this usage.

5.2   STEP-GROWTH AND CHAIN-GROWTH POLYMERIZATIONS

A more useful distinction is based on polymerization mechanisms, rather than polymer structures [2]. Most polymerizations can be classified as *step-growth* or *chain-growth* processes. A few reactions possess some characteristics of both mechanisms but the great majority of polymer syntheses can be characterized conveniently into one class or the other. We will see later that the nature of the polymerization is very different in the two cases, and the general operations required to produce high-molecular-weight polymers of good quality and in good yield are quite different for step-growth and chain-growth syntheses.

The distinguishing features of step- and chain-growth mechanisms are listed below. A subsequent example is provided to illustrate the differences, which may not be entirely clear from this bald summary of the characteristics of each reaction type.

The following features characterize step-growth polymerizations:

1.   The growth of polymer molecules proceeds by a stepwise intermolecular reaction. Only one reaction type is involved in the polymerization.

2.   Monomer units can react with each other or with polymers of any size. Polymer molecules grow over the course of the whole reaction, and such growth is in a series of fits and starts as the reactive end of a monomer or polymer encounters other species with which it can form a link.

3.   The functional group on the end of a monomer is usually assumed to have the same reactivity as that on a polymer of any size.

4. A high conversion of functional groups is required in order to produce high-molecular-weight products. Average polymer molecular weight rises steadily during the course of the polymerization.

5. Many step-growth polymerizations involve an equilibrium between reactants on the one hand and macromolecular products and eliminated small molecules on the other [cf. reactions (1-6) and (1-7)]. In such cases, all molecular species are present in a calculable distribution, and the course of the polymerization is statistically controlled. High polymer cannot coexist with much monomer in equilibrium systems. Such step-growth polymerizations are evidently reversible and also involve interchange reactions (Section 5.4.1) in which terminal functional groups react with linking units in other molecules to produce changes in molecular weight distributions.

6. Condensation polymers by the above definition are usually produced by step-growth polymerizations but not all step-growth syntheses are condensation reactions. Thus there is no elimination product in polyurethane synthesis from a diol and a diisocyanate (cf. reaction (1-12)):

$$\text{HOROH} + \text{OCN}-\text{R}'-\text{NCO} \longrightarrow \left[\text{O}-\text{R}-\text{O}-\underset{\underset{\text{O}}{\|}}{\text{C}}-\overset{\text{H}}{\underset{}{\text{N}}}-\text{R}'-\overset{\text{H}}{\underset{}{\text{N}}}-\underset{\underset{\text{O}}{\|}}{\text{C}}\right]_x \quad (5\text{-}3)$$

7. Step-growth reactions of the type of Eq. (5-3) are essentially irreversible. Interchange reactions are not significant. The reactions are usually very fast, and high degrees of polymerization can be realized, depending on the stoichiometric balance of the reactants. The polymers may be capable of further growth in size if additional reactants are added after the nominal end of the polymerization reaction. (This is illustrated in Section 1.5.4 and Problem 3-15.)

Chain-growth polymerizations have the following distinguishing features:

1. Each polymer molecule increases in size at a rapid rate once its growth has been started. When the macromolecule stops growing it cannot generally react with more monomers (barring side reactions).

2. Growth of polymer molecules is caused by a kinetic chain of reactions. (The name chain-growth reflects the existence of a chain reaction. Unfortunately, macromolecules are often also called chains because they are composed of linked identical entities. There is no necessary connection between the two usages. Some polymer chains are made by chain-growth polymerizations, and some are made by step-growth reactions.)

3. Chain-growth polymerization involves the reaction of monomers with active centers that may be free radicals, ions, or polymer-catalyst bonds.

4. In chain-growth polymerizations the mechanisms and rates of the reactions that initiate, continue, and terminate polymer growth are different.

5. Chain-growth polymerization is usually initiated by some external source (energy, highly reactive compound, or catalyst), and the reaction is allowed to proceed under conditions in which monomers cannot react with each other without the intervention of an active center.

6. Polymers made by chain-growth reactions are often addition polymers by Carothers's definition. The most common polymers made by these processes have only carbon–carbon links in their backbones.

An example of step- and chain-growth reaction mechanisms is provided by alternative pathways to poly(tetramethylene oxide) (**1-31**), which can be made by the self-condensation of tetramethylene glycol:

$$n\,HO-CH_2CH_2-CH_2-CH_2-OH \xrightarrow{H^+} \left[OCH_2CH_2CH_2CH_2\right]_n + n\,H_2O \qquad (5\text{-}4)$$

and by the acid-catalyzed ring opening polymerization of tetrahydrofuran:

$$n\;\underset{O}{\bigcirc} \xrightarrow{HClO_4} \left[OCH_2CH_2CH_2CH_2\right]_n \qquad (5\text{-}5)$$

Equation (5-4) represents a step-growth polymerization in which the overall reaction is a succession of etherifications:

$$HOCH_2CH_2CH_2CH_2OH + HOCH_2CH_2CH_2CH_2OH \xrightarrow{-H_2O} HOCH_2CH_2CH_2CH_2OCH_2CH_2CH_2CH_2OH$$

$$\xrightarrow[-H_2O]{HOCH_2CH_2CH_2CH_2OH} HOCH_2CH_2CH_2CH_2OCH_2CH_2CH_2CH_2OCH_2CH_2CH_2CH_2OH \longrightarrow \;\longrightarrow$$

$$HO \left[CH_2CH_2CH_2CH_2 - O\right]_x CH_2CH_2CH_2CH_2OH \qquad (5\text{-}6)$$

Each monomer addition involves the reaction between two hydroxyl groups, and a monomer can react as readily with a functional group on the end of a polymer as with another monomer. Initially, however, a monomer is far more likely to encounter another monomer than any other reactive molecule because these are the most prevalent species. The first stages of the reaction produce dimers which are likely in turn to yield trimers or tetramers by reacting with monomers or other dimers. There is thus a gradual increase in the average size of the species in the reaction mixture, and the molecular weight and yield of polymer both increase as the reaction proceeds.

The chain-growth polymerization of Eq. (5-5) actually represents a sequence of monomer reactions which is initiated by a small concentration of a strong acid:

$$HClO_4 + \underset{O}{\bigcirc} \longrightarrow HO\boxed{\oplus}\; ClO_4^{\ominus} \qquad (5\text{-}7)$$

The kinetic chain is propagated by successive additions of monomer to the active site generated in reaction (5-7).

$$HO\overset{\oplus}{\bigcirc} ClO_4^{\ominus} + \bigcirc_O \longrightarrow HOCH_2CH_2CH_2CH_2 - \overset{\oplus}{O}\bigcirc ClO_4^{\ominus} \longrightarrow \longrightarrow$$

$$\fpart{O \fpart{CH_2}_4}_n \overset{\oplus}{O}\bigcirc ClO_4^{\ominus} \tag{5-8}$$

The reactive site is regenerated on the oxygen atom of each new monomer as it is added to the growing macrocation, and this polymerization will continue until the existence of the propagating cation is terminated by reaction with adventitious impurities. The reaction rate of monomers with each other is negligible compared to the rate of addition of monomer to the cationic chain end. Polymers can grow to very large sizes in the presence of much monomer. The amount of polymer increases as the reaction proceeds but its molecular weight can be essentially constant as long as the initiator and monomer concentrations are not seriously depleted. [In practice, low-molecular-weight versions of this polymer are made by cationic polymerization of tetrahydrofuran and used mainly as flexible segments in polyurethanes (p. 19).]

We turn now to a review of step-growth polymerizations. Chain-growth polymerizations are the subject of subsequent chapters.

5.3   REQUIREMENTS FOR STEP-GROWTH POLYMERIZATION

A great many step-growth reactions are theoretically available to produce a given polymer. Figure 5-2 provides a partial list of reactions of bifunctional monomers which could be used to produce poly(ethylene terephthalate). Reaction (c) is the fastest of those listed and proceeds very quickly at room temperature. It can be carried out by dissolving the diacid chloride in an inert organic solvent and mixing this solution with an aqueous basic solution of the glycol. The polymer will form quickly at the interfaces between the two phases (cf. Section 5-5 on interfacial polymerization). It must, however, be freed of residual solvent and salts, dried, and densified before it can be fed into downstream equipment which will melt the polymer and shape it into fiber, film, or other products. Densification normally involves melting, extrusion, and chopping into granules that will flow efficiently in the hoppers of equipment used for fiber spinning or film extrusion. This operation and other required postpolymerization processes are expensive, as

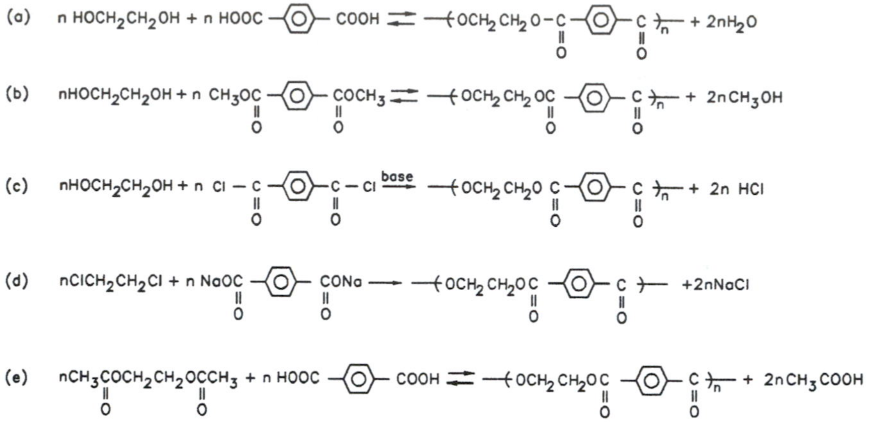

Fig. 5-2.   Some possible syntheses of poly(ethylene terephthalate) by step-growth reactions of bifunctional monomers.

are the costs of the diacid chloride monomer and recovery of the organic solvent. As a result, the net costs of process (c) are higher than those of either (a) or (b).

Process (a) is the preferred synthetic route. It involves a melt polymerization which is finished at high temperatures (about 275°C) and low pressures (about 1 mm Hg) to strip out the water produced as a condensation product and drive the equilibrium depicted to the polymer side. The product is a molten polymer which is suitable for immediate formation of fibers or films or for granulation. The acid monomer is less expensive than the diester in reaction (b). Thus, while the actual polymerization of route (a) is slower and more expensive than that of (c), the overall costs of producing a finished polymer are less.

Reaction (b) was once the preferred method for making poly(ethylene terephthalate) because dimethyl terephthalate can be readily purified to the quality necessary for production of this polymer. It should be remembered in this connection that any impurities or structural imperfections which are produced during the polymerization process will be left in the final polymer. An important side reaction in manufacture of this polymer results from the formation of diethylene glycol ($HOCH_2CH_2OCH_2CH_2OH$), the ether produced by dehydration of two molecules of ethylene glycol. This impurity combines randomly in the polyester. It lowers the polymer melting temperature compared to that of the pure homopolymer and affects its rate of crystallization. The net results are adverse influences on fiber spinning and heat setting to stabilize fiber dimensions. Also, inadvertent production of color-forming centers detracts from the appearance of textile fibers made from this polymer.

The only practical means for ensuring the desired polymer quality is to use scrupulously pure monomers. Purification of a polymer after it is synthesized would be prohibitively expensive, because these materials are sparingly soluble and are often difficult or impossible to crystallize or free of solvent. The overall least expensive route to good quality poly(ethylene terephthalate) was therefore through the dimethyl ester of terephthalic acid as shown in reaction (b). The by-product methanol was recovered to generate more diester from the acid. In more recent years, methods have been developed to produce the diacid with satisfactory purity, and reaction (a) is now the preferred route to this polymer because the esterification step with methanol can be eliminated. Reactions (d), (e), and others, which the reader may be able to write, will be more expensive in the final analysis for the various reasons mentioned above.

The foregoing discussion illustrates one of the reasons why the great majority of step-growth polymerizations which can be written on paper are not used in fact: the expenses involved in the purchase of monomers, carrying out the reaction, or preparing the polymer for further use. The overall process which produces a final product of required quality at the lowest cost will be chosen.

Another reason why many theoretical step-growth syntheses are not employed is because many of these polymerization reactions are not efficient or do not give a good yield of polymer with the desired molecular weight.

A satisfactory step-growth polymerization reaction must satisfy the following requirements:

1.   The reaction must proceed at a reasonably fast rate.
2.   The polymerization must be free of side reactions which produce cyclic or otherwise undesirable products.
3.   The monomers which are employed must be free of deleterious impurities.
4.   It must be possible to drive the process almost to complete reaction of the functional groups.
5.   Many important step-growth polymerizations are run under conditions in which the reverse reaction between the polymer and condensation products is significant [as in reactions (a) and (b) in Fig. 5-2]. In these cases, the stoichiometry of the reactants must be carefully controlled, and a close balance is needed between the concentrations of functional groups of opposite kinds.
6.   It must be possible to control the reaction to produce target average molecular weights and molecular weight distributions.

It is necessary to understand these requirements in order to be able to carry out and control step-growth polymerizations. The requirements listed above are considered in more detail in the following sections of this chapter.

5.3.1   Speed of Step-Growth Polymerizations

It can be seen from the preceding examples that the same macromolecular structure can be produced by alternative step-growth polymerizations of different monomers. The rate of a polymerization reaction will depend on the reactivity of the particular monomers under the experimental conditions. Most of the least expensive monomers used for the production of large-volume polymers by step-growth syntheses react very slowly at room temperature even when the reaction responds to catalysis. The examples given also illustrate the general rule that it is usually more practical to increase the reaction rate by moving to high temperatures than by switching to more reactive monomers which will be more expensive *per se* and may require extra finishing operations for the polymer. Hot reaction temperatures also facilitate removal of volatile condensation products and help drive the polymerization reaction to completion.

This conclusion is widely applicable. Where possible, step-growth polymerizations are carried out with the least expensive monomers and are accelerated by resort to high reaction temperatures and catalysts, when these are available. Further consideration of reaction engineering in such polymerizations is given in Chapter 10.

5.3.2   Side Reactions in Step-Growth Polymerization

Side reactions present a particular problem in step-growth polymerizations because these syntheses are often carried out at high temperatures compared to the comparable reactions in conventional organic chemistry. Thus, the acid-catalyzed esterification of ethanol with acetic acid is performed commercially at about 70°C whereas the polyesterification of ethylene glycol and terephthalic acid must be finished at about 275°C to obtain a high conversion of functional groups and a product with satisfactory molecular weight. Side reactions which are of negligible importance in conventional esterifications can become very significant at these higher temperatures.

The difficulties with side reactions are particularly important when linear polymers are being synthesized from bifunctional monomers. When branched thermosetting polymers are being produced from polyfunctional monomers, the functionality of the polymer increases as it grows in size, and some wastage of functional groups is not too serious as long as the polymer can still be cross-linked to convert it to its final, thermoset stage. Because of their application as adhesives and filled materials, the clarity and color purity of thermoset polymers is usually of less importance than that of linear, thermoplastic materials. For this reason also, lower purity monomers and side reactions during polymerization can sometimes be tolerated to a greater extent when thermosetting polymers are being manufactured.

An example of such side reactions occurs during the attempted self-polymerization of α-hydroxyacids like lactic acid (**5-1**) which dehydrates readily at temperatures near 250°C.

$$CH_3 - \overset{\overset{\displaystyle H}{|}}{\underset{\underset{\displaystyle OH}{|}}{C}} - COOH$$

5-1

It is also possible to produce low-molecular-weight cyclic compounds in preference to linear polymers. An example is provided by the homologous series of amino acids, $H_2N—(—CH_2)_x—COOH$[3].

$$2\ H_2N - CH_2 - COOH \longrightarrow O = C \underset{\underset{\displaystyle H}{|}}{\overset{\displaystyle CH_2 - N}{\underset{\displaystyle N - CH_2}{}}} C = O + 2H_2O + \text{NO POLYMER} \qquad (5\text{-}9)$$

$$H_2N - CH_2 - CH_2 - COOH \longrightarrow NH_3 + CH_2 = \overset{\overset{\displaystyle H}{|}}{C} - COOH + \text{NO POLYMER} \qquad (5\text{-}10)$$

$$H_2N \overset{}{\underset{3}{-(CH_2)-}} COOH \longrightarrow CH_2 \overset{\displaystyle CH_2}{\underset{\displaystyle CH_2}{\diagdown}} \overset{\displaystyle C = O}{\underset{\displaystyle NH}{}} + H_2O + \text{NO POLYMER} \qquad (5\text{-}11)$$

$$H_2N \overset{}{\underset{4}{-(CH_2)-}} COOH \longrightarrow \overset{\displaystyle CH_2}{\underset{\displaystyle CH_2}{\overset{\diagup CH_2 \diagdown}{}}} \overset{\displaystyle C = O}{\underset{\displaystyle NH}{}} + H_2O + \text{NO POLYMER} \qquad (5\text{-}12)$$

$$n\ H_2N \overset{}{\underset{5}{-(CH_2)-}} COOH \longrightarrow_y \overset{\displaystyle CH_2 \diagup CH_2 \diagdown C = O}{\underset{\displaystyle CH_2 \diagdown CH_2 \diagup}{\overset{}{|}}} \overset{\displaystyle N - H}{\underset{\displaystyle CH_2}{}} + \overset{}{\underset{n-y}{-(N-(CH_2)_5-\overset{O}{\overset{\|}{C}})-}} \qquad (5\text{-}13)$$

\qquad\qquad ~25% \qquad\qquad\qquad\qquad ~75%

Whether or not the polymerization proceeds will depend on the size of the ring which can be formed by intermolecular or intramolecular cyclization.

Five- and six-membered rings are thermodynamically stable and will be produced to the exclusion of polymer if the spacing between the coreactive functional groups is appropriate. Smaller all-carbon rings are relatively unstable because of bond angle distortions while the stability of rings of 8-12 carbons is adversely affected by crowding of hydrogens or other substituents inside the ring. The potential ring size increases with the distance between the coreactive ends of a growing polymer. Large rings will be stable if they are formed but the probability of the two reactive ends meeting is also diminished sharply, particularly if alternative reactions are available with other molecules in the reaction mixture. As an example, the equilibrium concentration of cyclic trimer in the polymerization of poly(ethylene terephthalate) (Fig. 5-2a) is about 1.5%. This material, with molecular weight 575, can be detected as a small peak in the SEC chromatogram of poly(ethylene terephathalate). It can affect dyeing properties of fibers made from this polymer.

5.4   POLYMER SIZE AND EXTENT OF CONVERSION OF FUNCTIONAL GROUPS IN EQUILIBRIUM STEP-GROWTH POLYMERIZATIONS

This section develops the relation between the number average size of polymers produced in a step-growth polymerization and the fraction of functional groups which have been reacted at any point in the process. The basic equation which will be developed, Eq. (5-19), is called the *Carothers equation*. It illustrates the fundamental principles which underlie the operation of such polymerizations to produce good yields of polymers with the desired molecular weights.

5.4.1   Basic Assumptions

The basic assumptions involved in deriving the Carothers equation are reviewed in this section.

(i)   A major assumption is that the reactivity of all functional groups of the same kind is equivalent and independent of the size of the molecule to which the functional group is attached. In the last stages of polymerization, the rate of polymerization will be low because the concentration of reactive groups is small, but the specific rate constant for reaction of the given group will remain constant if this assumption is true.

Equal reactivity of functional groups has been demonstrated by measurements of reaction rates of several series of reactants which differ in molecular weight. Further evidence is provided by the occurrence of interchange reactions under appropriate conditions. In these reactions the terminal functional group of one molecule reacts with the linking unit in another polymer.

$$\begin{array}{c}
\text{—(O—R—O—C—R'—C—)}_n\text{O—R—OH}\\
\quad\quad\quad\quad\; \| \quad\; \|\\
\quad\quad\quad\quad\; \text{O} \quad\; \text{O}\\[2mm]
\text{—(C—R—C—O—R—O—)}_m\text{C—R'—C—O—R—O}\\
\quad\;\| \quad\; \|\quad\quad\quad\quad\;\; \| \quad\; \|\\
\quad\;\text{O} \quad\; \text{O}\quad\quad\quad\quad\;\; \text{O} \quad\; \text{O}
\end{array}
\rightleftharpoons
\begin{array}{c}
\text{—(O—R—O—C—R'—C—)}_n\text{O—R}\\
\quad\quad\quad\quad\; \| \quad\; \|\\
\quad\quad\quad\quad\; \text{O} \quad\; \text{O}\\
\quad\quad\quad\quad\quad\quad\quad\quad\quad\;\; |\\
\text{—(C—R'—C—O—R—O—)}_m\text{C—R'—C—O}\\
\quad\;\| \quad\; \|\quad\quad\quad\quad\;\; \| \quad\; \|\\
\quad\;\text{O} \quad\; \text{O}\quad\quad\quad\quad\;\; \text{O} \quad\; \text{O}
\end{array}$$

$$+$$

$$\text{HO—R—O—C—R'—C—O—R—O} \\ \quad\quad\quad\quad\; \| \quad\; \| \\ \quad\quad\quad\quad\; \text{O} \quad\; \text{O}$$

$$(5\text{-}14)$$

It has been established that random interchange reactions occur with polyesters and polyamides at the high temperatures at which these thermoplastics are normally polymerized and subsequently extruded or molded. Other condensation polymers probably undergo similar shufflings of repeating units under the proper conditions. This implies that the equilibrium constant for condensation reactions like that in reaction (a) of Fig. 5-2 is independent of the sizes of the molecules which are reacting.

Note that interchange reactions do not change the number of molecules in the system. Therefore the number average degree of polymerization of the polymer will not be altered unless condensation products are also being removed from the system. The weight average degree of polymerization can change drastically, however, as the sizes of the macromolecules become randomized through this process.

Interchange reactions represent an approach to an equilibrium distribution of molecular sizes. Polymerizations that use very reactive monomers can proceed at room temperature [as in reaction (c) of Fig. 5-2]. They are kinetically controlled and will not produce random molecular weight distributions. Polymers from such syntheses will undergo interchange reactions, however, if they are subsequently heated to temperatures at which the equilibria such as that of reaction (5–14) can be reached in a reasonable time.

(ii)   All the functional groups in an n-functional reactant are usually assumed to be equally reactive. This has been demonstrated for dibasic aliphatic acids but is not true for 2,4-toluene diisocyanate (**5-2**) where the *para* isocyanate reacts

5-2

more readily than the *ortho* group. It is also thought that the secondary hydroxyl in glycerol is less reactive than the primary alcohol groups. (If such differences are known quantitatively they can be incorporated into the subsequent calculations by multiplying the concentration of less reactive groups by a fraction equal to the ratio of its reactivity to that of the more reactive functional groups.) Further, we take it for granted that the reactivity of a functional group does not change when other functional groups in the same molecule have reacted. This seems to be a valid assumption, by and large, with some exceptions, perhaps, in diisocyanates.

(iii)   The following calculations presuppose a homogeneous, single-phase reaction mixture. This condition is unlikely to be realized exactly in practical step-growth polymerizations. The consequences of deviations from this assumption in syntheses of branched polymers are considered briefly at the end of Section 5.4.2.

5.4.2   Number Average Degree of Polymerization in Step-Growth Reactions

Calculations of the number average degree of polymerization as a function of extent of reaction are very useful in designing and controlling step-growth polymerizations. Some examples of applications of these calculations are given later in this section.

The reactions we are considering can be summarized as in Fig. 5-3 where monomers carrying A-type functional groups react with B-type monomers. To be completely general, each functional group need not be confined to a single monomer (e.g., a mixture of acids could be used in a polyamidation reaction), the functionalities of the various monomers may differ, and the functional groups of opposite kinds need not be present in equivalent quantities.

The method we use is that of Carothers. Functionality is defined (p. 7) as the number of positions in the monomer that is available for reaction under the specified conditions, and f_i is the symbol for functionality of monomeric species i. We now define the average functionality f_{av} as the average number of functional groups per monomer molecule. That is,

$$f_{av} \equiv \frac{\sum N_i f_i}{\sum N_i} \tag{5-15}$$

where N_i is the number of moles (either per unit volume or in the whole reaction vessel) of species i.

The definition of Eq. (5-15) holds strictly when functional groups of opposite kinds are present in equal concentrations. If there is an excess of functional groups of one kind, the monomers carrying these groups will be able to react only until the

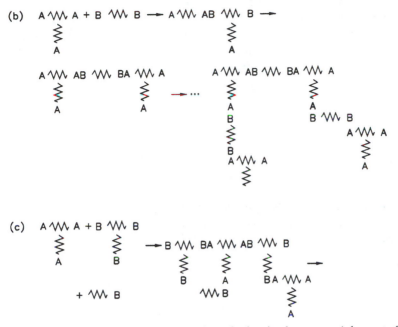

Fig. 5-3. Some step-growth polymerizations of *n*-functional monomers (where $n \geq 1$ and any number of different monomers may be present). (a) (b) and (c) represent different possible combinations of monomers.

opposite functional groups are consumed. In such nonstoichiometric mixtures the excess reactant does not enter the polymerization in the absence of side reactions and should not be counted in calculating f_{av}. Consider a polymerization which forms AB links and in which $n_A < n_B$, where n_i is the number of equivalents of functional groups of type i. In this case the number of B equivalents which can react cannot exceed n_A, and therefore

$$f_{av} = 2n_A \Big/ \sum N_i = \frac{2n_A}{N_A + N_B} \qquad (5\text{-}16)$$

The initial number of monomers is $\sum N_i = N_0$, and $N_0 f_{av}$ is the total number of useful equivalents of functional groups of all kinds that are present at the start of the reaction. We define p as the extent of reaction equal to the fraction of functional groups in deficient concentration which have reacted. Obviously $0 \leq p \leq 1$, and p in stoichiometric mixtures is the fraction of functional groups of either kind or of both kinds which have reacted. Also, N is the total number of moles of molecules (monomer plus polymers of all sizes) when the reaction has proceeded to an extent p.

Neglecting intramolecular linkages, every time a new linkage is formed the reaction mixture will contain one less molecule. Therefore, when the number of molecules has been reduced from N_0 to N moles, the number of linkages which have been formed is equal to $N_0 - N$ moles. It takes two functional groups to form a linkage and so $2(N_0 - N)$ moles of functional groups will have been lost in forming these $N_0 - N$ moles of linkages.

By definition,

$$p = \frac{\text{no. functional groups used}}{\text{no. functional groups initially}} = \frac{2(N_0 - N)}{N_0 f_{av}} \tag{5-17}$$

whence,

$$N = \frac{1}{2}(2N_0 - N_0 p f_{av}) \tag{5-18}$$

The number average degree of polymerization *of the reaction mixture* is $\overline{X}_n$, which is equal to the initial number of monomer units divided by the number of remaining molecules. That is,

$$\overline{X}_n = \frac{N_0}{\frac{1}{2}(2N_0 - N_0 p f_{av})} = \frac{2}{2 - p f_{av}} \tag{5-19}$$

This is the Carothers equation. Note that $\overline{X}_n$, is the number average degree of polymerization *of the reaction mixture* and not just of the polymer which has been formed. Note also that $\overline{X}_n$ is not necessarily equal to the degree of polymerization $\overline{DP}_n$, defined in Section 1.2 as the average number of repeating units per polymer molecule. In the present context a single monomer has $\overline{X}_n = 1$, a molecule containing two monomers has $\overline{X}_n = 2$, and so on. Thus for structure **5-3**, $\overline{X}_n = \overline{DP}_n = 100$, but for **5-4**, $X_n = 200$ while $\overline{DP}_n = 100$, since each

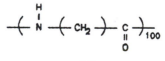

5-3

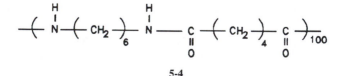

5-4

TABLE 5-1

Simplified Alkyd Recipe

Ingredient	Functionality	Moles	Equivalents[a]
(i) Fatty acid[b]	1	1.2	1.2
Phthalic anhydride	2	1.5	3.0
Glycerol	3	1.0	3.0
1,2-Propaneglycol	2	0.7	1.4
		4.4	8.6
(ii) Fatty acid	1	0.8	0.8
Phthalic anhydride	2	1.8	3.6
Glycerol	3	1.2	3.6
1,2-Propaneglycol	2	0.4	0.8
		4.2	8.8

[a] The number of equivalents equals the number of functional groups here.
[b] If the fatty acid were an unsaturated species like linoleic acid the alkyd

$$(CH_3-)(-CH_2-)-_4CH=CH-CH_2-CH=CH-(-CH_2-)-_7COOH)$$

would air dry to form cross-linked films.

repeating unit shown contains the residues of two monomers. The degree of polymerization derived in this section is generally applicable, even to branched polymers formed from a mixture of monomers, in which a regular repeating unit cannot be identified. (Such mixtures are illustrated in Table 5-1). It is unfortunate that these two somewhat different concepts of degree of polymerization are both current in polymer technology, but the intended meaning is usually clear from the context in which the term is used.

If only bifunctional monomers are present and there are equivalent concentrations of functional groups of opposite kinds, then $f_{av} = 2$. This is the case, for example, when a high-molecular-weight linear polymer is being synthesized. Under these conditions Eq. (5-19) becomes

$$\overline{X}_n = \frac{2}{2-2p} = \frac{1}{1-p} \qquad (5\text{-}20)$$

When the reaction is 95% complete, $X_n = (1 - 0.95)^{-1} = 20$. If hexamethylene diamine and adipic acid were being reacted, the number average molecule would have structure **5-4** with 10 repeating units. This is an oligomer but would not yet be considered to be a polymer. At $p = 0.98$, 98% of the functional groups will have been reacted, $X_n = 50$, and the nylon-6,6 of **5-4** would have a molecular weight that is barely high enough for fiber formation. At $p \simeq 0.995$, $\overline{M}_n \simeq 20,000$ and the polymer melt is becoming too viscous for fiber spinning. Still higher molecular

TABLE 5-2

Effect of Impurity on Development of Molecular Weight in Equilibrium
Step-Growth Polymerizations

	Moles	Functionality	Equivalents[a]
$H_2N-(CH_2-)_6NH_2$	1	2	2
$HOOC-(-CH_2-)-COOH$	0.99	2	1.98
$CH_3-(-CH_2-)-_4COOH$	0.01	1	0.01
	2.0		3.99

[a] Total amine equivalents = 2 and total acid equivalents = 1.99.

weights and degrees of conversion are needed, however, when the same polymer is

$$\therefore f_{av} = \frac{\text{no. useful equivalents}}{\text{total no. moles}} = \frac{2(1.99)}{2} = 1.99$$

At $p = 0.99$,

$$\overline{X}_n = \frac{2}{2 - pf_{av}} = \frac{2}{2 - (0.99)(1.99)} = 67$$

injection molded or extruded into articles that are relatively unoriented compared to fibers (cf. Section 1.8).

We can see from this simple example why step-growth polymerizations require very high conversions of functional groups compared to syntheses of the same linkages in conventional organic reactions.

Suppose now that the acid monomer for this polymerization consisted of 0.01 mol $CH_3(CH_2)_4COOH$ for every 0.99 mol of adipic acid. The situation would then be as shown in Table 5-2. It can be seen that $\overline{X}_n$ would be only 67 at $p = 0.99$ compared to 100 if the monomers were pure.

The same large decline in $\overline{X}_n$ would result if there were 0.99 mol of one bifunctional monomer for every mole of the coreactive monomer in this case.

It can be seen that it is relatively easy to carry out step-growth polymerizations inefficiently or entirely ineffectively if the requirements listed in Section 5.3 are not fully appreciated.

As another practical example consider how the molecular weight ($\overline{M}_n$) of nylon-66 can be controlled to a target value, say, 50,000. The formula weight of the repeating unit shown in structure **5-4** is 226 g mol^{-1}, and the number of repeating units $\overline{DP}_n = 50,000/226 = 221$ in this case; $\overline{X}_n$ then equals 442, by definition, since each repeating unit in nylon-66 contains residues of two monomers. Substitution of $f_{av} = 2$ and $\overline{X}_n = 442$ into Eq. (5-19) indicates that the target conversion p for this number average molecular weight is $1 - 1/442 = 0.9977$. It is not practical to try to control a factory-scale step-growth polymerization to this level of conversion. If p is 0.002 below this target value, $\overline{X}_n$ would be only 237, while a p that is 0.002 higher than that calculated would produce a product with $\overline{X}_n = 4708$.

An alternative procedure would involve feeding the polymerization reactor with a deliberate imbalance of diacid and diamine to limit $\overline{M}_n$ to 50,000 even if p were effectively forced close to unity. This is not attractive in this particular case because hexamethylene diamine and adipic acid react to form a zwitterion salt (**5-5**) which can be recrystallized from methanol and used to provide a pure feedstock with a stoichiometric balance of coreactive functional groups.

$$H_3\overset{\oplus}{N} \!-\!\!\left(CH_2\right)_{\!6}\overset{\oplus}{N}H_3$$

$$O\!=\!\underset{\underset{\ominus}{\overset{|}{O}}}{C} \!-\!\!\left(CH_2\right)_{\!4}\underset{\overset{|}{O}\;\ominus}{\overset{|}{C}}\!=\!O$$

5-5

The most general solution to the problem involves adding a monofunctional reactant such as acetic acid to the reaction mixture to control the product molecular weight. The composition of the polymerization recipe is calculated as follows. From Eq. (5-19) with $X_n = 442$ and $p = 1$, $f_{av} = 2 - 2/442 = 1.9955$. (*Note:* Do *not* round off to $f_{av} = 2.0$.) If we keep the adipic acid and ethylene diamine in stoichiometric balance and add y mol acetic acid for every mole of diacid, the initial reaction mixture is shown in the accompanying tabulation. There are $(2+y)$ acid groups for every two amine groups, so only two of these acid groups can react. From Eq. (5-16)

$$f_{av} = (2 \times 2)/(2 + Y) = 1.9955$$

The solution is then $y = 0.00445$ mol acetic acid for each mole of adipic acid.

	Moles	f	Equivalents
Adipic acid	1.0	2	2
Hexamethylene diamine	1.0	2	2
Acetic acid	y	1	y

(As a check on the results of this calculation we can make use of the fact that the number of molecules, N, left in the system at an extent of conversion p is equal to the original number of moles of monomer minus the number of linkages formed. From a mass balance, the number of linkages that have been formed is p times the number of moles of deficient groups and equals $2p = 2$ in the present example. Then $N = 2 + y - 2 = 0.0045$. As before, the number of structural units is N_0 = the initial number of moles of monomer = $2 + y = 2.0045$. $X_n = N_0/N = 2.0045/0.0045 = 445$. This is a close enough check. The difference between 445 and the target value of 442 reflects the effects of rounding-off decimals.)

The following paragraphs include sample calculations which illustrate the practical application of the Carothers equation to step-growth polymerizations which yield branched polymers.

Consider the simple alkyd recipe shown in Table 5-1, part (i). Alkyds are polyesters produced from polyhydric alcohols and polybasic and monobasic acids. They are used primarily in surface coatings. The ingredients of these polymers contain polyfunctional monomers and it is possible that such polymerizations could produce a thermoset material during the actual alkyd synthesis. This is of course an unwanted outcome, and calculations based on the Carothers equation can be used to adjust the polymerization recipe to produce a finite molecular weight polymer in good yield. The recipe can also be adjusted to provide other desirable characteristics of the product, such as an absence of free acid groups which may react adversely with some pigments.

In the example of Table 5-1, part (i), there are 4.2 equivalents of acid groups and 4.4 equivalents of alcohol. Therefore, from Eq. (5-16), $f_{av} = 2(4.2)/(4.4) = 1.91$. Then, with Eq. (5-19) at $p = 0.9$, $X_n = 2/[2 - (1.91)(0.9)] = 7.2$ and at $p = 1.0$, $X_n = 2/(2 - 1.91) = 22$. There are four monomers which can conceivably be incorporated into this polymer and it is clearly impossible to specify a regular repeating group, as was done in Chapter 1 for linear polymers.

Since $\overline{X}_n$ is finite at $p = 1$, the recipe listed in Table 5-1, part (i), will not produce cross-linked polymer (i.e., it will not "gel"). Note in this connection that the limit of $\overline{X}_n$ for polymerization of bifunctional monomers is infinite at $p = 1$ (Eq. 5-20). This simply means that the whole reaction mixture would be reduced to a single molecule if all the functional groups could actually be reacted.

The simple recipe we have been discussing is modified in Table 5-1, part (ii), by increasing the relative concentrations of the phthalic anhydride and glycerol. The coreactive functional groups are now in balance and $f_{av} = 8.8/4.2 = 2.10$. From Eq. (5-19), when the reaction mixture has gelled $\overline{X}_n$ is infinite and thus p at the gel point is given by the equality: $2 - pf_{av} = 0$, from which the limiting value of $p = 0.95$. According to this model, the reaction cannot be taken past 95% conversion of functional groups without producing an intractable, insoluble product.

Note that substitution of $p = 1$ into Eq. (5-19) with $f_{av} > 2$ yields a negative value for X_n. The Carothers equation obviously does not hold beyond the degree of conversion at which X_n is infinite.

It is pertinent to ask at this point how well the foregoing predictions work. In fact, there is a basic error in the Carothers model because an infinite network forms when there is one cross-link per weight average molecule (Section 7.9). That is to say, gelation occurs when $\overline{X}_w$ rather than $\overline{X}_n$ becomes infinite. Since $\overline{X}_w \geq \overline{X}_n$ (Section 2.7), reaction mixtures should theoretically "gel" at conversions lower than those predicted by the Carothers equation.

A method to calculate the molecular weight distribution (and $\overline{X}_w$) during the course of an equilibrium step-growth polymerization is described in Section 5.4.3. Such calculations are of great value in understanding the course of such reactions but they are not generally any more effective than the simpler Carothers equation in the design of practical polymerization mixtures. The major reasons for lack of agreement between calculated and observed gel points probably include intramolecular reactions, loss of functional groups through side reactions, unequal reactivities of functional groups of the same ostensible type, and the possibility that estimations correspond to initial formation of three-dimensional polymer whereas most experimental observations reflect the presence of massive amounts of gel.

Because of these factors, molecular weight calculations are used mainly for systematic modifications of formulations which have unsatisfactory property balances rather than for accurate predictions of gel points under practical operating conditions. The design of branched condensation polymers still relies heavily on $\overline{X}_n$ estimates, since these are less complicated than the theoretically more accurate $\overline{X}_w$ method described in the next section of this chapter.

5.4.3 Molecular Weight Distribution in Equilibrium Step-Growth Polymerizations

The calculations described in this section yield estimates of the molecular weight distribution of the reaction mixture during a step-growth polymerization which is proceeding according to the assumptions outlined in Section 5.4.1.

It will be helpful first to review some very simple principles of probability before calculating the relation between the degree of conversion and molecular weight distribution.

If an event can happen in a ways and fail to happen in b ways then the probability of success (or happening) in a single trail is $u = a/(a + b)$. Thus the probability of heads in a single coin toss is $1/2$. Similarly, the probability of failure in a single trial is $v = b/(a + b)$. If $u = 0$ the event is impossible, and if $u = 1$ the event is a certainty.

If u_1 is the probability of event E_1 happening and u_2 is the probability of event E_2, the likelihood that E_2 occurs after E_1 happened is $u_1 u_2$ if E_2 depends on E_1 (i.e., if the happening of E_1 affects the probability of the occurrence of E_2).

Some events are mutually exclusive. (For example, a coin toss may produce a head or a tail. The occurrence of one excludes the happening of the other event.) In that case the probability that either event happens is $u_1 + u_2$.

The coin toss example given here involves *a priori* probabilities. These are probabilities whose magnitudes can be calculated ahead of the event from the number of different ways in which a given event may occur or fail to happen. We are concerned here with actuarial probabilities, which are similar to estimates of chances used by insurance companies to calculate life insurance premiums. It is obviously impossible, for example, to estimate the life expectancy of a person who is now 21 years old by calculating the number of ways in which he or she may live or die. The life expectancy of an individual is equated to that experienced in the past by a large number of similar people.

In arithmetic, this is equivalent to noting that if S successes have already been experienced in n trials then the relative frequency of success is S/n and the $\lim_{n \to \infty}(S/n)$ is taken to be the probability of success in a single trial. This is the kind of probability we shall use in the following paragraphs. The rules for combination of such probabilities are as given earlier.

The concepts we use to develop relations between the degree of conversion and molecular weight distribution of the reaction mixture in equilibrium step-growth polymerizations are most clearly illustrated with reference to the self-polymerization of a monomer which contains two coreactive groups. An example would be a hydroxy acid that can undergo self-polymerization according to

$$n\ HO - R - \underset{\underset{O}{\|}}{C} - OH \longrightarrow H -\!\!\left(\!O - R - \underset{\underset{O}{\|}}{C}\!\right)_{\!n}\!\!- OH + (\,n-1\,)\,H_2O \quad (5\text{-}21)$$

This is a so-called AB-type monomer. The final equations are very similar to those which can be developed for the reaction AA- and BB-type monomers, as in Fig. 5-3, for example. The derivation in the latter case, however, requires some steps that are beyond the scope of this text. The reader is referred to the original source [2] for a more complete description of the ideas involved.

We are assuming, as before, that the probability of reaction and the reaction rate of either group in Eq. (5-21) is independent of the size of the molecule to which this group is attached.

Imagine that we can reach into the reaction mixture at time t when the extent of reaction is p and select an unreacted functional group of either kind. The probability that the *other* end of the structural unit we have picked has reacted is p, since a fraction p of all functional groups of that type will have reacted. If this other end has reacted, then the unreacted end we picked is attached to at least one other monomer residue. That means that the probability that an unreacted end is part of a molecule that contains *at least* two monomer units is p. The other group on the second monomer from the end we picked could also have reacted, however, and the probability that this single event may have happened is still equal to p. The probability that the *two* similar groups we have mentioned will have reacted equals

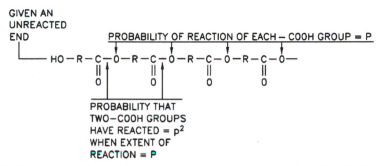

Fig. 5-4. Concepts involved in deriving relations between molecular weight distribution and extent of reaction for self-condensation of a bifunctional monomer.

p^2, and this equals the probability that our initial end is attached to a molecule that contains *at least* three monomer residues. By extension, the probability that the molecule in question contains *at least* i monomer residues will be p^{i-1}. The concepts involved here are summarized also in Fig. 5-4.

We now wish to know the probability that our initial unreacted end is attached to a molecule which contains exactly i monomer residues. This requires that there will have been $i - 1$ reactions, with a net probability of p^{i-1}, followed a nonreaction. The probability that the last functional group has *not* reacted is $1 - p$ (since the probability that it has either reacted *or* has *not* reacted must equal 1). The combined probability that the initial random selection of a molecule found a macromolecule with i monomer residues is evidently $p^{i-1}(1 - p)$.

During such a random choice of a molecule the chances that a polymer containing i structural units (an i-mer) will be selected will depend directly on how many such molecules there are in the reaction vessel. (To illustrate: If a box contains nine red balls and one black ball, the probability that a red ball will be selected in a blind choice is $9/10$.) In other words, if one molecule is selected the probability that it will be an i-mer equals the mole fraction x_i of i-mers in the reaction mixture. We have just calculated this probability and we see then that

$$x_i = (1 - p)p^{i-1} \tag{5-22}$$

Equation (5-22) is the differential number distribution function for equilibrium step-growth polymerizations in homogeneous systems.

We now derive an expression for the differential weight distribution. The total number, N, of molecules remaining at an extent of reaction p, is

$$N = N_0(i - p) \tag{5-23}$$

[This is the same as Eq. (5-18) with $f_{av} = 2$ in this particular case.] The mole fraction of i-mers x_i is

$$x_i = N_i/N_0(1 - p) = (1 - p)p^{i-1} \tag{5-24}$$

where N_i is the number of moles of i-mers. Thus

$$N_i = N_0(1 - p)^2 p^{i-1} \tag{5-25}$$

If the formula weight of a monomer that has reacted at both ends equals M_0, then the molecular weight of M_i of an i-mer is iM_0, and the weight of i-mers in the reaction mixture is iM_0N_i. The total weight of all molecules equals N_0M_0 (neglecting unreacted ends). Hence the weight fraction w_i of i-mers is

$$w_i = \frac{N_i(iM_0)}{N_0M_0} = \frac{N_0(1 - p)^2 p^{i-1}(iM_0)}{N_0M_0}$$

$$= i(1 - p)^2 p^{i-1} \tag{5-26}$$

This is the differential weight distribution functions.

Figure 5-5a shows the number distribution of degrees of polymerization in a linear ($f = 2$) step-growth polymerization for several extents of reaction p. This is a plot of Eq. (5-22). The corresponding weight distribution function (Eq. 5-26) is plotted in Fig. 5-5b. The former figure shows that monomer is the most prevalent species on a mole basis at any extent of reaction. The corresponding weight fraction of monomer is negligible, however, at any $p \geq 0.99[w_1/x_1 = (1 - p)$ from Eqs. (5-22) and (5-26)].

The weight distribution in Fig. 5-5b exhibits a peak at a particular degree of polymerization, i, for any given extent of reaction. This peak moves to higher i and flattens out as p increases. The value of i at which this peak occurs is obtained from the derivative of w_i with respect to i:

$$dw_i/di = (1 - p)^2(p^{i-1})(1 + i \ln p) \tag{5-27}$$

When $dw_i/di = 0$, the curve has a maximum and $i = 1/\ln p$. A series expansion of $-1/\ln p$ gives $p/(p-1)$ as the first term and this fraction approaches $1/(p-1)$ as p approaches 1. This is the value of $\overline{X}_n$ in linear step-growth polymerizations (Eq. 5-20). Thus, as a first approximation, the peak in the weight distribution of high conversion linear step-growth polymers is located at $\overline{M}_n$ of the polymer if the synthesis was carried out under conditions where interchange reactions and molecular weight equilibration could occur.

It is also easy to calculate how much polymer has been made at a given degree of conversion. Defining polymer as any species with $i \geq 2$, the weight fraction of polymer must be equal to the total weight fraction of all species minus the weight

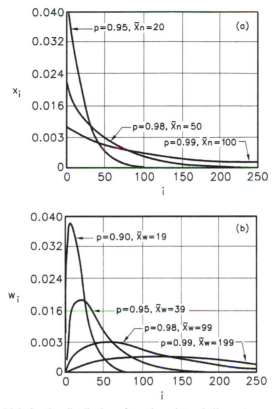

Fig. 5-5.   (a) Mole fraction distribution of reaction mixture in linear step-growth polymerization for several extents of reaction. (b) Weight fraction distribution of reaction mixture in linear step-growth polymerization for several extents of reaction [2].

fraction of monomer. That is,

$$\sum_{i=2}^{\infty} w_i = \sum_{i=1}^{\infty} w_i - w_1 \tag{5-28}$$

Substituting for w_1 from Eq. (5-26) and setting $\sum_{w=1}^{\infty} w_i = 1$

$$\sum_{i=2}^{\infty} w_i = 1 - 1(1 - p)^2 p^0 = 2p - p^2 \tag{5-29}$$

Similarly from Eq. (5-24), the mole fraction of the reaction mixture that is polymer at any degree of conversion p is

$$\sum_{i=2}^{\infty} x_i = \sum_{i=1}^{\infty} x_i - x_1 = 1 - (1 - p)p^0 = p \tag{5-30}$$

Since we know the distribution functions, we can calculate average molecular weights. To accomplish this, use must be made of the following series summations:

$$\sum_{y=1}^{\infty} ya^{y-1} = \frac{1}{(1-a)^2} \tag{5-31a}$$

$$\sum_{y=1}^{\infty} y^2 a^{y-1} = \frac{1+a}{(1-a)^3} \tag{5-31b}$$

$$\sum_{y=1}^{\infty} y^3 a^{y-1} = \frac{1+4a+a^2}{(1-a)^4} \tag{5-31c}$$

From Eqs. (2-6) and (5-24):

$$\overline{X}_n = \sum_i x_i i = \sum_i i(1-p)p^{i-1} \tag{5-32}$$

Terms in p alone can be removed from the summation since this calculation is at fixed degree of conversion. Then

$$\overline{X}_n = (1-p)\sum_{i=1}^{\infty} ip^{i-1} = \frac{1-p}{(1-p)^2} = \frac{1}{1-p} \tag{5-33}$$

(invoking the series sum of Eq. 5-31a). Of course, this calculation gives the same result as the Carothers equation for step-growth polymerization with $f_{av} = 2$ (Eq. 5-20).

The reader is reminded of Eq. (1-1), which reads as follows in the current context:

$$\overline{M}_n = \overline{X}_n M_0 = M_0/(1-p) \tag{5-34}$$

where M_0 is the formula weight of the repeating unit which is the residue of a single monomer in this case.

The weight average degree of polymerization X_w can be derived similarly by using the series summation of Eq. (5-31b). From Eqs. (2-13) and (5-26),

$$\overline{X}_w = \sum_i iw_i = \sum_i i^2(1-p)^2 p^{i-1}$$

$$= (1-p)^2 \sum_i i^2 p^{i-1} = \frac{(1-p)^2(1+p)}{(1-p)^3} = \frac{1+p}{1-p} \tag{5-35}$$

The breadth of the number distribution in equilibrium step-growth polymerization of linear polymers is indicated by

$$\overline{X}_w/\overline{X}_n = \overline{M}_w/\overline{M}_n = 1 + p \tag{5-36}$$

from Eqs. (5-34) and (5-35). In standard statistical terms, since $\overline{M}_w = (1+p)\overline{M}_n$ from the last equation, then the variance of the number distribution of molecular weights is

$$s_n^2 = \overline{M}_w\overline{M}_n - \overline{M}_n^2 = p(\overline{M}_n)^2 \qquad (5\text{-}37)$$

(cf. Eq. 2-33) and the standard deviation of the number distribution is

$$s_n = p^{1/2}\overline{M}_n \qquad (5\text{-}38)$$

In step-growth polymerizations of commercial linear polymers the degrees of conversion are close to unity. Thus, for linear nylons and polyesters, $\overline{M}_w/\overline{M}_n$ is close to 2 as indicated by Eq. (5-36) with $p = 1$.

Equations (5-24) and (5-26) describe a random distribution of molecular sizes. This distribution is also known as a *Flory–Schulz distribution* or a *most probable distribution*. Note that it is most probable only for linear step-growth polymers made under conditions which satisfy the assumptions in Section 5.4.1. If such polymers are made under other conditions (cf. Section 5.5), the molecular weight distributions of the products will not conform to the relations we have derived. However, when such polymers are melted and shaped, these processing conditions usually facilitate interchange reactions and randomization of the molecular weight distribution.

For example, while poly(ethylene terephthalate) is made from ethylene glycol and terephthalic acid according to the reaction of Fig. 5-2a, the very viscous nature of the reaction medium at high conversions limits the molecular weights which can be attained economically. The chain lengths of such polyesters can be further enhanced by a subsequent "solid-state polymerization" process in which finely divided polymer is held under reduced pressure at temperatures just below its softening point. Under these conditions, glycol is removed and the polymer molecular weight is increased:

$$(5\text{-}39)$$

Interchange reactions do not occur readily under such conditions, and the molecular weight distribution of the product may differ significantly from the random one derived above. When the polymer is melted subsequently for spinning or extrusion, the molecular weight distribution will tend to equilibrate toward the predicted "most probable" type, however. The rate at which this equilibrium distribution is approached will depend on the time and temperature conditions during melt processing and will be enhanced by the presence of residual polymerization catalyst or water. The $\overline{M}_n$ of the polymer will not change during this randomization because neither the number of structural units nor the number of molecules can change in a closed spinning system. The $\overline{M}_w$ will, however, tend to approach $2\overline{M}_n$ in accordance with Eq. (5-36) at $p = 1$.

Molecular weight distributions have also been derived for polymerizations of multifunctional monomers, and the original work [2] should be consulted for details of these calculations which are beyond the scope of this text.

5.5   INTERFACIAL AND SOLUTION POLYMERIZATIONS OF ACID CHLORIDES AND OTHER REACTIVE MONOMERS

If step-growth polymerizations are carried out with very reactive monomers, the polymer synthesis conditions may be mild enough that interchange reactions between the polymer formed and low-molecular-weight reaction products do not take place at a significant rate. Then the distribution of macromolecular sizes depends on the kinetics of the particular polymerization system. It is not determined statistically according to the concepts developed in the preceding section, because no equilibrium is reached between the various species in the reaction mixture at a given time. An example has been given in Section 5.3, in connection with reaction (c) of Fig. 5-2. This particular synthesis would not be used commercially for the reasons given earlier but such reactions are very useful generally in laboratory-scale polymerizations and in selected larger scale step-growth polymerizations.

Interfacial polymerizations rely on reactions of diacid chlorides and active hydrogen compounds. Other examples include production of a polyurethane from a diamine and a bischloroformate:

$$n\text{H}_2\text{N}-\text{R}-\text{NH}_2 + n\text{Cl}-\underset{\underset{\text{O}}{\|}}{\text{C}}-\text{O}-\text{R}'-\text{O}-\underset{\underset{\text{O}}{\|}}{\text{C}}-\text{Cl} \xrightarrow[\text{BASE}]{} \left[\!\!\left[-\text{N}-\text{R}-\overset{\overset{\text{H}}{|}}{\text{N}}-\underset{\underset{\text{O}}{\|}}{\text{C}}-\text{O}-\text{R}'-\text{O}-\underset{\underset{\text{O}}{\|}}{\text{C}}-\right]\!\!\right]_n + 2n\text{HCl}$$

$$(5\text{-}40)$$

and reaction of a diamine and a disulfonic acid chloride to yield a polysulfonamide:

$$n\text{H}_2\text{N}-\text{R}-\text{NH}_2 + n\text{Cl}-\text{SO}_2-\text{R}'-\text{SO}_2\text{Cl} \xrightarrow[\text{BASE}]{} \left[\!\!\left[-\text{N}-\text{R}-\overset{\overset{\text{H}}{|}}{\text{N}}-\underset{\underset{\text{O}}{\|}}{\overset{\overset{\text{O}}{\|}}{\text{S}}}-\text{R}'-\underset{\underset{\text{O}}{\|}}{\overset{\overset{\text{O}}{\|}}{\text{S}}}-\right]\!\!\right]_n + 2n\text{HCl}$$

$$(5\text{-}41)$$

The reactants are dissolved in a pair of immiscible liquids, usually water and a hydrocarbon or chlorinated hydrocarbon. The aqueous phase contains the diol, diamine, dimercaptan, or other active hydrogen compound, an acid acceptor like sodium hydroxide, and perhaps a surfactant, while the organic phase dissolves the acid chloride. Polymerization takes place very rapidly at the interface between the two phases, and the rate of reaction is controlled by the rate at which the monomers can diffuse to the reaction site. The number of growing polymer chains is restricted because the incoming monomers find it difficult to diffuse through the polymer layer at the interface without being captured by a coreactive end.

For this reason, it is possible to obtain higher molecular weight products than are normally formed in bulk step-growth polymerizations in which the high viscosity of the molten product limits the efficiency of stirring. Since interchange reactions are not significant under the mild conditions of these reactions there is no necessity for high conversions characteristic of equilibrium step-growth processes. Further, since the polymerization is diffusion controlled, one does not need to start with an exact balance of coreactive equivalents. The molecular weight distributions which are obtained are also quite different from the random one which is found for reversible systems, and the distributions of interfacial polymers may vary depending on the details of the particular interfacial reaction. As explained above, however, interfacial polyesters and polyamides will tend to randomize during subsequent melt processing.

The choice of organic solvents for interfacial polycondensation has an important effect on the molecular weight of the polymer that is made. Liquids which swell the polymer permit the formation of higher molecular weight products than media which are effectively nonsolvents for the polymer. If the polymer is insoluble in either phase, the migration of acid chloride will be hindered enough to limit the polymer yield and molecular weight.

Interfacial polymerizations are generally characterized by the following features [4]: (1) Reactive monomers are used; (2) the polymerization is irreversible at the reaction temperature; (3) the reaction rate is determined by the rates of encounter of complementary species; (4) the growing polymer is dissolved or highly swollen during the reaction; and (5) the aqueous phase contains a base to remove by-product acid from the polymerization zone.

Interfacial polycondensations can also be carried out in vapor–liquid systems. Reaction takes place at the interface between an aqueous solution of a bifunctional active hydrogen compound and the vapor of diacid chloride. Interfacial condensation is commercially important in the synthesis of polycarbonates (**1-52**). Polymerizations based on diacids are always less expensive than those that use diacid chlorides. In the polycarbonate case, however, the parent reactant, carbonic acid, is not suitable and the derived acid chloride, phosgene ($COCl_2$), must be used.

Similar reactions to those used in interfacial polymerizations are sometimes carried out in solution and are employed to prepare some polymers which yield ultra-high-strength high modulus fibers. These species typically contain *para*-linked aromatic rings and amide or ester linkages in the polymer backbone.

Aromatic polyamides are generally made by low-temperature reactions of aromatic diamines and aromatic diacid chlorides in special solvents such as a 1:3 molar mixture of hexamethylphosphoramide: N-methylpyrrolidone, as in reaction (4-50). Intensive stirring is required to attain high molecular weights because the polymer precipitates. These macromolecules are very rigid and rodlike. They form oriented liquid crystalline arrays in solution and require little postspinning orientation to produce extremely strong and stiff fibers. The polymer would not be made in the melt because it is infusible. It must be synthesized and handled in solution, and this requires the use of reactive precursors.

5.6   STEP-GROWTH COPOLYMERIZATIONS

Step-growth copolymerization involves the use of three or more monomers which do not ordinarily all react with each other. Examples include mixtures of acids and polyols in the synthesis of alkyds, as illustrated in the recipes in Table 5-1. Such polymers will contain a random distribution of monomer residues if they are synthesized under conditions in which the polymerization is reversible and the molecular weight distribution is random. Polymers like alkyds are intended to be homogeneous products with properties which represent an average of those of all the component monomers. The copolymerization of linoleic acid in the recipe in Table 5-1 would confer air-drying properties on all the macromolecules in which it is incorporated.

Copolymerization can be employed in a similar fashion to modify the properties of the homopolymer of p-hydroxybenzoic acid (**5-6**). Poly(p-hydroxybenzoic acid) is an infusible polymer which can be shaped only by compression sintering. A melt processable variation of this high modulus, thermally stable material can be made, however, by copolymerizing an ester of **5-6** with equimolar quantities of terephthalic acid (**5-7**) and biphenol (**5-8**) to produce an aromatic polyester which can be fabricated at temperatures near 400°C but still retain many useful properties at 300°C.

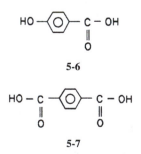

5-6

5-7

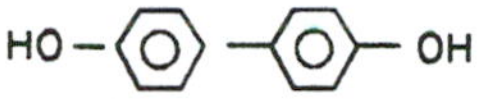

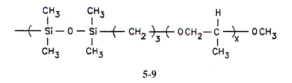

5-8

Not all copolymers which are produced by step-growth processes are random in nature. Block copolymers are also of major interest. An example of the synthesis of elastic polyurethane fibers was given in Section 1.5.4. Block and graft copolymers of polysiloxane-poly(alkylene ethers) with segments like **5-9** are used as surfactants in the production of polyurethane foams with uniform cell sizes.

$$
\underset{\substack{|\\CH_3}}{\overset{\substack{CH_3\\|}}{{-}(Si}} - O - \underset{\substack{|\\CH_3}}{\overset{\substack{CH_3\\|}}{Si}} \underset{}{-(CH_2)_3}\underset{}{(OCH_2 - \underset{\substack{|\\CH_3}}{\overset{\substack{H\\|}}{C}})_x} OCH_3
$$

5-9

Step-growth polymerizations at high temperatures produce nearly random copolymers because of end-group interchange reactions like (5–14) between macromolecules. Interfacial and low-temperature solution polycondensations are conducted under essentially irreversible conditions, by contrast. In these cases the average copolymer composition and blocklike character of the product may depend on the reaction conditions and relative reactivity of the functional groups involved in the polymerization.

PROBLEMS

5-1. Polybenzimidazoles are made by a two-stage step-growth melt polymerization:

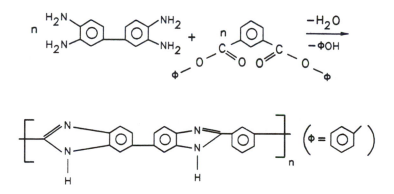

What extent of reaction is needed to produce $n = 200$ in the above formula? Carry four decimal places in your answer.

5-2.   Polyamide A has the structure

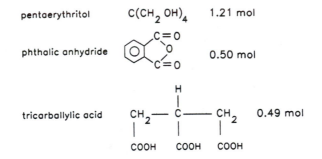

$$\overline{M}_n = 60{,}000 \text{ and } \overline{M}_w = 120{,}000.$$ Polyamide B has the structure

$\overline{M}_n = 90{,}000$ and $\overline{M}_w = 180{,}000$. The two polyamides are mixed in equal quantities by weight and are held at 280°C for 7 h. Interchange reactions can come to equilibrium in this time but no further polymerization could take place because condensation products were not removed from the system. What are $\overline{M}_n$ and $\overline{M}_w$ of the final, equilibrated system?

5-3.   Can the following alkyd recipe be carried to "complete" conversion without gelling?

pentaerythritol	$C(CH_2 OH)_4$	1.21 mol
phthalic anhydride		0.50 mol
tricarballylic acid		0.49 mol

5-4.   In the polymerization of $H_2N(CH_2)_{10} COOH$ to form Nylon-11, what weight fraction of the reaction mixture has the structure

when 99% of the functional groups have reacted?

5-5. Following is a simplified alkyd recipe:

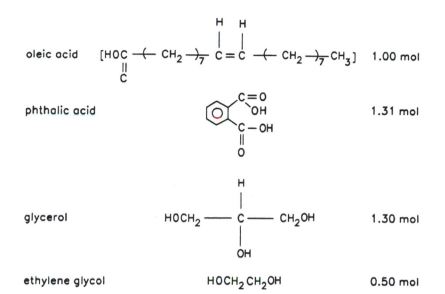

oleic acid		1.00 mol
phthalic acid		1.31 mol
glycerol		1.30 mol
ethylene glycol		0.50 mol

(a) Calculate the number average degree of polymerization when the esterification reaction is complete ($p = 1$).

(b) An operator makes up this mixture and forgets to add the ethylene glycol. He tries to run the reaction to completion. At what extent of conversion will he notice the results of this omission?

5-6. Nylon-11 is poly(11-aminoundecanoic acid). This polymer has a crystal melting point around 190°C and has lower water absorption than nylon-6,6 or nylon-6. It can be used to make mechanical parts, packaging films, bristles, monofilaments, and sprayed and fluidized coatings.

$$nH_2N \xleftarrow{\hspace{0.3cm}} CH_2 \xrightarrow{\hspace{0.3cm}}_{10} COOH \xrightarrow{220°C} \xleftarrow{\hspace{0.3cm}} N \xleftarrow{\hspace{0.3cm}} CH_2 \xrightarrow{\hspace{0.3cm}}_{10} \underset{O}{C} \xrightarrow{\hspace{0.3cm}}_n + nH_2O$$

In the step-growth polymerization of this monomer

(a) how much monomer (in terms of weight fraction of the reaction mixture) is left when 90% of the functional groups have reacted?

(b) Calculate $\overline{M}_n$, $\overline{M}_w$ and $\overline{M}_z$ of the reaction mixture at this stage.

5-7. Poly(ethylenoxy benzoate) is a fiber-forming polymer produced by the following sequence of reactions:

$$HO-\langle O\rangle + CO_2 \xrightarrow{KOH} HO-\langle O\rangle-COOH \qquad CH_2\overset{O}{\overbrace{}}CH_2$$

$$HOCH_2CH_2O-\langle O\rangle-COOH \xrightarrow{CH_3OH} HOCH_2CH_2O-\langle O\rangle-\underset{O}{\overset{}{C}}-OCH_3 \qquad (1)$$

$$n\ HOCH_2CH_2O-\langle O\rangle-\underset{O}{\overset{}{C}}-OCH_3 \xrightarrow{\Delta} -\left(OCH_2CH_2O\langle O\rangle-\underset{O}{\overset{}{C}}\right)_n + nCH_3OH\uparrow \quad (2)$$

The polymer synthesis is a step-growth polymerization. The fiber is melt-spun at 250–270°C and has silklike properties.

(a) When the extent of reaction (2) is 60%, what is the number average degree of polymerization *of the reaction mixture*?

(b) What weight fraction of the reaction mixture is polymer under these conditions?

(c) What is the weight average degree of polymerization *of the polymer*?

5-8. The polymer of Problem 5–7 becomes too viscous to spin conveniently when its number average molecular weight exceeds 50,000. Show a *practical* way to adjust the polymerization recipe so that $\overline{M}_n$ does not exceed this value.

5-9. When an equilibrium step-growth polymerization is 99.5% complete what fraction of the reaction mixture is still monomer

(a) on a mole basis?

(b) on a weight basis?

5-10. Ethylene glycol is polymerized with an impure sample of terephthalic acid, which contains 1.0 mol % benzoic acid. How does this impurity affect the limiting degree of polymerization at very high conversions?

REFERENCES

[1] W. H. Carothers, *J. Am. Chem. Soc.* **51**, 2548 (1929).

[2] P. J. Flory, "Principles of Polymer Chemistry." Cornell University Press, Ithaca, NY 1953.

[3] R. W. Lenz, "Organic Chemistry of Synthetic High Polymers." Wiley (Interscience), New York, 1967.

[4] P. W. Morgan and S. L. Kwolek, *J. Polym. Sci. Part A* **2**, 181 (1964).

Chapter 6

Free-Radical Polymerization

*Mathematics is all well and good but Nature keeps dragging us
around by the nose.*
—Albert Einstein, to Hermann Weyl, 1923

6.1 SCOPE

We turn our attention now to chain-growth polymerizations. The reader should recall that the features which distinguish chain-growth and step-growth polymerizations were summarized in Section 5.2. The present chapter is devoted to the basic principles of chain polymerizations in which the active centers are free radicals. Chain-growth reactions with active centers having ionic character are reviewed in Chapter 9.

Free-radical polymerization is the most widely used process for polymer synthesis. It is much less sensitive to the effects of adventitious impurities than ionic chain-growth reactions. Free-radical polymerizations are usually much faster than those in step-growth syntheses, which use different monomers in any case. Chapter 7 covers emulsion polymerization, which is a special technique of free-radical chain-growth polymerizations. Copolymerizations are considered separately in Chapter 8. This chapter focuses on the polymerization reactions in which only one monomer is involved.

6.2   POLYMERIZABILITY OF MONOMERS

A monomer must have a functionality greater than or equal to 2 in order for polymers to be produced from its reactions (Section 1.3). This functionability can be derived from (1) opening of a double bond, (2) opening of a ring, or (3) coreactive functional groups [1]. We considered the fundamentals of the reactions of monomers of type (3) in Chapter 5. This chapter is concerned with free-radicals reactions of monomers of type (1).

The most important functional groups that participate in chain-growth polymerizations are the carbon–carbon double bond in alkenes and the carbon–oxygen double bond in aldehydes and ketones. In such polymerizations the active species A adds to one atom of the double bond and produces a new active species on the other atom:

$$A^* + CH_2 = \overset{\overset{\displaystyle H}{|}}{\underset{\underset{\displaystyle X}{|}}{C}} \longrightarrow A - CH_2 - \overset{\overset{\displaystyle H}{|}}{\underset{\underset{\displaystyle X}{|}}{C^*}} \qquad (6\text{-}1)$$

6-1

or

$$A^* + \overset{\displaystyle R}{\underset{\displaystyle H}{>}} C = O \longrightarrow A - \overset{\overset{\displaystyle R}{|}}{\underset{\underset{\displaystyle H}{|}}{C}} - O^* \qquad (6\text{-}2)$$

The nature of the new active site on the residue of the monomer determines whether the polymerization mechanism is radical or ionic. Reaction (6-2) is not possible when A* is a free radical because of the difference in electronegativity of the C and O atoms. Aldehydes and ketones are not polymerized by radicals; they are enchained only by ionic or heterogeneous catalytic processes.

The alkene double bond can be polymerized in chain-growth reactions in which the active site is a radical, ion, or carbon–metal bond. The processes whereby a given alkene reacts depend on the inductive and resonance characteristics of the substituent X in the vinyl monomer **6-1**.

Electron-releasing substitutents

$$(R, RO-, R - \overset{\overset{\displaystyle H}{|}}{C} = \overset{\overset{\displaystyle H}{|}}{C} - \text{ and } \langle\!\bigcirc\!\rangle-)$$

increase the electron density of the double bond and facilitate addition of a cation. Thus monomers like isobutene, styrene, and vinyl ethers **(6-2)** all undergo cationic polymerization. Some of these vinyl monomers can also delocalize the positive charge, and this also facilitates reactions with cations:

$$A^{\oplus} + CH_2 = \underset{OR}{\overset{H}{\underset{|}{C}}} \longrightarrow A - CH_2 - \underset{OR}{\overset{H}{\underset{|}{C^{\oplus}}}} \longleftrightarrow A - CH_2 - \underset{\underset{\oplus}{O - R}}{\overset{H}{\underset{|}{C}}} \tag{6-3}$$

6-2

Styrene will act similarly but the effect of alkyl substitutents is almost entirely inductive.

Electron-withdrawing substituents

$$(- C \equiv N, - \underset{\overset{\|}{O}}{C} - R, - \underset{\overset{\|}{O}}{C} - OH \text{ or } - \underset{\overset{\|}{O}}{C} - OR)$$

promote attack of an anionic species on the double bond and may also stabilize the anion formed by delocalization of the charge:

$$A^{\ominus} + CH_2 = \underset{CN}{\overset{H}{\underset{|}{C}}} \longrightarrow A - CH_2 - \underset{CN}{\overset{H}{\underset{|}{C^{\ominus}}}} \longleftrightarrow A - CH_2 - \underset{\ominus : C = N}{\overset{H}{\underset{|}{C}}} \tag{6-4}$$

Phenyl and alkenyl ($-CH=CH_2$) substituents are electron releasing but they stabilize the product anions by resonance, and so styrene and butadiene can undergo both cationic and anionic polymerizations. Anionic mechanisms are more important, however, since they provide better control over the polymer structure (Chapter 9).

Radical reactions with the π bond of vinyl monomers are not nearly as selective as ionic attack, and free-radical initiators cause the polymerization of nearly all vinyl and vinylidene monomers. (Some of these polymerizations are not efficient because of side reactions. Propylene is a case in point as described in Section 6.8.5.) Resonance stabilization occurs to some extent with most vinyl monomers but it is important in radical polymerizations only when the monomers contain conjugated C—C double bonds as in styrene, 1,3-butadiene, and similar molecules:

$$A^{\bullet} + CH_2 = \overset{H}{\underset{|}{C}} \longrightarrow A - CH_2 - \overset{H}{\underset{|}{C^{\bullet}}} \longleftrightarrow A - CH_2 - \overset{H}{\underset{|}{C}} \longleftrightarrow A - CH_2 - \overset{H}{\underset{|}{C}} \longleftrightarrow A - CH_2 - \overset{H}{\underset{|}{C}}$$

$$\tag{6-5}$$

Steric hindrances prevent the polymerization of most 1,2-disubstituted ethylenes by any mechanism. However, 1,1-disubstituted monomers and vinylidene monomers usually polymerize more readily than the corresponding vinyl analogs.

6.3   OVERALL KINETICS OF RADICAL
POLYMERIZATION

Chain-growth polymerizations are so called because their mechanisms comprise chains of kinetic events. For successful polymerization, the sequence of reactions must first be initiated by some agent, and monomers must be added consecutively to a growing macromolecule. This chain of events may then be terminated by a reaction that is inherent in the system or by the action of impurities. In any case, we can usefully distinguish between at least three different reaction types in a kinetic polymerization chain. These are initiation, propagation, and termination reactions. (Recall that there is only one reaction involved in step-growth polymerizations where the monomers add to the end of a macromolecule without the intervention of an active center.)

It is not practical to conduct free-radical polymerizations under conditions where there is an equilibrium between polymerization and depolymerization processes. The polymer synthesis is effectively irreversible in normal radical polymerizations. The course of the reaction is then determined kinetically, and the molecular weight distribution cannot be predicted statistically as was done for equilibrium step-growth polymerizations described in Chapter 5.

This section develops the overall kinetics for an ideal free-radical polymerization. Various details of the individual reactions in the kinetic sequence are treated in subsequent sections of this chapter.

6.3.1   Initiation

Free radicals must be introduced into the system to start the reaction. There are many ways to accomplish this, but the most common method involves the use of a thermolabile compound, called an *initiator*, which decomposes to yield two free radicals at the temperature of the reaction mixture. That is,

$$I \xrightarrow{k_d} 2R \cdot \qquad (6\text{-}6)$$

Here the initiator I decomposes to yield two radicals $R \cdot$. (These are called *primary radicals*.) The specific rate constant for this reaction at the particular temperature is k_d. The initiation reaction *per se* follows if a radical $R \cdot$ adds to a monomer as in

$$
R^\bullet + CH_2 = \underset{\underset{C\ell}{|}}{\overset{\overset{H}{|}}{C}} \longrightarrow R - CH_2 - \underset{\underset{C\ell}{|}}{\overset{\overset{H}{|}}{C}}^\bullet \qquad (6\text{-}7)
$$

Reaction (6-7) may be abbreviated and generalized to

$$R^\bullet + M \xrightarrow{k_i} M_1^\bullet \qquad (6\text{-}8)$$

where M stands for monomer and $M_1\cdot$ denotes a monomer-ended radical (like that in the preceding reaction). There is only one monomer in this radical. The rate constant for reaction (6-8) is k_i.

Reaction (6-6) is a unimolecular decomposition, as written, so k_d will be a first-order rate constant. The magnitude of this decomposition rate constant is usually of the order of 10^{-4}–10^{-6} sec^{-1} at the temperatures at which such initiators are used.

The rate of radical production from reaction (6-6) is

$$d[R\cdot]/dt = 2k_d[I] \qquad (6\text{-}9)$$

since each molecular decomposition produces two radicals. The rate of initiation $R_i\cdot$ is the rate of reaction (6-8). This can be expressed in terms of the rate of radical production as

$$R_i\cdot = 2fk_d[I] \qquad (6\text{-}10)$$

where f, the fraction of all radicals generated that are captured by monomers, is called the *initiator efficiency*. This term is discussed in more detail in Section 6.5.5. Initiator efficiency is always ≤ 1.

The formulation of the rate of initiation in Eq. (6-10) contains k_d as the only rate constant. The other rate constant k_i of reaction (6-8) is not required if f and k_d can be measured.

6.3.2   Propagation

By definition, a propagation step in a chain reaction is one in which products are formed, and the site of the reactive center changes but the number of active centers is not changed. (This statement is qualified in [2].) There are two major propagation reactions under the conditions of most free-radical polymerizations. These are addition and atom transfer reactions.

(i)   Addition Reactions

Successive monomer additions after the initiation step of reaction (6-8) can be represented as

$$M_1^\bullet + M \longrightarrow M_2^\bullet \qquad (6\text{-}11a)$$

$$M_2^{\bullet} + M \longrightarrow M_3^{\bullet} \qquad (6\text{-}11\text{b})$$

$$M_i^{\bullet} + M \longrightarrow M_{i+1}^{\bullet} \qquad (6\text{-}12)$$

where M_i represents the radical $R\text{-}(M\text{-})_{i-1}M\cdot$. Each reaction in the sequence involves the addition of a monomer to a monomer-ended radical, and each is assigned the same rate constant k_p on the reasonable assumption that the rate of the addition reaction does not depend on the size of the participating macroradical. Values of the propagation rate constant k_p for most monomers are of the order of 10^2–10^3 liter/mol sec under practical polymerization conditions.

Reaction (6-12) is a bimolecular reaction as written, and k_p is therefore a second-order rate constant with units of $(\text{concentration})^{-1} (\text{time})^{-1}$.

The rate of propagation R_p is given by

$$R_p = k_p[\text{M}\cdot][\text{M}] \qquad (6\text{-}13)$$

where $[\text{M}\cdot]$ stands for the sum of the concentrations of all monomer-ended radicals in the system. This expression for R_p can be written as shown since the radical concentrations can be lumped together if k_p does not depend on the size of M_i.

(ii) Atom Transfer Reactions

Radicals can undergo other reactions as well as monomer addition. Atom abstraction reactions usually involve transfer of a hydrogen or halogen atom. An example from micromolecular chemistry involves the chlorination of hydrocarbons at about $200°\text{C}$ or during irradiation with light of wavelength less than 4875×10^{-10} m:

$$Cl_2 \longrightarrow 2\ Cl^{\bullet} \qquad (6\text{-}14\text{a})$$

$$Cl^{\bullet} + CH_4 \longrightarrow H\ Cl + {}^{\bullet}CH_3 \qquad (6\text{-}14\text{b})$$

$$CH_3^{\bullet} + Cl_2 \longrightarrow CH_3Cl + Cl^{\bullet} \qquad (6\text{-}14\text{c})$$

Reaction (6-14a) is the initiation step, while reactions (6-14b) and (6-14c) are atom abstraction propagation reactions. Atom abstraction reactions in free-radical polymerizations are called *chain transfer reactions*. They are discussed in some detail in Section 6.8.

(iii) Other Propagation Reactions

Radicals can undergo other reactions such as rearrangements and fragmentations. These are also propagation reactions by definition but they do not concern us here because free-radical polymerizations are not usually performed under conditions in which these processes are significant.

6.3.3   Termination

The sequence of monomer additions is terminated by the mutual annihilation of two radicals. Such termination reactions can occur if the radicals combine to form a paired electron bond as in

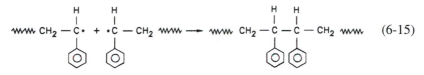

$$(6\text{-}15)$$

This process is called *termination by combination* and would be written in general terms as in

$$M_n \bullet \; + \; M_m \bullet \; \xrightarrow{\; k_{tc} \;} \; M_{n+m}$$

$$(6\text{-}16)$$

with k_{tc} as the corresponding rate constant.

Alternatively the two radicals can form two new molecules by a disproportionation reaction in which a hydrogen atom is transferred:

$$(6\text{-}17)$$

Generally for this case,

$$M_n \bullet \; + \; M_m \bullet \; \xrightarrow{\; k_{td} \;} \; M_m \; + \; M_n$$

$$(6\text{-}18)$$

where k_{td} is the disproportionation rate constant.

Termination may also occur by a mixture of disproportionation and combination. The rates of these reactions are additive for a given polymerization, because both terminations are bimolecular and have second-order rate constants. Thus we can write

$$M_n \cdot \; + M_m \cdot \; \xrightarrow{\; k_t \;} \text{dead polymer}$$

$$(6\text{-}19)$$

with the overall rate constant k_t given by

$$k_t = k_{tc} + k_{td}$$

$$(6\text{-}20)$$

The termination rates R_t corresponding to the different modes of termination are

$$R_{tc} = 2k_{tc}[\text{M}\cdot]^2$$

$$(6\text{-}21)$$

from Eq. (6-16)

$$R_{td} = 2k_{td}[M\cdot]^2 \tag{6-22}$$

from Eq. (6-18), and

$$R_t = 2k_t[M\cdot]^2 \tag{6-23}$$

from Eq. (6-20).

Typical termination rate constants are of the order of 10^6–10^8 liter/mol sec. These rate constants are much greater than k_p, but polymerization still occurs because the overall rate of polymerization is proportional to k_p and inversely proportional to $k_t^{1/2}$. This basic relation is derived in the following subsection.

6.3.4   Rate of Polymerization

When a free-radical polymerization is first started, the number of radicals in the system will increase from zero as the initiator begins to decompose according to reaction (6-6). The frequency of termination reactions will also increase from zero in the early stages of the polymerization because the rates of these reactions, (6-16) and (6-18), are proportional to the square of the total concentration of radicals in the system. Eventually the rate of radical generation will be balanced by the rate at which radicals undergo mutual annihilation, and the concentration of radicals in the system will reach a steady value. It can be shown that this steady state is reached very early in the reaction with the usual concentrations of initiator and monomer.

The assumption that the rate of initiation equals the rate of termination is called the *steady-state assumption*. It is equivalent to the two following statements:

$$R_i\cdot = R_t \text{ at steady state} \tag{6-24}$$

and

$$\frac{d[M\cdot]}{dt} = 0 \text{ at steady state} \tag{6-25}$$

Since the steady state is reached soon after polymerization starts, we can assume without significant error that it applies to the whole course of the polymerization.

Substituting Eqs. (6-10) and (6-23) into Eq. (6-24),

$$2fk_d[I] = 2k_t[M\cdot]^2 \tag{6-26}$$

Hence

$$[M\cdot] = [fk_d[I]/k_t]^{1/2} \tag{6-27}$$

This is an expression for the total concentration of monomer-ended radicals in terms of experimentally accessible quantities.

The rate of polymerization is taken to be the rate of disappearance of monomer, which is $d[M]/dt$. (The concentration of monomer $[M]$ decreases with time so $d[M]/dt$ is negative.) The two reactions listed that consume monomer are (6-8) (initiation) and (6-12) (propagation). Therefore,

$$-d[M]/dt = R_i + R_p \tag{6-28}$$

It can be shown that the initiation process accounts for a negligible amount of monomer if high-molecular-weight polymer is being produced. Then the rate of polymerization can be taken as equal to the rate of propagation. That is,

$$-d[M]/dt = R_p = k_p[M][M\cdot] = (k_p/k_{t^{1/2}})[M](fk_d[I])^{1/2} \tag{6-29}$$

The preceding relation is obtained by substituting Eq. (6-27) in Eq. (6-13). It is a differential equation and gives the rate of polymerization in moles of monomer per unit volume per unit time when the monomer concentration is $[M]$ and the initiator concentration is $[I]$. Since both these concentrations will decrease as the reaction proceeds, the amount of polymer formed in a given time is obtained by integrating Eq. (6-29) with respect to time (see Section 6.3.5).

Equation (6-29) was derived for the polymerization in which radicals are generated by the decomposition of an initiator in a homogeneous reaction mixture. More generally,

$$-d[M]/dt = (k_p/k_{t^{1/2}})[M]R_{M\cdot}^{1/2} \tag{6-30}$$

where $R_{M\cdot}$ is the rate of formation of monomer-ended radicals. Equation (6-30) can be used when initiation is by means summarized in Section 6.5 in which the relation between R_i and the initiator concentration may differ from that assumed in these paragraphs or in which the reaction may be started without an initiator *per se*.

Equations (6-29) and (6-30) show that the instantaneous rate of polymerization depends directly on the monomer concentration and on the square root of the rate of initiation.

The ratio $k_p/k_t^{1/2}$ will appear frequently in the equations we develop for radical polymerization. The polymerizability of a monomer in a free radical reaction is related to $k_p/k_t^{1/2}$ rather than to k_p alone. From Eq. (6-30) or (6-29), it can be seen that a given amount of initiator will produce more polymer from a monomer with a higher $k_p/k_t^{1/2}$ ratio. Thus, at 60°C, the k_p values for acrylonitrile and styrene are approximately 2000 and 100 liter/mol sec, respectively. The former monomer does not polymerize 20 times as fast as styrene under the same conditions at 60°, however, because the respective k_t values are 780×10^6 liter/mol sec. and 70×10^6 liter/mol sec. Then $k_p/k_t^{1/2}$ for acrylonitrile is 0.07 liter$^{1/2}$/mol$^{1/2}$ sec$^{1/2}$, which is just six times that of styrene. Styrene forms the less reactive radical in this case because the unpaired electron can be delocalized over the phenyl ring. It adds monomer more slowly as a consequence but it also terminates less readily because its radical is more stable than that of polyacrylonitrile.

6.3.5 Integrated Rate of Polymerization Expression

If the initiator decomposes in a unimolecular reaction the corresponding rate expression is first order in initiator [cf. reaction (6-6)]:

$$-d[I]/dt = k_d[I] \tag{6-31}$$

Integration of Eq. (6-31) between $[I] = [I]_0$ at $t = 0$ and $[I]$ at t gives

$$[I] = [I]_0 e^{-k_d t} \tag{6-32}$$

[For first-order reactions it is often convenient to integrate between $[I] = [I]_0$ at $t = 0$ and $[I] = [I]_0/2$ at $t = t_{1/2}$, the half-life of the initiator. The half-life is evidently the time needed for the initial concentration of initiator to decrease to half its initial value. It is related to the initiator decomposition rate constant k_d by

$$t_{1/2} = (\ln 2)/k_d \tag{6-33}$$

and is independent of the initial concentration. The conventional criterion for initiator activities is $t_{1/2}$.]

Equation (6-32) for $[I]$ can be inserted into Eq. (6-29) to obtain

$$-d[M]/[M] = (k_p/k_{t^{1/2}})(f k_d[I]_0)^{1/2} e^{-k_d t/2} \, dt \tag{6-34}$$

On integrating between $[M] = [M]_0$ at $t = 0$ and $[M]$ at t:

$$-\ln\frac{[M]}{[M_0]} = \frac{2k_p}{k_{t^{1/2}}} \left(\frac{f[I]_0}{k_d} \right)^{1/2} (1 - e^{-k_d t/2}) \tag{6-35}$$

This gives the amount of polymer (in terms of the moles of monomer converted) produced in time t at given temperature.

6.4 A NOTE ON TERMINATION RATE CONSTANTS

Much of our knowledge of free-radical polymerizations was pioneered by researchers in the United Kingdom and in the United States. Unfortunately, both groups used different conventions for termination rate constants, and the unwary reader may be misled unless the data and equations are self-consistent.

The American convention is followed in this text. In this case k_{tc} and k_{td} are defined by

$$-d[\text{M}\cdot]/dt \equiv 2k_{tc}[\text{M}\cdot]^2 + 2k_{td}[\text{M}\cdot]^2 \tag{6-36}$$

It follows then that

$$d[\text{polymer}]/dt = 2k_{td}[\text{M}\cdot]^2 + k_{tc}[\text{M}\cdot]^2 \tag{6-37}$$

since a single termination reaction produces one polymer molecule by combination or two if disproportionation is involved.

The British convention uses the following definitions:

$$-d[\text{M}\cdot]/dt = k_{tc}[\text{M}\cdot]^2 + k_{td}[\text{M}\cdot]^2 \tag{6-38}$$

and, hence,

$$d[\text{polymer}]/dt = \frac{1}{2}k_{tc}[\text{M}\cdot]^2 + k_{td}[\text{M}\cdot]^2 \tag{6-39}$$

Each convention is unobjectionable if used consistently but the two cannot be mixed because the termination rate constants quoted according to the American convention will be exactly half those measured by the British system. The same conclusions are reached when either usage is adopted because a compensating factor of 2 is present in the kinetic equations that use American rate constants and is absent in the corresponding British system.

All United States texts and data compilations adhere to the American system while most United Kingdom texts rely on the British method. The convention used is seldom explicitly specified, however, although it can usually be inferred from the particular context.

6.5   METHODS OF PRODUCING RADICALS

A large number of methods are available to bring about free-radical polymerizations. The more widely used initiation techniques are reviewed in this section.

6.5.1   Thermal Decomposition of Initiators

The thermal scission of a compound to yield two radicals was used for illustration in reaction (6-6), because this is the most common means of generating radicals. Thermal decomposition is ideally a unimolecular reaction with a first-order constant, k_d, which is related to the half-life of the initiator, $t_{1/2}$, by Eq. (6-33).

It is convenient in academic studies to select an initiator whose concentration will not change significantly during the course of an experiment. From experience it seems that a $t_{1/2}$ of about 10 h at the particular reaction temperature is a good rule-of-thumb in this regard. This corresponds to a k_d of 2×10^{-5} sec^{-1} from Eq. (6-33). More generally, initiators which are used in polymer production have k_d's under usage conditions which are as low as about 10^{-4} sec^{-1}. This corresponds to a 2-h half-life. Even faster decompositions are needed in the high-pressure polymerization of ethylene where reactor residence times are very short at temperatures between about 130 and 280°C.

In general, initiation should be as fast as is practical to produce as much polymer as possible per unit of reaction time. The reaction cannot be allowed to proceed more quickly than the rate at which the exothermic heat of polymerization can be removed from the system, however. The decomposition rates of free-radical initiators are very temperature sensitive (the $t_{1/2}$ of benzoyl peroxide drops from 13 h at 70°C to 0.4 h at 100°C), and a runaway reaction can result from overheating if the rate of initiation is not limited appropriately.

Many polymerizations are carried out at temperatures between 0 and 100°C. Initiation at the required rates under these conditions is confined to compounds with activation energies for thermal homolysis in the range 100–165 kJ/mol. If the decomposition process is endothermic, the activation energy can be considered to be approximately equal to the dissociation energy of the bond which is being split. It can be expected, then, that useful initiators will contain a relatively weak bond. (The normal C—C sigma bond dissociation energy is of the order of 350 kJ/mol, and alkanes must be heated to 300–500°C to yield radicals at the rates required in free-radical polymerizations.)

The major class of compounds with bond dissociation energies in the 100–165 kJ/mol range contain the O—O peroxide linkage. There are numerous varieties of compounds of this type, and some are listed in Table 6-1. Thermal decomposition always initiates by scission of the bond between the oxygen atoms:

$$R - C - O - O - C - R \longrightarrow 2R - C - O^{\bullet} \qquad (6\text{-}40)$$
$$\quad \| \qquad\qquad \| \qquad\qquad\quad \|$$
$$\quad O \qquad\qquad O \qquad\qquad\quad O$$

The acyloxy radicals produced in reaction (6-40) of diacyl peroxides can initiate polymerization or undergo side reactions as described in Section 6.5.5. Other peroxides behave similarly.

The temperature–half-life relations given in Table 6-1 may vary with reaction conditions, because some peroxides are subject to accelerated decompositions by specific promoters (Section 6.5.2) and are also affected by solvents or monomers in the system.

The choice of reaction conditions has much less effect on the behavior of azo initiators. The activation energies for decomposition of azo compounds are similar to those of peroxides although the azo initiators do not contain a weak bond like

TABLE 6-1

Some Peroxide Initiators for Radical Polymerizations [3]

Types	Example	Temp. for 10-h $t_{1/2}$(°C)
Diacyl peroxides	Dibenzoyl peroxide	73
	$(CH_3)_3 - C - C - O - O - C - C(CH_3)_3$ (with C=O double bonds) Di-tertbutyryl peroxide	21
Peroxy esters	$CH_3 - C - O - O - C(CH_3)_3$ (with C=O double bond) t-Butyl peracetate	102
Dialkyl peroxides	Dicumyl peroxide	115
Dialkyl peroxydicarbonates	Di-*sec*-butylperoxydicarbonate	45
tert-Alkylhydroperoxides[a]		158
Ketone peroxides[a]		and other structures 105

[a]Hydroperoxides and ketone peroxides are used mainly with activators in redox initiation systems (Section 6.5.2).

the O—O linkage. Decomposition of azodiisobutyronitrile (AIBN, 6-3), for example, proceeds because the nitrogen which is formed is a very stable gas and has a very high enthalpy of formation:

$$CH_3-\underset{\underset{CN}{|}}{\overset{\overset{CH_3}{|}}{C}}-N=N-\underset{\underset{CN}{|}}{\overset{\overset{CH_3}{|}}{C}}-CH_3 \overset{\Delta}{\longrightarrow} 2\ CH_3-\underset{\underset{CN}{|}}{\overset{\overset{CH_3}{|}}{C^{\bullet}}}+N_2 \qquad (6\text{-}41)$$

6-3

This initiator has a 10-h $t_{1/2}$ at 64°C.

Azo compounds are preferred for scientific investigations because their k_d values do not vary with the particular polymerization system as much as those of peroxides. Cost considerations generally favor the industrial use of peroxides.

6.5.2   Redox Sources

Oxidation–reduction reactions can also be used to initiate free-radical polymerizations. Such initiators include a metallic ion that can undergo a one-electron transfer:

$$ROOH + Co^{2+} \longrightarrow RO^{\bullet} + OH^{\ominus} + Co^{3+} \qquad (6\text{-}42)$$

$$S_2O_8^{2-} + Fe^{2+} \longrightarrow SO_4^{2-} + SO_4^{\cdot} + Fe^{3+} \qquad (6\text{-}43)$$

Redox reactions occur with hydroperoxides, peroxides, peresters, persulfates, hydrogen peroxide, and other peroxides. A wide variety of metal ions may be used as reducing agents. Ions which are commonly used include Co^{2+}, Fe^{2+}, Cr^{2+}, and Cu^+.

Transition metals can convert the radicals generated in reactions like (6-42) or (6-43) into ions:

$$RO^{\bullet} + Co^{2+} \longrightarrow RO^{\ominus} + Co^{3+} \qquad (6\text{-}44)$$

$$Co^{3+} + ROOH \longrightarrow ROO^{\bullet} + H^{\oplus} + Co^{2+} \qquad (6\text{-}45)$$

The radical destroyed in reaction (6-44) is not available for initiating polymerization. There is thus an optimum, low level of transition metal reducing agent for efficient generation of radicals.

The general redox reaction

$$A-B + X \longrightarrow A^{\bullet} + B^{\ominus} + X^{\oplus} \qquad (6\text{-}46)$$

could proceed with any molecule AB, provided the reducing agent X is strong enough to split the AB bond. Sodium would be effective with ethyl chloride, in which the C—Cl link has a bond dissociation energy of about 290 kJ/mol:

$$CH_3 - CH_2 - Cl + Na \longrightarrow CH_3CH_2\cdot + Cl^\ominus + Na^\oplus \qquad (6\text{-}47)$$

Such redox couples are not attractive because the reducing agent is readily oxidized in air and may attack some solvents in preference to the intended substrate. For practical purposes the A—B bond in redox systems must be relatively weak and this limits such materials to the peroxide compounds described in the preceding section.

Many redox initiators are water soluble and are widely used in emulsion polymerizations (Chapter 8) in which the radicals are generated in the aqueous phase. When metal ions are used as reducing agents in organic media they are commonly in the form of salts of carboxylic acids.

Redox initiation can be arranged to proceed quickly under mild reaction conditions and is particularly useful for low and ambient temperature radical polymerizations. The two components of the redox couple are stable when handled separately whereas a conventional thermal decomposition initiator with the same activity would be difficult to store and transport.

6.5.3 Photochemical Initiation

Radicals can be produced by ultraviolet irradiation of a monomer like styrene which absorbs sufficiently strongly in this wavelength region. Photochemical initiation may also be provided by thermal initiators or by compounds like benzoin ethers:

$$(6\text{-}48)$$

Current practice in the ultraviolet curing of paints, adhesives, and inks relies heavily on the use of photoinitiators of this type. An alternative procedure involves the use of photosensitizers which undergo excitation at an irradiation frequency which is not effective on the monomer or initiator of direct interest. Transfer of energy from the excited photosensitizer takes place at a frequency at which the recipient molecule can absorb and subsequently decompose. Benzophenone is a common photosensitizer in radical reactions.

6.5.4 Other Initiation Methods

Ionizing radiation like γ rays and fast electrons are sometimes used to initiate polymerizations and also to cross-link polymers. The interactions between such radiations and matter result in a complex series of chemical events that culminate generally in the production of free radicals. In scrupulously dry systems ionic polymerization may also take place along with or to the exclusion of radical processes.

Some pure monomers undergo initiation when heated. The subsequent polymerization is free radical in character. Styrene exhibits significant thermal initiation at temperatures of $100°C$ or more. Methyl methacrylate also self-initiates but at a slower rate. Low-molecular-weight vinyl polymers can often be made simply by heating the monomers, but the molecular weight control is not very close and initiation in some cases at least may be from thermal homolysis of impurities in the reaction mixture.

6.5.5   Initiator Efficiency

Initiators are not used efficiently in free-radical polymerizations. A significant proportion of the primary radicals that are generated are not captured by monomers, and the initiator efficiency f in Eq. (6-10) is normally in the range 0.2–0.7 for most initiators in homogeneous reaction systems. It will be lower yet in polymerizations in which the initiator may not be very well dispersed.

Two major reactions are involved in the wastage of primary radicals. These are induced decomposition of initiator by radicals and side reactions in the solvent cage. The mechanisms of induced decompositions depend on the structure of the initiator molecule. For benzoyl peroxide the reaction involved could be an S_N2 attack of propating macroradicals on the O—O bond:

$$M_n^{\cdot} + \bigcirc\!\!-\!COO-C\!\!-\!\!\bigcirc \longrightarrow M_n-O-C\!\!-\!\!\bigcirc$$

$$+\ \bigcirc\!\!-\!C-O^{\cdot} \qquad\qquad (6\text{-}49)$$

The radical concentration during the polymerization is not changed by reaction (6-49), but the initiator molecule involved has been wasted because its decomposition has not produced a net increase in the conversion of monomer to polymer. Induced decomposition reactions are negligible for azo initiators, but they can be very significant for some peroxides. Peroxydicarbonates, for example, have efficiencies which change greatly with reaction conditions.

The major cause of primary radical wastage involves "cage" reactions. When an initiator decomposes, the primary radicals are each other's nearest neighbors for about 10^{-10} sec. During this interval they are surrounded by a "cage" of solvent and monomer molecules through which they must diffuse to escape each other. Once one or the other radical leaves the cage it is extremely unlikely that the pair will encounter each other again. While they are in the cage, however, any reaction that takes place will be between the two primary radicals which can be expected to be colliding on the average once very 10^{-13} sec. (The vibrational frequency of a diatomic molecule at reaction temperatures is approximately 10^{13} sec^{-1}.) Direct recombination simply regenerates the original initiator molecule, but other reactions can also occur that waste the radicals for polymerization. To illustrate, the decomposition of acetyl peroxide could lead to the following events:

$$CH_3 - \underset{\underset{O}{\|}}{C} - O - O - \underset{\underset{O}{\|}}{C} - CH_3 \longrightarrow [CH_3 - \underset{\underset{O}{\|}}{C} - O^{\bullet} + CH_3 - \underset{\underset{O}{\|}}{C} - O^{\bullet}] \qquad (6\text{-}50)$$

$$\left[2\, CH_3 - \underset{\underset{O}{\|}}{C} - O^{\bullet} \right] \longrightarrow 2\, CH_3 - \underset{\underset{O}{\|}}{C} - O^{\bullet} \qquad (6\text{-}51)$$

$$CH_3 - \underset{\underset{O}{\|}}{C} - O^{\bullet} \longrightarrow CH_3{}^{\bullet} + CO_2 \qquad (6\text{-}52)$$

$$\left[2\, CH_3 - \underset{\underset{O}{\|}}{C} - O^{\bullet} \right] \longrightarrow \left[2\, CH_3 - \underset{\underset{O}{\|}}{C} - O - CH_3 + CO_2 \right] \qquad (6\text{-}53)$$

$$\left[2\, CH_3 - \underset{\underset{O}{\|}}{C} - O^{\bullet} \right] \longrightarrow \left[CH_3 - CH_3 + 2\, CO_2 \right] \qquad (6\text{-}54)$$

Here the brackets indicate caged radicals, and the last two reactions waste the initiator. Only reactions (6-51) and (6-52) yield radicals that could initiate polymerization. All initiators suffer cage wastage reactions. At 60°C, for example, f for AIBN is only 0.6 because of these events.

6.5.6 Measurement of k_d

Dead-end polymerization is a useful technique for assessing k_d. Equation (6-35) can be written

$$-\ln\frac{[M]}{[M_0]} = \left(2\frac{V_0}{k_d} \right) \left(1 - e^{-k_d t/2} \right) \qquad (6\text{-}35a)$$

where the parameter V_0 is defined by

$$V_0 \equiv k_p \left(\frac{f k_d}{k_t} \right)^{1/2} [I_0]^{1/2} \qquad (6\text{-}55)$$

A useful feature of equation (6-35a) is that k_d is separated from the other rate constants because it is the only kinetic parameter in the exponential function. If the starting initiator concentration $[I_0]$ was insufficient to cause polymerization of all the monomer, the reaction will reach a dead end (i.e., cease) after a long time t_∞. The corresponding limiting value of $[M]$ can be used to assess k_d if this concentration can be determined accurately and if the polymerization does not exhibit autoacceleration (Section 6.13.2) effects at high conversions [4]. To avoid these potential complications, it is convenient to expand the exponential function in Eq. (6-35a) in a power series and simplify the resulting expression to

$$-\frac{\ln[M]/[M_0]}{t} = V_0\left[1 - (1/2!)\left(\frac{k_d}{2}\right)t + (1/3!)\left(\frac{k_d}{2}\right)^2 t^2 + \cdots\right] \qquad (6\text{-}56)$$

If the t term is much larger than the t^2 term, the equation may be truncated to

$$-\frac{\ln[M]/[M_0]}{t} = V_0 - V_0\left(\frac{k_d}{4}\right)t \qquad (6\text{-}57)$$

A plot of the left-hand side of Eq. (6-57) versus time produces a straight line whose intercept at $t = 0$ is equal to V_0 and whose slope can be used to calculate k_d [5].

6.6 LENGTH OF THE KINETIC CHAIN AND NUMBER AVERAGE DEGREE OF POLYMERIZATION OF THE POLYMER

The kinetic chain length v is the average number of monomers that react with an active center from its formation until it is terminated. It is given by the ratio of the rate of polymerization to the rate of initiation and under steady-state conditions where $R_t = R_i$:

$$v = \frac{R_p}{R_i} = \frac{k_p[M][M\cdot]}{2fk_d[I]} = \frac{k_p[M][M\cdot]}{2k_t[M\cdot]^2} \qquad (6\text{-}58)$$

from Eqs. (6-10), (6-13), and (6-23). Therefore,

$$v = k_p[M]/2k_t[M\cdot] = k_p^2[M]^2/2k_t R_p \qquad (6\text{-}59)$$

by substituting for $[M\cdot]$ from Eq. (6-13).

If the polymerization is initiated by thermal homolysis of an initiator, Eq. (6-27) applies, and this expression for $[M\cdot]$ in Eq. (6-59) gives

$$v = k_p[M]/2(fk_d[I]k_t)^{1/2} \qquad (6\text{-}60)$$

The last two equations show that the average number of monomers converted to polymer per radical will be inversely proportional to the radical concentration or the polymerization rate and directly proportional to the monomer concentration.

The number average degree of polymerization (Section 1.2), $\overline{DP}_n$, is equal at any instant to the ratio of the rate of monomer disappearance to the rate at which completed polymer molecules are produced. That is,

$$\overline{DP}_n = -\frac{d[M]}{dt} \bigg/ \frac{d[\text{polymer}]}{dt} \tag{6-61}$$

$$\frac{d[\text{polymer}]}{dt} = k_{tc}[M\cdot]^2 + 2k_{td}[M\cdot]^2 \tag{6-62}$$

because each termination reaction by combination yields one polymer molecular while each disproportionation produces two macromolecules. Equations (6-13) and (6-62) can be substituted into Eq. (6-61) to yield

$$\overline{DP}_n = k_p[M]/(k_{tc} + 2k_{td})[M\cdot] \tag{6-63}$$

and with $[M\cdot] = R_p/k_p[M]$ from Eq. (6-13),

$$\overline{DP}_n = k_p^2[M]^2/R_p(k_{tc} + 2k_{td}) \tag{6-64}$$

Also, for initiation by thermal homolysis of an initiator, Eqs. (6-27) and (6-63) give

$$\overline{DP}_n = \frac{k_p[M](k_{tc} + k_{td})^{1/2}}{(k_{tc} + 2k_{td})(fk_d[I])^{1/2}} = \frac{k_p[M](k_{tc} + k_{td})^{1/2}}{(k_{tc} + 2k_{td})\left(\frac{1}{2}R_i\right)^{1/2}} \tag{6-65}$$

On comparing Eqs. (6-59) and (6-64) or Eqs. (6-60) and (6-65), it is evident that $\overline{DP}_n = v$ if termination is entirely by disproportionation ($k_{tc} = 0$). In the more common instance where termination is by combination only $k_{td} = 0$ and $\overline{DP}_n = 2v$.

(Recall that for vinyl polymers

$$\overline{M}_n = M_0\overline{DP}_n \tag{6-66}$$

where M_0 is the molecular weight of the monomer.)

The number average degree of polymerization is inversely proportional to the rate of polymerization at given monomer concentration and temperature (Eq. 6-64). It follows that $\overline{DP}_n$ also varies inversely with the square root of the rate of initiation. It is not possible to increase R_p and $\overline{DP}_n$ simultaneously in free-radical polymerizations.

6.7 MODES OF TERMINATION

It is believed that most macroradicals terminate in free-radical polymerizations predominantly or entirely by combination. Experimental measurements of polymer systems are scanty, however. It can be expected that disproportionation will be more important for tertiary than secondary macroradicals. The former have more β hydrogens available for transfer during disproportionation, and direct coupling of tertiary radicals is more hindered sterically. Thus the poly(methyl methacrylate) radical terminates by combination and disproportionation while combination is the only mode observed in polystyryl radical termination. The k_{tc}/t_{td} ratio in polymers which terminate by both processes is temperature sensitive with higher temperatures apparently tending to encourage disproportionation.

6.8   CHAIN TRANSFER

In many free-radical polymerizations, the molecular weight of the polymer produced is lower than that predicted from Eq. (6-64). This is because the growth of macroradicals in these systems was terminated by transfer of an atom to the macroradical from some other species in the reaction mixture. The donor species itself becomes a radical in the process, and the kinetic chain is not terminated if this new radical can add monomer. Although the rate of monomer consumption may not be altered by this change of radical site, the initial macroradical will have ceased to grow and its size is less than it would have been in the absence of the atom transfer process. These reactions are called *chain transfer processes*. They can be classified as varieties of propagation reactions (Section 6.3.2).

In general,

$$M_n^{\bullet} \ + \ TH \ \xrightarrow{\ k_{tr}\ } \ M_nH \ + \ T\cdot \tag{6-67}$$

Equation (6-67) is written for transfer of a hydrogen atom between a macroradical $M_n\cdot$ and a transfer agent TH. Other atoms, and particularly halogens, can be transferred in place of hydrogen and TH can be monomer, solvent, initiator, polymer, or any substance in the reaction mixture. The rate of transfer R_{tr} is

$$R_{tr} = k_{tr}[M\cdot][TH] \tag{6-68}$$

assuming as usual that the transfer rate constant k_{tr} is the same for all monomer-ended radicals and taking [M·] to be the concentration of all such species. The magnitude of k_{tr} will depend on the natures of $M_n\cdot$ and TH as well as the reaction temperature.

The new radical T· can reinitiate. The rate constant for addition of a particular monomer M to T· is k_r, so that the sequence of events in this process is

$$T\cdot \ + \ M \ \xrightarrow{\ k_r\ } \ M_1^{\bullet} \ \xrightarrow[M]{\ k_p\ } \ M_2^{\bullet} \ \xrightarrow{\ k_p\ } \ \cdots \ \xrightarrow{\ k_p\ } \ M_n^{\bullet} \tag{6-69}$$

where k_p is the propagation rate constant for this monomer.

Chain transfer always results in a lower polymer molecular weight than would occur in its absence. The effect on the rate of propagation varies however, and depends on the relative rates of the transfer and reinitiation steps (Eq. 6-69) compared to that of the normal propagation reaction. Several cases are conventionally distinguished. These are all instances of chain transfer, but they are usually given different names, as shown below, depending on the net effects on molecular weight and polymerization rate.

In effective (normal) chain transfer, the new radical formed by the transfer reaction is reactive enough to initiate growth of a new macroradical within about the same time period as that required for addition of a monomer to a monomer-ended radical in normal polymerization. In this case, then, $k_r \simeq k_p$ and the rate of polymerization R_p is not altered. If $k_p \gg k_{tr}$, the polymer molecular weight is reduced but the product is still macromolecular. Normal chain transfer may be a characteristic of the particular polymerization or it may be deliberately contrived to control the polymer molecular weight.

If $k_p \ll k_{tr}$ and $k_r \simeq k_p$, there will be a large number of transfer reactions compared to monomer additions, and only low-molecular-weight products will be made. This process, which is called *telomerization*, is illustrated by the radical reaction of ethylene and CCl_4 which yields waxy products of the general structure $Cl_3C(CH_2CH_2)_nCl$ with $n \leq 12$. The chain transfer reaction and reinitiation process in this case are

$$\text{\large www } CH_2 - CH_2^{\bullet} + CCl_4 \xrightarrow{k_{tr}} \text{\large www } CH_2 - CH_2 - Cl + {}^{\bullet}CCl_3 \qquad (6\text{-}70)$$

$$^{\bullet}CCl_3 + CH_2 = CH_2 \xrightarrow{k_r} Cl_3C - CH_2 - CH_2^{\bullet} \qquad (6\text{-}71)$$

If the reactivity of the radical formed by chain transfer is lower than that of propagating macroradicals, the rate of polymerization will be reduced along with the polymer molecular weight. Such cases are examples of retardation. If the reduction in polymerization rate and molecular weight are so severe as to make both effectively nil on the scale of polymer measurements, the process is labeled *inhibition*. Retardation and inhibition are reviewed separately in Section 6.9, because some retarders and inhibitors operate through reactions other than transfer. Degradative chain transfer, which results when the monomer itself is a true chain transfer agent that reinitiates poorly, is also more clearly explained by postponing its consideration to Section 6.8.5.

The ideal free-radical kinetics without chain transfer culminate in Eqs. (6-64) and (6-65) in which termination of the growth of polymeric radicals is accounted for only by mutual reaction of two such radicals. Chain transfer can also end the physical growth of macroradicals, and the polymerization model will now be amended to include the latter process. This can be easily done by changing Eq. (6-62) to include transfer reactions in the rate of polymer production, $d[\text{polymer}]/dt$. Combining Eqs. (6-62) and (6-68),

$$d[\text{polymer}]/dt = k_{tc}[\text{M}\cdot]^2 + 2k_{td}[\text{M}\cdot]^2 + k_{tr}[\text{M}\cdot][\text{TH}] \qquad (6\text{-}72)$$

in the presence of chain transfer. If we substitute Eq. (6-72) into (6-61) and invert the resulting expression (for ease in manipulation), we obtain

$$\frac{1}{\overline{DP}_n} = \frac{k_{tc}[M\cdot]}{k_p[M]} + \frac{2k_{td}[M\cdot]}{k_p[M]} + \frac{k_{tr}[TH]}{k_p[M]} \tag{6-73}$$

Then, putting $[M\cdot] = R_p k_p[M]$ (from Eq. 6-13):

$$\frac{1}{\overline{DP}_n} = \frac{k_{tc}R_p}{k_p^2[M]^2} + \frac{2k_{td}R_p}{k_p^2[M]^2} + \frac{k_{tr}[TH]}{k_p[M]} \tag{6-74}$$

The ratio k_{tr}/k_p will depend on the particular transfer agent and monomer, as well as the reaction temperature. The relation given can be generalized by breaking the last term on the right-hand side of Eq. (6-74) into a sum of the contributions from the various potential chain transfer agents (except polymer). Then

$$\frac{1}{\overline{DP}_n} = \frac{k_{tc}R_p}{k_p^2[M]^2} + \frac{2k_{td}R_p}{k_p^2[M]^2} + \frac{k_{tr,M}}{k_p}\frac{[M]}{[M]} + \frac{K_{tr,I}}{k_p}\frac{[I]}{[M]}$$
$$+ \frac{k_{tr,S}}{k_p}\frac{[S]}{[M]} + \frac{k_{tr}}{k_p}\frac{[Ta]}{[M]} \tag{6-75}$$

where $k_{tr,M}$, $k_{tr,I}$, $k_{tr,S}$ are the rate constants for the transfer reaction of reaction (6-67) with monomer, initiator, and solvent (S), respectively, and Ta stands for any chain transfer agent which is added deliberately for this purpose.

It is customary to define *a chain transfer constant C* for each substance as the ratio of k_{tr} for that material with a propagating radical to k_p for that radical. Thus,

$$C_M \equiv \frac{k_{tr,M}}{k_p}; \quad C_I \equiv \frac{k_{tr,I}}{k_p}; \quad C_S \equiv \frac{k_{tr,S}}{k_p}; \quad C \equiv \frac{k_{tr}}{k_p} \tag{6-76}$$

In this notation, Eq. (6-75) is written

$$\frac{1}{\overline{DP}_n} = \frac{k_{tc}R_p}{k_p^2[M]^2} + \frac{2k_{td}R_p}{k_p^2[M]^2} + C_M + C_I\frac{[I]}{[M]} + C_S\frac{[S]}{[M]} + C\frac{[Ta]}{[M]} \tag{6-77}$$

Equation (6-77) is not as formidable in practice as it appears on first glance. In the bulk polymerization of styrene, for example, $[S] = 0$, $k_{td} = 0$, $[I]$ is usually so small as to make transfer to initiator negligible, and the right-hand side of this equation is left with only three terms.

The following sections review the magnitudes of the various chain transfer constants, methods for measuring these parameters, and their significance in free-radical polymerizations.

6.8.1   Transfer to Initiator

Many peroxides are quite active chain transfer agents, but they are normally present in such low quantities that they do not have a significant effect on $\overline{DP}_n$. The usual values of [I] are 10^{-4}–10^{-2} M and thus the [I]/[M] term in Eq. (6-79) is of the order of 10^{-5}–10^{-3}. The C_I values vary from zero for azo compounds to figures that approach one for hydroperoxides.

Several methods can be used to determine C_I. If transfer to all species except initiator is negligible, Eq. (6-77) can be rearranged to

$$\left[\frac{1}{\overline{DP}_n} - \frac{k_{tc} R_p}{k_p^2 [M]^2} - \frac{2 k_{td} R_p}{k_p^2 [M]^2} - C_M \right] = C_I \frac{[I]}{[M]} \tag{6-78}$$

R_p and $\overline{DP}_n$ are measured at various values of [I] and [M], and a plot of the left-hand side of the above equation against [I]/[M] yields C_I if the ratio k_{td}/k_{tc} is known from other measurements.

A complication can arise at the higher initiator concentrations involved in such plots because radicals from the initiators can terminate polymer radicals. This process, called *primary radical termination*, reduces polymer molecular weight and reduces initiator efficiency.

6.8.2   Transfer to Monomer

Transfer of a hydrogen can occur either from the monomer to a propagating macro-radical [reaction (6-79)] or in the reverse direction [reaction (6-80)]. In either case, the active site is transferred to the monomer, and the growth of the polymer radical is terminated.

$$\text{(6-79)}$$

$$\text{(6-80)}$$

$$\text{(6-81)}$$

Reinitiation from the monomer radical in reaction (6-81) can produce a branched polymer because the vinyl end of the new polymer radical can react with another

growing radical to produce a structure which can grow by adding monomer from
the new radical in the interior of the macromolecule:

$$CH_2 = C - CH_2 - C \text{ wwww} + \text{wwww } CH_2 - \overset{H}{\underset{X}{C}} \cdot \longrightarrow \text{wwww } CH_2 - \overset{H}{\underset{X}{C}} - CH_2 - \overset{H}{\underset{X}{\overset{.}{C}}} - CH_2 - \overset{H}{\underset{X}{C}} \text{ wwww} \qquad (6\text{-}82)$$

Reaction (6-80), on the other hand, does not result in branched polymers because
the end C atoms in the unsaturated polymer end are both substituted and unreactive
toward vinyl polymerization (cf. Section 6.2).

Chain transfer reactions of monomers with α methyl groups are likely to proceed predominantly by reaction (6-79) because the new radicals which are produced are relatively stable:

$$\text{wwww } CH_2 - \overset{CH_3}{\underset{X}{C}} \cdot + CH_2 = \overset{CH_3}{\underset{X}{C}} \longrightarrow \text{wwww } CH_2 - \overset{CH_3}{\underset{X}{C}} - H + CH_2 \overset{CH_2}{\underset{X}{\overset{.}{C}}} \qquad (6\text{-}83)$$

Chain transfer can also occur with side groups of some monomers. With vinyl
acetate, for example,

$$\text{wwww } CH_2 - \overset{H}{\underset{\substack{O \\ | \\ C=O \\ | \\ CH_3}}{C}} \cdot + CH_2 = \overset{H}{\underset{\substack{O \\ | \\ C=O \\ | \\ CH_3}}{C}} \longrightarrow \text{wwww } CH_2 - \overset{H}{\underset{\substack{O \\ | \\ C=O \\ | \\ CH_3}}{CH_2}} + CH_2 = \overset{H}{\underset{\substack{O \\ | \\ C=O \\ | \\ \cdot CH_2}}{C}} \qquad (6\text{-}84)$$

This reaction also eventually produces branched macromolecules.

Monomer chain transfer constants are generally less than 10^{-4}. Reaction (6-79)
involves breaking the strong vinyl C—H bond and the products of reaction (6-80)
are not appreciably more stable than the reactants.

Transfer to monomer sets an upper limit to the molecular weight of the polymer that can be produced. This is not normally a problem, however, except
with allylic monomers like propylene (Section 6.8.5). C_M is about 10^{-4} for
styrene. The maximum $\overline{DP}_n$ of polystyrene that can be achieved is then 10^4 [from
Eq. (6-77) assuming very low [I] and vanishingly small R_p]. This corresponds to an
$\overline{M}_n$ of about 10^6. Commercial molding and extrusion grades of polystyrenes have

$\overline{M}_n$ of about $1 \times 10^5 - 2.5 \times 10^5$, which are considerably lower than this limiting value.

C_M can be measured from experiments based on Eq. (6-77) which reduces to

$$\frac{1}{\overline{DP}_n} = \frac{k_{tc} R_p}{k_p^2 [M]^2} + \frac{2 k_{td} R_p}{k_p^2 [M]^2} + C_M + C_I \frac{[I]}{[M]} \tag{6-85}$$

in the absence of solvent or other chain transfer agents. However,

$$[I] = \frac{k_t R_p^2}{k_p^2 [M]^2 f k_d} \tag{6-86}$$

from Eq. (6-29) for polymerizations initiated by thermal decomposition of an initiator. For such systems Eq. (6-85) becomes

$$\frac{1}{\overline{DP}_n} = \frac{k_{tc} R_p}{k_p^2 [M]^2} + \frac{2 k_{tc} R_p}{k_p^2 [M]^2} + C_M + C_I \frac{k_t R_p^2}{k_p^2 f k_d [M]^3} \tag{6-87}$$

R_p can be varied by changing [M] and [I] and corresponding values of $1/\overline{DP}_n$ plotted against R_p. The equation above is a quadratic in R_p, but the initial portion of the plot at low R_p is linear with intercept C_M at zero R_p. The slope of this linear portion gives $(k_{tc} + 2k_{td})/k_p^2 [M]^2$ from which $k_p/k_{tc} + 2k_{td})^{1/2}$ can be derived at known [M] [6].

The technique described relies implicitly on the assumption that termination rates are independent of $\overline{DP}_n$. More recent work has shown that this is not strictly true (see Section 6.13) and neglect of this effect may account for much of the scatter in C_M values reported in the polymer literature.

6.8.3   Transfer to Solvent and Chain Transfer Agents

The chain transfer agent in some laboratory polymerizations is the solvent. Industrial free-radical polymerizations are not normally run in solution, for economic reasons, and the chain transfer agents in these reactions are ingredients that are added deliberately to limit the molecular weight of the polymer. Although R_p and $\overline{DP}_n$ are inextricably linked in ideal radical polymerizations, the use of transfer agents permits the independent adjustment of these two parameters as long as the desired $\overline{DP}_n$ is less than that for the corresponding transfer-free reaction (Eq. 6-64).

To define a useful new symbol, consider a solvent–free-radical polymerization. The number average degree of polymerization of the polymer made at any instant in the absence of a transfer agent is given by Eq. (6-77) as

$$\frac{1}{\overline{DP}_n} = \frac{k_{tc}R_p}{k_p^2[M]^2} + \frac{2k_{td}R_p}{k_p^2[M]^2} + C_M + C_I\frac{[I]}{[M]} = \frac{1}{(\overline{DP}_n)_0} \tag{6-77a}$$

where $(\overline{DP}_n)_0$ is the number average degree of polymerization produced in the absence of any optional reactants like solvent or chain transfer agent. It depends only on the nature and concentrations of monomer and initiator as well as the reaction temperature.

There are a number of methods for measuring C (or C_s). The most readily obvious procedure involves the insertion of Eq. (6-77a) into (6-77) to yield

$$\frac{1}{\overline{DP}_n} = \frac{1}{(\overline{DP}_n)_0} + C\frac{[Ta]}{[M]} \tag{6-77b}$$

Experimentally, the polymerization is carried out under conditions where the rate of transfer to initiator is negligible (e.g., by using an azo initiator) and where $R_p/[M]^2$ is kept constant by adjusting [I] during the course of the reaction. $\overline{DP}_n$ is measured in a series of reactions, each with a different [Ta]/[M] ratio and the data are plotted as shown in Fig. 6-1, where styrene polymerization is taken as an example. Each plot yields a straight line with slope C.

Table 6-2 lists some transfer constants for various compounds in the polymerization of styrene, methyl methacrylate, and vinyl acetate. All transfer reactions involve a transfer agent and a radical and the reaction rates depend on

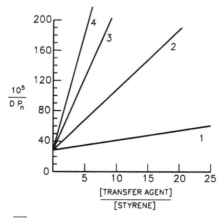

Fig. 6-1.   Plots of $(1/\overline{DP}_n)$ against [Ta]/[M] (Eq. 6-77b) for various chain transfer agents in the polymerization of styrene at 100°C [7]. Line 1, benzene; 2, toluene; 3, ethylbenzene; 4, isopropylbenzene. (The data points have been omittted for clarity.)

TABLE 6-2

Transfer Constants to Chain Transfer Agents^{a,b} ($C \times 10^4$ at 60°C)

Transfer Agent	Monomer		
	Styrene	Methyl methacrylate	Vinyl acetate
Benzene	0.03	0.04	1.5
Toluene	0.2	0.2	20
Isopropylbenzene	1	2	100
Isopropanol	3	1	45
Chloroform	0.5	1	150
Carbon tetrachloride	100	200	8,000
1-Butanethiol	200,000	-	480,000

^{a} These values are approximate averages of the data listed in the third edition of the "Polymer Handbook" [8]. There is a considerable spread in reported results, probably because of effects described in Section 6.13.

^{b} By convention, most free-radical kinetic data are measured at 60°C, but commercial polymerizations are performed at higher temperatures, where feasible, in order to complete the reaction faster.

the characteristics of both reactants. Propagation reactions in free-radical kinetic chains must be fast in order to exert a significant effect before termination events intervene. This means that they must have relatively low activation energies. They cannot be strongly endothermic, as a consequence, because the energy of activation for a reaction cannot be less than the difference in enthalpies of the reactants and products. The enthalpy change in a transfer reaction will depend on the nature of the bonds which are broken and formed and the relative stabilities of the radicals $M_n\cdot$ and $T\cdot$ in reaction (6-67). In general, a given transfer agent will be more reactive (C is greater) for a reactive radical like those in ethylene or vinyl chloride polymerizations than for a resonance-stabilized radical like that of styrene. Similarly, when a given monomer is being polymerized, aliphatic compounds that yield tertiary radicals will be more effective transfer agents than those that produce secondary radicals. Chain transfer activity is also enhanced by the possibility for resonance stabilization, and isopropyl benzene is a more active transfer agent than propane, for example. The slopes of the plots in Fig. 6-1 and the data in Table 6-2 conform to these general rules.

The most widely used chain transfer agents are compounds with one relatively weak bond like this thiols, disulfides, CCl_4, or CBr_4, but even hydrogen and propane can be used with the very reactive polyethylene radical. If C, as defined in Eq. (6-76), is relatively large, it can be expected that a chain transfer agent will be used up relatively early in a free-radical polymerization. For example, if $C = 1$, the transfer agent will be entirely consumed when not all of the monomer has been polymerized in a batch (all ingredients present in the reaction vessel before the process is started) reaction. In such cases it is advisable to meter the transfer agent into the vessel during the polymerization.

6.8.4   Chain Transfer to Polymer

Transfer to polymer yields a radical on the polymer chain. Polymerization of monomer from this site produces a polymer with a long branch:

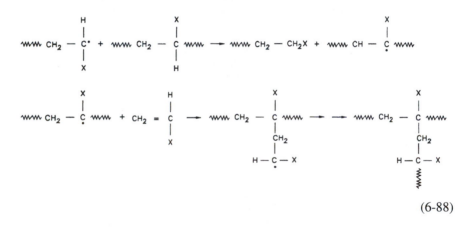

$$(6-88)$$

Transfer to polymer is important with very reactive radicals like those in the polymerizations of vinyl chloride, vinyl acetate, ethylene, and similar monomers in which significant resonance stabilization is absent. It is also most significant in high conversion reactions where the concentration of polymer in the system is relatively high.

Chain transfer to polymer is not included in Eq. (6-77). The occurrence of reaction (6-88) does not change the number of monomers which have been polymerized nor the number of polymer molecules over which they are distributed. It has no effect on $\overline{DP}_n$ for that reason. It will cause a change in the molecular weight distribution, however, and $\overline{DP}_w$ and higher averages are increased because the polymers which are already large are more likely to suffer transfer reactions and become yet bigger.

The existence of only a few long branches per molecule of radical-polymerized polyethylene produces a more compact structure at given overall molecular weight. Long branched macromolecules are less likely to become mutually entangled and exhibit lower melt elasticities and better processing properties for some applications than equivalent molecular weight polyethylenes without long branches, provided that the branches themselves are not long enough to entangle.

Reactive radicals like that of polyethylene can also undergo self-branching by a "backbiting" reaction in which a radical at the end of the polymer chain obstructs a hydrogen from a methylene unit in the same chain:

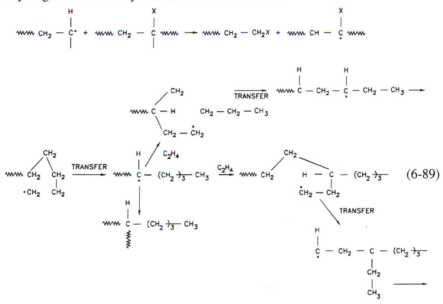

(6-89)

This is a very important process in the free-radical, high-pressure polymerization of this monomer. It produces short branches (ethyl, butyl, etc.) of a type and frequency that depend on the particular pressure and temperature in the reactor. Branched polyethylene from this process has lower crystallinity than its linear counterpart. As a consequence it tends to be less rigid and tougher and forms clearer films. Branches in free radical polyethylene vary more in length and are more clustered than those produced by copolymerization of ethylene and higher olefins by organometallic catalysts (Chapter 9). The conventional term for the latter materials is linear low density polyethylene (LLDPE), while the former type is called is called low density polyethyelene (LDPE). There are many varieties within each polymer type, but in general, LLDPE's are tougher than LDPE's. They can be extruded into thinner packaging films and normally require more power to extrude at the same output rates.

6.8.5   Allylic Transfer

Chain transfer to monomer was described in general terms in Section 6.8.2. Such reactions are particularly favored with allylic monomers which have the structure

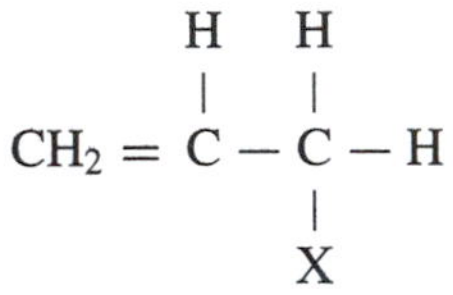

$$
\begin{array}{ccc}
 & H & H \\
 & | & | \\
CH_2 = & C - C - H \\
 & & | \\
 & & X
\end{array}
$$

6-4

with a C—H bond alpha to the double bond. The polymer radical is reactive, and the radical formed by transfer to the monomer is particularly unreactive. For allyl acetate, for example, transfer to monomer involves the reaction

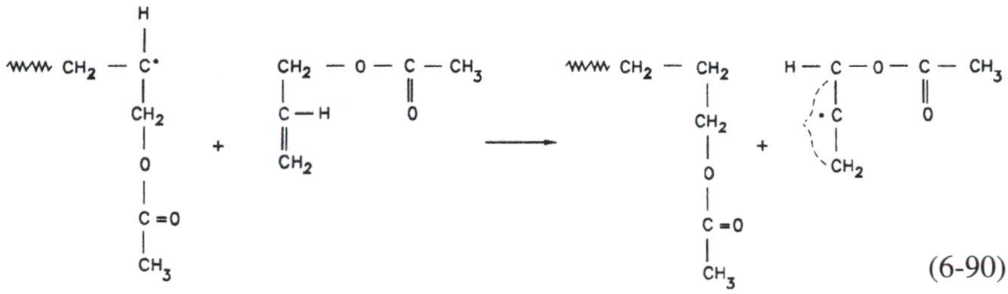

$$(6\text{-}90)$$

Allylic transfer is also variously named degradative chain transfer, autoinhibition, or allylic termination. The stable radical derived from the monomer by reactions like (6-90) are slow to reinitiate and prone to terminate. Low-molecular-weight products are therefore formed at slow rates and small concentrations of allyl monomers can inhibit or retard the polymerization of more reactive monomers.

The reluctance of olefins like propylene or isobutene to form high polymers in free radical reactions is a result of degradative chain transfer of allylic hydrogens.

Despite these problems useful polymers are made from allyl monomers. High initiator concentrations are needed because the kinetic chains terminate at low degrees of polymerization and multifunctional monomers are used to produce cross-linked structures even at low conversions.

Methyl methacrylate (Fig. 1-4) and methacrylonitrile (6-5) are allylic-type monomers that do yield high molecular weight polymers in free radical reactions. This is probably because the propagating radicals are conjugated with and stabilized to some extent by the ester and nitrile substituents. The macroradicals are less reactive than those in conventional allyl polymerizations and therefore have lower tendencies for transfer to monomer.

$$
\begin{array}{c}
CH_3 \\
| \\
CH_2\!=\! C \\
| \\
CN
\end{array}
$$

6-5

6.9   INHIBITION AND RETARDATION

Some substances retard or suppress free-radical polymerization by reacting with primary radicals or macroradicals to yield nonradical products or radicals that are too stable to add further monomer. Some inhibitors and retarders are chain transfer agents; others act by addition processes.

Inhibitors and retarders provide an alternative reaction path to propagating macroradicals:

$$
M_n^{\bullet} \left\{
\begin{array}{l}
\boxed{\text{MONOMER ADDITION}} \longrightarrow M_{n+1}^{\bullet} \qquad\qquad (6\text{-}91a) \\[1em]
\boxed{\text{CHAIN TRANSFER WITH TH}} \longrightarrow M_nH + T^{\bullet} \qquad (6\text{-}91b) \\[1em]
\boxed{\text{ADDITION TO Q}} \longrightarrow M_nQ^{\bullet} \qquad\qquad (6\text{-}91c)
\end{array}
\right.
$$

If the new radicals $T\cdot$ or $M_nQ\cdot$ do not react readily with more monomer, there will be a decrease in the concentration of reactive radicals and a concurrent reduction in the rate of polymerization. When the rate of reaction (6-91b) or (6-91c) is very much greater than that of reaction (6-91a) and the new radicals $T\cdot$ or $M_nQ\cdot$ do not add monomer then high-molecular-weight polymer will not be formed and R_p will be effectively zero. This is a case of inhibition. In retardation the polymerization is slowed but not entirely suppressed. This occurs if (1) the rate of either alternative process is close to that of the monomer addition reaction (6-91a) and the new radicals from steps (6-91b) or (6-91c) do not reinitiate, or (2) if the alternative processes are fast compared to ordinary monomer addition but the new radicals formed reinitiate slowly.

Retardation and inhibition differ only as to whether the production of high polymer is slowed or completely eliminated. When the retarder or inhibitor is a chain transfer agent it is distinguished from an effective transfer agent in that it

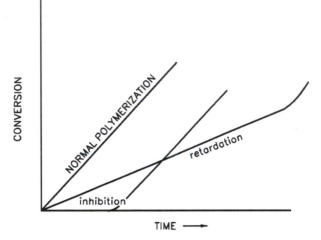

Fig. 6-2. Conversion time plots for normal, retarded and inhibited free radical polymerizations.

transfers an atom to the propagating radicals very readily to produce a new radical which reinitiates slowly or not at all.

Figure 6-2 compares the effects of inhibitors and retarders on the rate of free-radical polymerization. Polymerization at normal rates is resumed when and if the inhibitor or retarder has been consumed.

Inhibition can be achieved by three processes:

1.   Atom transfer to form a stable radical which does not reinitiate polymerization, as in the reaction of poly(vinyl acetate) radical and diphenylamine to yield a diphenyl nitrogen radical which will not add vinyl acetate but may terminate a macroradical.

2.   Stable radicals such as diphenyl picryl hydrazyl, which are reactive enough to terminate propagating radicals but not to initiate polymerization.

3.   Addition of primary or macroradicals to nonradical substances to yield free radicals that do not add monomer.

Oxygen is a prime example of such inhibitors. Polymerization can be initiated ultimately by the breakdown of hydroperoxides or peroxides that were formed by the oxidation of monomers or other species in the reaction mixture, but the initial effect of oxygen is to inhibit polymerization by forming unreactive peroxy radicals:

$$M\cdot + O_2 \rightarrow M—O—O\cdot \tag{6-92}$$

Oxygen is a general inhibitor for vinyl polymerizations and good practice requires the removal of air from the polymerization system before the reaction is started.

A monomer which forms a stable radical can be used to inhibit the polymerization of another monomer which yields a more reactive radical. Styrene inhibits the polymerization of vinyl acetate, for example.

Monomers are sold with added inhibitors (usually quite highly substituted phenols) to prevent their premature polymerization during storage and transportation. The inhibitors are either removed by washing with alkaline water before polymerization or else extra initiator is used to destroy the inhibitor. In that case the polymerization will exhibit an induction period as shown in Fig. 6-2.

Note that only a very small concentration of inhibitor is needed to suppress polymerization. Consider a reaction initiated by 10^{-3} M AIBN, for example [**6-3**, reaction (6-41)]. At 60°C, k_d is 1.2×10^{-5} sec^{-1} and f is close to 0.6. From Eq. (6-10) the rate of initiation will be 1.4×10^{-8} mol/liter/sec and the total number of radicals produced in 1 h in 1 liter of reaction mixture will be 5×10^{-5} mol. If every inhibitor accounts for one primary or monomer-ended radical, the inhibitor concentration need only be 5×10^{-5} M to suppress polymerization for an hour.

Retarders and inhibitors differ only in the frequency with which propagating radicals react with them rather than with monomer and possibly also in the ability of the radicals resulting from such reactions to reinitiate. It is to be expected, then, that a compound may not exert the same effect in the polymerization of different monomers. For example, aromatic nitro compounds that are inhibitors in vinyl acetate polymerizations are classified as retarders in polystyrene syntheses.

6.10   READILY OBSERVABLE FEATURES OF FREE-RADICAL POLYMERIZATIONS

The rate of polymerization $-d[\text{M}]/dt$ and the molecular weight of the polymer produced are easily measured experimental parameters. The rate of polymerization can be determined by a number of methods. A commonly used procedure involves quenching the polymerization with an inhibitor like hydroquinone or ferric chloride, precipitating the polymer with a nonsolvent, washing the polymer, and weighing it.

The polymerization of vinyl monomers is accompanied by considerable shrinkage (about 20% in the case of styrene), and the rate of the reaction can be followed from the change in volume of the reaction mixture if this relation is known for the particular monomer [9]. Gas-liquid chromatography is also a convenient technique for following the rate of polymerization because the polymer produced is not volatile and the amount of unused monomer can be measured with reference to solvent or other internal standard [10].

The $\overline{\text{M}}_n$ and $\overline{\text{DP}}_n$ of the polymer produced can be measured as outlined in Section 3.1.

Recall that corresponding instantaneous values of $\overline{DP}_n$ and R_p are linked by Eq. (6-64). If sufficient polymer can be collected for molecular weight measurements from a polymerization that goes to only 5 or 10% conversion, the average value of [M] can be inserted into this equation with the measured $\overline{DP}_n$ and R_p to estimate $k_p^2(k_{tc} + k_{td})$. If the ratio k_{td}/k_{tc} is known or assumed, then k_p^2/k_t is available.

Alternatively, when the initiator in these experiments is an azo compound like AIBN the initiator characteristics f and k_d are believed to be independent of the other constituents of the reaction mixture. (This cannot be assumed for peroxide or redox initiation.) Then, if f and k_d or their product is known from other measurements with the particular azo compound, $k_p/k_t^{1/2}$ can be calculated from polymerization rate measurements and Eq. (6-29) or (6-35). As well, $\overline{DP}_n$ alone provides an estimate of $k_p/k_t^{1/2}$ from Eq. (6-65) when fk_d and the mode of termination are already known.

Note that all these calculations except those that rely on Eq. (6-29) or (6-35) make the assumption that chain transfer is negligible. Estimations of $k_p/k_t^{1/2}$ that involve $\overline{DP}_n$ data can be rendered unreliable because of unsuspected chain transfer reactions.

The ratio $k_p/k_t^{1/2}$ has appeared again and again in the equations which were developed to this point in the text for free-radical polymerizations. It is not too difficult to measure this parameter from steady rate experiments, as shown in this section, and $k_p k_t^{1/2}$ is known for many systems of practical interest. Measurement of the separate values of k_p and k_t is more challenging, however, and methods pertaining to k_p and k_t are described in Section 6.12

6.11 RADICAL LIFETIMES AND CONCENTRATIONS

The average lifetime τ of the kinetic chain is given by the ratio of the steady-state radical concentration to the steady-state rate of radical disappearance:

$$\tau = \frac{[M\cdot]}{2(k_{tc} + k_{td})[M\cdot]^2} = \frac{1}{2(k_{tc} + k_{td})[M\cdot]} \tag{6-93}$$

Substituting for [M·] from Eq. (6-13),

$$\tau = k_p[M]/2(k_{tc} + k_{td})R_p \tag{6-94}$$

Also, with Eq. (6-29) for R_p and Eq. (6-20) for k_t,

$$\tau = 1/2(fk_d[I])^{1/2}(k_{tc} + k_{td})^{1/2} \tag{6-95}$$

for initiation by thermal decomposition of an initiator. The average radical lifetime can also be related to the number average molecular weight of the polymer that is being produced by substituting in Eq. (6-94) for R_p from Eq. (6-64):

$$\tau = \overline{DP}_n(k_{tc} + 2k_{td})/2k_p[M](k_{tc} + k_{td}) \qquad (6\text{-}96)$$

If termination is by combination alone, $k_{td} = 0$ and Eq. (6-96) simplifies to

$$\tau = \overline{DP}_n/(2k_p[M]) = \frac{v}{k_p[M]} \qquad (6\text{-}97)$$

with the kinetic chain length v defined as in Section 6.6. Note that the last two equations which relate τ and $\overline{DP}_n$ are valid only in the absence of significant chain transfer.

As an example, consider the polymerization of bulk styrene at 60°C initiated by 1×10^{-3} M azodiisobutyronitrile. The density of liquid styrene is 0.909 g cm^{-3} at the reaction temperature. What is the average radical lifetime and what is the steady-state radical concentration?

The concentration of styrene [M] is obtained from

$$[M] = \frac{0.909 \text{ g}}{cm^3}\left(\frac{1000 \text{ cm}^3}{\text{liter}}\right)\left(\frac{\text{mol}}{104 \text{ g}}\right) = 8.74 \text{ } M$$

since the molecular weight of styrene is 104. For styrene, $k_{tc} = 1.8 \times 10^7 \, M^{-1} \sec^{-1}$, while k_d for AIBN is $0.85 \times 10^{-5} \sec^{-1}$ and $f = 0.6$. From Eq. (6-95).

$$\tau = (1/2)[(0.6)(0.85 \times 10^{-5})(10^{-3})(1.8 \times 10^7)]^{-1/2} = 1.6 \text{ sec}$$

Also, from Eq. (6-27),

$$[M\cdot] = \left(\frac{(0.6)(0.85 \times 10^{-5})(10^{-3})}{1.8 \times 10^7}\right)^{1/2} = 1.7 \times 10^{-8} \text{ } M$$

Multiplying by Avogadro's constant, we see that there are 10^{16} macroradicals per liter of reaction mixture, with an average lifetime of 1.6 sec per radical. The rate of polymerization is

$$R_p = -d[M]/dt = (84)(8.74)(1.7 \times 10^{-8})$$

$$= 0.26 \times 10^{-4} \text{ mol monomer/liter/sec}$$

$$= 1.4 \times 10^{-3} \text{ g polymer/liter/sec}$$

In this calculation (from Eq. 6-13), k_p for polystyrene is taken to be 84 $M^{-1} \sec^{-1}$.

This is a slow polymerization, since the calculated rate will result in less than 6 g of polymer per hour from 1 liter (909 g) of monomer.

6.12   DETERMINATION OF k_p AND k_t

A number of nonsteady polymerization rate techniques can be used to measure k_p [11]. The most widely used method involves pulsed-laser-induced polymeriza-tion in the low monomer conversion regime. Briefly, a mixture of monomer and photoinitiator (Section 6.5.3) is illuminated by short laser pulses of about 10 ns (10^{-8} sec) duration. The radicals that are created by this burst of ligh propagate for about 1 sec before a second laser pulse produces another crop of radicals. Many of the initially formed radicals will be terminated by the short, mobile radicals created in the second illumination. Analysis of the number molecular weight distribution of the polymer produced permits the estimation of k_p from the relation

$$\overline{DP}_n = k_p[M]t_f \tag{6-98}$$

where t_f is the "dark time" between pulses [12,13]. This technique has been applied successfully at low monomer conversions of about 2–3%.

At higher conversions and radical concentrations, carefully calibrated electron paramagnetic resonance spectroscopy (EPR) can be used in some cases to measure the concentration of propagating radicals directly [14]. Application of Eq. (6-13) then yields a value for k_p.

If k_p is determined, k_t can be estimated from the readily observable relation between k_p^2/k_t outlined in Section 6.10. Techniques for measuring k_t directly are summarized in specialized reports [11]. Note, however, that termination rates in free-radical polymerizations are always diffusion controlled (see Section 6.13.1) and the apparent value of k_t will depend on the conditions under which it has been measured.

6.13   DEVIATIONS FROM IDEAL KINETICS

The kinetic scheme that has been developed in this chapter provides a very use-ful framework for the organization of experimental results and for the systematic modification of polymerization conditions when the polymer properties or re-action are not entirely satisfactory. Most practical free-radical polymerizations will deviate to a greater or lesser extent from the standard relations outlined, however, either because the reaction conditions are not entirely as postulated here or because some of the assumptions that underly the kinetic scheme are not valid. This section is a review of the principal causes and results of such deviations.

Equation (6-30) predicts that the rate of polymerization should be propor-tional to the monomer concentration and to the square root of the rate at which monomer-ended radicals are formed. When initiation is by thermal decomposition of an initiator I, which yields two radicals, the rate of polymerization is ideally proportional to $[I]^{1/2}$. In initiation by photolysis of an initiator, R_i depends directly on the intensity Φ of the incident light, as well as on $[I]$, and R_p is therefore ideally proportional to $(\Phi[I])^{1/2}$. When a photosensitizer is used the concentration of this species takes the place of $[I]$ in this relation. In a redox polymerization, like reaction (6-42), R_p is predicted to be proportional to $([ROOH])(Co^{2+}))^{1/2}$.

Deviations from the predicted dependence of R_p on $[M]$ and $[I]^{1/2}$ are not unusual. The initiation rate and the initiator efficiency f may depend on $[M]$ if primary radicals escape from their solvent cage (Section 6.5.5) by reaction with the nearest monomer molecules. At high initiation rates, some of the primary radicals from initiator decomposition may terminate kinetic chains. This "primary termination" causes the observed R_p to depend on $[M]$ to a power greater than one and reduces the dependence of R_p on $[I]$ to less than the power 0.5.

These examples illustrate the point that initiation reactions may vary from the standard kinetic scheme that has been described in this chapter. Propagation reactions are believed to occur generally as postulated, but termination reactions are very clearly not as has been assumed to this point. In particularly, the rate of termination may depend on the size of the polymer being produced and on the extent of conversion. Both effects are discussed below.

6.13.1 Diffusion Control of Termination

The assumption that k_t is independent of the sizes of the radicals involved in the termination reaction is not true. The standard kinetic scheme has been developed to this point by making use of this assumption, however, because the presentation is simpler and more readily comprehended and because the errors involved are not large in most commercial free-radical polymerizations.

Termination reactions occur between two relatively large radicals, and termination rates are limited by the rates at which the radical ends can encounter each other. As a result, k_t is a decreasing function of the dimensions of the reacting radical. The segmental diffusion coefficient and the termination rate constant increase as the polymer concentration increases from zero. This initial increase is more pronounced when the molecular weight of the polymer is high and/or when the polymerization is carried out in a medium which is a good solvent for the polymer. For similar reasons, k_t is inversely proportional to the viscosity of reaction medium. A model has been proposed that accounts for these variations in k_t in low-conversion radical polymerizations [15,16].

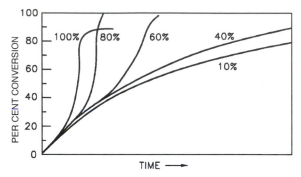

Fig. 6-3. Conversion-time plots for the polymerization of methyl methacrylate in benzene at 50°C. The labeled curves are for the indicated monomer concentrations [17].

6.13.2 Autoacceleration

Deviations resulting from the diffusion control of termination at low conversions of monomer to polymer are the relatively weak effects discussed in the preceding subsection. By contrast, changes in reaction rate resulting from hindered diffusion at high conversions are very important in most radical polymerizations. Figure 6-3 shows rate curves for the polymerization of methyl methacrylate in benzene at 50°C [17]. At monomer concentrations less than about 40 wt % in this case, the rate is approximately as anticipated from the standard kinetic scheme described in this chapter. R_p decreases gradually as the reaction proceeds and the concentrations of monomer and initiator are depleted.

An acceleration is observed at higher monomer concentrations, however. This is due to a decrease of k_t at higher conversions in these polymerizations. When the polymer concentration becomes high enough, the macroradicals will become entangled with segments of other polymer chains. As a result, the rate of diffusion of the polymeric radicals and the frequency of their mutual encounters will decrease. The rate of termination is reduced accordingly.

The rate of propagation is affected much less than the rate of termination. The propagation reaction involves the reaction of a large radical with a small monomer whose diffusion rate is not changed significantly, whereas the termination process involves two macroradicals whose ends have reduced mobility, because motion of their centers of mass has become restrained. The net result in this case is an increase in the effective $k_p/k_t^{1/2}$ ratio in Eq. (6-29) and an increase in the rate of polymerization.

Because vinyl polmerizations are exothermic, the increased polymer production associated with the autoacceleration effect can cause a temperature rise and faster initiator decomposition. Runaway reactions or explosions may result if the heat of reaction is not removed efficiently.

The polymer concentration at which autocatalytic effects are significant varies from system to system. When a polymer is insoluble in its own monomer, radicals can be occluded in the precipitated polymer-rich phase where they are prevented from terminating by their low mobilities. Bulk polymerization of vinyl chloride provides an example of this.

Autoacceleration is also known as the *gel effect* or as the *Tromsdorff effect* or *Norrish–Smith effect* after pioneering workers in this field.

Note in the bulk polymerization example in Fig. 6-3 that R_p increases rapidly at low conversions and then effectively ceases before all the monomer in the system has been consumed. This occurs when the polymerization mixture becomes glassy and even the propagation step becomes subject to diffusion control. Most monomers will solvate and soften their own polymers so that the system does not become rigid until a high proportion of the monomer has been converted.

6.14 MOLECULAR WEIGHT DISTRIBUTION

6.14.1 Low-Conversion Polymerization

It is convenient to focus separately on the cases in which termination is entirely by disproportionation or by combination. When both modes of termination occur, the size distribution of the polymer which is formed is described by a weighted average of the distribution functions for the individual modes.

We assume here that the concentrations of monomer and initiator remain sensibly constant during the polymerization, and that any dependence of termination rate constants on macroradical size and concentration or autoacceleration effects can be neglected. This means that the molecular weight distributions to be derived can be expected to apply to low-conversion polymers. Commercial macromolecules, whose polymerizations are often finished at high conversions, may have distributions that differ from those calculated here. Section 6.14.2 discusses the size distributions of such polymers.

A given monomer-ended radical may add monomer or undergo chain transfer or termination. The probability S that it will grow by monomer addition is

$$S = \frac{R_p}{R_p + R_{tr} + R_t} \tag{6-99}$$

Now

$$R_p = k_p[M][M\cdot] \tag{6-13a}$$

$$R_{tr} = C_I k_p[I][M\cdot] + C_M k_p[M\cdot][M] + C k_p[TH][M\cdot] \tag{6-100}$$

from Eqs. (6-76) and (6-77). Also,

$$R_t = 2(k_{tc} + k_{td}[M\cdot]^2 \tag{6-101}$$

from Eqs. (6-20) and (6-23). Insertion of the last three relations into Eq. (6-100) and simplification gives

$$\frac{1}{S} = 1 + C_I \frac{[I]}{[M]} + C_M + C \frac{[TH]}{[M]} + 2 \left[\frac{(k_{tc} + k_{td})}{k_p} \right] [M\cdot] \tag{6-102}$$

With $[M\cdot] = R_p/k_p[M]$ from Eq. (6-13),

$$\frac{1}{S} = 1 + C_I \frac{[I]}{[M]} + C_M + C \frac{[TH]}{[M]} + \frac{2(k_{tc} + k_{td})}{k_p^2[M]^2} R_p \tag{6-103}$$

Note that any chain transfer effects of solvent are included in the general term for transfer agent TH and that transfer to polymer is not included in this expression.

We first consider the polymerization where each kinetic chain yields one polymer molecule. This is the case for termination of the growth of macroradicals by disproportionation and/or chain transfer ($k_{tc} = 0$). The situation is completely analogous to that for linear, reversible step-growth polymerization described in Section 5.4.3. If we randomly select an initiator residue at the end of a macromolecule, the probability that the monomer residue which was captured by this primary radical has added another monomer is S and the probability that this end is attached to a macromolecule which contains *at least i* monomers is S^{i-1}. The probability that this macromolecule contains *exactly i* monomers equals the product of S^{i-1} and the probability of a termination or transfer step. The latter probability must be equal to $(1 - S)$ since it is certain that the last monomer under consideration will undergo one of these three reactions. That is, the probability that a randomly selected molecule contains i monomer units is $S^{i-1}(1 - S)$. Since such probabilities are equal to the corresponding mole fraction of this size molecule, x_i, we have the expression

$$x_i = (1 - S)S^{i-1} \tag{6-104}$$

for the number distribution function. The weight distribution function w_i is also given by direct analogy to that for linear, equilibrium step-growth polymerization as

$$w_i = i(1 - S)^2 S^{i-1} \tag{6-105}$$

The distribution described is a random one, with average degrees of polymerization (Section 5.4.3):

$$\overline{DP}_n = 1/(1 - S) \tag{6-106}$$

$$\overline{DP}_w = (1 + S)/(1 - S) \tag{6-107}$$

$$\overline{DP}_w/\overline{DP}_n = 1 + S \tag{6-108}$$

[As a check, note that in the absence of chain transfer, $\overline{DP}_n$ from Eqs. (6-106) and (6-103) equals the value from Eq. (6-64) with $k_{tc} = 0$.]

For most addition polymerizations, $R_p \gg R_t + R_{tr}$ (or high-molecular-weight polymer would not be formed). In that case $S \simeq 1$ and X_w/X_n will be $\simeq 2$ from Eq. (6-108).

There is an important difference between the distributions calculated for equilibrium, bifunctional step-growth polymerization in Chapter 5 and for the free-radical polymerizations with termination by disproportionation or chain transfer that are being considered here. The distribution functions in the step-growth case apply to the whole reaction mixture; in the free-radical polymerization this distribution describes only the polymer which has been formed. There is obviously a strong parallel between the probability S of this section and the extent of reaction p used in the step-growth calculations in Chapter 5. Many authors use the same symbol for both parameters. Different notations are used here, however, for clarity.

When termination occures by combination as well as disproportination, the initial instantaneous distribution of molecular weights is give by [18, 19]

$$w_i = F_i(1 - S)^2 S^{i-1} + (1 - A)i(i - 1)(1 - S)^3 S^{i-2}/2 \qquad (6\text{-}109)$$

where F is the fraction of product formed by chain disproportionation and/or transfer and

$$A = \frac{C_I[I] + C_M[M] + C[TH] + 2k_{td}R_p/k_{p^2}[M]}{C_I[I] + C_M[M] + C[TH] + 2(k_{tc} + k_{td})R_p/k_{p^2}[M]^2} \qquad (6\text{-}110)$$

Equation (6-109) reduces to Eq. (6-105) when $k_{tc} = 0$ ($F = 1$) and to Eq. (6-111) when $k_{td} = 0$ and transfer is negligible ($A = 0$):

$$w_i = \frac{1}{2}i(i - 1)(1 - S)^3 S^{i-2} \qquad (6\text{-}111)$$

In the latter case

$$\overline{DP}_n = 2/(1 - S) \qquad (6\text{-}112)$$

$$\overline{DP}_w = (2 + S)/(1 - S) \qquad (6\text{-}113)$$

and

$$\overline{DP}_w/\overline{DP}_n = \frac{1}{2}(2 + S) \qquad (6\text{-}114)$$

The ratio in Eq. (6-114) has a limiting value of 1.5 at high polymer molecular weights when S approaches 1. This is narrower than the instantaneous distribution produced in the absence of termination by coupling.

6.14.2   High-Conversion Polymerization

Molecular weight distributions in high-conversion polymerizations are not nearly as predictable as those in low-conversion reactions, and they will vary with the particular monomer and polymerization conditions.

A number of factors combine to make high conversion molecular weight distributions broader than those calculated in Section 6.14.1 Since autoacceleration results from a reduced termination rate, it is always accompanied by an increase in the average molecular weight and the breadth of the distribution of molecular weights. If chain transfer to polymer can occur (Section 6.8.4), this will be most significant at higher conversions when the polymer concentration is high. This results in a further skewing of the molecular weight distribution to higher molecular weights, because the larger polymer molecules are the most likely to suffer transfer reactions and then grow even larger.

Molecular weight distributions in commercial polymers are characterized by $\overline{M}_w/\overline{M}_n$ ratios of about 3 for substances like polystyrene in which transfer to polymer does not appear to be important. Where long branches can be formed by chain transfer to polymer, the molecular weight distribution will be even broader and $\overline{M}_w/\overline{M}_n$ ratios of 50 and more are observed in some polyethylenes made by free-radical syntheses.

Note that the molecular weight distributions of high-conversion polymers made under conditions where the growth of macromolecules is limited primarily by chain transfer will be random, as described in Section 6.14.1 for low-conversion cases. Then $\overline{M}_w/\overline{M}_n$ will be 2. An exception to this rule occurs when the chain transfer reactions which determine the polymer molecular weight are to monomer and can result in branching [as in reactions (6-79) or (6-84)]. The molecular weight distributions of the branched polymers that are produced will be broader than the random one, and bimodal distributions may also be observed.

Poly(vinyl chloride) made by suspension polymerization (Chapter 8) is a polymer in which molecular weight control is effectively by chain transfer—to monomer in this case. The $\overline{M}_w/\overline{M}_n$ ratio is slightly higher than the expected value of 2 because the polymerization is performed at rising temperatures, rather than isothermally, as assumed in the ideal kinetics discussed above, and possibly because of autoacceleration effects.

When chain transfer agents are used to control polymer molecular weight the molecular weight distribution will tend to narrow toward the random case from the characteristics it would have in the absence of transfer agents.

6.15   FREE-RADICAL TECHNIQUES FOR POLYMERS WITH NARROWER MOLECULAR WEIGHT DISTRIBUTIONS

Free-radical polymerization is the most economical process for use with vinyl monomers because the reaction mixture does not require the high-purity reactants and rigorous exclusion of moisture, air, and other impurities that is needed for successful operation of the alternative ionic or organometallic catalyses described in Chapter 9. The molecular weight distributions that are characteristic of free-radical reactions are, however, not optimum for some polymer applications, such as toners for reprographic processes [20]. Process variations have been developed that yield narrow molecular weight polymers from free-radical polymerizations. These differ from conventional free-radical polymerizations in that the growing macroradicals can be considered (somewhat inexactly) to be *living*. True "living" polymerizations are those without transfer or termination reactions. [They are discussed in Chapter 9, in connection with the ionic initiation systems for which they were first developed.] This implies that the end of the macrospecies is still active for the addition of more monomer and affords a means for the synthesis of block copolymers (Section 1.5.4) by addition of another monomer after the consumption of the preceding monomer.

Polymerization in the presence of stable nitroxide free radicals under specific conditions yields polymers with polydispersities ($\overline{M}_w/\overline{M}_n$ ratios) less than the theoretical values given in Section 6.14 for low-conversion polymerizations [20]. The polymerization kinetics differ from those summarized above because the nitroxide radicals form stable adducts with the macroradicals. "Pseudoliving" polymerization schemes may vary. One process consists first of forming an initiator-nitroxide adduct, which can be isolated, purified, and added to monomer to initiate polymerization by thermal homolysis of the adduct. A useful advantage of the method is the ability to produce AB block copolymers of controlled structure by adding preformed stable B macroradicals to designed, purified A-type macroradicals. An alternative process with the same objectives is "degenerative chain transfer" [21] in which conventional free-radical termination processes (Section 6.3.3) are swamped by chain transfer to an appropriate transfer agent. The growing macroradicals are capped by halogen atoms which are supplied by the transfer agent and removed by primary initiator radicals. The method differs from telomerization, reactions (6-70) and (6-71), in the choice of lower initiator levels and less reactive transfer agents. The techniques described in this paragraph use more expensive reactants and the reactions are slower than in conventional free-radical polymerizations. They are useful primarily for syntheses of polymers with structures that cannot be produced by more established processes.

6.16   EFFECTS OF TEMPERATURE

6.16.1   Rate and Degree of Polymerization

R_p (Eq. 6-29) and $\overline{DP}_n$ (Eq. 6-63) each depend on a combination of three rate constants: k_d, k_p, and k_t. [Actually k_t may itself be the sum of k_{tc} and k_{td}, Eq. (6-20), but experimental data on the temperature dependencies of each of these termination rate constants are scanty.] The influence of temperature on an individual rate constant can be expressed by an Arrhenius-type expression:

$$k = Ae - E/RT \qquad (6\text{-}115)$$

where A is a preexponential factor, E the Arrhenius activation energy, and T the absolute temperature. A plot of $\ln k$ versus $1/T$ yields A and E from the intercept and slope, respectively. The three separate Arrhenius expressions can be combined in a straightforward manner to obtain the temperatue dependence of the $k_p(k_d/k_t)^{1/2}$ ratio in the expression for R_p in Eq. (6-29):

$$\ln[k_p(k_d/k_t)^{1/2}] = \ln\left[A_p\left(\frac{A_d}{A_t}\right)^{1/2}\right] - \left[\frac{E_p + 1/2E_d - 1/2E_t}{RT}\right] \qquad (6\text{-}116)$$

The overall activation energy for the rate of polymerization E_{R_p} is

$$E_{R_p} = E_p + 1/2E_d - 1/2E_t \qquad (6\text{-}117)$$

and can be measured by plotting $\ln R_p$ against $1/T$. The units of the activation energies are

$$E_{R_p} = \text{kJ/mol of polymerizing monomer}$$

$$E_t = \text{kJ/mol of propagating radicals}$$

$$E_d = \text{kJ/mol of initiator.}$$

E_d is of the order of 126–165 kJ/mol for thermal decomposition of chemical initiators, E_p appears to be in the 20–40 kJ/mol range, while E_t is least at 8–20 kJ/mol. Thus the overall E_{R_p} is approximately 84 kJ/mol from Eq. (6-116). Each 10°C increase in reaction temperature will result in a two- to threefold increase in the rate of polymerization.

Note that the initiation step dominates the overall temperature dependence of the rate of polymerization. When the method of initiation varies, E_{R_p} will also change. For redox initiation, for example, E_d is of the order of 40–60 kJ/mol and E_{R_p} for redox polymerizations is about 40 kJ/mol. For photochemical or radiation-induced polymerizations, E_d is practically zero and the rate of polymerization in such cases does not change much with the reaction temperature.

For the sake of clarity, we consider the effects of temperature on $\overline{DP}_n$ in the simplest case, which implies an absence of transfer reactions and termination by combination alone. When the polymerization is started by thermal decomposition of an initiator, insertion of appropriate Arrhenius expressions into Eq. (6-65) produces

$$\ln(\overline{DP}_n) = \ln[A_p/A_dA_t)^{1/2}] + \ln[[M]/(f[I])^{1/2}] - E_{\overline{DP}_n}/RT \qquad (6\text{-}118)$$

where

$$E_{\overline{DP}_n} = E_p - 1/2E_d - 1/2E_t \qquad (6\text{-}119)$$

is the activation energy for the degree of polymerization of the polymer formed. With the activation energies for propagation, initiator decomposition, and termination quoted above $E_{\overline{DP}_n}$, is of the order of -60 kJ/mol.

The degree of polymerization decreases with increasing reaction temperature with every mode of initiation except photochemical, where E_d is close to zero.

In general, then, when the temperature of a free-radical polymerization is increased, the rate of polymerization is strongly enhanced and the molecular weight of the polymer is reduced. These effects are a consequence primarily of the strong temperature dependence of the rate at which chemical initiators decompose.

There is little quantitative information about the effects of temperature on chain transfer reactions. Activation energies for transfer reactions of polystyryl radical with its own monomer and with transfer agents like isopropylbenzene are in the range 40–60 kJ/mol at 60°C. The activation energy for the transfer constant C (Eq. 6-76) of polyethylene radical and chain transfer agents like propane or isobutane are about 16 kJ/mol at 1360 atm pressure and 130–200°C [22]. In general, transfer agents with low transfer constants have higher activation energies for chain transfer.

As a general rule, higher reaction temperatures results in lower polymer molecular weights because of higher initiation rates and enhanced rates of chain transfer.

Chain transfer to polymer increases with reaction temperatures. The backbiting reaction (6-89) results in the production of polyethylene with more short branches at higher polymerization temperatures. This reaction changes the polymer constitution but not its molecular weight.

6.16.2 Polymerization–Depolymerization Equilibrium

Most polymerizations are characterized by negative enthalpy (ΔH_p) changes since the reactions are exothermic and by negative entropy (ΔS_p) values because the total disorder of the monomer–polymer system is decreasing. Since the free energy of polymerization is

$$\Delta G_p = \Delta H_p - T \Delta S_p \qquad (6\text{-}120)$$

ΔG_p will become less negative as the positive contribution of the $T \Delta S_p$ term increases at higher temperatures. If side reactions due to chemical degradation do not intervene, a temperature T_c may be reached at which $\Delta G_p = 0$. It will not be possible to produce high-molecular-weight polymer at temperatures $> T_c$, just as liquids do not aggregate into crystals at temperatures above their melting point.

In a dynamic equilibrium situation, forward and reverse reactions proceed at equal rates. Thus reaction (6-12) should be written more generally as

$$M_i\cdot \;+\; M \;\; \underset{k_{dp}}{\overset{k_p}{\rightleftarrows}} \;\; M_{i+1} \qquad (6\text{-}121)$$

where k_{dp} is the reaction rate constant for depolymerization. Equation (6-13) is then replaced by

$$R_p = -\frac{d[M]}{dt} = k_p[M][M\cdot] - k_{dp}[M\cdot] \qquad (6\text{-}122)$$

Since the net rate of polymer production at $T_c = -d[M]/dt = 0$, then under these conditions

$$k_p/k_{dp} = K = 1/[M] \qquad (6\text{-}123)$$

where K is the equilibrium constant.

Analysis of polymerization–depolymerization equilibria makes use of the reaction isotherm

$$\Delta G_p = \Delta G_p^0 + RT \ln K \qquad (6\text{-}124)$$

where ΔG^0 is the Gibbs free-energy change for the polymerization with the monomer and polymer in appropriate standard states [23]. Further, since

$$\Delta G_p^0 = \Delta H_p^0 - T \Delta S_p^0 \qquad (6\text{-}125)$$

and since $\Delta G = 0$, at $T = T_c$ then

$$T_c = \frac{\Delta H_p^0}{\Delta S_p^0 + R \ln[M]_e} \qquad (6\text{-}126)$$

where $[M]_e$ is the equilibrium monomer concentration.

It is obvious from the last equation that T_c will depend on the monomer concentration in the system. Usually $[M]_e$ is taken as unit concentration and T_c, the *ceiling temperature*, is then that temperature above which it is not possible to form high polymer from unit or lower concentration. In other words, if polymerization were to be started at T_c it would proceed until [M] fell to the value of $[M]_e$. Conversely, if high polymer that is made at some lower temperature is warmed to T_c, depolymerization will ensue until concentration $[M]_e$ of monomer is established.

Note in this connection that these last predictions refer to thermodynamic equilibria and give no information as to how quickly the equilibrium monomer concentrations will be attained. Macromolecules may in fact be quite useful above their ceiling temperatures if depolymerization processes are kinetically hindered. T_c for poly(formaldehyde) is 126°C, for example, but the polymer can be made stable enough for melt processing at temperatures above 200°C. This is accomplished by esterifying or etherifying the thermolabile hydroxyl ends of the macromolecule, copolymerizing with small concentrations of ethylene oxide, and using basic additives as stabilizers. These expedients all retard the initiation or propagation steps of chain reactions that could cause the polymer to "unzip" to monomer.

The equations developed for T_c contain no reference to the mode of polymerization. Although they are presented here in a chapter devoted to radical polymerization, they are characteristic of the polymer and not of its method of synthesis. The same T_c applies to all polystyrenes of given molecular weight and tacticity, for example, regardless of whether they were polymerized by anionic, cationic, or free-radical initiation.

It is evident that T_c will be lower if ΔH_p^0 is small. It is found, in fact, that ΔS_p^0 varies little between monomers so that the ceiling temperature is dominated by the magnitude of ΔH_p^0.

Vinyl monomers with 1,1-disubstitution generally have lower standard enthalpies of polymerization and lower ceiling temperatures than the corresponding singly substituted analogs. Thus the ceiling temperatures of polystyrene and poly(alpha-methylstyrene) are 310 and 61°C, respectively, for high-molecular-weight polymer in equilibrium with pure liquid monomer. It is interesting that poly(alpha-methylstyrene) unzips to monomer if degradation is initiated by free radicals at temperatures above its ceiling temperature. Poly(methyl methacrylate) with $T_c \simeq 165$°C exhibits the same behavior, and depolymerization is a valuable method for recovering scrap quantities of this polymer as monomer. Polystyrene and most other polymers that have fairly high ceiling temperatures and contain hydrogen atoms bonded to tertiary carbon atoms do not produce major quantities of monomer when they are thermally degraded. This is because other modes of molecular scission occur at temperatures lower than T_c.

The concepts developed in this section can be used to calculate how much monomer will be in equilibrium with high-molecular-weight polymer at any temperature. This is useful information because many monomers are toxic or have offensive odors and it is often necessary to limit their concentrations in their polymers. Equation (6-126) is valid at any temperature since $[M]_e$ will vary along with

T. The equilibrium monomer concentration is given by

$$\ln[M]_e = \frac{\Delta H_p^0}{RT} - \frac{\Delta S_p^0}{R} \qquad (6\text{-}126a)$$

For vinyl chloride, $\Delta H_p = -96$ kJ/mol and ΔS_p can be taken to be $\simeq -100$ J/deg mol. From Eq. (6-126a), only a negligible concentration of monomer will be in equilibrium with high-molecular-weight polymer at 25°C. If unreacted monomer can be purged from the polymer, no significant concentration will develop thereafter at room temperature because of the equilibrium of reaction (6-121).

In contrast ΔH_p for methyl methacrylate is only -56 kJ/mol, with a ΔS_p of -117 J/deg mol. The concentration of monomer in equilibrium with high polymer at 25°C is 2×10^{-4} M and $[M]_e$ at 100°C is calculated to be approximately 0.05 M [from Eq. (6-126a) assuming no significant temperature dependence of ΔH_p or ΔS_p]. If a free-radical polymerization of methyl methacrylate were being carried out at 100°C it would not be possible to convert more than 95% $(1 - [M]_e)$ of the monomer to polymer. For reasons given above, however, one can synthesize the polymer at a lower temperature and use the stabilized product for a reasonable period at 100°C without generating significant concentrations of monomer.

6.17   FREE-RADICAL POLYMERIZATION PROCESSES

The kinetic schemes described in this chapter apply to free-radical polymerizations in bulk monomer, solution, or in suspension. Suspension polymerizations ([Section 10.4.2.(iii)]) involve the reactions of monomers which are dispersed in droplets in water. These monomer droplets contain the initiator, and polymerization is a water-cooled bulk reaction in effect. Emulsion systems also contain water, monomer and initiator, but the kinetics of emulsion polymerizations are different from those of the processes listed above. Chapter 8 describes emulsion polymerizations.

PROBLEMS

6-1. Rate constants for termination k_t may be of the order of 10^8 liter/mol sec in free-radical polymerizations. Consider the polymerization of styrene initiated by di-t-butyl peroxide at 60°C. For a solution of 0.01 M peroxide and 1.0 M styrene in benzene, the initial rate of polymerization is 1.5×10^{-7} mol/liter sec and $\overline{M}_n$ of the polymer produced is 138,000.
(a)   From the above information estimate k_p for styrene at 60°C.

(b)   What is the average lifetime of a macroradical during initial stages of polymerization in this system?

6-2.   For a particular application, the molecular weight of the polymer made in Problem 6-1 is too high. What concentration of t-butyl mercaptan should be used to lower the number molecular weight of the polymer to 85,000? For this, transfer agent C is 3.7 in the polymerization of styrene.

6-3.   When bulk styrene is heated to 120°C, polymerization occurs because of thermal initiation in the absence of an added initiator. It is observed that polystyrene with $\overline{M}_n = 200,000$ is produced under these conditions at a rate of 0.011 g polymer/liter/min. Using this information calculate the total initial rate of polymerization expected if an initiator with concentration 0.1 M, $k_d = 10^{-4}$ sec^{-1}, and $f = 0.8$ is added to this system at 120°C. (Molar mass of styrene = 104; $k_t = k_{tc}$; $k_{td} = 0$.)

6-4.   Free-radical polymerization can be initiated by a redox system involved Ce^{4+} and an alcohol:

$$RCH_2OH + Ce^{4+} \overset{k}{\to} Ce^{3+} + H^+ + R - \dot{C}HOH$$

The propagation reaction can be summarized as usual as

$$M_n \cdot + M \overset{k_p}{\to} M_{n+1}$$

where $M_n \cdot$ is a macroradical with degree of polymerization n. The major termination reaction involves the Ce^{+4} component of the redox system:

$$M_n \cdot + Ce^{4+} \overset{k_{t'}}{\to} Ce^{3+} + H^+ + \text{dead polymer}$$

Using the steady-state assumption, derive a useful expression for the rate of polymerization. Take the initiator efficiency $= f$ and assume that the propagation reaction is the only one which uses significant quantities of monomer.

6-5.   For acrylamide, $k_p^2/k_t = 22$ liter/mol sec at 25°C and termination is by coupling alone. At this temperature the half-life of isobutyryl peroxide is 9.0 h and its efficiency in methanol can be taken to be equal to 0.3. A solution of 100 g/liter acrylamide in methanol is polymerized with 10^{-1} M isobutyryl peroxide.
(a)   What is the initial steady-state rate of polymerization?
(b)   How much polymer has been made in the first 10 min of reaction in 1 liter of solution?

6-6.   One hundred liters of methyl methacrylate is reacted with 10.2 mol of an initiator at 60°C.
(a)   What is the kinetic chain length in this polymerization?
(b)   How much polymer has been made in the first 5 h of reaction?

$$k_p = 5.5 \text{ liter/mol sec} \qquad k_t = 25.5 \times 10^6 \text{ liter/mol sec}$$

$$\text{Density of monomer} = 0.94 \text{ g cm}^{-3}$$

$$t^{1/2} \text{ for this initiator} = 50 \text{ h} \qquad f = 0.3$$

6-7. Redox initiation is often used in polymerizations in aqueous systems. Thus H_2O_2 and Fe^{2+} ion can be used to initiate the polymerization of acrylamide in water. Derive an expression for the steady-state rate of polymerization in this case:

$$H_2O_2 + Fe^{2+} \xrightarrow{k} HO\cdot + OH^- + Fe^{3+}$$

6-8. (a)   In a free-radical polymerization in solution the initial monomer concentration is increased by a factor of 10:

(i)   How is the rate of polymerization affected quantitatively?

(ii)   What change takes place in the number average degree of polymerization?

(b)   If the monomer concentration was not changed but the initiator concentration was halved instead:

(i)   What is the change in the rate of polymerization?

(ii)   What is the change in the number of average degree of polymerization?

6-9. Consider the chain transfer reactions that occur in the radical-initiated homopolymerizations of isopropenyl acetate (**6-9-1**) and methacrylonitrile (**6-9-2**).

Which monomer yields high-molecular-weight polymers in conventional free radical polymerizations? Explain the difference.

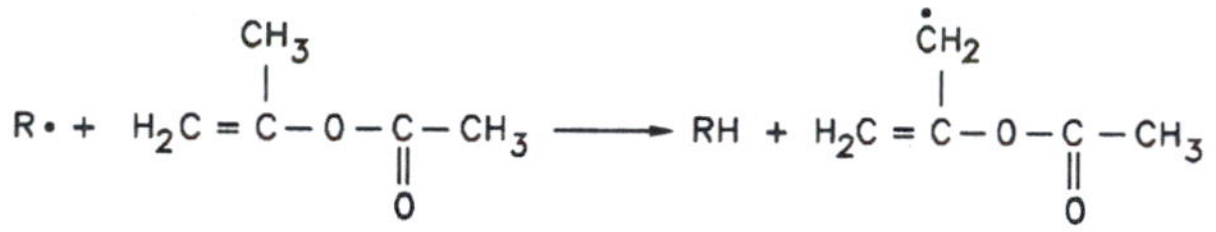

6-9-1

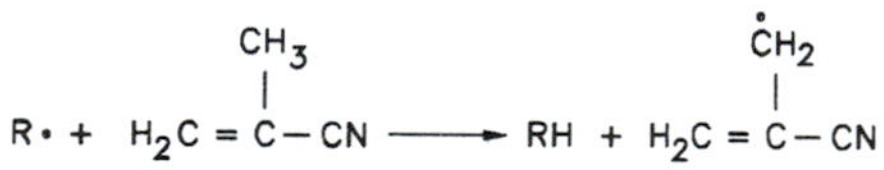

6-9-2

6-10. Vinyl acetate was polymerized in a free-radical reaction.   The initial monomer concentration was 1 mol/liter and its concentration after 1 h was 0.85 mol/liter. Chloroform was present as a chain transfer agent, with concentrations 0.01 mol/liter at time zero and 0.007 mol/liter after 1 h. What is the chain transfer constant C in this case?

6-11. An engineer is studying the azodiisobutyronile-initiated polymerization of butyl acrylate in solution. She wants to change the monomer and initiator concentrations so as to double the initial steady-state rate of polymerization without changing the number average degree of polymerization of the polymer or the reaction temperature. How should she proceed?

REFERENCES

[1] N. C. Billingham and A. D. Jenkins, *in* "Polymer Science" (A. D. Jenkins, ed.), Vol. 1. American Elsevier, New York, 1972.

[2] This is strictly true only for nonbranched chain reactions like polymerizations. In branching chain reactions, the number of active sites may increase during a propagation step. An example is $H \cdot + O_2 \rightarrow HO \cdot + O \cdot$ in hydrogen–oxygen flames. Such reactions could not be employed in a controllable polymerization and need not be considered here.

[3] C. S. Sheppard, *Encycl. Polym. Sci. Technol.* **11**, 1 (1987).

[4] A. V. Tobolsky, C. E. Rogers, and R. D. Brickman, *J. Am. Chem. Soc.*, **82**, 1277 (1960).

[5] S. C. Ng and K. K. Chee, *J. Polym. Sci., Polym. Chem. Ed.* **20**, 409 (1982).

[6] B. Baysal and A. V. Tobolsky, *J. Polym. Sci.* **8**, 529 (1952).

[7] R. A. Gregg and F. R. Mayo, *Discuss. Faraday Soc.* **2**, 328 (1947).

[8] J. Brandrup and E. Immergut, Eds., "Polymer Handbook," 3rd ed. Wiley (Interscience), New York, 1989.

[9] N. Bauer and S. Z. Lewin, *in* "Techniques of Chemistry" (A Weissberger and B. W. Rossiter, eds.), 3rd ed., Vol. 1, Part 4, Wiley (Interscience), New York, 1971.

[10] A. Rudin and S. S. M. Chiang, *J. Polym. Sci, Chem, Ed.* **12**, 2235 (1974).

[11] M. Buback, R. G. Gilbert, G. T. Russell, D. J. T. Hill, G. Moad, K. F. O'Driscoll, J. Shen, and M. A. Winnik, *J. Polym. Sci., Pt. A: Polym. Chem.* **30**, 851 (1992).

[12] O. F. Olaj, I. Bitai, and F. Hinkelmann, *Makromol. Chem.* **14**, 1689 (1987).

[13] K. F. O'Driscoll and M. E. Kuindersma, *Macromol. Theory Simul.* **3**, 469 (1994).

[14] T. G. Carswell, D. J. T. Hill, D. S. Hunter, P. J. Pomery, J. H. O'Donnell, and C. L. Winzor, *Eur. Polym. J.* **26**, 541 (1990).

[15] H. K. Mahabadi and K. F. O'Driscoll, *Macromolecules* **10**, 55 (1977).

[16] H. K. Mahabadi and A. Rudin, *J. Polym. Sci. Chem. Ed.* **17**, 1801 (1979).

[17] G. V. Schulz and G. Haborth, *Makromol. Chem.* **1**, 106 (1948).

[18] W. B. Smith, J. A. May, and C. W. Kim, *J. Polym. Sci. Part A-2* **4**, 365 (1966).

[19] P. J. Flory, "Principles of Polymer Chemistry," pp. 334–336. Cornell University Press, Ithaca, NY, 1953.

[20] M. K. Georges, R. P. Veregin, P. M. Kazmaier, and G. K. Hamer, *Trends Polym. Sci.* **2**, 66 (1994).

[21] D. Greszta, D. Madare, and K. Matyjaszewski, *Macromolecules* **27**, 638 (1994).

[22] P. Ehrlich and G. A. Mortimer, *Adv. Polym. Sci.* **7**, 386 (1970).

[23] F. S. Dainton and K. J. Ivin, *Quart. Rev. (London)* **12**, 61 (1958).

Chapter 7

Copolymerization

*It is a good morning exercise for a research scientist to discard a pet
hypothesis every day before breakfast. It keeps him young.*
—Konrad Lorenz, *The So-Called Evil, Chapter 2, (1966).*

7.1   CHAIN-GROWTH COPOLYMERIZATION

Copolymerization of two or more monomers during chain-growth polymerization
is an effective way of altering the balance of properties of commercial polymers.
Free-radical copolymerization of 20–35% of the relatively polar monomer acrylonitrile with the hydrocarbon styrene produces a transparent copolymer with better
oil and grease resistance, higher softening point, and better impact resistance than
polystyrene. The copolymer can be used for applications in which these properties of the styrene homopolymer are marginally inferior. Similarly, although
polyisobutene is elastomeric, the macromolecule is resistant to the cross-linking
reactions involved in sulfur vulcanization. Cationic copolymerization of isobutene
with 1–3 mol% isoprene at very low temperatures yields a polymer with sufficient
unsaturation to permit vulcanization by modified sulfur systems.

Reactivity of a monomer in chain-growth copolymerization cannot be predicted
from its behavior in homopolymerization. Thus, vinyl acetate polymerizes about
twenty times as fast as styrene in a free radical reaction, but the product is almost
pure polystyrene if an attempt is made to copolymerize the two monomers under the
same conditions. Similarly, addition of a few percent of styrene to a polymerizing
vinyl acetate mixture will stop the reaction of the latter monomer. By contrast,
maleic anhydride will normally not homopolymerize in a free-radical system under
conditions where it forms one-to-one copolymers with styrene.

These examples do not mean that copolymerization reactions defy understanding. The simple copolymer model described here accounts for the behavior of

many important systems and the entire process is amenable to statistical calculations which provide a great deal of useful information from few data. Thus, the composition of a copolymer of three or more monomers can be estimated reliably from a knowledge of the corresponding binary copolymerization reactions, and it is possible to calculate the distribution of sequences of each monomer in the macromolecule and the drift of copolymer composition with the extent of conversion of monomers to polymer.

7.2 SIMPLE COPOLYMER EQUATION

To predict the course of a copolymerization we need to be able to express the composition of a copolymer in terms of the concentrations of the monomers in the reaction mixture and some ready measure of the relative reactivities of these monomers. The utility of such a model can be tested by comparing experimental and estimated compositions of copolymers formed from given monomer concentrations. As a general rule in science, the preferred model is the simplest one which fits the facts. For chain-growth copolymerizations, this turns out to be the simple copolymer model, which was the earliest useful theory in this connection [1,2]. All other relations which have been proposed include more parameters than the simple copolymer model. We focus here on the simple copolymer theory because the basic concepts of copolymerization are most easily understood in this framework and because it is consistent with most copolymer composition and sequence distribution data.

In the copolymerization chain-growth reaction, we shall concentrate only on the propagation step in which a monomer adds to an active site at the end of a macromolecular species and the active site is transferred to the new terminal unit created by this addition.

$$M^* + M \rightarrow MM^* \tag{7-1}$$

Here M denotes a monomer and the asterisk means an active site which could be a radical, ion (with an appropriate counterion), or a carbon–metal bond. The reactivity of the active site is assumed to be determined solely by the nature of the terminal monomer residue which carries this site. Thus, for copolymerizations of monomers A and B, the two species AABAABA* and BBABAA* would be indistinguishable.

For simplicity, the simple copolymer equation will be developed for the case of free radical reactions. The four possible propagation reactions with monomers M_1 and M_2 are

$$M_1^\cdot + M_1 \xrightarrow{k_{11}} M_1^\cdot \qquad \text{rate} = k_{11}[M_1^\cdot][M_1] \tag{7-2}$$

$$M_1^{\cdot} + M_2 \xrightarrow{k_{12}} M_2^{\cdot} \qquad \text{rate} = k_{12}[M_1^{\cdot}][M_2] \tag{7-3}$$

$$M_2^{\cdot} + M_2 \xrightarrow{k_{22}} M_2^{\cdot} \qquad \text{rate} = k_{22}[M_2^{\cdot}][M_2] \tag{7-4}$$

$$M_2^{\cdot} + M_1 \xrightarrow{k_{21}} M_1^{\cdot} \qquad \text{rate} = k_{21}[M_2^{\cdot}][M_1] \tag{7-5}$$

In these expressions M_i stands for a radical of any size ending in a unit derived from monomer M_i and $[M_i]$ denotes the total concentration of all such radicals, regardless of molecular chain length or structure. Similarly k_{ij} is the propagation rate constant for addition of monomer M_j to radical M_i. (Then k_{ii} is the propagation rate constant k_p for hompolymerization of M_i under the given reaction conditions.)

If we now assume that the only significant changes in monomer concentrations result from propagation reactions (i.e., changes in $[M_i^{\cdot}]$ from initiation and transfer reactions are negligible), the rates of monomer disappearance according to the reaction schemes of (7-2)–(7-5) are

$$-d[M_1]/dt = k_{11}[M_1^{\cdot}][M_1] + k_{21}[M_2^{\cdot}][M_1] \tag{7-6}$$

$$-d[M_2]/dt = k_{22}[M_2^{\cdot}][M_2] + k_{12}[M_1^{\cdot}][M_2] \tag{7-7}$$

Similarly, the time dependence of the concentration of radical $M_1^{\cdot}$ is

$$-d[M_1^{\cdot}]/dt = -k_{12}[M_1^{\cdot}][M_2] + k_{21}[M_2^{\cdot}][M_1] \tag{7-8}$$

Under steady-state conditions $[M_1^{\cdot}]$ is sufficiently small that $d[M_1^{\cdot}]/dt$ is negligible compared to the rates of change of concentrations of the reactants. Hence, setting $d[M_1^{\cdot}]/dt = 0$

$$k_{12}[M_1^{\cdot}][M_2] = k_{21}[M_2^{\cdot}][M_1] \tag{7-9}$$

and

$$[M_1^{\cdot}]/[M_2^{\cdot}] = k_{21}[M_1]/k_{12}[M_2] \tag{7-10}$$

The relative rates of incorporation of the two monomers into the copolymer at any instant follow by dividing Eq. (7-6) by Eq. (7-7):

$$\frac{d[M_1]}{d[M_2]} = \frac{k_{11}[M_1^{\cdot}][M_1] + k_{21}[M_2^{\cdot}][M_1]}{k_{22}[M_2^{\cdot}][M_2] + k_{12}[M_1^{\cdot}][M_2]} \tag{7-11}$$

Now divide the right-hand side of Eq. (7-11) by $[M_2^{\cdot}]$, insert expression (7-10) for

$[M_1^\cdot]/[M_2^\cdot]$, and divide through by k_{21} to obtain

$$\frac{d[M_1]}{d[M_2]} = \frac{[M_1][(k_{11}/k_{12})[M_1] + [M_2]]}{[M_2][(k_{22}/k_{21})[M_2] + [M_1]]} \tag{7-12}$$

Define the reactivity ratios r_i as $r_1 \equiv k_{11}/k_{12}$ and $r_2 \equiv k_{22}/k_{21}$, so that the preceding equation becomes

$$\frac{d[M_1]}{d[M_2]} = \frac{[M_1](r_1[M_1] + [M_2])}{[M_2](r_2[M_2] + [M_1])} \tag{7-13}$$

An equivalent expression can of course be derived in terms of mole fractions rather than concentrations. If f_1 and f_2 are the respective mole fractions of monomers M_1 and M_2 in the reaction feed and F_1 and F_2 are the corresponding mole fractions in the copolymer formed from this mixture, then

$$F_1 = \frac{r_1 f_1^2 + f_1 f_2}{r_1 f_1^2 + 2f_1 f_2 + r_2 f_2^2} \tag{7-14}$$

and from the definition of mole fraction

$$F_2 = 1 - F_1 \tag{7-15}$$

Equations (7-13) and (7-14) are alternative versions of the simple copolymer equation. Measurements of corresponding feed and copolymer compositions should yield values of r_1 and r_2 which can then be used to predict the relative concentrations of monomer in copolymers formed from any other mixtures of the particular monomers. The equations given are differential expressions and define the composition of the copolymer formed at any instant during the polymerization. They may be integrated, as noted in Section 7.5, to follow reactions in which one monomer is consumed more rapidly than the other.

Assumptions are invoked whenever an attempt is made to reduce the complexities of the real world to a mathematically tractable model. Following are some of the assumptions which are implied in the simple copolymer model:

1.   Radicals $M_1^\cdot$ are formed by an initiation reaction as well as by propagation reaction (7-5) and are eliminated by termination reactions and by propagation step (7-3). [Reaction (7-2) has no effect on $[M_1^\cdot]$.] If the kinetic chains are long, initiation and termination reactions are rare compared to propagation events and the former may be ignored. If the kinetic chain is long, then so is the molecular chain length (Section 6.6). The effects of initiation and termination reactions and of side reactions can probably be neglected safely if the copolymer molecular weight is fairly high (say, $\overline{M}_n \geq \sim 100,000$).

2.   The equal reactivity hypothesis in this case assumes that the rates of the propagation and termination reactions are independent of the size of the macromolecular radical and depend on its composition only through the terminal unit bearing the actual site.

3.   Equations (7-13) and (7-14) are dimensionless; all units cancel on each side of the equality sign. Thus the reactivity ratios are indicated to be independent of dilution and of the concentration units used. The reactivity ratios for a particular monomer pair should be the same in bulk and in dilute solution copolymerizations.

4.   The reactivity ratios should likewise be independent of initiator concentration, reaction rate, and overall extent of monomer conversion, since no rate constants appear as such in the copolymer composition equation.

5.   Reactivity ratios should be independent of inhibitors, retarders, chain transfer reagents, or solvents. We see later (Section 7.12.3) that solvent independence is not universal, but the effect is small.

6.   Addition of complexing agents that might change the mechanism of the reaction or the reactivity of monomers will, of course, alter the reactivity ratios (Section 7.12.4).

7.   When the ratio of unreacted monomer concentrations is $[M_1]/[M_2]$, the increment of copolymer formed has the relative composition $d[M_1]/d[M_2]$. From Eq. (7-13), the copolymer composition will change continuously as the reaction proceeds unless $d[M_1]/d[M_2] = [M_1]/[M_2]$. Thus when emulsion SBR rubber is made from a mixture of 72 parts butadiene (M_1) with 28 parts styrene (M_2) the respective concentrations in the initial monomer droplets are about 9.4 and 1.9 M. Since $r_1 = 1.4$ and $r_2 = 0.8$, Eq. (7-13) indicates that the copolymer formed in the initial stages of the reaction will contain about 78% by weight of butadiene, compared to 72% of this monomer in the feed.

7.3   COPOLYMER STRUCTURE INFERENCES FROM REACTIVITY RATIOS

Before measurements of reactivity ratios are reviewed, it is useful to consider what the absolute magnitude of these parameters implies. This is most easily approached from the relation between the $r_1 r_2$ product and copolymer structure as summarized in Table 7-1. This table also includes a list of reactivity ratios that apply in some important free radical polymerizations.

7.3.1   Random Copolymers

According to our reaction scheme $M_1 M_1$ bonds are formed only by reaction (7-2). The probability that a radical ending in an M_1 unit adds an M_1 unit is equal to the rate of this reaction divided by the sum of the rates of all reactions available to this radical. This is the probability P_{11} that an M_1 unit follows an M_1 unit in the

TABLE 7-1

(a) Copolymer Structure and $r_1 r_2$ Product

r_1	r_2	$r_1 r_2$	Copolymer structure
$r_1 = 1/r_2$	$r_2 = 1/r_1$	1	Random (ideal)
$\ll 1$	$\ll 1$	$\to 0$	Alternating
$\gg 1$	< 1	< 1	Tends to be homopolymer of M_1

(b) Some Reactivity Ratios for Radical Copolymerizations

M_1	M_2	r_1	r_2	$r_1 r_2$
Butadiene	Styrene	1.4	0.8	1.1
Ethylene	Vinyl acetate	1.0	1.0	1.0
Vinyl chloride	Vinyl acetate	1.4	0.65	0.9
Vinylidene chloride	Vinyl chloride	3.2	0.3	1
Methyl methacrylate	Methacrylamide	1.5	0.5	0.8
Vinylidene chloride	Acrylonitrile	0.4	0.9	0.4
Styrene	Methyl methacrylate	0.5	0.5	0.3
Butadiene	Methyl methacrylate	0.7	0.3	0.2
Acrylonitrile	Vinyl chloride	2.7	0	0
Acrylonitrile	Styrene	0	0.4	0
Styrene	Maleic anhydride	0	0	0

copolymer, and since the only other reaction assumed important for this radical is (7-3),

$$P_{11} = \frac{k_{11}[M_1^{\cdot}][M_1]}{k_{11}[M_1^{\cdot}][M_1] + k_{12}[M_1^{\cdot}][M_2]} = \frac{r_1[M_1]}{r_1[M_1] + [M_2]} \qquad (7\text{-}16)$$

Similarly, the probability P_{21} that an M_1 unit follows an M_2 unit in the polymer is

$$P_{21} = \frac{k_{21}[M_2^{\cdot}][M_1]}{k_{21}[M_2^{\cdot}][M_1] + k_{22}[M_2^{\cdot}][M_2]} = \frac{[M_1]}{r_2[M_2] + [M_1]} \qquad (7\text{-}17)$$

If, however, $r_1 r_2 = 1$, then P_{11} and P_{21} defined above are equal. That is to say, the likelihood that an M_1 unit follows an M_1 unit equals the likelihood that it follows an M_2 unit in the product. The absolute value of this probability depends on the relative concentrations of monomers in the feed, but the equivalence of probabilities is independent of the feed and copolymer compositions. A similar equality for P_{22} and P_{12} can be shown by analogous reasoning, and this equivalence of probabilities is a necessary condition for a random distribution of monomer residues in the copolymer.

Random copolymers will be formed, or course, if each radical attacks either monomer with equal facility ($k_{11} = k_{12}$, $k_{22} = k_{21}$, $r_1 = r_2 = 1$). Free-radical copolymerization of ethylene and vinyl acetate is an example of such a system, but this is not a common case. Random monomer distributions are obtained more generally if k_{11}/k_{12} is approximately equal to k_{21}/k_{22}. That is to say, $r_1 \simeq 1/r_2$. This means that k_{11}/k_{22} and k_{21}/k_{22} will be simultaneously either greater or less than unity or in other words, that both radicals prefer to react with the same monomer.

If $r_1 r_2 = 1$, copolymer equation (7-13) reduces to

$$\frac{d[M_1]}{d[M_2]} = \frac{r_1[M_1]}{[M_2]} = \frac{[M_1]}{r_2[M_2]} \tag{7-18}$$

The copolymer and feed compositions in random copolymerizations are identical only in the rare case when both reactivity ratios equal unity. The copolymer composition curves in Fig. 7-1 are typical of copolymerizations which are effectively random ($r_1 r_2 \sim 1$). The curves show no inflection points; they do not cross the 45° line corresponding to equal feed and copolymer compositions. There is no appreciable range of monomer feeds over which both monomers enter random copolymers in significant quantities if *both* reactivity ratios are not near unity, and the difficulty of making such copolymers becomes more severe as the difference in absolute values of r_1 and r_2 increases.

Some commercially important examples of random free radical copolymerizations include the styrene–butadiene pair mentioned above, in which $r_1 r_2 = 1.1$

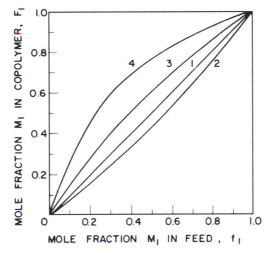

Fig. 7-1. Relation between instantaneous feed composition f_1 and corresponding copolymer composition F_1 for random copolymerizations. Curve 1, ethylene ($r_1 = 1$)–vinyl acetate ($r_2 = 1$); curve 2, styrene ($r_1 = 0.8$)–butadiene ($r_2 = 1.4$); curve 3, vinyl chloride ($r_1 = 1.4$)–vinyl acetate ($r_2 = 0.65$); curve 4, vinylidene chloride ($r_1 = 3.2$)–vinyl chloride ($r_2 = 0.3$).

and vinyl chloride ($r_1 = 1.4$)–vinyl acetate ($r_2 = 0.65$), for which $r_1 r_2 = 0.9$. In these products a given monomer is randomly linked to monomers of the same or different type. The relative amount of each monomer in the copolymer depends on the feed concentrations and reactivity ratios (Eqs. 7-13 and 7-14), and copolymers with different average compositions can be made by altering the feed ratios in batch reactions.

The vinylidene chloride (M_1)–vinyl chloride (M_2) system is an example of a random copolymerization in which $r_1 = 3.2$, $r_2 = 0.3$, and $r_1 r_2 \simeq 1$. Reasonably homogeneous copolymers with controlled concentrations of the less reactive component can be made in semibatch processes by controlling the monomer feed. In this case the more reactive ingredient, vinylidene chloride, is added intermittently or continuously in a proportional manner to a mixture to which all the vinyl chloride is charged initially.

7.3.2 Alternating Copolymers

If each radical prefers to add the monomer of the opposite type both reactivity ratios will tend to zero, and the copolymer equation becomes

$$d[M_1]/d[M_2] = 1 \qquad\qquad (7\text{-}19)$$

or

$$F_1 = F_2 = 0.5 \qquad\qquad (7\text{-}20)$$

The tendency to alternation increases as the $r_1 r_2$ product nears zero, as long as both r_1 and r_2 are less than unity. Such copolymerizations occur in free-radical systems when the two monomers have opposite polarities (Section 7.10.2). The styrene maleic anhydride copolymers mentioned in Chapter 1 are an example of a purely alternating system ($r_1 = r_2 = 0$), while styrene (M_1)–acrylonitrile (M_2) copolymers have a pronounced tendency to alternate monomer residues ($r_1 = 0.4$, $r_2 = 0$).

7.3.3 Long Sequences of One Monomer

When one reactivity ratio is greater than one and the other is less than one, either radical will prefer to add monomers of the first type. Relatively long sequences of this monomer will be formed if the reactivity ratios differ sufficiently. In that case the product composition will tend toward that of the homopolymer of the more reactive monomer. Such reactivity ratios reflect the existence of an impractical copolymerization. Styrene ($r_1 \simeq 50$)–vinyl acetate ($r_2 = 0$) is such a system and these copolymers cannot be made by free radical initiation.

7.3.4   Block Copolymers

If both reactivity ratios are greater than one, there will be a tendency for formation of sequences of uniform composition in the copolymers. Such reactivity ratio combinations are not known in free-radical copolymerizations when both monomers are present simultaneously in the reaction vessel, but they can be made in other systems.

7.4   AZEOTROPIC COMPOSITIONS

Binary copolymerization resembles distillation of a bicomponent liquid mixture, with a reactivity ratio corresponding to the ratio of vapor pressures of the pure components in the latter case. The vapor-liquid composition curves of ideal binary mixtures have no inflection points and neither do the polymer-composition curves for random copolymerizations, in which $r_1 r_2 \simeq 1$ (Fig. 7-1). For this reason, such comonomer systems are sometimes called *ideal*.

Distillation terminology is also borrowed for instances when the copolymer composition is the same as that of the comonomer feed from which it is derived. Such a feed composition is called an azeotrope by analogy with distillation. Under these conditions $d[M_1]/d[M_2] = [M_1]/[M_2]$, and hence Eq. (7-13) reduces to

$$\frac{r_1[M_1] + [M_2]}{[M_1] + r_2[M_2]} = 1 \tag{7-21}$$

From Eq.(7-21), an azeotropic feed composition is such that

$$\left(\frac{[M_1]}{[M_2]}\right)_{\text{azeotrope}} = \frac{1 - r_2}{1 - r_1} \tag{7-22}$$

or from Eq. (7-14)

$$(f_1)_{\text{azeotrope}} = \frac{1 - r_2}{2 - r_1 - r_2} \tag{7-23}$$

Note that f_1 is physically meaningful ($0 \leq f_1 \leq 1$) only if r_1 and r_2 are simultaneously both either greater or less than one. (If $r_1 = r_2 = 1$, all values of f_1 are azeotropic compositions.) Since the case of $r_1 > 1, r_2 > 1$ is unknown in free-radical systems, the necessary conditions for azeotropy in such copolymerizations is that $r_1 < 1, r_2 < 1$.

Figure 7-2 includes some representative copolymer-feed composition curves for r_1 varying with r_2 constant at 0.8. Azeotropic feed compositions containing appreciable quantities of both polymers can be achieved only when the reactivity ratio values are close to each other.

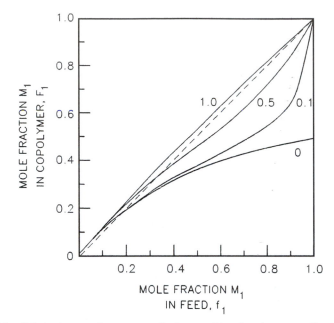

Fig. 7-2. Relation between instantaneous feed composition f_1 and corresponding polymer composition F_1 for indicated values of r_1 with $r_2 = 0.8$. Azeoptropic compositions exist at $r_1 = 0$, $f_1 = 0.17$; $r_1 = 0.1$, $f_1 = 0.18$; $r_1 = 0.5$, $f_1 = 0.29$; $r_1 = 1.0$, no azeotrope [Eq. (7-23)].

Equation (7-23) calculates the feed composition that yields an invariant copolymer composition as the conversion proceeds in a batch polymerization. Note that comonomer ratios that are near but not equal to the estimated azeotropic value may produce copolymers whose compositions are constant for all practical purposes. The permissible range of feed compositions for which this "approximate" azeotropy occurs is evidently greater the closer the two reactivity ratios are to each other.

7.5   INTEGRATED BINARY COPOLYMER EQUATION

If the feed composition is an azeotropic mixture or if $r_1 = r_2 = 1$, the feed and copolymer compositions will remain constant during the course of a batch copolymerization. More generally, however, both compositions will change with conversion, and it is important to be able to calculate the course of such changes.

An algorithm [3] suitable for manual or computer-assisted calculations proceeds as follows:

A batch copolymerization mixture initially contains m_1^0 mol monomer M_1 and m_2^0 mol monomer M_2. After a fraction p of the initial monomers have been polymerized, the unreacted monomers are, respectively, m_1 and m_2. Now, f_i and F_i are the mole fractions of monomer i in the feed and corresponding copolymer, respectively, and

$$m_1^0 + m_2^0 = m^0 \tag{7-24}$$

$$m_1^0 \equiv f_1^0 m^0 \tag{7-25}$$

$$m_2 \equiv f_2^0 m^0 \tag{7-26}$$

$$m_1 \equiv f_1(1-p)m^0 \tag{7-27}$$

$$m_2 \equiv f_2(1-p)m^0 \tag{7-28}$$

where the superscript 0 denotes initial values and the unsuperscripted symbols refer to values after an extent of reaction p.

Integration of the simple copolymer equation between the limits m_1^0, m_2^0 and m_1, m_2 yields [1]:

$$\log\left(\frac{m_2}{m_2^0}\right) = \frac{r_2}{1-r_2} \log\left(\frac{m_2^0 m_1}{m_1^0 m_2}\right)$$
$$- \frac{1-r_1 r_2}{(1-r_1)(1-r_2)} \log\left[\frac{(r_1-1)(m_1/m_2)-r_2+1}{(r_1-1)(m_1^0/m_2^0)-r_2+1}\right] \tag{7-29}$$

By substituting the above definitions,

$$(1-p)^{1+r_1 r_2 - r_1 r_2} = \left(\frac{f_1}{f_1^0}\right)^{r_2 - r_1 r_2} \left(\frac{f_2}{f_2^0}\right)^{r_1 - r_1 r_2}$$
$$\times \left[\frac{(1-r_2)f_2^0 - (1-r_1)f_1^0}{(1-r_2)f_2 - (1-r_1)f_1}\right]^{1-r_1 r_2} \tag{7-30}$$

[Note that Eq. (7-30) cannot be used when $r_1 = 1.0$, $r_2 = 0.5$. In that case make calculations with $r_1 = 0.95$ or 1.05. This will make very little difference to the calculated values. Alternatively, one can use both suggested r_1 values and interpolate, if absolutely necessary.]

Since F_1 is the ratio of the number of moles of M_1 converted divided by the total number of moles of M_1 and M_2 polymerized in the same interval,

$$F_1 = \frac{m_1^0 - m_1}{(m_1^0 - m_1) + (m_2^0 - m_2)} = \frac{f_1^0 - f_1(1-p)}{p} \tag{7-31}$$

Equation (7-30) gives the feed composition after a fraction p of the initial monomer mixture has been reacted. The composition of the copolymer made during this

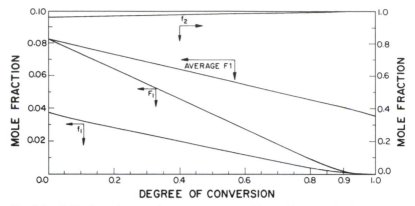

Fig. 7-3. Drift of copolymer and comonomer compositions with conversion in the copolymerization of glycidyl methacrylate

$$(H_2C = C - C - O - CH_2 - C - CH_2 .$$

(with H_3C and O substituents shown below)

$(M_1, r_1 = 0.5)$ with styrene $(M_2, r_2 = 0.4)$ with $f_1^0 = 0.037$ (5 wt% of the methacrylate). Azeotropic composition is at $f_1^0 = 0.55$.

interval of reaction is given by Eq. (7-31). For given r_1, r_2 and f_1^0, one assumes an f_1 and estimates p from Eq. (7-30). (It is necessary to remember that f_1 will decrease if M_1 is the more reactive monomer, and vice versa.) The corresponding value of F_1 is obtained from Eq. (7-31) with the p figure calculated as described. The F_1 values for a series of stepwise f_1 levels are calculated by repeating this sequence with f_1 in one step becoming f_1^0 in the next estimate. The cumulative average copolymer composition can also be calculated in a straightforward manner by entering Eq. (7-31) with the cumulative value of p and the initial value of f_1^0. (As a check, the average value of F_1 at $p = 1$ must equal f_1^0.)

Figure 7-3 records the changes of monomer feed and copolymer compositions with conversion in the case of glycidyl methacrylate and styrene. This copolymerization would produce an essentially styrenic polymer which is cross-linkable through the pendant epoxy groups of the methacrylate residues. The last 10% of copolymer formed is practically pure polystyrene. In the styrene-butadiene copolymerization depicted in Fig. 7-4, the product composition is almost constant for the first 70% of the reaction where this polymerization would normally be halted anyway (Section 7.2.3).

The variation of copolymer composition during the course of a batch polymerization can be reduced by conducting the reaction as a so-called "semibatch" process. This is a starved feed operation in which part of the charge is fed to the reaction vessel and polymerization is started. The remainder of the monomer feed is pumped in continuously or intermittently at a rate sufficient to keep the

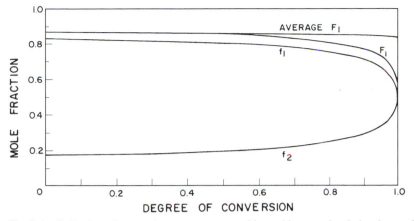

Fig. 7-4. Drift of copolymer and comonomer compositions with conversion during the copolymerization of butadiene (M_1, $r_1 = 1.4$) and styrene (M_2, $r_2 = 0.8$) with $f_1^0 = 0.832$ and $f_2^0 = 0.168$.

copolymerization going at the desired rate. (In effect this could be at a rate sufficient to keep the temperature of the reaction mixture contents in a small range established by experience; these are exothermic reactions.) The degree of conversion of the comonomer mixture *in the reactor* is always high and relatively invariant, and the value of F_1 in Eq. (7-31) remains sensibly constant since p is almost the same for reaction of most of the total monomer charge.

7.6 DETERMINATION OF REACTIVITY RATIOS

The measurement of reactivity ratios appears to be straightforward provided the equation linking feed and copolymer compositions fits the data obtained by analyzing the compositions of copolymers formed from several different concentrations of monomers. If a differential form of the copolymer equation (Eq. 7-13 or 7-14) is used with initial feed composition values, it is necessary to keep the total conversion to polymer in each experiment less than about 5% so as to minimize the drift of copolymer makeup. Ten or more percent of the monomers can be converted to polymer in a single run without significant calculation error if the arithmetic averages of the final and initial monomer concentrations are used in these differential copolymer equations [4]. The extent of reaction at which this procedure becomes unreliable depends on the relative magnitudes of the reactivity ratios, except of course, for the case of azeotropic feed mixtures. This expedient is safe in general so long as the concentration of each unreacted monomer is linearly related to reaction time [5]. Alternatively, it is quite feasible to fit the experimental feed and polymer compositions to an integrated form of the copolymer equation, although the calculations are slightly more cumbersome than with the differential form.

The fitting of corresponding feed and copolymer compositions to the copolymer equation to obtain reactivity ratio values is not without pitfalls. Many of the available r_1 and r_2 values in the literature are defective because of unsuspected problems which were involved in estimation procedures, use of inappropriate mathematical models to link polymer and feed compositions, and experimental or analytical difficulties.

Several procedures for extracting reactivity ratios from differential forms of the copolymer equation are mentioned in the following paragraphs. These methods are arithmetically correct, but they do not give reliable results because of the nature of the experimental uncertainties in reactivity ratio measurements.

The *method of intersections* [1] has been widely used for computing reactivity ratios from data fitted to the differential copolymer equation. In this procedure, Eq. (7-13) is recast into the form

$$r_2 = \frac{[M_1]}{[M_2]} \left[\frac{d[M_2]}{d[M_1]} \left[1 + \frac{[M_1]}{[M_2]} r_1 \right] - 1 \right] \tag{7-32}$$

Corresponding experimental values of $[M_1]$, $[M_2]$, $d[M_1]$, and $d[M_2]$ are substituted into Eq. (7-32), and r_2 is plotted as a function of assumed values of r_1. Each experiment yields one straight line in the $r_1 r_2$ plane and the intersection region of such lines from different feed composition experiments is assumed to give the best values of r_1 and r_2. The same basic technique may be applied to the integrated form of the copolymer equation. The intersection point which corresponds to the "best" values of r_1 and r_2 is selected imprecisely and subjectively by this technique. Each experiment yields a straight line, and each such line can intersect one line from every other experiment. Thus n experiments yield $(n(n-1)/2$ intersections and even one "wild" experiment produces $(n-1)$ unreliable intersections. Various attempts to eliminate subjectivity and reject dubious data on a rational basis have not been successful.

Alternatively [6], the simple copolymer equation can be solved in a linear graphical manner by substituting $x = [M_1]/[M_2]$, $z = d[M_1]/[M_2]$, so that Eq. (7-13) becomes

$$z = x(1 + r_1 x)/(r_2 + x) \tag{7-33}$$

Equation (7-33) can be linearized in the alternative forms

$$G = x(z - 1)/z = r_1 H - r_2 \tag{7-34}$$

and

$$G/H = -r_2/H + r_1 \tag{7-35}$$

where $H = x^2/z$. Linear least squares fits to Eq. (7-34) or (7-35) yield one reactivity ratio as the intercept and the other as the slope of the plotted line. The experimental data are, however, unequally weighted by these equations and the values obtained at low $[M_2]$ in Eq. (7-34) or low $[M_1]$ in Eq. (7-35) have the greatest influence on the slope of a line corresponding to these equations. Equations (7-34) and (7-35) are not symmetrical in r_1 and r_2. The same set of experimental data can yield different r_1, r_2 sets depending on which monomer is indexed as M_1 and which is M_2. This procedure (the Fineman–Ross method) has been widely used, because it is simple and can be treated graphically or by a linear least squares regression on the data. The latter procedure is statistically unsound, however. Equation (7-13) is not linear in r_1 and r_2. Its transformation into the linear form of Eq. (7-34) or (7-35) is algebraically correct, but the error structure in the copolymer composition (F_1 and F_2) is also changed, so that the errors in G and H do not have constant variance and zero mean. This means that Eqs. (7-34) and (7-35) do not meet the statistical requirements for linear least squares computations [7].

More recent linear graphical methods are invariant to the inversion of monomer indexes. One procedure [8] uses the form

$$r_2/H^{1/2} = r_1/H^{1/2} - z^{1/2} + z^{-1/2} \tag{7-36}$$

Another method [9] involves recasting the copolymer equation in the form

$$\eta = (r_1 + r_2)/\alpha)\xi - r_2/\alpha \tag{7-37}$$

or

$$\eta = r_1\xi - (r_2/\alpha)(1 - \xi) \tag{7-38}$$

where

$$\eta(\text{eta}) = G/(\alpha + H) \tag{7-39}$$

$$\xi(x_i) = H/(\alpha + H) \tag{7-40}$$

and α (alpha) is defined from

$$\alpha = (H_{\max} H_{\min})^{1/2} \tag{7-41}$$

with $H_{\max}$ the maximum and $H_{\min}$ the lowest values of $([M_1]/[M_2])^2 \cdot d[M_2]/d[M]_1$ in the series of measurements.

Other procedures [10,11] rely on computerized adjustments of r_1 and r_2 to minimize the absolute values of deviations between predicted and measured copolymer compositions.

The last four methods cited [8-11] are statistically inexact, in that they cannot give good estimates of the reactivity ratio values, but they can provide good estimates of r_1 and r_2 if the copolymerization experiments are suitably designed.

The copolymerizations should not be carried out over a random range of feed compositions, but the available effort should be devoted to replications of copolymerizations at two monomer feeds f_1' and f_1'' given by

$$f_1' = 2/(2 + r_1^*)$$

(7-42)

and

$$f_1'' = r_2^*(2 + r_2^*)$$

(7-43)

where r_1^* and r_2^* are approximated values of the reactivity ratios [12]. (A convenient method for approximating reactivity ratios is given in Section 7.11.)

The most generally useful methods and the only statistically correct procedures for calculating reactivity ratios from binary copolymerization data involve nonlinear least squares analysis of the data or application of the "error in variables (EVM)" method. Effective use of either procedure requires more iterations than can be performed by manual calculations. An efficient computer program for nonlinear least squares estimates of reactivity ratios has been published by Tidwell and Mortimer [13]. The EVM procedure has been reported by O'Driscoll and Reilly [14].

Another important recent contribution is the provision of a good measurement of the precision of estimated reactivity ratios. The calculation of independent standard deviations for each reactivity ratio obtained by linear least squares fitting to linear forms of the differential copolymer equations is invalid, because the two reactivity ratios are not statistically independent. Information about the precision of reactivity ratios that are determined jointly is properly conveyed by specification of joint confidence limits within which the true values can be assumed to coexist. This is represented as a closed curve in a plot of r_1 and r_2. Standard statistical techniques for such computations are impossible or too cumbersome for application to binary copolymerization data in the usual absence of estimates of reliability of the values of monomer feed and copolymer composition data. Both the nonlinear least squares and the EVM calculations provide computer-assisted estimates of such joint confidence loops [15].

The copolymer composition can be estimated usefully in many cases from the composition of unreacted monomers, as measured by gas-liquid chromatography. Analytical errors are reduced if the reaction is carried to as high a conversion as possible, since the content of a given monomer in the copolymer equals the difference between its initial and final measured contents in the feed mixture. The uncertainty in the copolymer analysis is thus a smaller proportion of the estimated quantity, the greater the magnitude of the decrease in the monomer concentration in the feed. It may seem appropriate under these circumstances to estimate reactivity ratios by fitting the data to an integrated form of the copolymer equation.

7.7 MULTICOMPONENT COPOLYMERIZATIONS

The composition of the copolymer formed from n monomers in addition polymerization can be expressed in terms of the monomer feed composition and $n(n-1)$ binary reactivity ratios. Thus, for terpolymerization [16],

$$\frac{d\mathrm{M}_1}{d\mathrm{M}_3} = \frac{\mathrm{M}_1(\mathrm{M}_1 r_{23} r_{32} + \mathrm{M}_2 r_{31} r_{23} + \mathrm{M}_3 r_{32} r_{21})(\mathrm{M}_1 r_{12} r_{13} + \mathrm{M}_2 r_{13} + \mathrm{M}_3 r_{12})}{\mathrm{M}_3(\mathrm{M}_1 r_{12} r_{23} + \mathrm{M}_2 r_{13} r_{21} + \mathrm{M}_3 r_{12} r_{21})(\mathrm{M}_3 r_{31} r_{32} + \mathrm{M}_1 r_{32} + \mathrm{M}_2 r_{31})}$$

(7-44)

$$\frac{d\mathrm{M}_2}{d\mathrm{M}_3} = \frac{\mathrm{M}_2(\mathrm{M}_1 r_{32} r_{13} + \mathrm{M}_2 r_{13} r_{31} + \mathrm{M}_3 r_{12} r_{31})(\mathrm{M}_2 r_{21} r_{23} + \mathrm{M}_1 r_{23} + \mathrm{M}_3 r_{21})}{\mathrm{M}_3(\mathrm{M}_1 r_{12} r_{23} + \mathrm{M}_2 r_{13} r_{21} + \mathrm{M}_3 r_{12} r_{21})(\mathrm{M}_3 r_{31} r_{32} + \mathrm{M}_1 r_{32} + \mathrm{M}_2 r_{31})}$$

(7-45)

where the M_i are the molar concentrations of monomer $i (i = 1, 2, 3)$ and $r_{ij} (i \neq j)$ are the appropriate binary reactivity ratios. [In Eqs. (7-44) and (7-45) the usual symbol of [M_i] for a concentration is abbreviated for ease of typesetting. The symbols for reactivity ratios now require two indexes each, since there are three monomers. The reactivity ratios defined in the derivation of Eq. (7-13) would be called r_{12} and r_{21} in this modified notation.] The preceding differential copolymer equations can be integrated numerically [17] if the reaction volume can be assumed to remain approximately constant. For example, if 0.001 mol changes are assumed in M_3, the corresponding changes in M_1 and M_2 can be calculated from Eqs. (7-44) and (7-45), and the final monomer concentrations ($\mathrm{M}_i - d\mathrm{M}_i$) used as the initial values in the next iterative step. About 50 iterations account for 15 wt% conversion of monomers. The mean copolymer composition is given by

$$\overline{C}_{i\mu} = 100 P_i \sum_{v=1}^{\mu} \delta\mathrm{M}_{iv} \left/ \sum_{i=1}^{3} \sum_{v=1}^{\mu} \mathrm{M}_{oi} \delta\mathrm{M}_{iv} \right. \qquad i = 1, 2, 3 \qquad v = 1, 2$$

(7-46)

where $\overline{C}_{i\mu}$ is the weight fraction of component i in the copolymer produced during μ iterative calculations, M_{oi} is the formula weight of this monomer, and $\delta\mathrm{M}_{iv}$ is the decrease in concentration of monomer i during the vth iteration [from Eqs. (7-44) and (7-45)].

The foregoing equations are derived on the assumption that the binary reactivity ratios in Eqs. (7-13) and (7-14) also apply to multicomponent polymerizations. Experimental tests of this point are rather scanty, but the weight of such evidence seems to support this assumption.

It is possible to analyze multicomponent feed and copolymer composition data directly to determine the reactivity ratios that apply to a particular system [18].

Azeotropic feed compositions can exist in terpolymerizations if S, Z and R [Eq. (7-47a–c)] have the same sign and are different from zero. The azeotropic feed composition is given by Eq. (7-48).

$$S = \left(1 - \frac{1}{r_{13}}\right)\left(\frac{1}{r_{32}} - 1\right) - \left(\frac{1}{r_{12}} - \frac{1}{r_{13}}\right)\left(\frac{1}{r_{13}} - 1\right) \qquad (7\text{-}47a)$$

$$Z = \left(\frac{1}{r_{21}} - \frac{1}{r_{23}}\right)\left(\frac{1}{r_{12}} - \frac{1}{r_{13}}\right) - \left(1 - \frac{1}{r_{23}}\right)\left(1 - \frac{1}{r_{13}}\right) \qquad (7\text{-}47b)$$

$$R = \left(\frac{1}{r_{31}} - 1\right)\left(1 - \frac{1}{r_{23}}\right) - \left(\frac{1}{r_{32}} - 1\right)\left(\frac{1}{r_{21}} - \frac{1}{r_{23}}\right) \qquad (7\text{-}47c)$$

$$
\begin{aligned}
[M_1] : [M_2] : [M_3] = \ & R\left(\frac{Z}{r_{13}r_{32}} + \frac{S}{r_{12}r_{23}} + \frac{R}{r_{12}r_{13}}\right) \\
: \ & S\left(\frac{S}{r_{21}r_{23}} + \frac{Z}{r_{23}r_{31}} + \frac{R}{r_{21}r_{13}}\right) \\
: \ & Z\left(\frac{Z}{r_{31}r_{32}} + \frac{S}{r_{21}r_{32}} + \frac{R}{r_{12}r_{31}}\right) \qquad (7\text{-}48)
\end{aligned}
$$

7.8   SEQUENCE DISTRIBUTION IN COPOLYMERS

We have already derived expressions for P_{11} and P_{21} in Eqs. (7-16) and (7-17). These are the respective probabilities that M_1M_1 and M_2M_1 sequences exist in the copolymer. (The assumption implicit here, as in the simple copolymer equations in general, is that the molecular weight of the polymer is fairly large.) The probabilities P_{22} and P_{12} can be derived by the same reasoning, and all four can be expressed in terms of mole fractions f_i, in place of the concentrations used to this point:

$$P_{11} = \frac{r_1 f_1}{r_1 f_1 + f_2} = \frac{r_1([M_1]/[M_2])}{r_1([M_1]/[M_2]) + 1} \qquad (7\text{-}49)$$

$$P_{12} = \frac{f_2}{r_1 f_1 + f_2} = \frac{1}{r_1([M_1]/[M_2]) + 1} \qquad (7\text{-}50)$$

$$P_{22} = \frac{r_2 f_2}{f_1 + r_2 f_2} = \frac{r_2}{[M_1]/[M_2] + r_2} \qquad (7\text{-}51)$$

$$P_{21} = \frac{f_1}{f_1 + r_2 f_2} = \frac{[M_1]/[M_2]}{[M_1]/[M_2] + r_2} \qquad (7\text{-}52)$$

To determine the distribution of sequence lengths of each monomer in the polymer, an M_1 unit in the copolymer is selected at random. If this unit is part of a sequence of $n_i M_1$ units, reaction (7-2) would have to have been repeated

$(n_i - 1)$ times. Since the probability of one such event is P_{11}, the probability that it occurs $(n_i - 1)$ times is $(P_{11})^{n_i-1}$. If the M_1 sequence is exactly n_i units long the $(n_i - 1)$ reactions of M_1 with M_i must be followed by reaction (7-3). This placement has the probability $P_{12} = 1 - P_{11}$. [Reactions (7-2) and (7-3) represent the only alternatives available to radical M_i and therefore $P_{11} + P_{12}$ must equal 1.] We conclude, then, that the probability that the original M_1 unit was part of a sequence of n_i such units is $P_{11}^{n_i-1}(1 - P_{11})$. But the probability that the sequence contains $n_i M_1$ units is also the fraction of all M_1 sequences which contain n_i units. That is to say, it is the number distribution function $N(M_1, n_i)$ for M_1 sequence lengths:

$$N(M_1, n_i) = P_{11}^{n_i-1}(1 - P_{11}) = P_{12}P_{11}^{n_i-1} \tag{7-53}$$

Similarly, the fraction of all M_2 sequences that contain exactly n_j units is

$$N(M_2, n_j) = P_{22}^{n_i-1}(1 - P_{22}) = P_{22}^{n_j-1}P_{21} \tag{7-54}$$

The number average length of M_1 sequences $\overline{N}_1$ is

$$\overline{N}_1 = \sum_{n_i=1}^{\infty} N(M_1, n_i)n_i \tag{7-55}$$

[This is completely analogous to the definition of number average molecular weight $\overline{M}_n$ in Eq. (2-6). A number average of a quantity is always the sum of products of values of that quantity times the corresponding fraction of the whole sample which is characterized by the particular value.] Substituting Eq. (7-53) into Eq. (7-55),

$$\overline{N}_1 = \sum_{n_i=1}^{\infty} n_i P_{11}^{n_i-1} P_{12} = \frac{P_{12}}{P_{11}} \sum_{1}^{\infty} n_i P_{11}^{n_i} \tag{7-56}$$

The summation in Eq. (7-56) is of the same form as the series $[(1-x)/x]\cdot\sum_{n=1}^{\infty} nx^n$, which equals $1/(1 - x)$ if x (here P_{11}) < 1. Thus,

$$\overline{N}_1 = \frac{1}{1 - P_{11}} = \frac{1}{P_{12}} \tag{7-57}$$

and similarly

$$\overline{N}_2 = \frac{P_{21}}{P_{22}} \sum_{n_j=1}^{\infty} n_j P_{22}^{n_j} = \frac{1}{1 - P_{22}} = \frac{1}{P_{21}} \tag{7-58}$$

Every M_1 sequence is joined to two M_2 sequences in the interior of the copolymer. The number of sequences of each type cannot differ by more than one, therefore. For long polymer molecules, which will contain a large number of sequences of each type, this difference is negligible and the number of M_1 sequences effectively equals the number of M_2 sequences. Then the ratio of M_1 units to M_2

units in the copolymer, $d[M_1]/d[M_2]$ in our previous notation, must equal the ratio of the respective average sequence lengths. That is to say,

$$\frac{d[M_1]}{d[M_2]} = \frac{\overline{N}_1}{\overline{N}_2} = \frac{1/P_{12}}{1/P_{21}} \qquad (7\text{-}59)$$

Substituting Eqs. (7-56) and (7-58) into Eq. (7-59) and rearranging, we obtain the simple copolymer equation (7-14) without specific reference to steady-state approximations.

Since considerations of sequence distributions can be used to derive the simple copolymer equation, it is not surprising that measured values of triad distributions in binary copolymers [by ^1H or ^{13}C NMR analyses] can be inserted into the copolymer equation to calculate reactivity ratios [19].

If one of the reactivity ratios, say, r_2, is zero then $P_{22} = 0$ and Eq. (7-54) becomes

$$N(M_2, n_j) = P_{22}^{n_j-1}(1 - P_{22}) = \begin{cases} 0 & \text{for } n_2 \neq 1 \\ 1 & \text{for } n_2 = 1 \end{cases} \qquad (7\text{-}60)$$

Thus, all M_2 units are present as isolated sequences.

When one reactivity ratio, say, r_1, is greater than 1 the average sequence length of M_1 units increases, since

$$\overline{N}_1 = 1/P_{12} = r_1([M_1]/[M_2]) + 1 \qquad (7\text{-}61)$$

Calculations of sequence lengths as a function of the $r_1 r_2$ product and the monomer feed concentration ratio are available in tabular form [20]. The calculations above refer, strictly speaking, to infinitely long copolymer molecules. There is still the possibility that different copolymer molecules of finite length will vary in composition even though the overall composition will be as calculated. This likelihood has been analyzed [21], and it is expected theoretically that the variation of composition about the mean value calculated from the simple copolymer equation will be quite narrow and will shrink as the mean molecular weight of the copolymer increases. If the reaction medium is homogeneous (this is not always true), this conclusion indicates that most practical copolymerizations from a given feed yield a product in which all macromolecules have close to the same composition at any instant.

7.9   GEL FORMATION DURING COPOLYMERIZATION AND CROSS-LINKING[22]

If one of the monomers in a copolymerization is a divinyl compound or any other entity with a functionality greater than 2, a branched polymer can be formed and it is possible for the growing branches to interconnect and form infinite molecular weight products known as *gels*. It is useful to be able to predict the conditions under which such gel formation will occur. The criteria for this condition are applicable also to cross-linking of preformed polymers which occurs in radiation-induced cross-linking, to vulcanization with addition of other reagents, and to chain-growth and step-growth polymerizations of polyfunctional monomers.

Consider first a sample containing discrete polymer molecules that can be interconnected either during further polymerization or by a separate reaction on the macromolecules. Suppose some of the molecules are cross-linked by linkages formed between randomly selected monomer units. We choose a cross-link at random. The probability that the monomer unit on this cross-link resides in a primary molecule that contains y monomer units equals the fraction of all monomer units that are in y-mers. That is, this probability P_y is

$$P_y = yN_y \bigg/ \sum_{y=1}^{\infty} yN_y \qquad (7\text{-}62)$$

where N_y is the number of y-mer molecules in the sample. However, if $\overline{M}_0$ is the mean formula weight of monomeric units in this sample

$$P_y = yN_y\overline{M}_0 \bigg/ \overline{M}_0 \sum_1^{\infty} yN_y = w_y \qquad (7\text{-}63)$$

where w_y is the weight fraction of y-mers in the sample.

If a fraction q of all monomer units in the sample forms parts of cross-links and these cross-links are randomly placed, then an additional $q(y-1)$ monomer units in our original y-mer are also cross-linked, on the average. That is to say, the probability that an arbitrarily selected cross-link is attached to a primary chain which contains y monomer units is w_y, and it is expected that $q(y-1)$ of these y monomer units are also cross-linked. Thus, the initial, randomly chosen cross-link leads through the primary molecules and other cross-links to $w_yq(y-1)$ additional primary molecules. Since y can have any positive nonzero value, the expected number of additional cross-links ϵ in a molecule that already contains one arbitrarily chosen cross-link is

$$\epsilon = q \sum_{y=1}^{\infty} w_y(y-1) = q\left(\overline{y}_{\mathrm{w}} - \sum w_y\right) = q(\overline{y}_{\mathrm{w}} - 1) \qquad (7\text{-}64)$$

where $\overline{y}_{\mathrm{w}}$ is the weight average degree of polymerization of the primary chains in the sample. This molecule can be part of an infinite cross-linked network only if ϵ is at least 1. The critical value of q, q_{cr}, for this condition (which is sometimes called gelation) is then given by

$$1 = q_{\mathrm{cr}}(\overline{y}_{\mathrm{w}} - 1) \qquad (7\text{-}65)$$

and

$$q_{cr} = \frac{1}{\bar{y}_w - 1} \simeq \frac{1}{\bar{y}_w} \qquad (7\text{-}66)$$

since $\bar{y}_w \gg 1$.

Since gelation occurs when a fraction $1/\bar{y}_w$ of all monomer units is cross-linked, a completely gelled polymer sample contains one cross-link per weight average molecule.

If all functional groups in a polymerization are equally reactive and some monomers have two functional groups, gel may be formed during the polymerization reaction. This "gel point" occurs at a degree of conversion to monomer of polymer, $p(0 < p < 1)$ such that

$$p = 1/\rho_0 \bar{y}_w^0 \qquad (7\text{-}67)$$

where ρ_0 is the fraction of reactive groups that are part of a multifunctional cross-linking agent and $\bar{y}_w^0$ is the weight average degree of polymerization that the polymer would have had if all monomers contained one functional group.

Cross-linking can occur in free-radical polymerizations because of chain transfer to polymer (Section 6.8.4) or monomer (Section 6.8.2), as well as the presence of multifunctional monomers. When it is desireable to retard or suppress crosslinking, chain transfer agents are added to the polymerization mixture. Equation (7-67) shows why this is helpful.

Consider the copolymerization of bifunctional monomer A with a monomer BB that contains two functional groups. If all the reactive groups are equally reactive, Eq. (7-67) is equivalent to

$$p = \frac{[A] + 2[BB]}{2[BB]\bar{y}_w^0} = \frac{[A] + [B]}{[B]\bar{y}_w^0} \qquad (7\text{-}68)$$

where the [A] and [B] are concentrations of functional groups and $\bar{y}_w^0$ is the weight average degree of polymerization that would be observed in the homopolymerization of monomer A under the particular reaction conditions. For example, if styrene were being polymerized under conditions such that $\bar{y}_w$ of the polystyrene being made were 1000, the product would be completely gelled at full monomer conversion if the reaction mixture contained 0.5×10^{-3} mol fraction of divinyl benzene and the vinyl groups in the latter monomer were equally as reactive as those in styrene. (Their reactivities are not exactly equal [23].)

Equation (7-67) indicates that the extent of reaction at which gelation occurs can be increased by reducing the concentration of divinyl monomer, by reducing the weight average chain length (increase initiator concentration or add chain transfer agents), or by using a divinyl monomer in which the second vinyl group is less reactive than those in the monovinyl comonomer.

Note that the foregoing analysis assumes that no cross-links are wasted. The calculations thus function best at low diene concentrations. In practice, gelation will be observed at higher values of conversion than predicted, and the error will increase with increasing diene concentration and hence with increasing wastage by cross-linking different parts of the same molecule.

7.10   REACTIVITIES OF RADICALS AND MONOMERS

The reactivities of various chemical species are usually assessed by comparing rate constants for selected reactions. This is not a convenient procedure in free-radical polymerizations, however, because absolute rate constant measurements are rare. More convenient and plentiful parameters in free-radical systems are functions of more than one rate constant as in the $(k_p/k_t^{1/2})$ factor, reactivity ratios, and chain transfer constants.

The relative reactivities of various monomers towards a given radical can be computed from the reciprocals of the reactivity ratios. Thus, if the reference monomer is M_1 its reactivity ratio with each comonomer is $r_1 = k_{11}/k_{12}$, with k_{11} constant for the particular series. Relative reactivities of test monomers towards radical $M_1^{\cdot}$ are given by comparing $1/r_1 = k_{12}/k_{11}$. Similar considerations apply to transfer constants, such as $C_S = k_{tr,S}/k_p$, for different transfer agents in the homopolymerization of a particular monomer.

The relative reactivities of different radicals towards the same monomer in copolymerization or toward the same chain transfer agent in homopolymerization can be assessed by comparing values of $k_{12} = k_{11}/r_1$ and $k_{tr,S} = k_p C_S = k_{11} C_S$, respectively.

When such comparisons are made it becomes clear that the reactivities of radicals, monomers, or transfer agents depend on the particular reaction being considered. It is not possible to conclude, for example, that poly(vinyl acetate) radical will always react x times more rapidly than polystyrene radical in addition reactions or y times as rapidly in the atom abstraction reactions involved in chain transfer. Similarly the relative order of efficiency of chain transfer agents will not be the same for all radical polymerizations. This is because resonance, steric, and polar influences all come into play and their effects can depend on the particular species involved in a reaction.

7.10.1   Resonance Effects

As a general rule, the controlling factor in reactivity of a vinyl monomer toward radical homopolymerization appears to be the stability of the radical formed by addition of the monomer to the initial radical.

Monomers that yield radicals in which the unpaired electron is extensively delocalized have ground state structures that are themselves resonance stabilized. The
important factor is the relative stability of the product radical, however, because a
single electron is more easily delocalized than one in a C=C double bond. Thus
resonance stabilization causes as increase in monomer reactivity and a decrease in
reactivity of the resulting polymer radical. Styrene is more reactive toward polymerization than vinyl acetate, for example, and the propagation rate in the former
polymerization is much slower than in the radical synthesis of poly(vinyl acetate).

Nonpolar radicals follow a consistent pattern, in which the basic reactivity
decreases with more extensive delocalization of the unpaired electron. Efficient
copolymerization between nonpolar monomers occurs only when both or neither
are resonance stabilized. The double bonds in styrene and butadiene are conjugated and the unpaired electrons are extensively delocalized in both corresponding
radicals. These monomers copolymerize but neither reacts with vinyl chloride to
any appreciable extent, because the addition of the latter monomer to the styryl
or butadiene radical is energetically unprofitable. Similarly, the chain transfer
constants of cyclohexane and of toluene are several hundredfold greater in vinyl
acetate than in styrene polymerizations. Also, since abstraction of a hydrogen
atom from the side group of isopropylbenzene produces a more stable radical than
that resulting from the same reaction on the methyl group of toluene, the former
is a more generally reactive chain transfer agent in free-radical polymerizations.

Basic reactivity is controlled by the extent of delocalization of the unpaired
electron. When the radical is significantly polar, however, its behavior during
a propagation reaction is no longer exclusively determined by the potential for
resonance stabilization.

7.10.2 Polar Effects

Free radicals and vinyl monomers are neutral, but variations in the reactivities of
both species can be rationalized and predicted by considering that the transition
states in their reactions may have some polar character. Appropriate substituents
may facilitate or hinder a particular reaction because of their influence on the
polarity of the reaction site.

An example from micromolecular chemistry involves the selectivity of hydrogen and chlorine atoms in abstraction of hydrogen from propionic acid (**7-1**).
Hydrogen atoms attack at carbon atom 2 more rapidly than at carbon 3, in accordance with the relative strengths of the

$$\overset{3}{C}H_3 - \overset{2}{C}H_2 - \overset{|}{C}OOH$$

7-1

bonds involved. Polar effects are not involved in these reactions. Electrophilic chlorine atoms exhibit the reverse selectivity, however. This can be attributed to the repulsion between the electron-seeking chlorine and carbon atom 2, which is rendered relatively electron deficient by the carbonyl group in the aliphatic acid.

Similarly, electrophilic CH_3 radicals add faster to ethylene than to tetrafluorethylene, in which the inductive effects of the fluorine substituents make the double bond somewhat electron deficient. The reverse selectivity is shown by CH_3 radicals, which are nucleophilic in character [24]. This parallels the observations in organic chemistry that carbonium ion formation is facilitated by replacement of hydrogens by alkyl groups and hindered by CF_3 groups. There is a polarization of σ electrons towards the trifluoromethyl group, whereas these electrons can be more effectively released by alkyl substituents.

Studies primarily of effects of substituents on reactions on phenyl rings permit the assignment of nucleophilic (electron-releasing) character to various groups. These include alkyls, vinyl, hydroxyl, ether, phenyl, and

$$-O-\underset{\underset{O}{\|}}{C}-R$$

species. Electrophilic (electron-withdrawing) substituents include halogen, nitro, cyanide, carboxyl, and carbonyl groups. Thus the CH_2 site in styrene is relatively electron rich while that in acrylonitrile has the opposite character, and the radicals derived from these monomers (**7-2** and **7-3**, respectively) also share the same tendency. Note that the isomers vinyl acetate (**7-4**) and methyl acrylate (**7-5**) differ in their selectivity. The double bond in vinyl acetate is weakly

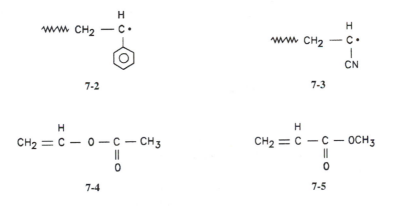

electron rich whereas that in methyl acrylate is electron poor, because it is conjugated with a carbonyl group on which slight excess charge can be stabilized.

Monomers with opposite polarities will tend to copolymerize in alternating fashion. Thus styrene and acrylonitrile copolymerize normally with an $r_1 r_2$ product effectively equal to zero. With more polar copolymerizing pairs, the use of a solvent with higher dielectric constant increases the alternation tendency and reduces the r_1r_2 product to a small, but significant extent. Similarly, complexing of one of the monomers increases alternation (Section 7.12.4).

The reactivity of a polar monomer can be considerably enhanced in copolymerization with a species of the opposite polarity. Maleic anhydride does not homopolymerize under normal free-radical reaction conditions but it forms 1:1 copolymers with styrene under the same conditions and even reacts with stilbene (**7-6**), which itself will not homopoloymerize.

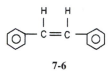

7-6

7.10.3   Steric Effects

Steric influences may retard some radical polymerizations and copolymerizations. Double bonds between substituted carbon atoms are relatively inert (unless the substituents are F atoms) and 1,2-substituted ethylenes do not homopolymerize in normal radical reactions. Where there is some tendency of such monomers to enter into polymers, the *trans* isomer is more reactive. When consideration is restricted to monomers that are doubly substituted on one carbon atom, it is usually assumed that steric effects can be neglected and that the influence of the two substituents is additive. Thus vinylidene chloride is generally more reactive in copolymerizations than is vinyl chloride.

7.11   ANALYSIS OF REACTIVITY DATA

Several attempts have been made to codify the relations between monomer reactivity ratios and structures. These approaches are essentially empirical but they are useful for predictions of reactivity ratios.

The $Q–e$ scheme [25] assumes that each radical or monomer can be classified according to its "general reactivity" and its polarity. The general reactivity of radical $\sim M_i^{\cdot}$ is represented by B_i while the corresponding factor for monomer M_j is Q_j. Polarity is denoted by e_i and e_j, with the e values for a monomer and its resulting radical being assumed to be equal. The rate constant for reaction (7-3) would then be expressed as $k_{12} = B_1 Q_2 \exp(-e_1 e_2)$. From analogous expressions for the three other propagation reactions, one obtains the following expressions for the reactivity ratios:

$$r_1 = (Q_1/Q_2) \exp[-e_1(e_1 - e_2)] \tag{7-69}$$

$$r_2 = (Q_2/Q_1) \exp[-e_2(e_2 - e_1)] \tag{7-70}$$

and

$$r_1 r_2 = \exp\left[-(e_1 - e_2)^2\right] \tag{7-71}$$

Experimental reactivity ratios provide values for (Q_1/Q_2) and $(e_2 - e_1)$ for a given comonomer pair. In order to obtain numerical values of the four Q and e parameters from two reactivity ratios, two of the former are assigned arbitrary values. Originally, the scheme was anchored by selecting styrene as the reference monomer with $Q = 1.0$ and $e = -0.8$. Later modifications have broadened the calculation base to include styrene copolymerization data of other well-researched monomers [26].

Extensive lists of Q–e values are given in handbooks and texts on copolymerization. Table 7-2 contains a brief list. As a general rule monomers with electron-rich double bonds have more negative e values and those which are extensively resonance stabilized have higher Q numbers.

The Q–e numbers are obtained by fitting to experimental reactivity ratios, many of which are not very accurate. It is not surprising, then, that this prediction scheme is not quantitatively reliable. It is nevertheless very simple and convenient to use and it is at least qualitatively reasonable. Monomers with very different e values will evidently react with higher absolute values of $(e_1 - e_2)^2$ values (Eq. 7-71). They will also be more likely to be influenced in their behavior by

TABLE 7-2

Q–e Values[26]

Monomer	e	Q
Butadiene	−0.50	1.70
Styrene	−0.80	1.00
Vinyl acetate	−0.88	0.03
Ethylene	0.05	0.01
Vinyl chloride	0.16	0.06
Vinylidene chloride	0.34	0.31
Methyl methacrylate	0.40	0.78
Methyl acrylate	0.64	0.45
Acrylic acid	0.88	0.83
Methacrylonitrile	0.68	0.86
Acrylonitrile	1.23	0.48
Methacrylamide	−0.05	0.40
Maleic Anhydride	3.69	0.86

complexing agents (Section 7.12.4). An alternative approach is the *patterns of reactivity* method [27] in which the rate constant for the reaction of a radical and any given substrate is expressed in a three-parameter form. This approach has been used to correlate other radical reactions as well as copolymerizations. It lacks the extensive parameter tabulations that have been made for the $Q-e$ scheme.

7.12   EFFECT OF REACTION CONDITIONS

7.12.1   Temperature[28]

The effect of temperature on reactivity ratios in free radical copolymerization is small. We can reasonably assume that the propagation rate constants in the reactions (7-2)–(7-5) can be represented by Arrhenius expressions over the range of temperatures of interest, such as

$$k_{11} = A_{11} \exp(-E_{11}/RT) \qquad (7\text{-}72)$$

where A_{11} is a temperature-independent *preexponential factor* and E is the activation energy. Then a reactivity ratio will be the ratio of two such expressions as in

$$r_1 = \frac{A_{11} \exp(-E_{11}/RT)}{A_{12} \exp(-E_{12}/RT)} \qquad (7\text{-}73)$$

Now, according to the transition-state theory of chemical reaction rates, the preexponential factors are related to the entropy of activation, ΔS_{ii}, of the particular reaction [$A_{ii} = (\kappa T/h)e^{\Delta S_{ii}/R}e^{\Delta n}$ where κ and h are the Boltzmann and Planck constants, respectively, and Δn is the change in the number of molecules when the transition state complex is formed.] Entropies of polymerization are usually negative, since there is a net decrease in disorder when the discrete radical and monomer combine. The range of values for vinyl monomers of major interest in connection with free radical copolymerization is not large (about -100 to -150 J K^{-1} mol^{-1}) and it is not unreasonable to suppose, therefore, that the A_{ii} values in Eq. (7-73) will be approximately equal. It follows then that

$$r_1 \simeq \exp(-E_{11} - E_{12})/RT \qquad (7\text{-}74)$$

and the temperature dependence of r_1 can therefore be approximated by

$$\frac{d \ln r_1}{d(1/T)} = -\frac{(E_{11} - E_{12})}{R} = T \ln r_1 \qquad (7\text{-}75)$$

Similarly,

$$\frac{d \ln r_2}{d(1/T)} = T \ln r_2 \qquad (7\text{-}76)$$

The activation energy for r_1, from the slope of an Arrhenius plot of $\ln r_1$ against $1/T$, (Eq. 7-74), will be equal to $(-RT \ln r_1)$. Similar expressions hold for r_2 and the product $r_1 r_2$. The absolute value of the logarithm of a number is a minimum when this number equals unity, and so a strong temperature dependence of r_1 will be expected only if either $r_i \gg 1$ or $r_i \ll 1$.

This reasoning predicts that a reactivity ratio or an $r_1 r_2$ product greater than unity will decrease with increasing temperature and vice versa. The tendency for random polymerization will increase and the tendency for monomer alternation will decrease with increasing reaction temperature, so long as the same copolymerization mechanism predominates over the experimental temperature range.

These predictions are essentially confirmed by experience. Most free-radical reactivity ratios are measured by convention at temperatures near 60°C, and the effect of changes in conditions in the range 0–90°C is usually assumed to be negligible, compared to the experimental difficulties in detecting the effects of slight variations in r_1 or r_2.

7.12.2 Pressure

In transition-state theory, the temperature dependency of a rate constant k on pressure P can be expressed as

$$\left(\frac{\partial \ln k}{\partial P} \right)_T = -\frac{\Delta V^*}{RT} \qquad (7\text{-}77)$$

where the activation volume ΔV^* represents the volume change that occurs when the transition state is formed from the reactants. The pressure dependence of a reactivity ratio is then of the form

$$-RT \partial \ln r_i / \partial P = \Delta V_{ii}^* - \Delta V_{ij}^* \qquad (7\text{-}78)$$

and reflects the differences between activation volumes for reaction of radical M_i with monomer M_i or M_j. This difference is usually not large, and the variation of reactivity ratios with pressure appears to be very small in the systems that have been studied.

7.12.3   Medium

The copolymer composition equation is written in terms of monomer concentrations at the locus of reaction. The same reactivity ratios should apply in principle whether the polymerization is carried out in bulk, solution, suspension, or emulsion systems. In general, the only concentration values available to the experimenter are the overall bulk figures. Deviations of copolymer composition can be expected, therefore, if the concentrations at the polymerization sites differ from these figures. This can occur in emulsion systems, for example, if the monomers differ appreciably in aqueous solubility and diffusion rates.

The occurrence of a homogeneous reaction system is also implicit in the derivation of the copolymer composition equation. Some polymers, like poly(vinylidene chloride), are insoluble in their own monomer and are not highly swollen by monomer. In emulsion copolymerizations of such reactants the relative concentrations of the comonomers in the polymerizing particles will be influenced by the amounts that can be adsorbed on the surface or absorbed into the interior of these polymerization loci.

Similarly the copolymerization behavior of acidic or basic monomers or of ionizable monomers in aqueous solutions may be affected by the pH of the reaction medium.

Early work indicated that the nature of the reaction medium had no effect on the course of free radical copolymerizations in homogeneous reaction systems. More recent studies have not always supported this conclusion and it has been suggested that a "bootstrap effect" may be operating whereby there is a partitioning of the comonomers between the bulk of the reaction medium and the polymerization locus (i.e., the macroradical end) [29].

7.12.4   Effects of Complex Formation

Monomers with electron-rich double bonds produce one-to-one copolymers with monomers having electron-poor double bonds in reaction systems that also contain certain Lewis acids. These latter are halides or alkyl halides of nontransition metal elements, including $AlCl_2$, $ZnCl_2$, $SnCl_4$, BF_3, $Al(CH_2CH_3)Cl_2$, alkyl boron halides, and other compounds. The acceptor monomer generally has a cyano or carbonyl group conjugated to a vinyl double bond. Examples are acrylic and methacrylic acids and their esters, acrylonitrile, vinyl ketones, maleic anydride, fumaric esters, vinylidene cyanide, sulfur dioxide, and carbon monoxide. The variety of donor molecules is large and includes various olefins, styrene, isoprene, vinyl halides and esters, vinylidene halides, and allyl monomers [30].

The donor molecules usually have *e* values (Section 7.11) of less than 0.5. Monomers with negative *e* values are especially effective. Acceptor molecules have higher *e* values. Copolymerization conditions can be fairly mild for conjugated donor monomers but use of nonconjugated donors requires more careful selection of reactants and conditions.

In general, an alternating copolymer is formed over a wide range of monomer compositions. It has been reported that little chain transfer occurs, and in some cases, conventional free radical retarders are ineffective. Reaction occurs with some combinations, like styrene–acrylonitrile, when the monomers are mixed with a Lewis acid, but addition of a free-radical source will increase the rate of polymerization without changing the alternating nature of the copolymer. Alternating copolymerizations can also be initiated photochemically and electrochemically. The copolymerization is often accompanied by a cationic polymerization of the donor monomer.

One practical disadvantage of these systems lies in the fact that relatively high concentrations of Lewis acids are needed to achieve alternation. Thus methyl methacrylate and styrene alternate perfectly with azodiisobutyronitrile initiation at 50°C when the molar ratio of $ZnCl_2$ to methyl methacrylate is 0.4. Alternation is less exact, however, when this ratio is 0.25. Alkyl boron halides like ethyl boron dichloride are effective at lower concentrations and act also as initiators if oxygen is present.

7.13 RATES OF FREE-RADICAL COPOLYMERIZATIONS

The simple copolymer model, with two reactivity ratios for a binary comonomer reaction, explains copolymer composition data for many systems. It appears to be inadequate, however, for prediction of copolymerization rates. (The details of various models that have been advanced for this purpose are omitted here, in view of their limited success.) Copolymerization rates have been rationalized as a function of feed composition by invoking more complicated models in which the reactivity of a macroradical is assumed to depend not just on the terminal monmomer unit but on the two last monomers in the radical-ended chain. This is the penultimate model, which is mentioned in the next Section.

At present, the kinetic parameters for prediction of copolymerization rates are scanty, except for a few low conversion copolymerizations of styrene and some acrylic comonomers. Engineering models of high conversion copolymerizations are, however, overdetermined, in the sense that the number of input parameters (kinetic rate constants, activation energies, enthalpies of polymerization, and so on) outnumber the output parameters (such as copolymer conversion and composition). Any number of copolymerization rate models may be found to fit the experimental data adequately, because the model designer is forced to rely on some adjustable parameters at some point in the exercise. This does not mean that such models may not be useful, for training reactor operators, for example, but it does indicate that any such model will be more reliable for interpolation of data between than for insights into the mechanism of the copolyermization.

7.14 ALTERNATIVE COPOLYMERIZATION MODELS

A series of judgments, revised without ceasing, goes to make up the incontestable progress of science.

—Duclaux

The extensive reactivity ratio data in the literature exhibit a wide scatter for many monomer pairs. This is partly due to errors in copolymer analysis and computational methods that result in larger uncertainties in r_1 and r_2 than was realized when the results were reported. Another factor reflects the frequent reliance on an inadequate number of data points because copolymerization experiments tend to be tedious and time consuming.

Deviations from the behavior of the simple copolymer model have been noted for various systems and have prompted the development of alternative models, all of which use more parameters than the two reactivity ratios in Eq. (7-13). Such models will often fit particular sets of copolymerization data better than the simple copolymer model. It appears in retrospect, however, that many of the apparent deviations from this model may be accounted for by large uncertainties in reactivity ratio values. The inadequacy of the simple copolymer theory can be established only if deviations between calculated and observed copolymer compositions are shown to be systematic as the feed composition or monomer dilution is varied. Random errors do not necessarily show that the basic model is inapplicable.

The simple copolymer model is a first-order Markov chain in which the probability of reaction of a given monomer and a macroradical depends only on the terminal unit in the radical. This involves consideration of four propagation rate constants in binary copolymerizations, Eqs. (7-2)–(7-4). The mechanism can be extended by including a penultimate unit effect in the macroradical. This involves eight rate constants. A third-order case includes antepenultimate units and 16 rate coefficients. A true test of this model is not provided by fitting experimental and predicted copolymer compositions, since a match must be obtained sooner or later if the number of data points is not saturated by the adjustable reactivity ratios.

A much better test is provided by comparing sequence distributions of monomer units in the copolymers. Such data are unfortunately very sparse at this time, and the reality of the penultimate unit model has not been proved.

An alternative hypothesis to account for deviations from simple copolymer behavior invokes reactions in which one or more of the propagation steps is significantly reversible under the reaction conditions. For example, if reaction (7-4) is not important, k_{22} and r_2 will both be zero and Eq. (7-13) will reduce to

$$d[M_1]/d[M_2] = 1 + r_1[M_1]/[M_2] \tag{7-79}$$

This is the case, for example, in the copolymerization of carbon monoxide and ethylene where the CO will not add to itself but does copolymerize with the olefin monomer. General theoretical treatments have been developed for such cases, taking into account temperature and penultimate effects. Again, the superiority of these more complicated theories over the simpler copolymer model is not proved for all systems to which they have been applied.

PROBLEMS

7-1. (a) Calculate the copolymer composition (in mole percent) formed at an early stage of the reaction of methyl methacrylate (monomer 1) at 5 mol/liter and 5-ethyl-2-vinyl pyridine at 1 mol/liter concentration. Reactivity ratios are $r_1 = 0.40$ and $r_2 = 0.69$.
(b) What molar ratio of monomers in the feed produces a copolymer composition which is the same as the feed composition?

7-2. Calculate the composition (mole fractions) of the initial terpolymer which would be formed from the radical polymerization of a feed containing 0.414 mol fraction methacrylonitrile (MAN), 0.424 mol fraction styrene (S), and 0.162 mol fraction alpha-methylstyrene (AMS). Reactivity ratios are:

$$\begin{array}{lll} MAN(M_1)/S: & r_1 = 0.44, & r_2 = 0.37 \\ MAN(M_1)/AMS: & r_1 = 0.38, & r_2 = 0.53 \\ S(M_1)/AMS: & r_1 = 1.124, & r_2 = 0.627 \end{array}$$

7-3. When 0.3 mol fraction methacrylonitrile is copolymerized with styrene in a radical reaction, what is the average length of sequences of each monomer in the copolymer?

7-4. Rank the following monomers in order of their increased tendency to alternate in copolymerization with butadiene and explain your reasoning: vinyl acetate, styrene, acrylonitrile, and methyl methacrylate, (*Hint:* Use $Q-e$ values if reactivity ratios are not readily available.)

7-5. (a) Acrylonitrile (monomer 1, $r_1 = 0.9$) is copolymerized with 0.25 mol fraction vinylidene chloride (monomer 2, $r_2 = 0.4$). What fraction of the acrylonitrile sequences contain 3 ore more acrylonitrile units? (Polyacrylonitrile is used to make "acrylic" fibers. Copolymerization with vinylidene chloride reduces the flammability of such products. Such copolymers form the basis of "mod-acrylic" fibers.)

(b) What is the feed composition for copolymerization of vinylidene chloride and acrylonitrile such that the copolymer composition does not vary with the conversion of monomers to polymer?

7-6. The extent of reaction at which gelation occurs in the copolymerization of a vinyl monomer and a divinyl monomer can be increased by increasing the concentration of free-radical initiator, at fixed polymerization temperature and comonomer feed composition. Explain why this should be so.

7-7. A thermosetting appliance enamel consists of a terpolymer comprising about 72 parts of vinyl toluene (70/40 meta/para) with about 20 parts of ethyl acrylate (to reduce brittleness of the copolymer) and 8 parts of an acidic vinyl comonomer. The acid is incorporated in the copolymer to provide sites for subsequent cross-linking with a diepoxide. It seems reasonable to expect that grease and stain resistance of the cross-linked enamel will be enhanced if the cross-links are not clustered and almost all initial polymer molecules contain at least one or a few cross-linking sites. To achieve this in a batch copolymerization, what are the best reactivity ratios (approximately) of the major component (vinyl toluene) and the vinyl acid comonomer? Show you reasoning.

7-8. It has been suggested that free-radical polymerization would be a useful way to react an equimolar mixture of allyl acetate and methyl methacrylate. Is this a good idea? (Allyl acetate: $e = -1.07$, $Q = 0.24$; methyl methacrylate: $e = 0.40$, $Q = 0.78$).

7-9. The copolymerization of ethylene and propylene is found to be essentially random ($r_1 r_2 \cong 1$) with $(C_2H_5)_2AlCl/VO(OC_2H_5)_3$ catalyst in chlorobenzene at 30°C. (Such polymerizations are discussed in Chapter 9.) The control of such systems is frequently on monomer concentration in the gas phase over the reaction mixture. This is because gas phase concentrations vary less with temperature, pressure, and solvent. What monomer composition in the gas phase is needed to produce a copolymer containing 30 mol% propylene. The reactivity ratio of ethylene (r_1) has been found to be 5, based on gas phase concentrations.

7-10. In the copolymerization of vinyl chloride and vinyl acetate, what monomer feed composition is needed to produce a copolymer containing 5 mol % vinyl acetate? (The reactivity ratios are listed in Table 7-1.)

7-11. Copolymers of methyl methacrylate and styrene (and other monomers) are made with acrylamide or methacrylamide and then post-reacted with formaldehyde and alcohols to yield the N-alkoxymethyl derivative of the acrylamide unit:

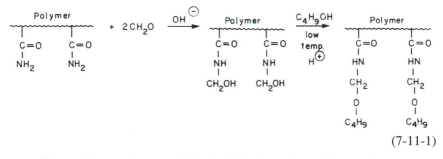

$$(7\text{-}11\text{-}1)$$

The product can be cross-linked with hydroxyl-containing polymers:

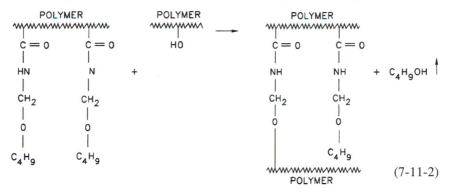

$$(7\text{-}11\text{-}2)$$

Such materials are used in baking finishes. Assume that a manufacturer wishes to make a polymer based on methyl methacrylate or styrene and about 15 mol% acrylamide. If the following reactivity ratios are correct, is there any advantage in terms of evenly spaced cross-links in using either styrene or methyl methacrylate as the major component?

$$\text{Methylmethacrylate}(M_1): \quad r_1 = 1.65$$
$$\text{Acrylamide}(M_2): \quad r_1 = 0.5$$
$$\text{Styrene}(M_1): \quad r_1 = 1.13$$
$$\text{Acrylamide}(M_2): \quad r_1 = 0.60$$

7-12. Following is a sample polycondensation recipe:

Ethylene glycol	2 mol
Adipic acid	1.90 mol
Tricarballylic acid (1,2,3-propanetricarboxylic acid)	0.051 mol

(a) At what degree of conversion (p) will the mixture gel according to the concepts described in Section 7.9. [*Hint:* In bifunctional condensation polymerizations the weight average degree of polymerization of the reaction mixture is $(1 + p)/(1 - p)$ (Eq. 5-35)].

(b) How does this estimated gel point differ from that calculated with the aid of the Carothers equation (Section 5.4.2)? What is the reason for the difference in the two estimates?

REFERENCES

[1]   F. R. Mayo and F.M. Lewis, Jr., *J. Am. Chem. Soc.* **66**, 1594 (1944).
[2]   T. Alfrey, Jr., and G. Goldfinger, *J. Chem. Phys.* **12**, 322 (1944).
[3]   R. K. Kruse, *J. Polym. Sci. Part B* **5**, 437 (1967).
[4]   R.M. Joshi, *J. Macromol. Sci. Chem.* **A7**, 1231 (1973).
[5]   A. Rudin, S. S. M. Chiang, H.K. Johnston, and P. D. Paulin, *Can. J. Chem.* **50**, 1757 (1972).
[6]   M. Fineman and S. D. Ross, *J. Polym. Sci.* **5**, 259 (1950).
[7]   P. W. Tidwell and G. A. Mortimer, *J. Macromol. Sci.-Rev.* **C4**, 281 (1970).
[8]   A. I. Yezreelev, E. L. Brokhina, and Y.S. Roskin, *Polym, Sci. (USSR)* **11**, 1894 (1969).
[9]   T. Kelen and F. Tudos, *J. Macromol. Sci.-Chem.* **A9**, 1 (1975).
[10] H. K. Johnston and A. Rudin, *J. Paint Technol.* **42** (547), 429 (1970).
[11] D. Braun, W. Brendlein, and G. Mott, *Eur. Polym. J.* **9**, 1007 (1973).
[12] R. C. McFarlane, P. M. Reilly, and K. F. O'Driscoll, *J. Polym. Sci. Chem. Ed.* **18**, 251 (1980).
[13] P. W. Tidwell and G. A. Mortimer, *J. Polym. Sci. Part A* **3**, 369 (1965).
[14] K. F. O'Driscoll and P. M. Reilly, *Makromol. Chem., Macromol. Symp.* **10/11**, 355 (1987).
[15] P. J. Rossignoli and T. A. Duever, *Polym. React. Eng.* **3**, 361 (1995).
[16] C. Walling and E. R. Briggs, *J. Am. Chem. Soc.* **67**, 1774 (1945).
[17] L. Crescentini, G. B. Gechele, and A. Zanella, *J. Appl. Polym. Sci.* **9**, 1323 (1965).
[18] T. A. Duever, K. F. O'Driscoll, and P. M. Reilly, *J. Polym. Sci. Polym. Chem. Ed.* **21**, 2003 (1983).
[19] A. Rudin, K. F. O'Driscoll, and M. S. Rumack, *Polymer* **22**, 740 (1981).
[20] C. Tosi, *Adv. Polym. Sci.* **5**, 451 (1968).
[21] W.H. Stockmayer, *J. Chem. Phys.* **13**, 199 (1945).
[22] P. J. Flory, "Principles of Polymer Chemistry." Cornell University Press, Ithaca, NY, 1953.
[23] C. D. Frick, A. Rudin, and R. H. Wiley, *J. Macromol. Sci. Chem. A* **16**, 1275 (1981).
[24] J. M. Tedder, *in* "Reactivity, Mechanism and Structure in Polymer Chemistry" (A. D. Jenkins and A. Ledwith, Ed.). Wiley, New York, 1974.
[25] T. Alfrey, Jr. and C. C. Price, *J. Polym. Sci.* **2**, 101 (1947).
[26] R. Z. Greenley, pII/267 *in* "Polymer Handbook," 3rd ed. (J. Brandrup and E. H. Immergut, Eds.). John Wiley and Sons, New York, 1989.
[27] C. H. Bamford and A. D. Jenkins, *Trans. Faraday Soc.* **59**, 530 (1963).
[28] K. F. O'Driscoll, *J. Macromol. Sci. Chem.* **A3**, 307 (1969).
[29] H. J. Harwood, *Makromol. Chem. Macromol. Symp.* **10/11**, 331 (1987).
[30] M. Hirooka, H. Yobiuchi, S. Kawasumi, and K. Nakaguchi, *J. Polym. Sci. Part A-1* **11**, 128 (1973).

Chapter 8

Dispersion and Emulsion Polymerizations

"Science" means simply the aggregate of all the recipes that are always successful. All the rest is literature.
—Paul Valéry, *Moralités* (1932)

The discussion of free-radical polymerizations in Chapters 6 and 7 focused primarily on homogeneous reaction systems, in which monomer, polymer, and any solvent were all miscible. This conventional presentation makes it much easier to grasp the fundamentals of free-radical polymerizations. In fact, however, many large-scale processes are carried out in heterogeneous systems, because these offer advantages over alternative procedures. Their overall importance is such as to justify this chapter describing the effects of process conditions on polymer properties.

There are two basic types of heterogeneous polymerizations:

1.   The initial reaction medium comprises several phases and polymerization occurs in a heterogeneous system, as in emulsion and suspension reactions.

2.   The reaction medium is initially homogeneous and the polymer forms a separate phase as the polymerization proceeds. Examples are dispersion reactions, described below, and polymerization of acrylonitrile. Various polyolefin processes discussed in Chapter 9 are other examples.

Description of suspension polymerizations fits most appropriately in Chapter 10, where polymer reaction engineering is the topic. This chapter focuses on dispersion and emulsion polymerizations.

277

8.1   DISPERSION POLYMERIZATION

In this process, the monomer and initiator are soluble in the continuous phase and the polymer particles, which precipitate as they are produced, are stabilized against coagulation by dispersants that comprise different segments that are respectively soluble and insoluble in the continuous phase. Dispersion polymerizations have been used successfully as an alternative to solution polymerization of vinyl polymers for application as surface coatings. In that case the diluents are usually aliphatic hydrocarbons, and the process acronym is NAD [for nonaqueous dispersion].

Before considering process details, a short digression is worthwhile to compare alternative means for applying an organic coating, say, an acrylic copolymer, to a metal substrate. Save for powder coatings, which are outside the scope of this text, the film former must be diluted before application, which will be by spraying on an industrial scale. Solution coatings have viscosities dependent on polymer molecular weight and concentration. By contrast, the viscosities of dispersion coatings are independent of polymer molecular weight (viscosities of suspensions depend on the size and concentration of the particles, Chapter 3) and are lower than those of solutions with practical polymer levels. A third alternative involves use of aqueous emulsions. Here again, there are advantages in terms of low application viscosities and high polymer concentrations. The emulsion system has the further benefit of containing low levels of volatile organic materials. Emulsion-based coatings contain higher polymer concentrations than NADs, where the solute levels are limited by the formation of some soluble low-molecular-weight polymer. The cost of drying the coating is in favor of the NAD, however, since the boiling point and heat of vaporization of a low-molecular-weight hydrocarbon are appreciably lower than those of water. In both the NAD and emulsion systems, the polymer is made in the diluent in which it will be applied; solution-polymerized acrylics require the use of expensive solvents like ketones and esters. In industrial spray applications, NADs have the advantages of rapid release of diluent between the substrate and the spray nozzle, better resistance to sagging (the polymer contains less solvent and has a lower tendency to flow) and fewer problems with solvent popping (caused by release of the retained diluent during drying). In summary, then, while the chemical nature of the coating will be approximately the same in all three cases, the choice of polymerization process can have profound effects on the subsequent performance of the material.

During dispersion polymerization polymer particles are formed from an initially homogeneous reaction mixture by polymerization in the presence of a polymeric steric stabilizer. The process is applicable to monomers which yield polymers that are insoluble in a solvent for the monomer. Styrene has been polymerized in alcohols, with steric stabilizers such as poly(N-vinylpyrrolidone) (see Fig. 1-4 for monomer structure) or hydroxypropyl cellulose. Hydrocarbon solvents are used for the polymerization of methacrylic esters, using steric stabilizers like poly(12-hydroxystearic acid), polyisobutene, poly(dimethylsiloxane) (**1-44**) and other polymers. The dispersions are suitable for use as surface coatings and the polymer particles themselves have applications as toners in xerography and as chromatographic packing materials. Particle sizes can be produced with narrow distributions, when necessary, and diameters between 0.1 and about 15 μm are attainable. Other features of NADs include their ability to make dispersions of polymers like polyacrylonitrile that are insoluble in common commercial solvents. By controlling the monomer feed in a semicontinuous polymerization certain monomers may be concentrated toward the interior or surfaces of the polymer particles, for example, so that particle fusion is favored during the baking cycle after a coating has been applied to a substrate. This is illustrated in the two-stage polymerization of an acrylonitrile/alkyl acrylate copolymer. High acrylonitrile contents are used in the first-stage polymerization, and the higher acrylate composition of the second-stage reaction gives an NAD with superior film properties, since the acrylate-rich copolymer on the particle surfaces fuses readily.

Early work on production of stable polymer dispersions in hydrocarbon diluents is summarized in Ref. [1]. More recently, publications in this field have been directed at polymerizations of nonpolar monomers in polar media, with the aim of producing large particles with narrow size distributions for use such as toners and packings for solid phase peptide syntheses [2, 3].

Stabilizers are generally polymers and are added as preformed graft or block polymers or as a precursor polymer which grafts *in situ* during the polymerization. The polymerization of styrene in ethanol, with poly(vinyl pyrrolidone) (PVP) stabilizer serves as a useful example here. The starting mixture, containing an initiator like azoisobutyronitrile, is a single-phase solution. When the reaction mixture is heated to decompose the initiator both PS homopolymer and PVP–PS graft copolymers are formed. The growing PS polymers soon reach their solubility limit in ethanol and associate into unstable nuclei, which collapse together to reduce the polymer/alcohol interfacial area. This coagulation process continues until the graft copolymer has been adsorbed in sufficient surface concentration to inhibit further coalescence. This marks the end of the nucleation period in the reaction. Once the initial crop of particles precipitates, the reaction can be continued so that the particles grow without fresh nucleation. This is accomplished by control of the dispersant concentration at the level needed for the particle size that is wanted. Insufficient stabilizer results in particle coalescence. In practice, such control is achieved by establishing empirical relations for particular polymerization systems of interest. The end result is particles with insoluble (in the diluent) PS cores, shielded by anchored surface layers of PVP–PS graft copolymers in which the PVP segments are dissolved in the ethanol medium.

Coagulation of the polymer particles is hindered by *steric stabilization*. The nature of the soluble segment of the stabilizer polymer is not important as long as it is miscible with the reaction medium, while the counterpart of the stabilizer and the polymer particle are effectively insoluble. Polymer particles are then surrounded by a shroud of attached, solvated polymeric fibrils. When two particles with attached soluble segments on their surfaces approach, the number of conformations available to the soluble dispersant polymer is reduced and their entropy decreases (ΔS for this compression is negative). Since the corresponding change in Gibbs free energy is positive (recall that $\Delta G = \Delta H - T \Delta S$), the close approach of sterically shielded polymer particles is energetically unprofitable. In effect, the higher concentration of dissolved polymer segments in the interparticle solvent region generates an osmotic pressure (cf. Eq. 2-68), which tends to force the particles apart. The same phenomenon operates in emulsion polymerization systems, which are the next topic in this chapter. In the latter case, however, the dispersion medium is usually water, and steric stabilization typical of NAD reactions is supplemented by ionic repulsions.

Most successful dispersants are based on block or graft polymers which are physically adsorbed on the polymer particle surfaces. For a copolymer segment to be sufficiently insoluble to function as an anchor group the molecular weight is usually >1000. The soluble portion of the dispersant is also at least about the same size. The chemical nature of the soluble portion is not as important as the requirement that it be freely soluble in the diluent. The same soluble segment has been used to stabilize dispersions of widely different polymers. A case in point is the use of poly(12-hydroxystearic acid) in aliphatic hydrocarbon continuous phases. The anchor component of the dispersant is, of course, specific to the particular polymer that is being synthesized, and can be provided by graft polymerization of the disperse polymer during its synthesis. Such graft copolymerizations in free-radical systems are initiated by the creation of radicals on the backbone of the soluble dispersant portion, by atom abstraction. The mechanism is exactly as in chain transfer to polymer (Section 6.8.4). The precursor soluble component may be modified, if necessary, to facilitate free-radical chain transfer.

Covalent links or acid–base interactions are alternatives to physical adsorption of the anchor segments of the dispersants. Such procedures are valuable in special cases, where other factors compensate for added complication of the polymerization process.

Mention of several possible operating variations may help the reader's understanding of the dispersion polymerization process. It is important to note, first, that dispersion polymerizations are usually conducted so that the monomers and other appropriate ingredients are metered into the vessel during the course of the reaction. (That is, these are so-called "semibatch" operations, which are considered in more detail in Chapter 10.) In polymerizations in a nonpolar diluent like an aliphatic hydrocarbon, the introduction of a small amount of a highly polar comonomer at the start of the reaction reduces the solubility of both the disperse polymer and the anchor group. As a result, anchoring is stronger and the precipitated polymer particles are finer. Conversely, to obtain a coarser particle size product a small quantity of strong solvent for the polymer may be added at the beginning of the polymerization. Alternatively, the process may be started with an increased amount of monomer in the reaction vessel, since monomers are usually solvents for their own polymers.

8.2   EMULSION POLYMERIZATION

An emulsion consists of a discontinuous liquid phase dispersed throughout a different, continuous liquid phase. Milk and the sap of the rubber tree are examples of naturally occurring emulsions. The term *latex* is used also to denote aqueous dispersions of polymers.

Most emulsion polymerizations are free-radical reactions. The main difference from alternative free-radical polymerizations, such as those in bulk, solution, and suspension systems, is that the propagating macroradicals in emulsion reactions are isolated from each other. Encounters between macroradicals are hindered as a consequence, and termination reactions are less frequent than in comparable systems in which the reaction mixture is not subdivided. Emulsion polymerizations thus often yield high-molecular-weight products at fast rates when suspension or bulk reactions of the same monomers are inefficient.

Both the emulsion and suspension processes use water as a heat sink. Polymerization reactions are easier to control in both these processes than in bulk or solution systems because stirring is easier and removal of the exothermic heat of polymerization is facilitated.

The free-radical kinetics described in Chapter 6 hold for homogeneous systems. They will prevail in well-stirred bulk or solution polymerizations or in suspension polymerizations if the polymer is soluble in its monomer. Polystyrene suspension polymerization is an important commercial example of this reaction type. Suspension polymerizations of vinyl chloride and of acrylonitrile are described by somewhat different kinetic schemes because the polymers precipitate in these cases. Emulsion polymerizations are controlled by still different reaction parameters because the growing macroradicals are isolated in small volume elements and because the free radicals which initiate the polymerization process are generated in the aqueous phase. The emulsion process is now used to make large tonnages of styrene–butadiene rubber (SBR), latex paints and adhesives, PVC "paste" polymers, and other products.

Emulsion polymerizations vary greatly, and no single reaction mechanism accounts for the behavior of all the important systems. The kinetics and mechanism of emulsion polymerizations are reviewed in detail in Ref. [4]. It is important to note that the nature of the products made by emulsion reactions are highly dependent on the details of the process whereby they were produced. (Incidentally, this is an advantage industrially, because, although the component monomers in complicated structures can be identified fairly readily, competitors cannot easily deduce the polymerization process by examining the final product.) Emulsion polymerization is described in a qualitative format in this introductory text, partly in order to emphasize the versatility of the process. Mathematical features of the reaction kinetics have not been covered because emulsion polymerizations are controlled most effectively by the rates and modes of reactant addition.

Emulsion polymerization is a particularly attractive route for the production and control of polymer structures on a size scale from a few hundred nanometers to several microns. This section is then the first instance in this text in which polymerization reactions are considered on a scale larger than molecular.

The essential ingredients in an emulsion polymerization are the water, a monomer which is not miscible with water, an oil-in-water emulsifier, and a compound or compounds which release free radicals in the aqueous phase. Other ingredients which may be used in practical recipes are mentioned briefly later. Typical proportions (by weight) are monomers 100, water 150, emulsifier 2-5, and initiator 0.5, although these ratios may vary over a wide range.

Figure 8-1 is a schematic of a typical laboratory apparatus for emulsion polymerization. Industrial reactors are usually large-scale versions of this basic

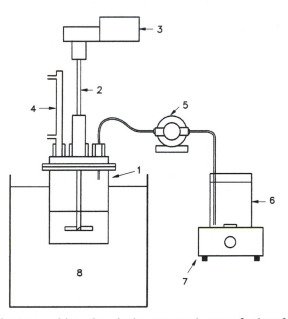

Fig. 8-1.   Laboratory emulsion polymerization apparatus. 1, reactor; 2, stirrer; 3, motor; 4, reflux condenser; 5, pump; 6, monomer feed; 7, magnetic stirrer; 8, water bath.

arrangement, with the modifications that the reactor will have a bottom discharge valve and a jacket for temperature control.

Most emulsifiers are either anionic surfactants, including salts of fatty acids or alkane sulfonic acids, or nonionics based on polyether groupings or sugar derivatives. A typical emulsifier will have a molecular weight around 300, so that there will be about 6×10^{19} surfactant molecules per milliliter of water if the bulk concentration of soap is 30g/liter. Emulsifiers are molecularly dissolved in the water at very low concentrations, but they form aggregates called micelles at their critical micelle concentrations (cmc). At concentrations higher than the cmc all the surfactant in excess of this concentration will be micellar. If the emulsifier has very low solubility in water, the course of the reaction will be affected by micellar soap. The cmc of the commonly used surfactant sodium dodecyl sulfate [also known as sodium lauryl sulfate, $Na^+ \ ^-OSO_3CH_2(CH_2)_{10}CH_3$] is about 6×10^{-3} mol/liter at 80°C.

Micelles are approximately spherical aggregates of surfactant molecules with their nonpolar tails in the interior and their hydrophilic ends oriented towards the aqueous medium. They are some 50–100 Å in diameter. The bulk concentration of surfactant is usually around 0.1 M and this corresponds to approximately 10^{18} micelles per milliliter of aqueous phase, since there are typically about 50–100 emulsifier molecules per micelle. The apparent water solubility of organic molecules is enhanced by micellar surfactants, because the organic molecules are absorbed into the micelle interiors. The extent of this "solubilization" of organic molecules depends on the surfactant type and concentration, the nature of the solubilized organic substance, and the concentration of electrolytes in the aqueous phase. As an example, there will be about an equal number of styrene molecules and potassium hexadecanoate (palmitate) molecules in a micelle of the latter material. In this case about half the volume of the micelle interior is occupied by solubilized monomer, and the concentration of styrene is approximately 4.5 M at this site. Thus radical polymerization starts very rapidly in the interior of a micelle once it is initiated there.

In batch emulsion polymerizations all ingredients are present in the reactor at the start of the reaction and much more monomer is present than can be incorporated into the available micelles. When monomer is being added to an aqueous phase containing a surfactant, the onset of turbidity signals saturation of the micelles. About 99% of the monomer will be located initially in droplets with diameters in the range of 10^{-4} cm. There are typically some 10^{12} soap covered droplets per milliliter of water.

Before polymerization starts there will thus be some monomer "solubilized" inside the micelles, more monomer in soap covered large droplets, and perhaps a small amount of monomer in true solution in the water. Emulsifier will also be located in the micelles, in aqueous solution, and on the surfaces of the monomer droplets. Most of this soap will be located in the micelles. The concentrations of dissolved monomer and soap are, however, not negligible. Styrene is regarded as insoluble in water, for example, since its solubility is limited to about $2.10^{-3} M$. This corresponds to a concentration of about 10^{18} molecules per cm^{-3}. Butyl acrylate is five times as soluble as styrene, while methyl methacrylate is about a 100-fold more soluble in water. Free radicals produced in the water phase are thus likely to find monomer molecules in their close proximity, except for exceptionally insoluble monomers like alkyl derivatives of styrene.

It is characteristic of emulsion polymerizations that free radicals are generated in the aqueous phase. (If radicals were generated in the monomer droplets the reaction would behave like a suspension polymerization.) The most widely used initiators are water soluble salts of persulfuric acid. The decomposition of these chemicals at pH values usually encountered in emulsion systems proceeds according to

$$S_2O_8^{2-} \rightarrow 2\,^{\bullet}OSO_3^- \tag{8-1}$$

$$\cdot SO_4^- + H_2O \rightarrow HSO_4^- + \cdot OH \tag{8-2}$$

$$2 \cdot OH \rightarrow H_2O + \frac{1}{2}O_2 \tag{8-3}$$

Each initiator molecule yields two primary radicals which can be sulfate ion or hydroxyl radicals. Carboxylate soaps and other ingredients of the reaction mixture may accelerate the rate of decomposition of persulfates. Their useful temperature range is generally between 40 and 90°C.

Redox systems are used for polymerizations at lower temperatures. Many of these redox initiator couples were developed for the emulsion copolymerization of butadiene and styrene, since the 5–10°C "cold recipe" yields a better rubber than the "hot" 50°C emulsion polymerization.

Common redox systems include the persulfate–bisulfite couple:

$$S_2O_8^{2-} + HSO_3^- \rightarrow SO_4^{2-} + SO_4^{\cdot -} + HSO_3 \tag{8-4}$$

and ferrous ion reducing agents combined with persulfate or hydroperoxide oxidizing agents:

$$S_2O_8^{2-} + Fe^{2+} \rightarrow Fe^{3+} + SO_4^{2-} + SO_4^{\cdot -}$$

$$ROOH + Fe^{2+} \rightarrow Fe^{3+} + OH^- + RO\cdot \tag{8-5}$$

The rate of radical generation by an initiator is greatly accelerated when it is coupled with a reducing agent. Thus, an equimolar mixture of $FeSO_4$ and $K_2S_2O_8$ at 10°C produces radicals about 100 times as fast as an equal concentration of the persulfate alone at 50°C. Redox systems are generally used only at lower temperatures. The activation energy for the decomposition reaction is usually 12–15 kcal/mol so that it is very difficult to obtain a controlled and sustained generation of radicals at an acceptable rate at high temperatures.

It is possible to prevent a too rapid depletion of the initiator system by making repeated additions of the oxidizer and reducer or by having one component soluble in the monomer and the other in the aqueous phase. Thus, organic peroxides like cumyl hydroperoxide (**8-1**) are soluble in the monomer droplets but the redox reaction itself occurs in the aqueous phase where the hydroperoxide encounters the water soluble reducing agent. The rate of production of radicals is then controlled by the diffusion of the oxidizing component from the organic phase.

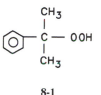

8-1

During polymerization there is often an induction period of variable length caused by traces of oxygen (see Section 6.9), followed by an accelerating rate.

8.2.1   Harkins–Smith–Ewart Mechanism

Emulsion polymerization first gained industrial importance during World War II when a crash research program in the United States resulted in the production of styrene-co-butadiene [SBR] synthetic rubber. The Harkins–Smith–Ewart model [5-6] summarized the results of early research, which focussed on this and similar systems. Current thinking is not entirely in accord with this mechanism. It is still worthwhile to review it very briefly here, however, because it is still widely referenced in the technical literature and because some aspects of the model provide valuable insights into operating procedures.

Strictly speaking, the Smith–Ewart model applies only to the batch polymerization of a completely water-insoluble monomer in the presence of micellar soap. (The terms *soap, surfactant*, and *emulsifier* are used interchangeably in this technology.) Its predictions do in fact apply neatly to the case of styrene. The polymerization reaction, after the induction period, can be classified conveniently into three stages, as shown schematically in Fig. 8-2.

Interval I is a region of accelerating rate of conversion and proceeds until all the micelles are consumed. Only 10–15% or less of the total monomer is consumed in this period. In interval II the rate of polymerization is constant. It is assumed that this period begins when all the micelles disappear and is complete when no monomer droplets remain. About one-third of the monomer is consumed in this period. In interval III, the rate of polymerization begins to decrease. This interval

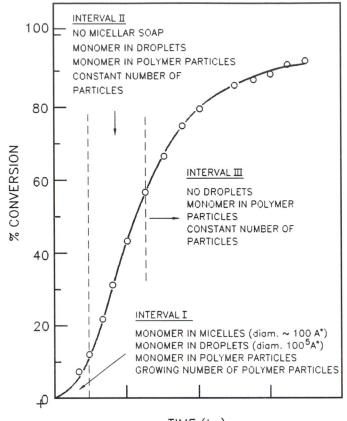

Fig. 8-2. Course of a batch emulsion polymerization of a water-insoluble monomer in the presence of micellar surfactant.

begins when the monomer droplets disappear and ends when the polymerization reaction is completed.

In the ideal case in which both the monomer and emulsifier have negligible water solubility, polymerization will begin in the micellar phase although radicals are generated only in the aqueous phase. Most of the monomer is initially present in the relatively large droplets, but there are normally about one million times as many micelles as droplets. Thus the micellar surface area available to capture radicals from the aqueous phase is much greater than that of the larger monomer droplets. (The soap/water interfacial area is 8×10^5 cm^2/cm^3 aqueous phase if there are 10^{18} micelles in this volume, and each micelle has a diameter of only 50 Å. The surface of 10^{12} monomer droplets, each with diameter 10^4 Å, is 3×10^4 cm^2.) The droplets act mainly as reservoirs to supply the polymerization loci by migration of monomer through the aqueous phase.

The foregoing considerations of relative capture surfaces and the available experimental evidence rule out the monomer droplets as the polymerization loci. The most realistic alternative assumption to micellar initiation is that the first addition of monomer to primary radicals occurs in the aqueous phase. The oligomeric free radicals which are generated would be $^-O_3SOMM\cdots M\cdot$, if persulphate initiators are used. These oligomeric species contain a polar end and a hydrophobic tail. According to this mechanism, they could be expected to add monomer units until they encounter phase interfaces (current ideas include alternative fates for these oligomers, as described later). For reasons mentioned, such interfaces are likely to be those between the water and the monomer-swollen micelles. Oligomeric radicals would be readily incorporated in a micelle and would initiate polymerization of the monomer which is already solubilized therein. The micelle is said to be "stung" when it is first entered by such an oligomeric radical or a primary radical derived from initiator decomposition. Polymerization proceeds rapidly inside a stung micelle because the monomer concentration there is large, as noted earlier. The micelle is rapidly transformed into a small monomer-swollen polymer particle. All the monomer which is needed for this growth process is not contained within the initial micelle but diffuses to it from the droplets.

The flux of monomers diffusing from the droplets to the particles will decrease as the former reservoirs are depleted, but reasonable calculations lead to the expectation that the arrival rate at the particle–water interface will exceed the usage rate even when more than 90% of the monomer has already left the droplets.

If the monomer is a good solvent for the polymer, the latex particles might be assumed to expand indefinitely because of inhibition of monomer. An equilibrium monomer concentration and swelling equilibrium is reached, however, because the free energy decrease due to mixing of polymer and monomer is eventually balanced by the increase in surface free energy which accompanies expansion of the particle volume.

Once a micelle is stung, polymerization proceeds very rapidly. The particle can accommodate more monomer as its polymer content increases and the water–polymer interfacial surface increases concurrently. The new surface adsorbs emulsifier molecules from the aqueous phase. This disturbs the equilibrium between micellar and dissolved soap, and micelles will begin to disintegrate as the concentration of molecularly dissolved emulsifier is restored to its equilibrium value. Thus the formation of one polymer particle leads to the disappearance of many micelles. The initial latex will usually contain about 10^{18} micelles per milliliter water, but there will be only about 10^{15} particles of polymer in the same volume of the final emulsion. When all the micelles have disappeared, the surface tension of the system increases because there is little surfactant left in solution. Any tendency for the mixture to foam while it is being stirred decreases at this time.

The transfer of free radicals out of the particles is assumed to be negligible. Also, the rate of mutual termination of two radicals inside a particle is taken to be very much faster than the free-radical entry rate, as is reasonable for small particles.

Then each particle will contain either zero or one radical. This radical will add monomer until another radical enters the particle and terminates the kinetic chain. The particle will be dormant until the cycle is started again by the entry of another radical. Thus, on the average, each particle will be active for half the time, or half the particles will be active and half dormant at any instant. Since the number of particles is fixed in interval II, addition of more initiator then will have no effect on the polymerization rate, but the average lifetime of each growing macroradical and the mean molecular weight will be reduced. The average number of radicals per particle will be equal to 0.5, if chain transfer can be neglected. Transfer reactions are important in emulsion polymerizations, however, because small free-radicals, e.g., from transfer to monomer, are mobile enough to diffuse out of the polymer particles. Hence, the mean number of radicals per particle is <0.5 for vinyl acetate and other monomers with significant C_M values.

The foregoing mechanism is amenable to mathematical analysis, with the salient results that during ideal interval II polymerization, the rate of reaction is proportional to $[I]^{2/5}$ and $S^{3/5}$, while $\overline{DP}_n$ depends on $S^{3/5}$ and $[I]^{-3/5}$. (Here [I] is the initiator molar concentration and S is the weight concentration of surfactant.) In conventional solution, suspension, or low-conversion bulk free-radical reactions, the rate of polymerization depends on $[I]^{1/2}$ while $\overline{DP}_n$ is proportional to $[I]^{-1/2}$. In these cases $\overline{DP}_n$ cannot be increased at given [M] without decreasing R_p. In emulsion polymerization, however, both R_p and $\overline{DP}_n$ can be changed in parallel by controlling the soap concentration.

The Harkins–Smith–Ewart theory predicts that the number of polymer particles formed, N, will be proportional to $S^{3/5}$ and $[I]^{2/5}$. This is observed for some batch polymerizations, as mentioned. In general, $N \propto S^x$, but the value of the exponent depends on the range of soap concentrations and the monomer solubility in water. This topic is of more academic than practical interest, however, because most useful polymerizations are not batch operations.

8.2.2 Role of Surfactants

Surfactants are employed in emulsion polymerizations to facilitate emulsification and impart electrostatic and steric stabilization to the polymer particles. Steric stabilization was described earlier in connection with nonaqueous dispersion polymerization; the same mechanism applies in aqueous emulsion systems. Electrostatic stabilizers are usually anionic surfactants, i.e., salts of organic acids, which provide colloidal stability by electrostatic repulsion of charges on the particle surfaces and their associated double layers. (Cationic surfactants are not commonly used in emulsion polymerizations.)

The presence of free surfactant is not preferred in some applications of polymer emulsions. It may, for instance, result in water sensitivity of dried latex coatings or affect the properties of particles used as substrates in solid phase peptide syntheses. Colloidal stability can be conferred by so-called "copolymerizable surfactants" in cases when it is not feasible to add surfactants. These "surfactants" are actually comonomers that are intended to provide surface activity. It is desirable that they be located preferentially on the particle surfaces and this can be accomplished by methods outlined below in the section on Core-Shell polymerization. For example, incorporation of methacrylic acid (in copolymers with styrene or acrylics) or vinyl sulfonic acid (in copolymers with vinyl acetate) provides electrostatic stabilization under the alkaline conditions used to store most emulsions (acid storage favors corrosion of metal containers). Copolymerizable steric stabilizers include esters of methacrylic acid and poly(ethylene glycol), which can be made by transesterification of methyl methacrylate.

8.2.3 Surfactant-Free Emulsion Polymerizations

In view of the foregoing discussion, it is not surprising that many emulsion polymerizations can be carried out without the addition of free surfactants. This is accomplished by use of ionic initiators, like the widely used persulfates, with the additional assistance of copolymerizable surfactants, if these are needed. It should be realized, however, that emulsion polymerization itself generates surfactants, and the process cannot therefore be entirely surfactant free.

The persulfate initiator resides in the aqueous phase and it is there that it first encounters monomers, which are added to growing oligomeric radicals:

$$\bullet OSO_3^- + M \rightarrow \bullet MOSO_3^- \tag{8-6}$$

$$\bullet OSO_3^- + M \rightarrow \bullet MOSO_3^- \rightarrow \rightarrow \rightarrow \bullet M_y OSO_3^- \tag{8-7}$$

Here, M stands for monomer, as usual. The oligomers produced by reaction (8-7) could meet several fates. If such a radical encounters another radical in the aqueous phase it may undergo termination by combination or disproportionation (Section 6.3.3) to yield a molecule with one or two sulfate ion ends, respectively. The end groups may also reflect whether the second reactant in the termination reaction is a hydroxyl radical (reaction 8-2) or sulfate ion radical (reaction 8-10), or results from hydrolysis, as in:

$$M_i OSO_3^- + H_2O \rightarrow M_i OH + H + HSO_4^- \tag{8-8}$$

Since the product here contains ionic and nonionic groups, it will be an anionic surfactant. Such materials, which are always formed in persulfate-initiated emulsion polymerizations, have been termed *in situ surfactants*. Their nature has not been studied extensively. In one study, of the polymerization of a 64 : 36 (w/w) methyl methacrylate : butyl acrylate copolymer in the presence of a chain transfer agent, more than 50% of the oligomeric radicals containing sulfate end groups produced surfactants. The surfactant-free polymerization of methyl methacrylate, with ammonium persulfate initiation, produced *in situ* surfactants with a mean degree of polymerization of 8-9 [7]. Oligomers from styrene would be expected to be shorter and those from vinyl acetate would be longer, in line with the relative water solubilities of the monomers [8].

We turn now to the other possible destinies of the oligomeric radicals produced by aqueous phase initiation. The mechanism of "surfactant-free" emulsion polymerizations is believed to include at least four stages: initiation (reaction 8-7), nucleation, coagulation, and particle growth. These intervals may ovelap in practice, but it is convenient to consider them separately. As the oligomeric radicals grow they become progressively less soluble in water and eventually collapse upon themselves to form so-called "primary particles." Such particles can increase their surface charge density and colloidal stability through adsorption of *in situ* surfactant and by coagulation, assuming plausibly that all the charged groups remain on the surface of the growing particle. Incorporation of an ionogenic monomer will also contribute to the particle stability.

Another reaction path involves the micellization of *in situ* surfactant. Both the micelles and primary particles can absorb monomer. Particle growth occurs by polymerization fed by entry into these loci of oligomers generated in the water. Figure 8-3 is a schematic representation of these mechanisms, the details of which are still debatable [9–13].

8.2.4   Mechanism of Emulsion Polymerization

Emulsion polymerizations are normally performed in the presence of micellar surfactants at concentrations higher than their cmcs. The most important nucleation mechanisms are then likely to be as sketched in Fig. 8-3, with the important qualification that the micelles mainly comprise the added soaps. One or other reaction path may be favored in different emulsion systems; experimental data are sometimes consistent with either the micellar entry or the homogeneous nucleation route. Stated generally, an aqueous phase monomer-ended radical may either undergo termination in the water phase, participate in the formation of a new particle or enter a polymer particle. Only the first two fates are involved in particle nucleation. When more hydrophobic monomers are polymerized oligomeric growth in the aqueous phase will tend to stop earlier, thus favoring particle nucleation by polymerization in micelles. Monomers which are more water soluble will produce oligomers that continue to grow in the continuous medium until the polymer chain collapses out of solution, and homogeneous nucleation is then more likely [14].

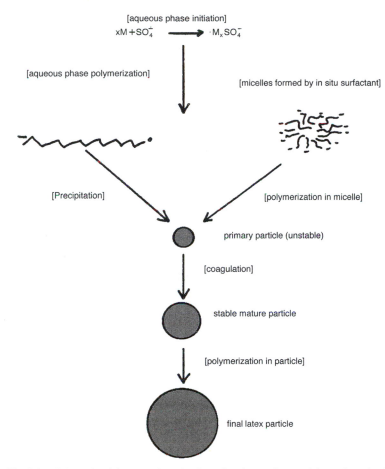

Fig. 8-3.   Schematic of the general mechanism of surfactant-free emulsion polymerization.

Macroradicals are isolated from each other in emulsion polymerizations because they grow in particles, which can accommodate either one or zero radicals at any instant. The distinguishing feature of the kinetics of such reactions is that the polymerization rate and polymer molecular weight are proportional to the number of particles, as distinct from free-radical polymerizations in bulk, solution or suspension. An interesting consequence is that the rate of polymerization will be inversely proportional to the particle size. This holds *at fixed final polymer content*, which is the way such reactions are usually performed. Polymer molecular weight may also be affected by particle size under the same conditions.

When a mixed monomer feed is used, the course of the copolymerization may be sensitive to the volume ratio between the monomer and water phases, if the monomers differ in water solubility. This is a factor especially at high monomer : water phase ratios, where the relative concentration of the more soluble monomers in the aqueous reaction medium will greatly exceed that in the bulk organic phase.

8.2.5   Heterogeneous Polymer Particles [15, 16]

The copolymer composition may drift during the course of an emulsion copolymerization because of differences in monomer reactivity ratios or water solubilities. Various techniques have been developed to produce a uniform copolymer composition. The feed composition may be continuously or periodically enriched in a particular monomer, to compensate for its lower reactivity. A much more common procedure involves pumping the monomers into the reactor at such a rate that the extent of conversion is always very high [>about 90%]. This way, the polymer composition is always that of the last increment of the monomer feed.

This section, by contrast, focuses on techniques to synthesize latex particles whose composition varies radially. Such materials are industrially important and this topic is also of interest because its consideration provides useful insights into emulsions technology. Particles with nonuniform structures are key ingredients in surface coatings, adhesives, opacifiers, impact improvers for plastics, waxes, chromatographic packings, components of polymer alloys, Xerographic toners, substrates for polypeptide syntheses, and other applications. Their sizes may be controlled in a range from <100 nm to several microns. These products are often called core-shell polymers, but their structures may vary from a "simple" core and shell to a multilayer onion-like structure. Particle shapes are usually spherical, but myriad other contours have also been observed.

Inhomogeneous particles can be made by dispersion polymerization and other methods, but the most versatile processes for production of such controlled structures are variations of emulsion polymerization techniques. In essence, the desired particle morphology is produced by sequential polymerizations of layers of designed compositions. The aim in staged polymerizations that are intended to produce particles with nonuniform structures requires that the polymers produced in later stages envelope (or are enveloped by) polymer made in earlier process steps. The whole operation is defeated if later stage polymers form separate particles. For this reason, the amount of surfactant must be regulated in order avoid providing enough free emulsifier to stabilize the formation of such new particles. At the same time, there must be enough surfactant to prevent coagulation of the emulsion, in which the interfacial area between polymers and water phases is growing. This is most conveniently accomplished by adding the later stage monomers as emulsions containing the required surfactant at the calculated concentration.

Practical applications do not always require tight control of particle size distributions, but the necessity for avoiding generation of new particles at any stage after the first polymerization process may necessitate this precaution. Analysis of the particle size distribution after successive polymerizations is a simple expedient to ensure that later stages have produced polymer that is attached to the previously existing particles.

Simple sequential processes frequently do not yield particles with the planned architectures. This is because of the complexity of emulsion polymerizations and because a system in which different polymers coexist with water will tend to rearrange toward the composition with the lowest overall surface energy. Theoretical descriptions of such phenomena [17–19] are based on the concept that the final state of the system consisting of polymer 1, polymer 2, and water (labeled phase 3) depends on the three interfacial tensions γ_{12}, γ_{23} and γ_{23}, and the corresponding interfacial areas A_{ij}. The equilibrium state of the three phases is determined by the minimum value of the surface free energy, G_s:

$$G_s = \sum \gamma_{ij} A_{ij}. \tag{8-9}$$

On a less mathematical, but still very useful format, various research groups have presented guidelines for determining which configuration will be the most stable, in multistage emulsion polymerizations [20–22]. These are primarily in terms of the relative hydrophilicities of the various polymers. The more hydrophilic polymer will tend to be situated at the interface with water, if it is mobile enough to get there. Thus, if the first-stage polymer is the more hydrophilic, the final two-stage structure is more likely to be inverted if the second-stage monomer can dissolve in the first-stage polymer and plasticize it. Also, with persulfate initiation by ion radicals, many of the polymer chain ends will be ionic (reaction 8-7) and can be expected to be located preferentially at the water interface.

Although the thermodynamics of the products of multistage polymerization determine which structure will be the most stable, it is possible, and frequently necessary, to produce particle structures that appear nominally to be thermodynamically forbidden. This is achieved either by changing the surface characteristics of a polymer phase from those of the bulk material or by employing kinetic factors to produce and anchor energetically unfavored morphologies. Both types of methods are summarized below.

(i)   Alteration of Particle Surface Hydrophilicity

Note that the nature of the particle surface need not necessarily be the same as that of the bulk polymer. (In fact, it seldom is.) This makes it possible to produce thermodynamically stable species in which the more polar (hence, more hydrophilic) polymers form the encapsualted phases. The most direct approach is to use surfactants to modify polymer surface characteristics. This is a preferred method, when it works, but several cautions must be observed. Remember that the surfactant that is added to the reaction mixture is not the only one that is present, as pointed out in Section 8.2.3. The *in situ* surfactant competes with added surfactants for adsorption sites on polymer surfaces and may augment or detract from the effectiveness of the latter soaps. The surfaces of earlier and subsequent stage polymers may also compete for adsorbed surfactant. If the later stage polymer is more hydrophobic than the preformed polymer, some of the surfactant adsorbed on the latter may desorb and stabilize the formation of particles of the later stage polymer. This effect, which is consistent with the thermodynamic factors mentioned above, results in the production of some second crop particles of the later stage material.

One of the difficulties is that there is no convenient theory for selecting the appropriate surfactants *a priori*. The simplest procedure is to try to match the HLB (hydrophile-lipophile balance) character of the surfactants to the polarity of the particular polymers. Tabulated HLB values are available [23, 24] (the higher the HLB number the more hydrophilic is the soap), but the rating is essentially empirical and effective use requires some experience and intuition.

Mention has been made of the fact that the polar character of polymer surfaces is strongly affected by the ionic polymer end groups that are residues of initiator-derived ion radicals, when persulfates are used in emulsion polymerizations. Variation of the initiator type between those that yield ionic and nonionic end groups is an effective way to control particle stability and avoid complications due to migration of surfactant from one polymer surface to another [25]. This method can also be supplemented by copolymerization with polar monomers to affect surface hydrophilicity.

Poly(methyl methacrylate) (PMMA)/polystyrene (PS) latex particles made using an ionic initiator in the first (PMMA) stage and a nonionic initiator in the second (PS) polymerization have a strong tendency to invert to a PS core inside a PMMA shell, because of the PMMA is more hydrophilic than PS and as a result of the presence of ionic groups from the initiator on the PMMA surface. The core-shell structure for this system can be made more stable by using a nonionic initiator, like an azo compound, in the first stage PMMA polymerization and an ionic initiator in the second stage, PS polymerization. This procedure may, however, necessitate the presence of an anionic surfactant in the first-stage reaction, for latex stability.

None of the factors that affect particle morphology operate alone. In particular, the mode of monomer addition is an interacting factor. This is illustrated by procedures used to produce core-and-shell polymers for use in architectural paints [26]. Polymers used for this purpose are primarily copolymers of butyl acrylate with either vinyl acetate or methyl methacrylate. The goal here was to make particles with conventional film-forming polymer shells and cores comprised of less expensive monomers than were used in the shells. In practice, this could mean poly(vinyl acetate) cores inside vinyl acetate/butyl acrylate shells or the same or poly(methyl methacrylate) cores surrounded by methyl/methacrylate/butyl acrylate shells. In all cases, the core polymers are more hydrophilic than the shell materials. The procedure used to sidestep this difficulty illustrates many of the principles discussed in this section.

Thermodynamically stable structures were made by modifying the surface characteristics of the two polymer phases. To illustrate with respect to one of the systems, the intended vinyl acetate/butyl acrylate shell polymer was made in the first polymerization stage. The surfaces of these particles were adjusted so as to have a lower interfacial tension against water by copolymerizing with an ionic comonomer, vinyl sulfonic acid. The core polymer, poly(vinyl acetate), was made in the second-stage polymerization and its surface was rendered less hydrophilic than in normal persulfate-initiated polymers by using a nonionic initiator. Such an initiator would not have produced a stable poly(vinyl acetate) latex and so the second-stage core polymer was made by performing a batch polymerization after adding all the vinyl acetate monomer to the preformed shell latex polymer. A parallel method was used for the methyl methacrylate-based particles, except that the ionic comonomer used in the shell polymer was methacrylic acid, which copolymerizes better with the other monomers used here.

It can be shown theoretically that the relative amounts of the two monomers in a two-stage emulsion polymerization can affect the final particle structure [19], with core-shell morphology being favored thermodynamically, as the amount of second-stage polymer is increased.

(ii) Influence of Polymerization Kinetics on Latex Particle Structure

Although the lowest free-energy structure is the most stable phase arrangement in heterogeneous polymer particles, kinetic factors are also important because the polymeric phases have limited mobility and the most favorable thermodynamic state may not be achieved during the course of the polymerization or subsequent storage. Note, in this connection, that thermodynamically unprofitable structures are prone to rearrange on storage for long periods or under warmer conditions.

The manner in which the second-stage monomer is added to the polymerization system has a strong influence on the final structure of the composite particles. Semibatch (or semicontinuous) operation is preferable to batch operation when it is desired to have the second-stage polymer on the outside of the final particles. In a batch second-stage polymerization all the monomer is added to the first-stage particles at the beginning of the reaction. The second-stage monomer has much higher mobility than the second-stage polymer; the second-stage monomer is more likely to polymerize with the first-stage particles if this produces an energetically favored state. Also, the plasticizing effect of the second-stage monomer in the first-stage polymer enhances the ability of the latter molecules to rearrange to the exterior of the particles, if this should reduce the overall surface energy of the system.

When second-stage monomer is added gradually to the first-stage polymer particles the level of monomer can be kept to a minimum during the polymerization. Thus, if the energetics of the system favor the second-stage polymer in a core arrangement inside the first-stage polymer, batch polymerization of the shell monomers will decrease the kinetic barrier to this inversion, while semibatch or, better even, starve–feed polymerization imposes a higher kinetic barrier. On the other hand, when it was desired to produce inverted structures, as in [25], the second-stage polymer was produced by batch polymerization in the presence of the first.

Similarly, when intermingling of the core and shell polymers is required, in order to provide mechanical interlocking of one polymer with the other, then it is advantageous to let some of the second monomer soak into the first-stage particles before starting polymerization. This may be followed by semibatch polymerization of more second-stage monomer in order to produce the desired outer shell [27].

It is possible to lock the polymer phases into a particular core-shell configuration by introducing a sufficient level of cross-linking. This is not always feasible, of course, if this results in other undesired changes in polymer properties. Conversely, the inclusion of solvents or plasticisers will facilitate inversion of second-stage polymer into the first-stage material, if this produces a lower energy state. Not all such phase inversions are clear-cut. In many cases this phenomenon will be observed as penetration of shell polymers into the core polymer. Unreacted monomer from stage one polymerization will act as a plasticizer to lower the kinetic barrier to such rearrangements. Use of a chain transfer agent which lowers polymer molecular weight and viscosity also facilitates phase rearrangements.

An approach to raising the kinetic barrier to stage 2 inversion involves lowering the polymerization temperature of the second stage. This will reduce the mobility of the first-stage polymer and retard phase mixing.

8.3   OTHER INGREDIENTS IN EMULSION RECIPES

The essential ingredients in an emulsion polymerization are the water, monomer, surfactant, and free-radical source. Other ingredients are frequently added for a variety of reasons. Stabilizers, which are usually water-soluble high polymers or carbohydrate gums, are employed to control latex viscosity and freeze–thaw stability of products that are used in the latex form.

Chain transfer agents are used to control the molecular weights of polymers like butadiene copolymers, which are subsequently isolated and processed. Molecular weight control is not normally needed for latexes which are used directly as surface coatings or adhesives. The most important class of transfer agents are aliphatic mercaptans, and these are used particularly with hydrocarbon monomers that do not transfer readily with the corresponding macroradicals. The mercaptans used are C_8–C_{16} species which are soluble in the monomer. The rate of consumption of the transfer agent depends partly on its reactivity with the particular polymeric free radical and partly on the rate of diffusion of the mercaptan from the monomer droplets to the polymer particles where polymerization is taking place. The hydrocarbon chain length of the mercaptan transfer agent is selected to control this diffusion rate at a desirable level. If this is not done most of the mercaptan would be consumed in early portions of the polymerization.

8.4   EMULSION POLYMERIZATION PROCESSES

Continuous processes will generally not be used for acrylic surface coating latexes, adhesives, and other low tonnage products. Semibatch processes are frequently used in which not all the ingredients are added initially. Continuous systems are favored for large-volume polymers like SBR in order to increase the reactor output and reduce fluctuations in product properties. Such continuous reactor trains usually consist of a series of batch reactors each connected with bottom inlets and top outlets and operated with continuous overflow.

Mixtures of anionic and nonionic surfactants are usually employed. The anionic emulsifiers are the less water soluble and control the number and size of the particles. The nonionic surfactants are often ethylene oxide condensates of alkyl phenols; their water solubility is proportional to the degree of polymerization of the poly(ethylene oxide) component. Their function is primarily to provide colloidal stability against electrolytes, mechanical shearing, and freezing.

The target polymerization temperature will usually be chosen to optimize production rates or product quality. "Cold" SBR, which is made near 5°C, is an interesting case in this regard. The cold product is superior as a rubber to hot (60°C emulsion polymerization) SBR, because it contains less low-molecular-weight polymer which cannot be reinforced with carbon black. There is also less branching and more *trans*-1,4 units in the cold SBR. Hot SBR is easier to mill and extrude because of its low-molecular-weight fraction and is used mostly for adhesive applications while cold SBR, which is made mainly for tires, accounts for about 90% of all production of this polymer.

The completion of the polymerization can be slow when the monomer concentration in the aqueous phase is low and the radicals which reach the polymer particles are mainly fragments of the primary initiator and not oligomers mentioned earlier in this chapter. Since water-soluble initiators yield hydrophilic radicals these may be reluctant to enter the polymer particles. This difficulty can be circumvented by using an initiator in the late stages of the reaction with a more favorable partition coefficient toward the polymer phase. Alternatively, a comonomer that has some water solubility may be added in the last intervals to generate oligomers that will enter the polymer particles. Vinyl acetate is sometimes added in the last stages of vinyl chloride emulsion polymerization, for example.

PROBLEMS

8-1.   (a)   What happens to the rate of emulsion polymerization if more monomer is added to the reaction mixture during interval II polymerization?
(b)   What happens to the number average degree of polymerization?

8-2.   A particular emulsion polymerization yields polymer with $\overline{M}_n = 500,000$. Show quantitatively how you would adjust the operation of a semibatch emulsion process to produce polymer with $\overline{M}_n = 250,000$ in interval II without changing the rate of polymerization, reaction temperature, or particle concentration.

8-3.   "Normal" behavior is observed in the emulsion polymerization of methyl methacrylate at low conversions and generally when the polymer particle sizes are small. The rate of polymerization is observed to accelerate, however, at high conversion levels when the polymer particles are large. Explain briefly.

REFERENCES

[1]   K. Barrett, Ed., "Dispersion Polymerisation in Organic Media." Wiley, New York, 1975.
[2]   A. J. Paine, *Macromolecules* **23**, 3109 (1990); *J. Coll. Interf. Sci.* **138**, 157 (1990).
[3]   B. Thomson, A. Rudin, and G. Lajoie, *J. Appl. Polym. Sci.* **33**, 345 (1995).
[4]   R. G. Gilbert, "Emulsion Polymerization." Academic Press, San Diego, CA (1995).
[5]   W. D. Harkins, *J. Chem. Phys.* **13**, 381 (1945); **14**, 47 (1946); *J. Am. Chem. Soc.* **69**, 1428 (1947); W. V. Smith and R. H. Ewart, *J. Chem. Phys.* **16**, 592 (1948).
[6]   J. L. Gardon, *Rubber Chem. Technol.* **43**, 74 (1970).
[7]   Z. Wang, A. J. Paine, and A. Rudin, *J. Polym. Sci. Part A: Polym. Chem.* **33**, 1597 (1995); B. Thomson, Z. Wang, A. J. Paine, and A. Rudin, *ibid.* **33**, 2297 (1995).
[8]   I. A. Maxwell, B. R. Morrison, R. G. Gilbert, D. F. Sangster, and D. H. Napper, *Makromol. Chem. Macromol. Symp.* **53**, 233 (1992).
[9]   F. K. Hansen and J. Ugelstad, *J. Polym. Sci. Part A, Polym. Chem.* **16**, 1953 (1978).
[10]   R. M. Fitch and C. H. Tsai, in "Polymer Colloids" (R. M. Fitch, Ed.). Plenum, New York, 1971.
[11]   J. W. Goodwin, J. Hearn, C. C. Ho, and R. H. Ottewill, *Brit. Polym. J.* **5**, 347 (1973).
[12]   P. J. Feeney, D. H. Napper, and R. G. Gilbert, *Macromolecules* **20**, 2922 (1987).
[13]   Z. Song and G. W. Poehlein, *J. Polym. Sci. Part A, Polym. Chem.* **28**, 2359 (1990).
[14]   J. W. Vanderhoff, *J. Polym, Sci. Polym. Symp.* **72**, 161 (1985).

[15]   A. Rudin, *Makromol. Chem. Macromol. Symp.* **92,** 53 (1995).

[16]   N. Sutterlin, *Makromol. Chem. Suppl.* **10/11,** 403 (1985).

[17]   D. Sundberg, B. Kronberg, and J. Berg, *J. Microencap.* **6,** 327 (1989).

[18]   Y. C. Chen, V. Dimonie, and M. S. El-Aasser, *J. Appl. Polym. Sci.* **42,** 1049 (1991).

[19]   J. A. Waters, European Patent Application, EP 327 199 (1989).

[20]   M. S. Silverstein, Y. Talmon, and M. Narkis, *Polymer (London)* **30,** 416 (1980).

[21]   J. C. Daniel, *Makromol. Chem. Suppl.* **10/11,** 359 (1985).

[22]   D. I. Lee and T. Ishikawa, *J. Polym. Sci. Part A, Polym. Chem.* **21,** 147 (1983).

[23]   W. C. Griffen, *J. Soc. Cosmetic Chemists*, **1,** 311 (1949).

[24]   "McCutcheon's Emulsifiers and Detergents," Vol. 1. M. C. Publishing Co., Glen Rock, NJ.

[25]   A. J. Paine, K. J. O'Callaghan, and A. Rudin, US Patent 5,455,315 (1995).

[26]   G. A. Vandezande and A. Rudin, *J. Coat. Technol.* **66,** (828) 99, (1994).

[27]   D. G. Cook, A. Plumtree, and A. Rudin, *J. Appl. Polym. Sci.* **48,** 75 (1993).

Chapter 9

Ionic and Coordinated Polymerizations

If an elderly but distinguished scientist says that something is possible
he is almost certainly right, but if he says that it is impossible he is
very probably wrong.
—Arthur C. Clarke, *New Yorker,* (August 9, 1969)

9.1 COMPARISON OF IONIC AND FREE-RADICAL POLYMERIZATIONS

The active site in chain-growth polymerizations can be an ion instead of a free-radical. Ionic reactions are much more sensitive than free-radical processes to the effects of solvent, temperature, and adventitious impurities. Successful ionic polymerizations must be carried out much more carefully than normal free-radical syntheses. Consequently, a given polymeric structure will ordinarily not be produced by ionic initiation if a satisfactory product can be made by less expensive free-radical processes. Styrene polymerization can be initiated with free radicals or appropriate anions or cations. Commercial atactic styrene polymers are, however, all almost free-radical products. Particular anionic processes are used to make research-grade polystyrenes with exceptionally narrow molecular weight distributions and the syndiotactic polymer is produced by metallocene catalysis. Cationic polymerization of styrene is not a commercial process.

As mentioned in Section 6.2, free-radical reactions are not selective and most olefinic monomers can be polymerized by such processes although the control of product molecular weight and stereoregularity may not always be what is desired. Ionic polymerizations, by contrast, are restricted to monomers whose structures enhance the stability of the particular ion that is involved in the process. Some examples are reviewed in Sections 9.2.1 and 9.3.1, which deal with monomers for anionic and cationic processes, respectively.

The characteristics of the active centers in free-radical polymerizations depend only on the nature of the monomer and are generally independent of the reaction medium. This is not the case in ionic polymerizations because these reactions involve successive insertions of monomers between a macromolecular ion and a more or less tightly attached counterion of opposite charge. The macroion and counterion form an organic salt which may exist in several forms in the reaction medium. The degree and nature of the interaction between the cation and anion of the salt and the solvent (or monomer) can vary considerably.

If we represent the organic ionic species as $U^{\oplus}V^{\ominus}$, then at least four forms must be considered in principle:

$$\text{UV} \rightleftharpoons U^{\oplus}V^{\ominus} \rightleftharpoons U^{\oplus} // V^{\ominus} \rightleftharpoons U^{\oplus} + V^{\ominus} \tag{9-1}$$

| covalent bonding | contact ion pair | solvent separated ion pair | free solvated ions |

The forms which are actually important in a given polymerization will depend on the natures of the species U and V, the solvating ability of the medium, and the temperature. It is not unusual to find two of these forms coexisting in significant quantities in a given polymerization. In general, more polar media favor solvent-separated ion pairs or free solvated ions. Free solvated ions will not exist in hydrocarbon media, where other equilibria may occur between ion pairs and clusters of ions.

It is well known in organic chemistry that the reactivities of these various forms of organic ions and counterions may differ significantly [1], and it is thus not surprising that the same influences are detectable in ionic polymerizations. In general, propagation rates are higher the more the macroion and its counterion are separated. Steric control of the polymer microstructure is greater, however, when the macroion and the counterion are associated.

Rates of ionic polymerizations are by and large much faster than in free-radical processes. This is mainly because termination by mutual destruction of active centers occurs only in free-radical systems (Section 6.3.3). Macroions with the same charge will repel each other and concentrations of active centers can be much higher in ionic than in free-radical systems. Rate constants for ionic propagation reactions vary but some are higher than those in free-radical systems. This is particularly true in media where the ionic active center is free of its counterion.

The high reactivity of ionic active centers which yields fast propagation rates also results in a greater propensity toward side reactions and interference from trace impurities. Low polymerization temperatures favor propagation over competing reactions, and ionic polymerizations are often performed at colder temperatures than those used in free-radical processes, which would be impossibly slow under the same conditions.

In this chapter we first review anionic polymerizations and then discuss cationic processes. A final section describes coordinated polymerizations.

9.2 ANIONIC POLYMERIZATION

9.2.1 Overview

Anionic polymerizations are chain-growth processes in which the active center to which successive monomers are added is a negative ion that is associated with a positive counterion. The degree of interaction between the macromolecular anion and its counterion depends on the nature of the respective ions and the medium in which the polymerization is proceeding.

Anions which are active enough to cause polymerization through reactions of carbon–carbon double bonds are strong bases. However, if the carbanion end of the growing chain is an extremely strong base, it will be converted to the corresponding hydrocarbon by transfer of a proton from solvent or any other species in the reaction mixture which is less basic. The kinetic chain would be terminated in such cases since the new anions produced by these transfer reactions would not be basic enough to reinitiate the propagation sequence. (Recall that a propagation reaction is one in which monomers are consumed, the location of the active site is changed, but the number of active sites remains the same. Monomer addition is the main propagation reaction.) Practical anionic polymerizations therefore require the use of macromolecular carbanions which are relatively stable under accessible reaction conditions.

Monomers that can be polymerized anionically include those in which some anion stability is realized by delocalization of the negative charge over a larger region of the molecule. Examples include styrene and butadiene in which the C=C double bonds are conjugated and cyclic heteroatom compounds in which the negative charge can be delocalized onto atoms that are more electronegative than carbon. The latter group includes ethylene oxide, propylene sulfide, caprolactam, and siloxane ring compounds. Vinyl monomers with electron-withdrawing substituents like cyano and nitro groups (Section 7.10.2) are also suitable for anionic polymerizations, because attack of an anion is facilitated on the electron-poor unsubstituted carbon of the double bond.

If the monomers contain reactive groups that could be attacked by carbanions they will not be suitable for anionic polymerizations. Halogen-containing vinyl monomers are difficult to polymerize in these systems because of the elimination of alkyl halides. Interactions between many initiators and the carbonyl groups of methacrylates or acrylates necessitate the use of special reaction conditions, like very low temperatures, for the anionic polymerization of these monomers.

Protection of some reactive functional groups is possible during polymerization, using known organic chemical methods, such as silanation of alcohol or phenol groups. This expedient and the subsequent deprotection step will, of course, complicate the overall polymerization.

Solvents with high dielectric constants are generally not employed in anionic polymerizations. This is because protic liquids like alcohols or amines are acidic enough to destroy carbanionic active centers, and highly polar aprotic solvents may form strong complexes with the anions and hinder addition of monomers. Anionic polymerizations are performed as a consequence in fairly nonpolar solvents with dielectric constants in the range 2–10. (At room temperature the dielectric constant of benzene is 3 while that of water is 78.5. Recall that the attractive force between opposite charges is inversely proportional to the dielectric constant of the medium in which they are immersed. Dissociation of ion pairs is facilitated in solvents with higher dielectric constants.)

The propagating anion and its counterion exist in relatively nonpolar solvents mainly in the form of associated ion pairs. Different kinds of ion pairs can be envisaged, depending on the extent of solvation of the ions. As a minimum, an equilibrium can be conceived between intimate (contact) ion pairs, solvent-separated ion pairs, and solvated unassociated ions. The nature of the reaction medium and counterion strongly influences the intimacy of ion association and the course of the polymerization. In some cases the microstructure of the polymer that is produced from a given monomer is also influenced by these variables. In hydrocarbon solvents, ion pairs are not solvated but they may exist as aggregates. Such intermolecular association is not important in more polar media where the ion pairs can be solvated and perhaps even dissociated to some extent.

Anionic chain growth polymerizations are particularly distinguished from free-radical polymerizations in the following respects.

1. Initiation is the slow reaction in the initiation–propagation–termination sequence in free radical reactions whereas selected initiation reactions in anionic systems can be very rapid compared to the subsequent propagation reaction. This facilitates the preparation of anionic polymers with narrow molecular weight distributions, if the polymerization is conducted carefully.

2. Spontaneous termination reactions are effectively absent in a number of anionic polymerizations. Some block copolymers can therefore be produced by adding a second monomer to the active anionic end of a polymer made from a different, first monomer. (Macromolecular species in which the propagation reaction can be interrupted for lack of monomer and then resumed with the same or a selected different monomer are called "living polymers." Living polymerizations are reactions in which only the initiation and propagation steps are significant.) Alternatively, polymers with specific end groups can be formed by deliberately adding suitable terminating agents to the system.

3. The choice of initiator has no effect on the propagation reactions in free-radical polymerizations but it can influence ionic propagations because the reactivity of the active center is partly determined by the nature of the counterion that is derived from the initiator.

4. The choice of reaction medium is much more significant in anionic and cationic reactions than in free-radical polymerizations because the character of the growing chain end is altered if the ion pair is more or less solvated.

5. Geometrical isomerism of polymers made from conjugated diolefins can be regulated in some anionic polymerizations. The tacticity of vinyl polymers is, however, not always controlled in anionic reactions. The products of anionic vinyl polymerizations are usually atactic, as in free-radical syntheses.

9.2.2 Anionic Initiation

Anionic initiators are all electron donors of varying base strengths. The initiator type required for a particular polymerization depends on the ease with which an anion can be formed from the monomer, which acts as an electron acceptor in anionic polymerizations. The relatively weak base sodium methoxide ($NaOCH_3$) can polymerize acrylonitrile. This monomer contains the electronegative —CN group which decreases the electron density on the C=C double bond and enhances reactivity toward anions. Vinylidene cyanide carries two —CN groups on the same carbon atom and be polymerized by even weaker bases like water and amines. Similarly, cyanoacrylates have two electron-withdrawing groups on the same carbon [—CN and —C=O(OCH_3)] and are polymerized by water; they can be used to glue human tissue together in oral surgery procedures. By contrast, polymerization of nonpolar monomers such as conjugated diolefins requires initiation by very strong bases like metal alkyls.

The two principal anionic initiation processes are nucleophilic attack on the monomer and electron transfer. Nucleophilic attack is essentially addition of a negatively charged entity to the monomer and involves mainly alkali metal alkyls, "living polymers," metal alkoxides, metal amides, and Grignard reagents. The general initiation process is

$$Z - B + CH_2 = \underset{\underset{X}{|}}{\overset{\overset{H}{|}}{C}} \longrightarrow B - CH_2 - \underset{\underset{X}{|}}{\overset{\overset{H}{|}}{C}}{}^{\ominus} Z^{\oplus} \qquad (9\text{-}2)$$

where the nucleophilic part of the initiator is denoted B and the counterion is formed from the metal Z. Alkali metal alkyls and living polymers are the most important initiators of this type for vinyl monomers.

The group of electron transfer initiators includes alkali metals and complexes of these metals with hydrocarbons. The general reaction is

$$Z + CH_2 = \overset{\overset{\displaystyle H}{|}}{\underset{\underset{\displaystyle X}{|}}{C}} \longrightarrow \bullet CH_2 - \overset{\overset{\displaystyle H}{|}}{\underset{\underset{\displaystyle X}{|}}{C}}{}^{\ominus} Z^{\oplus} \tag{9-3}$$

The more important anionic initiators are reviewed briefly in the following paragraphs.

(i)   Alkali Metal Alkyls

These initiators produce anionic propagating species by attack on the C=C double bond of vinyl and diene monomers. A common example of this type is *n*-butyl lithium. The C—Li bond is not ionic in hydrocarbon media where the initiator molecules exist as aggregates. The unaggregated form is more active for initiation. Butyl lithium is usually available as a solution in hexane. Addition of tetrahydrofuran to this solvent increases the concentration of the unaggregated initiator by forming a 1:1 complex with this compound. This accelerates the rate of initiation of styrene:

$$C_4H_9Li + CH_2 = \overset{\overset{\displaystyle H}{|}}{\underset{\underset{\displaystyle \bigcirc}{|}}{C}} \longrightarrow C_4H_9 - CH_2 - \overset{\overset{\displaystyle H}{|}}{\underset{\underset{\displaystyle \bigcirc}{|}}{C}}{}^{\ominus} Li^{\oplus} \tag{9-4}$$

which is fairly slow in hydrocarbon media. Alkyl lithium initiators yield stereoregular polymers of conjugated dienes if the polymerization is carried out in hydrocarbon solvents (Section 9.2.7). Addition of tetrahydrofuran or other more polar solvents changes the microstructure of the polymers which are produced.

There is no inherent termination step in organolithium polymerizations of hydrocarbon monomers, and this method of initiation yields living polymers. Living polymerizations are defined as those in which there is no inherent termination reaction (as described in Section 6.3.3 for free-radical polymerizations) and in which the macrospecies continue to grow as long as monomer is supplied.

Other metal alkyls, like triphenylmethyl sodium and potassium benzyl, are sometimes used. They have lower solubilities in organic solvents than lithium alkyls because of the greater ionic character of the Na—C and K—C bonds.

(ii) Living Polymers

A polymer with an active carbanionic end will initiate the polymerization of a different monomer if the initial macroion is more nucleophilic than the anion formed from the second monomer in the particular solvent/counterion environment. Thus alpha-methylstyryl anion (**9-1**) will initiate the polymerization of methyl methacrylate (Fig. 1-4) but the poly(methyl methacrylate) carbanion will not initiate polymerization of alpha-methylstyrene.

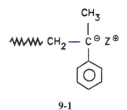

9-1

The relative ease of anionic polymerization can be correlated with the base strengths of the respective anions or crudely with the e values of the $Q-e$ scheme (Section 7.11) as shown in Table 9-1. Polymerization of a given monomer can generally be initiated by a carbanion from any monomer higher in the list, but the reverse reaction does not occur.

There are exceptions to this guideline, however. Thus, α-methylstyryl initiation of styrene or styryl initiation of isoprene are both faster than the reverse initiations. If inductive effects only are considered, then positive e numbers favor anionic polymerizability.

TABLE 9-1

Initiation of Anionic Polymerization by
Living Polymer Ends

Monomer	e value
α-Methylstyrene	−0.81
Styrene	−0.80
Butadiene	−0.50
Isoprene	−0.55
Methacrylate esters	0.20–0.40
Ethyl acrylate, butyl acrylate	0.55–0.85
Acrylonitrile	1.23
Ethylene oxide, propylene oxide	—
α-Cyanoacrylates	—

(iii)   Metal Amides

Styrene and other monomers can be polymerized by potassium amide in liquid ammonia. The dielectric constant of the solvent is quite high ($\sim$22) and this is one of the few anionic systems in which the active centers behave kinetically as free ions:

$$KNH_2 \rightleftharpoons K^{\oplus} + NH_2^{\ominus} \qquad (9\text{-}5a)$$

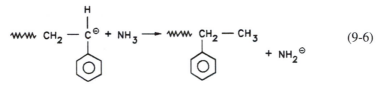

$$(9\text{-}5b)$$

Also because of the unusual nature of this solvent, chain transfer is important in this system:

$$(9\text{-}6)$$

(iv)   Alkali Metals

The initiation step in this case involves a one-electron transfer from a Group IA metal to the monomer. A radical ion is formed:

$$(9\text{-}7)$$

The radical ion may dimerize to give a dianion:

$$(9\text{-}8)$$

Further electron transfer from another alkali metal atom can also produce a dianion:

$$(9\text{-}9)$$

The eventual result of the initiation process is a species capable of propagating at both of its ends.

Historically, the most important application of this initiation step involved the production of stereoregular diene rubbers by lithium metal initiation. Other alkali metals are not as attractive for this purpose because polymers with the higher overall 1, 4 contents and molecular weights are provided by lithium initiation.

Direct attack of the monomer on the metal is a heterogeneous reaction. The lithium was used as a fine dispersion with a large surface area to speed up the initiation reaction, and the process was carried out in hydrocarbon solvents because polar solvents increase the generally undesired vinyl side chain content of the product polymer (cf. Section 9.2.7).

(v) Alkali Metal Complexes

Polycylic aromatic compounds can react with alkali metals in ether solution to produce monomeric radical ions with an extra electron in the lowest unoccupied π orbital of the hydrocarbon. For sodium and naphthalene, for example,

$$\text{(9-10)}$$

Tetrahydrofuran is a useful solvent for such reactions. This fairly polar solvent (dielectric constant $= 7.6$ at room temperature) promotes transfer of the $3s$ electron from the sodium to the aromatic compound and stabilizes the resultant complex. The stability of such complexes depends on the solvent, alkali metal counterion, and the nature of the aromatic compound.

The complexes can initiate the polymerization of conjugated monomers by a very rapid electron transfer process:

$$\text{(9-11)}$$

The styryl radical anion may dimerize to a dianion that is capable of growing at both ends:

$$\text{(9-12)}$$

This initiation process is similar to alkali metal initiation in this respect. The monomer in these systems often has a lower electron affinity than the polycyclic hydrocarbon, but dimerization of the monomeric radical anion (reaction 9-12) drives the equilibrium of reaction (9-11) to the right.

9.2.3   Anionic Propagation

Anionic polymerizations are generally much faster than free-radical reactions although the k_p values are of the same order of magnitude for addition reactions of radicals and solvated anionic ion pairs (free macroanions react much faster). The concentration of radicals in free-radical polymerizations is usually about 10^{-9}–10^{-7} M while that of propagating ion pairs is 10^{-3}–10^{-2} M. As a result, anionic polymerizations are 10^4–10^7 times as fast as free-radical reactions at the same temperature.

(i)   Heterogeneous Initiation/Homogeneous Propagation

In polymerizations by alkali metals or insoluble organometallics, the initiation step occurs at a phase interface while subsequent propagation reactions may occur in a homogeneous medium. The overall kinetics of such reactions are often very complex and specific to the particular systems. Useful generalizations are more likely to be provided at the present time by systems in which the initiation and propagation processes are both homogeneous. Such polymerizations are discussed next.

(ii)   Homogeneous Initiation and Propagation

Propagation reactions can be studied intensively in homogeneous systems because conditions can be arranged so that initiation is very fast and there are virtually no termination or chain transfer reactions.

In the absence of side reactions the number average degree of polymerization will be $d[M]/d[I]$ if initiation is by nucleophilic attack on the monomer or $2\,d[M]/d[I]$ if initiation is by electron transfer followed by dimerization of the monomeric radical anions ($d[M]$ and $d[I]$ are the reacted concentrations of monomer and initiator, respectively). If the rate of initiation is very rapid compared to the propagation rate and the initiator is mixed very rapidly and efficiently into the reaction mixture, then all macroions should start growing at almost the same time and should add monomer at equal rates. The active centers can be terminated deliberately and simultaneously since there are no spontaneous termination reactions under appropriate experimental conditions. Polymers made in such reactions have molecular weight distributions which approximate the Poisson distribution. That is, the number fraction x_i of I-mers is

$$x_i = e^{-\mu}\mu^i/i! \tag{9-13}$$

where μ equals the number of moles of monomer reacted per mole of initiator. The ratio of weight to number average chain lengths is

$$\overline{DP}_w/\overline{DP}_n = 1 + 1/\overline{DP}_n \tag{9-14}$$

and the molecular weight distribution is narrow if the degree of polymerization $\gtrsim 100$. (Recall also Section 2.5.)

A major interest in narrow distribution polymers is for research and molecular weight calibrations in gel permeation chromatography. Narrow-molecular-weight polystyrenes are made by initiation with alkali metal alkyls that are particularly effective in this application but that only polymerize conjugated monomers like styrene or butadiene.

The detailed kinetics of homogeneous anionic propagation reactions differ in hydrocarbon solvents and in media like ethers and amines which can solvate the metal counterion. These systems are discussed separately below.

Solvating Solvents. In these media the rate of polymerization is found to be directly proportional to the monomer concentration as would be expected for a propagation reaction like

$$\text{\Large \sim\!\!\!\sim\!\!\!\sim MMM}^{\ominus}\text{//}Z^{\oplus} + M \longrightarrow \text{\Large \sim\!\!\!\sim\!\!\!\sim MMM}^{\ominus}\text{//}Z^{\oplus} \tag{9-15}$$

The dependence of the propagation rate on initiator concentration is more complex, however, and can be explained as reflecting the existence of more than one kind of active center in media that can solvate the counterion. The simplest situation, which is used here for illustration, corresponds to an equilibrium between free ions and ion pairs. [It is likely that various kinds of ion pairs exist (cf. Eq. 9-1) but these ramifications can be neglected in this simple treatment.] The reactions involved in the actual propagation steps in the polymerization of a monomer M by an alkyl lithium compound RLi can then be represented as

$$
\begin{array}{ccc}
\text{RM}_j^{\ominus}\,//\,\text{L}_i^{\oplus} & \overset{K}{\underset{}{\rightleftharpoons}} & \text{RM}_j^{\ominus} + \text{L}_i^{\oplus} \\[4pt]
k_p^{\mp}\downarrow\,M & & M\downarrow k_p^{-} \\[4pt]
\text{RM}_{j+1}^{\ominus}\,//\,\text{L}_i^{\oplus} & \overset{K}{\underset{}{\rightleftharpoons}} & \text{RM}_{j+1}^{\ominus} + \text{L}_i^{\oplus} \\[4pt]
\text{solvated} & & \text{solvated free} \\
\text{ion pairs} & & \text{ions}
\end{array}
\tag{9-16}
$$

Here $k_p^{\mp}$ and k_p^{-} are rate constants for ion pair and free ion propagation, respectively, and K is the equilibrium constant for dissociation of ion pairs into solvated free ions.

$$K = \frac{\left[\sum_j RM_j^{\ominus}\right][Li^{\oplus}]}{\left[\sum_j RM_j^{\ominus}//Li^{\oplus}\right]} = \frac{\left[\sum_j RM_j^{\ominus}\right]^2}{\left[\sum_j RM_j^{\ominus}//Li^{\oplus}\right]} \tag{9-17}$$

since $[Li^{\oplus}]$ must $= [\sum_j RM_j^{\ominus}]$ for electrical neutrality. If the polymer molecular weight is high, the consumption of monomer in initiation reactions will be negligible and the overall rate of reaction can be expressed as

$$-\frac{d[M]}{dt} = R_p = k_p^{\mp}\left[\sum_j RM_j^{\ominus}//Li^{\oplus}\right][M] + k_p\left[\sum_j RM_j^{\ominus}\right][M] \tag{9-18}$$

Then, combining Eqs. (9-17) and (9-18),

$$\frac{R_p}{[M]} = k_p^{\mp}\left[\sum_j RM_j^{\ominus}//Li^{\oplus}\right] + k_p^{-}K^{1/2}\left[\sum_j RM_j^{\ominus}//Li^{\oplus}\right]^{1/2} \tag{9-19}$$

If there is little dissociation $[\sum_j RM_j^{\ominus}//Li^{\oplus}]$ can be set equal to the total initiator concentration [RLi] so that Eq. (9-19) becomes

$$\frac{R_p}{[M][RLi]} = k_p^{\mp} + \frac{k_p^{-}K^{1/2}}{[RLi]^{1/2}} \tag{9-20}$$

A plot of the left-hand side of Eq. (9-20) against $[RLi]^{1/2}$ yields a straight line with intercept $k_p^{\mp}$ and slope k_p^{-}. K can be determined independently by measuring the conductivity of solutions of low-molecular-weight living polymers and k_p^{-} can therefore also be estimated [1]. Such experiments show that free ions are generally present only in concentrations about 10^{-3} those of the corresponding ion pairs. The free ions are, however, responsible for a significant proportion of the polymerization, since k_p^{-} values are of the order of 10^4–10^5 liter/mol sec compared to $k_p^{\mp}$ magnitudes $\simeq 10^2$ liter/mol sec which are of the same order as free radical k_p's.

Hydrocarbon Solvents. Hydrocarbon solvents do not solvate the metal counterions. The rate of polymerization depends directly on the monomer concentration as in the case of solvating media, but the rate dependence on initiator concentration is complicated and variable. Some of this complexity is due to the association of growing polymer chains and their counterions into dimers and larger aggregates. The detailed mechanisms of these reactions still require clarification.

Initiation by soluble organolithium compounds in hydrocarbons is a slow process and is related to the existence of aggregates of initiators alone as well as aggregates which include both initiators and macromolecular species. The propagation reactions are also complex and not well understood at this time.

9.2.4 Termination Reactions

Anionic polymerizations must be carried out in the absence of water, oxygen, carbon dioxide, or any other impurities which may react with the active polymerization centers. Glass surfaces carry layers of adsorbed water which react with carbanions. It is necessary to take special precautions, such as flaming under vacuum, to remove this adsorbed water in laboratory polymerizations. Anionic reactions are easier to carry out on a factory scale, however, because the surface-to-volume ratio is much less in large reactors.

Polymers with specific end groups can be prepared by deliberately introducing particular reagents that "kill" living polymers. Thus in the anionic polymerization of butadiene with bifunctional initiators, carboxyl end groups are produced by termination with CO_2:

$$
\text{Na}^{\oplus}\text{CH}_2^{\ominus} - \underset{\underset{H}{|}}{C} = \underset{\underset{H}{|}}{C} - \text{CH}_2 \, \text{WWW} \, \text{CH}_2 - \underset{\underset{H}{|}}{C} = \underset{\underset{H}{|}}{C} - \text{CH}_2^{\ominus}\text{Na}^{\oplus} + 2\,CO_2 \longrightarrow
$$

$$
\text{Na}^{\oplus}\,\,\overset{\ominus}{\text{O}}\text{OC} - \text{CH}_2 - \underset{\underset{H}{|}}{C} = \underset{\underset{H}{|}}{C} - \text{CH}_2 \, \text{WWW} \, \text{CH}_2 - \underset{\underset{H}{|}}{C} = \underset{\underset{H}{|}}{C} - \text{CH}_2 - C\,\text{OO}^{\ominus}\,\text{Na}^{\oplus}
$$

$$(9\text{-}21)$$

Hydroxyl end groups are provided by termination with ethylene oxide:

$$
\text{WWW}\,\text{CH}_2^{\ominus}\text{Li}^{\oplus} + \text{CH}_2 - \text{CH}_2 \longrightarrow \text{WW} \, \text{CH}_2 - \text{CH}_2 - \text{CH}_2\text{O}^{\ominus}\,\text{Li}^{\oplus} \xrightarrow{\text{CH}_3\text{OH}}
$$

$$
\text{WWW}\text{CH}_2 - \text{CH}_2 - \text{CH}_2\text{OH} + \text{LiOCH}_3
$$

$$(9\text{-}22)$$

Low-molecular-weight (3,000–10,000) versions of such elastomers are used to produce liquid rubbers which can be shaped more easily than the conventional high-viscosity elastomers. The liquid rubbers can be vulcanized by reactions of their specific end groups. Hydroxyl-ended polybutadienes can be caused to grow in molecular size and to cross-link, for example, by reaction with isocyanates with functionalities > 2 (cf. reaction 1-12).

9.2.5 Temperature Effects

The direction of temperature effects in anionic polymerizations is conventional, with increased temperature resulting in increased reaction rates. Observed activation energies are usually low and positive. This apparent simplicity disguises complex effects, however, and the different ion pairs and free ions do not respond equally to temperature changes. Overall activation energies for polymerization will be influenced indirectly by the reaction medium because the choice of solvent shifts the equilibria of Eq. (9-1).

9.2.6   Anionic Copolymerization Reactions

(i)   Statistical Copolymerization

Copolymerizations analogous to free-radical reactions occur between mixtures of monomers which have more or less the same *e* values [Table 9-1]. The copolymerizations of styrene and dienes have been particularly studied in this connection. The simple copolymer equation (Eq. 7-13) applies to most of these systems, but the reactivity ratios will vary with the choice of solvent and positive counterion because these factors influence the nature of the propagating ion pair.

Note that styrene and conjugated dienes can be copolymerized to yield statistical or block copolymers. The latter process, which involves additions of one monomer to a living polymer of the other monomer, is described in the following section.

The styrene/methyl methacrylate pair contains monomers with different relative reactivity levels in Table 9-1. Polystyryl anion will initiate the polymerization of methyl methacrylate, but the anion of the latter monomer is not sufficiently nucleophilic to cross-initiate the polymerization of styrene. Thus the anionic polymerization of a mixture of the two monomers yields poly(methyl methacrylate) while addition of methyl methacrylate to living polystyrene produces a block copolymer of the two monomers.

(ii)   Block Copolymers

Anionic polymerizations are particularly useful for synthesizing block copolymers. These macromolecules contain long sequences of homopolymers joined together by covalent bonds. The simplest vinyl-type block copolymer is a two segment molecule illustrated by structure (**9-2**). This species is called an AB block copolymer, because it is composed of a poly-A block joined to a long sequence of B units. Other common block copolymer structures are shown as (**9-3**)–(**9-6**).

$$(M_1 M_1 M_1 \cdots)_n (M_2 M_2 M_2 \cdots)_m \quad \text{AB block copolymer}$$

9-2

$$(M_1 M_1 M_1 \cdots)_x (M_2 M_2 M_2 \cdots)_y (M_1 M_1 M_1 \cdots)_z \quad \text{ABA block copolymer}$$

9-3

$$(M_1 M_1 M_1 \cdots)_i (M_2 M_2 M_2 \cdots)_j (M_3 M_3 M_3 \cdots)_k \quad \text{ABC block copolymer}$$

9-4

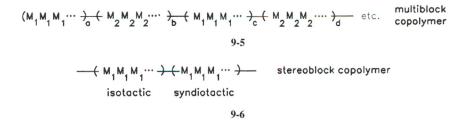

9-5

$$-\!\!\!(\; M_1 M_1 M_1 \cdots \;)\!\!-\!\!\!(\; M_1 M_1 M_1 \cdots \;)\!\!-\qquad \text{stereoblock copolymer}$$

isotactic   syndiotactic

9-6

The properties of block copolymers differ from those of a mixture of the corresponding homopolymers or a statistical copolymer with the same net composition. An important example is the ABA-type styrene/butadiene/styrene thermoplastic elastomers. Normal elastomers are cross-linked by covalent bonds [p. 10] to preserve the shape of stressed articles. A rubber tire, for example, would flow eventually into a pancake shape under the weight of an automobile if the elastomer were not vulcanized. Cross-linked rubbers are thermosets. When such materials are heated, chemical degradation and charring occurs at temperatures at which the rubbers are still not soft enough to permit reshaping. Solid thermoplastic styrene–butadiene triblock elastomers consist of glassy, rigid, polystyrene domains linked together by rubbery polybutadiene segments (**9-7**). The polystyrene regions serve as effective cross-links and stabilize the structure against moderate stresses. They can be softened sufficiently at temperatures over 100°C, however, that these products can be molded and remolded many times. When they are cooled the rigid polystyrene zones reform, and the polybutadiene links confer a rubbery character on the solid product.

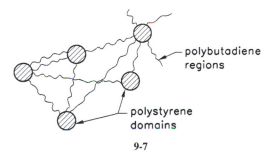

polybutadiene
regions

polystyrene
domains

9-7

Although some block copolymers can be made by other techniques, anionic polymerizations are particularly useful in this application. This is mainly because of the absence of an inherent termination step in some anionic systems and because anions with terminal monomer units of one type can be used to initiate the polymerization of other selected monomers. The different anionic reaction sequences that are employed include sequential monomer addition, coupling reactions, and termination with reactive groups.

(iii)   Sequential Monomer Addition

Monofunctional Initiators. AB, ABA, and multiblock copolymers can be synthe-sized by initiation of one monomer with a monofunctional initiator like *n*-butyl lithium. When the first monomer has been reacted, a second monomer can be added and polymerized. This monomer addition sequence can be reversed and repeated if the anion of each monomer sequence can initiate polymerization of the other monomer. The length of each block is determined by the amount of the corresponding monomer which was provided. Styrene–isoprene–styrene block copolymers can be made by this method by polymerizing in benzene solution and adding the styrene first. Addition of a small amount of ether accelerates the slow attack of dienyl lithium on styrene.

Bifunctional Initiation. The bifunctional initiators like alkali metal com-plexes of polycyclic aromatic compounds can be used to produce ABA triblock copolymers even when the A anion is not sufficiently basic to initiate polymeriza-tion of B monomers. In these cases polymerization would be started with monomer B to produce a polymeric dianion which could initiate polymerization of the A monomer which is added later. These initiators can be prepared only in aliphatic ethers, however. This precludes their use for the synthesis of useful styrene-diene ABA copolymers because polydienes made anionically in such solvents have low 1, 4 contents and are not good rubbers.

Coupling Reactions. In this technique, a living AB block copolymer is made by monofunctional initiation and is then terminated with a bifunctional coupling agent like a dihaloalkane. ABA copolymers can be made by joining AB polymeric anions:

$$2\,(M_1 M_1 \cdots)_i (M_2 M_2 M_2 \cdots)_j M_2^{\ominus} Li^{\oplus} + Br\,(CH_2)_K\,Br \longrightarrow$$

$$(M_1 M_1 M_1 \cdots)_i (M_2 M_2 M_2 \cdots)_{j+1} (CH_2)_K (M_2 M_2 M_2 \cdots)_{j+1} (M_1 M_1 M_1 \cdots)_i + 2\,Li\,Br \qquad (9\text{-}23)$$

Reaction (9-23) could be used to produce an ABA copolymer of styrene and methyl methacrylate even though the latter anion is too weakly basic to initiate the polymerization of styrene. If the presence of a linking group, like the $—(—CH_2—)_k$ in the above reaction, is undesirable a coupling agent like I_2 can be used.

Star-shaped block copolymers can be made by using polyfunctional linking agents, like methyltrichlorosilane or silicon tetrachloride, to produce tribranch or tetrabranch polymers.

9.2.7   Stereochemistry of Anionic Diene Polymerization

Anionic polymerization proceeds through only one of the two available double bonds in dienes like butadiene or isoprene (Fig. 1-4). Possible structures that could be produced from enchainment of isoprene have been listed earlier (in Section 4.1.1). In addition, the polymer segments attached to internal C—C double bonds may be in *cis* or *trans* configurations. The number of possible isomers of a polymer of isoprene is clearly very large. The preferred structure for elastomeric applications is that in natural rubber produced by the Hevea Brasiliensis tree. This polymer, which is sometimes called "synthetic natural rubber" can be made by anionic and Ziegler–Natta polymerizations.

The character of the counterion and the solvent both affect the microstructure of polymers made anionically from dienes. In general, the proportion of 1, 4 chains is highest for Li and decreases with decreasing electronegativity and increasing size of the alkali metals in the order: Li > Na > K > Rb > Cs. A very high (>90%) 1,4 content is achieved only with lithium alkyl or lithium metal initiation in hydrocarbon solvents. The properties of polymers of conjugated diolefins tend to be like those of thermoplastics if the monomer enchainment is 1,2 or 3,4 [reactions (4-3) and (4-4)]. Elastomeric behavior is realized from 1,4 polymerization and particularly if the polymer structure is *cis* about the residual double bond.

Addition of polar solvents like ethers and amines causes an increase in side chain vinyl content resulting from 1,2 or 3,4 polymerization. This effect is particularly marked in polymerizations with lithium alkyls, which are the only alkali metal alkyls that are soluble in bulk monomer or hydrocarbon solutions. Polar media also tend to increase the proportion of 1,4 units with *trans* configurations.

The molecular structures produced are influenced by: (1) the relative stabilities of the two conformations about the double bond in the terminal unit of the macroion; (2) the conformation of the diene monomer when it adds to the macroion; and (3) in the case of 1,4 addition, the relative rates of isomerization between *cis* and *trans* terminal units compared to the rates at which new monomers are added.

The double bond between carbons 3 and 4 of isoprene (**9-8**) is the more reactive in anionic polymerizations because the electron-donating methyl group makes carbon 1 more electron rich than carbon 4. The general reaction for anionic attack on isoprene is thus given by

$$\text{wwww } CH_2^{\ominus} \ Z^{\oplus} + \underset{4}{H_2}C = \underset{3}{\overset{|}{C}} - \underset{2}{\overset{|}{C}} = \underset{1}{CH_2} \longrightarrow \text{wwww } \underset{4}{CH_2} - \underset{3}{CH_2} - \underset{2}{\overset{|}{C}} - \underset{1}{\overset{|}{C}} - CH_2$$

9-8   9-9   (9-24)

Structure **9-9** is that of an allyl carbanion which can reach at carbon 1 or carbon 3. When the metal is lithium, which is unsolvated in hydrocarbon media, it is attached to the chain end at carbon 1, as shown in **9-10**. *Cis*-1,4 chain units can result from a concerted attack of isoprene on the bond between carbon 1 and Li while the incoming monomer is held in a *cis* conformation by coordination to the small $Li^\oplus$ ion.

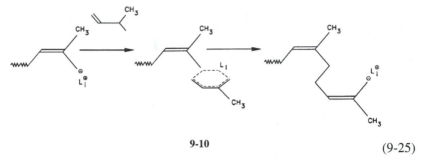

9-10   (9-25)

If isomerization can occur, then the configuration of end units is finally fixed only when a new monomer is added because the only units which can isomerize are terminal ones. *Trans* isoprenyl forms are more stable than *cis* terminal isomers in hydrocarbon media, but the rate of isomerization from the initial *cis* form shown above is not very fast and the *cis* configuration can be fixed in the polymer structure if the new monomers are added sufficiently quickly to the terminal unit. The *cis*-1, 4 content in the polymer decreases as the number of active centers is increased or the monomer concentration is lowered.

Butadiene reacts generally slower than isoprene, and the difference in propagation rates between *cis* and *trans* terminal units is less. Thus the *cis* configuration is not as favored in 1,4-polybutadiene as it is in polyisoprene.

A variation of the sequential monomer addition technique described in Section 9.2.6(i) is used to make styrene–diene–styrene triblock thermoplastic rubbers. Styrene is polymerized first, using butyl lithium initiator in a nonpolar solvent. Then, a mixture of styrene and the diene is added to the living polystyryl macroanion. The diene will polymerize first, because styrene anions initiate diene polymerization much faster than the reverse process. After the diene monomer is consumed, polystyrene forms the third block. The combination of $Li^\oplus$ initiation and a nonpolar solvent produces a high *cis*-1,4 content in the central polydiene block, as required for thermoplastic elastomer behavior.

9.3   GROUP TRANSFER POLYMERIZATION

Although group transfer polymerization does not involve ionic reactions, it is reviewed in this chapter because it bears many practical similarities to anionic polymerizations and is an alternative route to (meth-)acrylic polymers and block copolymers. This process is applicable to vinyl monomers with electron-withdrawing substituents, like acrylates and methacrylates.

The initiator is a silyl ketene acetal (**9-11**) which can be synthesized from methyl methacrylate and trimethyl- or tetramethylsilane:

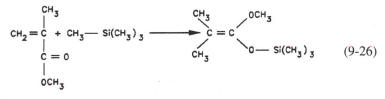

$$(9\text{-}26)$$

9-11

Monomer addition, which is catalyzed by anions like $HF_2^\ominus$, $F^\ominus$, $CN^\ominus$ or selected Lewis acids, proceeds by Michael addition in which the silyl group is transferred to the new monomer unit to renew the terminal silyl ketene acetal. (A Michael reaction, in general, is the addition of an enolate to an α, β-unsaturated carbonyl compound.)

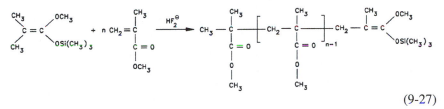

$$(9\text{-}27)$$

Unlike anionic polymerizations, the reaction sites are not ion pairs; the catalyst is believed to facilitate transfer of the trimethylsilyl group by dipolar interaction with the Si atom. As in anionic polymerizations, however, the reactive end group is deactivated by compounds carrying labile hydrogens. Group transfer polymerizations therefore must be carried out under anhydrous conditions.

Group transfer processes have some of the characteristics of living polymerizations. The degree of polymerization is determined by the ratio of monomer to silyl ketene initiator [cf. Section 9.2.3(ii)] and the molecular weight distribution of the product is narrow. In the absence of terminating impurities, the terminal silylketene groups remain active and block copolymers can be produced by introduction of another polymerizable monomer. Functional terminal groups can be introduced by reacting the silylketene acetal with an appropriate electrophile, such as aldehydes or acryl halides and active macrospecies can be coupled with dihalides. Also, terminal functional groups can be introduced as substituents in the silylketene initiator.

Currently, this polymerization technique is applied primarily for the synthesis of fairly low-molecular-weight functionalized polymers and copolymers of methyl methacrylate, for automotive finishes.

9.4   CATIONIC POLYMERIZATION

9.4.1   Overview

The active center in these polymerizations is a cation, and the monomer must therefore behave as a nucleophile (electron donor) in the propagation reaction. The most active vinyl monomers are those with electron-releasing substituents or conjugated double bonds in which the positive charge can be delocalized.

Figure 9-1 illustrates inductive influences in cationic polymerizations. The electron-releasing inductive influence of alkyl groups causes isobutene to polymerize very quickly at low temperatures where propylene reacts inefficiently and ethylene is practically inert. For similar reasons, α-methylstyrene (**9-12**) is more reactive than styrene, and substitution of an electron-withdrawing halogen for an *ortho*- or *para*-hydrogen, decreases the monomer reactivity still further. As a corollary, *ortho*- and *para*-electron-releasing substituents (RO—, RS—, aryl) increase cation stability and monomer reactivity.

Conjugated olefins, like styrene, butadiene, and isoprene, can be caused to polymerize by cationic and anionic as well as by free-radical processes because the active site is delocalized in all cases. The most practical ionic polymerizations for these species are anionic, because such reactions involve fewer side reactions and better control of the diene polymer microstructure than in cationic systems. Free-radical polymerization of styrene is preferred over ionic processes, however, for cost reasons.

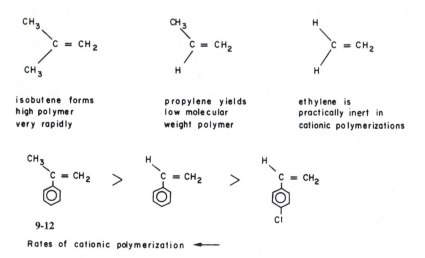

Fig. 9-1.   Inductive effects in cationic polymerization of vinyl monomers.

For efficient cationic polymerization of vinyl monomers, it is necessary that the carbon–carbon double bond be the strongest nucleophile in the molecule. Thus vinyl acetate would be classed as an electron-donor-type monomer (Section 7.10.2) but it cannot be polymerized cationically because the carbonyl group complexes the active center. (It is polymerized only by free radicals; anionic initiators attack the ester linkage.)

Other monomers that are suitable for cationic polymerization include cyclic ethers (like tetrahydrofuran), cyclic acetals (like trioxane), vinyl ethers, and *N*-vinyl carbazole. In these cases the hetero atom is bonded directly to the electron deficient carbon atom, and the respective carboxonium ion (**9-13**) and immonium ion (**9-14**) are more stable than the corresponding carbocations.

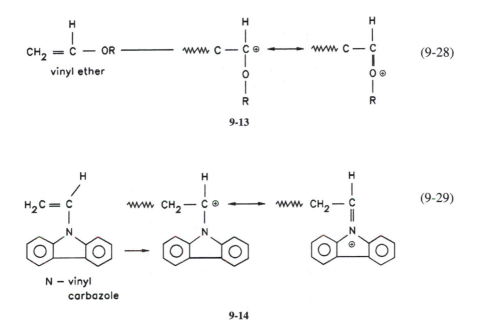

Because of this difference in stability monomers which yield onium ions will not copolymerize cationically with olefins like isobutene or styrene. (A similar difference in radical stabilities accounts for the reluctance of styrene to copolymerize with vinyl chloride in free-radical reactions, Section 7.10.1.)

Cationic polymerizations of vinyl monomers differ from other chain-growth polymerizations particularly as follows:

1.   Cationic reactions often involve complex initiation and preinitiation equilibria that may include the rate-determining steps in the polymerization sequence.

2.   The reaction medium in cationic polymerizations is usually a moderately polar chlorinated hydrocarbon like CH_3Cl (dielectric constant $= 12.6$ at $-20°C$). A greater proportion of the macroions are free of their counterions in cationic than in anionic polymerizations in the usual solvents for the latter processes. Cationic polymerizations are characterized by extremely fast propagation rates.

Chlorinated aliphatic solvents are useful only for cationic reactions, because they would be attacked by the strong bases used to initiate anionic polymerizations. Conversely, tetrahydrofuran is a useful solvent in some anionic processes but is polymerized by appropriate cationic initiators.

3.   The growth of vinyl polymer molecules in cationic reactions is limited by transfer rather than by true termination processes. Transfer to monomer is often the most important of these processes.

4.   Some monomers isomerize during cationic polymerization. The net result is that the structure of the polymer repeating unit is not the same as that of the monomer. This occurs if the carbocation which is formed by initial attack on the monomer can isomerize to a more stable form. Isomerization is more prevalent in cationic than in anionic or free radical polymerizations because carbenium or oxonium ions rearrange easily.

9.4.2   Initiation of Polymerization of Vinyl Monomers

Initiation is usually by attack of a cation on a monomer to generate a carbocation which can propagate the kinetic chain by adding further monomers. (We shall refer to this *sp²-hydridized trigonal species* as a carbenium ion. It is also called a *carbonium ion*.) The initiator is itself usually a proton or a carbenium ion. The actual initiation process is often preceded by a reversible reaction or series of reversible reactions between the initiator and other species. An example is

$$AIR_2^1 Cl + RCl \rightleftharpoons R^{\oplus} \left[AI\, R_2^1\, Cl_2 \right]^{\ominus} \qquad (9\text{-}30)$$

$$R^{\oplus} \left[AI\, R_2^1\, Cl_2 \right]^{\ominus} + CH_2 = \underset{\underset{CH_3}{|}}{\overset{\overset{CH_3}{|}}{C}} \longrightarrow R - CH_2 - \underset{\underset{CH_3}{|}}{\overset{\overset{CH_3}{|}}{C}}\!\oplus \left[AI\, R_2^1\, Cl_2 \right]^{\ominus} \qquad (9\text{-}31)$$

The preinitiation reaction sequence may be quite complex and attainment of the equilibria involved is often slower than subsequent initiation or propagation reactions. The following paragraphs review a few of the more important initiators for vinyl monomers.

(i)   Bronsted (Protonic) Acids

The general reaction between an olefin and a protonic acid HA can be written as

$$H^\oplus A^\ominus + CH_2 = C\underset{R'}{\overset{R}{\diagup}} \longrightarrow CH_3 - \underset{R'}{\overset{R}{C^\oplus}} \: A^\ominus \qquad (9\text{-}32)$$

It is necessary for practical polymerizations that the anion A not be a strong nucleophile or else it will react with the carbenium ion to form the nonpropagating covalent compound

$$CH_3 - \underset{R}{\overset{R'}{\underset{|}{\overset{|}{C}}}} - A$$

Acids such as H_2SO_4, $HClO_4$, and H_3PO_4 are better initiators than hydrogen halides because their anions are larger and less nucleophilic.

Initiation by protonic acids is relatively inexpensive. The major applications of such processes are for reactions of simple olefins like propylene or butenes or olefinic derivatives of aromatics such as coumarone (**9-15**) and indene (**9-16**). The products of these reactions have very low molecular weights because of transfer reactions which are discussed below (Section 9.4.4). The low polymers of olefins are used as lubricants and fuels while the coumarone-indene polymers are employed in coatings and as softeners for rubbers and bitumens.

9-15   9-16

Heterogeneous acidic materials are used widely as cracking and isomerization catalysts in the petroleum industry. These materials consist of acidic clays or synthetic, porous, crystalline aluminosilicates (molecular sieves). They appear to function as protonic acids. The counterion in these cases is the catalyst surface, which is a very poor nucleophile.

Initiation by proton addition is usually accomplished by generating the protons from the interaction of Lewis acids and auxiliary, hydrogen-donating substances. These processes are discussed in the following section.

(ii) Lewis Acids

These are halides and alkyl halides of Group III metals and of transition metals in which the d electron shells are incomplete. This is the most generally useful group of initiators and includes compounds like BF_3, $SnCl_4$, $AlCl_3$, AlR_2Cl, $SbCl_5$, and so on.

Initiation is facilitated by interaction of the Lewis acid with a second compound (called a *cocatalyst*) that can donate a proton or carbenium ion to the monomer. Typical cocatalysts are water, protonic acids, and alkyl halides. Examples of Lewis acid–cocatalyst interactions are given in reactions (9-29) and (9-32). The general reaction for

$$BF_3 + H_2O \rightleftharpoons H^\oplus BF_3 OH^\ominus \qquad (9\text{-}33)$$

initiation by a Lewis acid ZX_n and a cocatalyst BA is

$$ZX_n + BA \rightleftharpoons \left[ZX_n B \right]^\ominus_A{}^\oplus \qquad (9\text{-}34)$$

$$\left[ZX_n B \right]^\ominus_A{}^\oplus + CH_2 = C{\overset{R}{\underset{R'}{\diagup}}} \longrightarrow A - CH_2 - C{\overset{R'}{\underset{R}{\diagdown}}}^\oplus \left[ZX_n B \right]^\ominus \qquad (9\text{-}35)$$

here Z is a metal, X is generally a halogen or an organic entity, and BA is an ionizable compound. The anionic fragment of the complex between the Lewis acid and the cocatalyst adds to the monomer to initiate polymerization. The monomer cation that is formed does not form a covalent link with the complex anion, because the size and solvation of the latter reduce its nucleophilicity.

The use of cocatalysts is desirable and possibly absolutely necessary in many Lewis acid systems. The concentration of cocatalyst must be carefully controlled, however, and optimum Lewis acid/cocatalyst concentration ratios can be established for particular polymerizations. This is because the cocatalyst must be more basic than the monomer; otherwise the Lewis acid would react preferentially with the monomer. If excess co-catalyst BA is present, however, it can compete with the monomer for reaction with the primary Lewis acid/cocatalyst complex. For example,

$$BF_3 + H_2O \rightleftharpoons H^\oplus BF_3OH^\ominus \overset{H_2O}{\longrightarrow} H_3O^\oplus BF_3OH^\ominus \qquad (9\text{-}36)$$

where excess water transforms the bare proton of reaction (9-33) into the weaker acid $H_3O^\oplus$ which cannot initiate polymerization of olefins.

In many practical systems the aprotonic acids are difficult to purify and substances may be present that act as unidentified cocatalysts. The dependence of the rate of polymerization on the concentrations of Lewis acid and ostensible cocatalyst cannot then be elucidated.

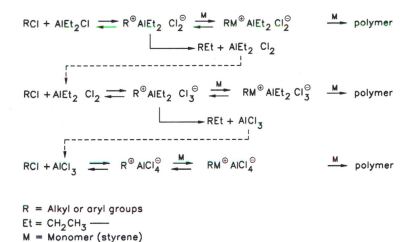

R = Alkyl or aryl groups
Et = CH_2CH_3 ——
M = Monomer (styrene)

Fig. 9-2. Equilibria in initiation by alkyl aluminum halides and alkyl halides [2].

Alkyl halides are widely used as cocatalysts in combination with aluminum alkyl halides or aluminum halide Lewis acids. The reaction scheme in Fig. 9-2 illustrates the complicated equilibria which may affect the initiation process. Each carbenium ion can initiate polymerization or remove an ethyl group from the counterion to produce a saturated hydrocarbon, REt, and a new more acidic Lewis acid. The propagating macrocarbenium ions can also terminate by the same process to produce ethyl-capped polymers and new Lewis acids. Thus, even though the initiator is ostensibly diethylaluminum chloride there may be major contributions to the polymerization from ethyl aluminum dichloride or aluminum chloride.

(iii) Stable Organic Cation Salts

Some organic cation salts can be isolated as crystalline solids. Stability of carbenium ions is enhanced if the electron-deficient carbon is conjugated with olefin or aromatic groups or with atoms with unshared electron pairs (O, N, S). The positive charge is diffused over a larger region as a consequence. Examples of such initiators are hexachloroantimonate ($SbCl_6^{\ominus}$) salts of triphenyl methyl (($C_6H_5)_3C^{\oplus}$) and cycloheptrienyl ($C_7H_7^{\oplus}$) carbenium ions.

These initiators are primarily of academic interest because their initiation processes are fairly straightforward compared to the complicated equilibria that can exist in other systems (e.g., Fig. 9-2). This simplifies the study of the kinetics of the propagation and other processes involved in cationic polymerization. However, since these cations are stable, their use is limited to the initiation of more reactive monomers like N-vinylcarbazole and alkyl vinyl ethers.

9.4.3   Propagation Reactions

Determination of propagation rate constants in cationic (and in anionic) systems is complicated by the simultaneous occurrence of different types of propagating sites. In olefin polymerizations, some portion of the active centers may exist as free ions and others as ion pairs of varying degrees of solvation. In the solvents in which cationic polymerizations are normally carried out, the polymerization is mainly due to free ions. In low dielectric constant media like benzene or hydrocarbon monomers, however, ion pairs will dominate the reaction.

The initiation and propagation processes are influenced by equilibria between various degrees of association of the active center and its counterion. As a minimum, it is necessary to conceive of the existence of contact (associated) ion pairs, solvent-separated ion pairs, and free solvated ions. A simplified reaction scheme [3] is presented in reaction (9-37).

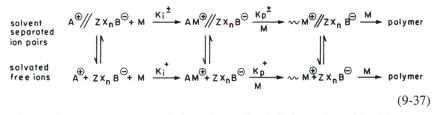

$$\tag{9-37}$$

The existence of contact pair ions (as in Eq. 9-1) is neglected in this representation because the dielectric constants of the solvents usually used for cationic polymerizations are high enough to render concentrations of intimate ion pairs negligible compared to those of solvated ion pairs.

The observed k_p in these simplified reactions will be composed of contributions from the ion pairs and free ions:

$$k_{p_{obs}} = \alpha k_p^+ + (1 - \alpha)\, k_p^\pm \tag{9-38}$$

where α is the degree of dissociation of ion pairs into free solvated ions and the k_p values are defined in (9-37). In solvents like CH_2Cl_2 (dielectric constant $\sim$10–12), the k_p^+ values for olefins are of the order of 10^3–10^5 liter/mol sec. These rate constants are similar to the k_p^- values for anionic polymerizations with free ions [Section 9.2.3(ii)]. The anionic reactions are carried out in solvents with lower dielectric constants, however, so the comparison is not strictly valid. In media of low polarity, like bulk monomer, the few k_p^+ values that have been measured for cationic olefin polymerizations are of the order of 10^6–10^9 liter/mol sec. It is not surprising that carbenium ions are more reactive than carbanions under more or less equivalent conditions. A carbenium ion has vacant bonding orbitals while those in the propagating carbanion and monomer are already filled, and the formation of an anionic transition state requires use of the antibonding orbitals of the monomer.

In general, both k_p^+ and $k_p^\pm$ will decrease with increasing solvent polarity. The transition-state theory of chemical reactions suggests that this is because the initial state (monomer plus ion or ion pair) is more polar than the activated complex in which the monomer is associated with the cation, and the charge is dispersed over a larger volume. More polar solvents will tend, then, to stabilize the initial state at the expense of the transition complex and reduce k_p^+ and $k_p^\pm$.

Note, however, that more polar solvents enhance the rate of polymerization even though they reduce the specific rate constants for propagation. This apparent conflict reflects the greater degree of dissociation of ion pairs into free ions in higher dielectric constant media.

The $k_p^\pm$ values are generally at least 100 times as great as the corresponding k_p^+ figures for olefin monomers. In many cationic polymerizations in media with dielectric constants greater than about 10, it is likely that the concentration of free ions is such that the contribution of ion pairs to the overall rate of polymerization is negligible.

9.4.4 Termination and Transfer Processes

True termination reactions interrupt the growth of a macrocation without generating a new active center which can add more monomer. This can occur through the formation of a stable cation:

$$(9\text{-}39)$$

Termination reactions are unlikely in very pure cationic systems but slight traces of impurities can exert significant effects because of the great reactivity of the propagating species.

The following termination reaction is important in the polymerization of isobutene at room temperature:

$$(9\text{-}40)$$

The kinetic chain is interrupted here, but the catalyst-cocatalyst complex is regenerated and can initiate new kinetic chains. In the production of butyl rubber (as distinguished from polyisobutene) isobutene is copolymerized with about 3%

(w/w) isoprene, to facilitate sulfur vulcanization [p. 10]. The major chain transfer reaction to monomer occurs at the isopropeneyl end of the macrocation. Thus the molecular weight of the copolymer is inversely proportional to the concentration of isoprene in the monomer feed mixture.

The most important processes which limit molecular weights in cationic vinyl polymerizations are transfer reactions to monomer (9-41) or residual water (9-42):

$$\text{\textasciitilde\textasciitilde\textasciitilde CH}_2 - \overset{\overset{\displaystyle H}{|}}{\underset{\underset{\displaystyle R}{|}}{C}} \oplus X^{\ominus} + \text{CH}_2 = \overset{\overset{\displaystyle H}{|}}{\underset{\underset{\displaystyle R}{|}}{C}} \longrightarrow \text{\textasciitilde\textasciitilde\textasciitilde CH} = \overset{\overset{\displaystyle H}{|}}{\underset{\underset{\displaystyle R}{|}}{C}} + \text{CH}_3 - \overset{\overset{\displaystyle H}{|}}{\underset{\underset{\displaystyle R}{|}}{C}} {}^{\oplus} X^{\ominus} \qquad (9\text{-}41)$$

$$\text{\textasciitilde\textasciitilde\textasciitilde CH}_2 - \overset{\overset{\displaystyle H}{|}}{\underset{\underset{\displaystyle R}{|}}{C}} \oplus X^{\ominus} + \text{H}_2\text{O} \longrightarrow \text{\textasciitilde\textasciitilde\textasciitilde CH}_2 - \overset{\overset{\displaystyle H}{|}}{\underset{\underset{\displaystyle R}{|}}{C}} - \text{OH} + \text{H}^{\oplus}X^{\ominus} \xrightarrow{\text{monomer}} \text{CH}_3 - \overset{\overset{\displaystyle H}{|}}{\underset{\underset{\displaystyle R}{|}}{C}} \oplus X^{\ominus}$$

$$(9\text{-}42)$$

Facile chain transfer to polymer is thought to account for the failure of cationic reactions of α-olefins like propylene to yield high-molecular-weight products. Reaction (9-43) produces a relatively stable tertiary carbenium ion from a more reactive, propagating secondary ion.

$$\text{\textasciitilde\textasciitilde\textasciitilde CH}_2 - \overset{\overset{\displaystyle H}{|}}{\underset{\underset{\displaystyle CH_3}{|}}{C}} \oplus X^{\ominus} + \text{\textasciitilde\textasciitilde\textasciitilde CH}_2 - \overset{\overset{\displaystyle H}{|}}{\underset{\underset{\displaystyle CH_3}{|}}{C}} - \text{CH}_2 \text{\textasciitilde\textasciitilde\textasciitilde} \longrightarrow \text{\textasciitilde\textasciitilde\textasciitilde CH}_2 - \text{CH}_2 - \text{CH}_3$$

$$+ \quad {}^{\oplus}X^{\ominus}$$
$$\text{\textasciitilde\textasciitilde\textasciitilde CH}_2 - \overset{\oplus}{\underset{\underset{\displaystyle CH_3}{|}}{C}} - \text{CH}_2 \text{\textasciitilde\textasciitilde\textasciitilde}$$

$$(9\text{-}43)$$

9.4.5   Kinetics of Cationic Polymerization of Olefins

Cationic polymerizations differ from free-radical and homogeneous anionic syntheses of high polymers in that the cationic systems have not so far been fitted into a generally useful kinetic framework involving fundamental reactions like initiation, propagation, and so on. To explain the reasons for the peculiar problems with cationic polymerizations we will, however, postulate a conventional polymerization reaction scheme and show where its inherent assumptions are questionable in cationic systems.

An "ideal" reaction scheme is shown in Fig. 9-3. Note that the Lewis acid ZX_n is referred to as a catalyst. This is the common terminology for such polymerizations but it will be realized that ZX_n is really an initiator and the compound BA, which is often called a cocatalyst, is really a coinitiator.

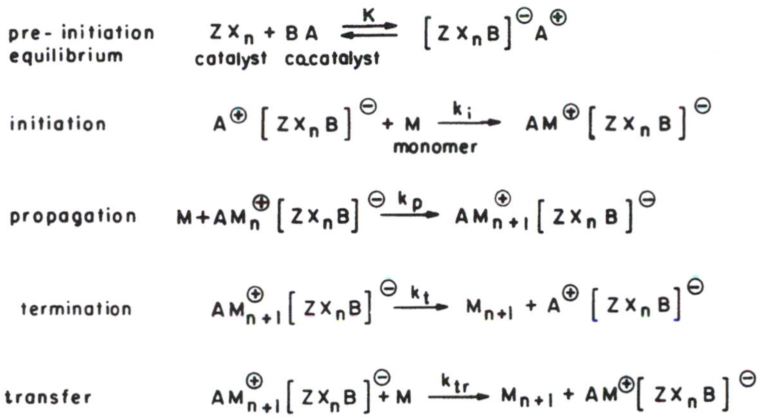

pre-initiation equilibrium	$ZX_n + BA \xrightleftharpoons[\quad]{K} \left[ZX_nB\right]^{\ominus} A^{\oplus}$
	catalyst cocatalyst

initiation $\qquad A^{\oplus}\left[ZX_nB\right]^{\ominus} + M \xrightarrow{k_i} AM^{\oplus}\left[ZX_nB\right]^{\ominus}$

monomer

propagation $\quad M + AM_n^{\oplus}\left[ZX_nB\right]^{\ominus} \xrightarrow{k_p} AM_{n+1}^{\oplus}\left[ZX_nB\right]^{\ominus}$

termination $\qquad AM_{n+1}^{\oplus}\left[ZX_nB\right]^{\ominus} \xrightarrow{k_t} M_{n+1} + A^{\oplus}\left[ZX_nB\right]^{\ominus}$

transfer $\qquad AM_{n+1}^{\oplus}\left[ZX_nB\right]^{\ominus} + M \xrightarrow{k_{tr}} M_{n+1} + AM^{\oplus}\left[ZX_nB\right]^{\ominus}$

Fig. 9-3. Simplified reaction scheme for cationic polymerizations.

Referring to Fig. 9-3, we see that the reaction sequence is greatly oversimplified to begin with, because we have ignored the existence and equilibria between free solvated ions and ion pairs of various degrees of intimacy. Thus all the rate constants that are listed are actually composite values (cf. Eq. 9-38) which can vary with the nature of the medium and counterion. The observed k_p values can even vary with the total concentration of reactive species, because organic ion pairs tend to cluster in the more nonpolar environment, and the reactivity of aggregated and individual ion pairs is not generally equal.

We can detect a further complication from observations that some cationic polymerizations exhibit bimodal (two peaks) molecular weight distributions. This can happen if different active species (say, free ions and some form of ion pairs) engage in propagation or transfer reactions faster than they can come to equilibrium with each other. Under these circumstances there can be two effectively independent processes that govern the size of the macromolecules that are produced. If we ignore all these important complications we can write the following expressions for the rates of initiation (R_i), propagation (R_p), termination (R_t), and transfer to monomer ($R_{tr,M}$):

$$R_i = (K[ZX_n][BA])k_i[M] \tag{9-44}$$

$$R_p = k_p[M][M^{\oplus}] \tag{9-45}$$

$$R_t = k_t[M^{\oplus}] \tag{9-46}$$

$$R_{tr,M} = k_{tr}[M^{\oplus}][M] \tag{9-47}$$

where $[M^{\oplus}]$ represents $\sum_{i=1}^{\infty}[AM_i^{\oplus}]$ and it is assumed that an overall rate constant can represent the reactions of all species of carbocations and their counterions.

It is necessary to invoke the steady state assumption ($R_i = R_t$; $d[M^\oplus]/dt = 0$) to make this model mathematically tractable. With this assumption,

$$[M^\oplus] = \frac{Kk_i[ZX_n][BA][M]}{k_t} \tag{9-48}$$

and

$$R_p = \frac{k_p[M]^2[ZX_n][BA]Kk_i}{k_t} \tag{9-49}$$

If the molecular growth is controlled by chain transfer to monomer, then

$$\overline{X}_n = R_p/R_{tr} = k_p/k_{tr} \tag{9-50}$$

This is more likely under practical polymerization conditions than the alternative case in which termination reactions limit the size of the macromolecules and

$$\overline{X}_n = R_p/R_t = k_p[M]/k_t \tag{9-51}$$

We pause here to note that the steady-state assumption that is so helpful in simplifying the analysis of free-radical kinetics (Section 6.3.4) will not apply to many cationic polymerizations of vinyl monomers, because propagation through free carbenium ions is so much faster than any of the other reactions in the kinetic chain.

Despite all these qualifications, Eq. (9-49) gives a form with which the kinetics of cationic systems can perhaps be studied. This is

$$-\frac{d[M]}{dt} = R_p = k_{obs}[M]^a[\text{catalyst}]^b[\text{cocatalyst}]^c \tag{9-52}$$

where the superscripts are determined experimentally. It is obvious from the preceding discussion, however, that the composite nature of k_{obs} in Eq. (9-52) and the rate constants in Eq. (9-49) make it possible to postulate many mechanisms to fit any single set of kinetic data.

In summary, cationic polymerizations are much more variable and complex than homogeneous free-radical or anionic chain-growth polymerizations. No convincing general mechanism has been provided for cationic reactions, and each polymerization system is best considered as a separate case.

9.4.6   Temperature Effects

The effects of temperature on cationic polymerizations can be described with caution by an Arrhenius expression like

$$k = Ae^{-E/RT} \tag{9-53}$$

where E is the activation energy for the particular process. From Eqs. (9-49) and (9-50), the activation energies for the rate of polymerization (E_{R_p}) and the number average degree of polymerization $(E_{\overline{X}_n})$ would be expected to be

$$E_{R_p} = E_i + E_p - E_t \tag{9-54}$$

and

$$E_{\overline{X}_n} = E_p - E_{tr} \tag{9-55}$$

where the Es are the activation energies for the processes identified by the particular subscripts in the reaction scheme in Fig. 9-3.

Kinetic data for cationic polymerizations are not usually reliable enough to establish the activation energies for the various processes very well. In general, it is expected, however, that energies for reactions that involve free ions will approach zero and those of other species will be positive.

The net activation energy for cationic polymerizations is low (<10 kcal/mol) and may even be negative. In the latter case one observes a rate of polymerization that increases with decreasing temperature. This is very probably because the proportion of free ions increases as the temperature is lowered. If the equilibria

$$R^{\oplus}X^{\ominus} \; \rightleftarrows \; \underset{\substack{\text{solvent separated} \\ \text{ion pair}}}{R^{\oplus} /\!/ X^{\ominus}} \; \overset{K_d}{\rightleftarrows} \; R^{\oplus} + X^{\ominus} \tag{9-56}$$

$$\underset{\substack{\text{contact} \\ \text{pair}}}{}$$

are considered, it can be shown [4] that

$$K_d = [R^{\oplus}][X^{\ominus}]/[R^{\oplus}X^{\ominus}] = \exp(-z^2/a\epsilon\kappa T) \tag{9-57}$$

where z is the charge of the ions, a is the sum of the van der Waals ionic radii, and κ is Boltzmann's constant. A common solvent for cationic polymerization is CH_2Cl_2. Its dielectric constant, which is 17.0 at $-100°C$, rises to 13.4 at $-60°C$ [5]. From these values and Eq. (9-57) it is clear that the proportion of free ions will increase strongly as the temperature of the reaction mixture falls. The net observed k_p value will also increase because free carbenium ions react so much more quickly than paired carbocations. The rate of polymerization will then increase as the reaction temperature is lowered.

The activation energy for degree of polymerization $E_{\overline{X}_n}$ is always negative and the degree of polymerization decreases as the temperature is raised. This is because E_{tr} is greater than E_p (as would be expected if E_p is zero for propagation with free ions). Thus low-molecular-weight grades of polyisobutene are made by $AlCl_3$ or BF_3 initiation at temperatures between 0 and $-40°C$ while high-molecular-weight polymers are made at -85 to $-100°C$. High-molecular-weight butyl rubber can only be made at low temperatures, because the comonomer isoprene is also a chain transfer agent.

9.4.7 Polymerization of Vinyl and Cyclic Ethers

The ether oxygen is a Lewis base (electron donor), and polymerization of vinyl and cyclic ethers can be initiated by reaction with an ion pair comprising an acidic cation and a weakly nucleophilic base. These monomers do not polymerize by free-radical or anionic processes. Thio ethers behave similarly.

The general structure of cyclic ethers may be written as **9-17** with $x \geq 2$.

9-17

Molecules with these structures polymerize because this process involves a ring-opening reaction which relieves strain. Four-member rings (oxetanes, $x = 3$) are subject to angle strain and to crowding of hydrogens on adjacent carbons. Tetrahydrofuran ($x = 4$) also polymerizes because of steric repulsion between adjacent hydrogens. Six-membered ring ethers like tetrahydropyran ($x = 5$) do not polymerize. However, trioxane (the cyclic trimer of formaldehyde) can be reacted cationically to produce polyformaldehyde. The reactivity of the trioxane ring is perhaps a result of the stability of the propagating carbenium ion **9-18** in which the charge is delocalized onto the adjacent oxygen:

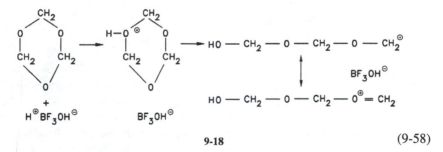

9-18   (9-58)

Epoxides (1,2 epoxides are also called oxiranes) can be polymerized by anionic, cationic, and coordinated anionic processes. These are the only cyclic ethers that polymerize through mechanisms that are not cationic. This propensity for ring opening reflects the high ring strain, because the bond angles in three-membered rings are distorted from the equilibrium tetrahedral angle of carbon. Cationic polymerization is a complex chain-growth process which is difficult to control, and anionic reactions are preferred for the production of polymers of ethylene and propylene oxide. These processes, which are important in the synthesis of nonionic detergents and polyols for urethane reactions, involve catalysis by alkali metal alcoholates, in alcohol solvents.

9.4.8   Living Cationic Polymerizations [6]

As illustrated earlier, the inherent problem in controlled cationic polymerizations is the instability of the macrocarbocations. However, living polymerizations can be realized by stabilizing the growing carbenium ions with a suitable nucleophilic counterion (B⊖), **9-19**, or an added Lewis base (X) containing a weakly nucleucleophilic counteranion (B⊖), **9-20** [7]. That is:

$$9\text{-}19 \qquad\qquad\qquad 9\text{-}20$$

Both methods spread the charge on the growing macrocation and render the β-proton less likely to transfer to monomer (as in reaction 9-41). The first method is typified by initiation with HI/I_2 in which the nucleophilic counterion is the $I\ominus/I_2$ anion. The second primarily involves combinations of a cation-generator, like a tertiary alkyl halide, with a Lewis acid, such as $EtAlCl_2$. A number of initiator combinations of the latter type have been reported for living cationic polymerization of isobutene [8].

Polymers may be made with functionalized end-groups, leading to block copolymers with controlled structures, in parallel with the anionic systems described in more detail in Section 9.2.6.2. Also, as in living anionic polymerizations, $\overline{M}_n$ of the polymer is directly proportional to the monomer conversion, and the polymerization may be restarted by adding more monomer after the initial monomer charge has been consumed.

9.5   COORDINATION POLYMERIZATION

In polymerizations of this type, each monomer is inserted between the growing macromolecule and the initiator. Complexing of the monomer to the initiator frequently precedes the insertion process and this polymerization is therefore often called coordination polymerization. The most important group of initiators in this category are Ziegler–Natta catalysts.

A different group consisting generally of supported oxides of transition metal elements from Groups V–VII of the periodic table are used to produce linear polyethylene.

9.5.1   Ziegler–Natta Catalysts

Ziegler–Natta catalysts consist of a combination of alkyls or hydrides of Group I–III metals with salts of the Group IV–VIII metals. The most generally efficient catalyst combinations are those in which an aluminum alkyl derivative is interacted with titanium, vanadium, chromium or zirconium salts. The most important application of these catalysts is in the polymerization of olefins and conjugated dienes. Not every catalyst combination is equally effective in such polymerizations. As a general rule, Ziegler–Natta combinations that will polymerize 1-olefins will also polymerize ethylene, but the reverse is not true.

The Ziegler–Natta catalysts that are used most industrially are solids that are suspended in the reaction medium. The polymerization reaction that is favored depends on the catalyst components and on their state of aggregation and other details of their preparation. Heterogeneous catalytic systems appear to be necessary for the production of isotactic polyolefins, but soluble catalysts have been used for the synthesis of polyethylene and syndiotactic polypropylene.

Although there are two metals in the Ziegler–Natta catalyst, the weight of current evidence indicates that polymerization takes place at the transition metal–carbon bond. The mechanism is illustrated here with reference to polymerization by $TiCl_3/Al(CH_2CH_3)_3$ catalyst complex. The normal geometry for Ti atoms is octahedral, and the eatalyst site, as shown in **9-21**, is believed to be a coordinately unsaturated Ti bonded to four Cl's [which in turn are bridged to two other Ti's] and to an alkyl group, derived from the aluminum alkyl component.

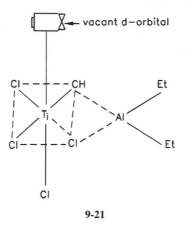

9-21

The olefin is assumed to become coordinated to the transition metal at a vacant octahedral site through overlap of the olefin π-electrons with the vacant d-orbitals of the metal [9, 10]. The bond between the transition metal and the alkyl group R is weakened by this coordination and propagation takes place by insertion of the complexed olefin between the metal and alkyl group via a four-membered cyclic transition state. The unsubstituted carbon atom of the olefin becomes attached to the metal during the opening of the double bond, which is always a cis addition. The alkylated olefin shown attached to Ti in the final product of reaction (9-59) is the R group for attack of the next monomer molecule.

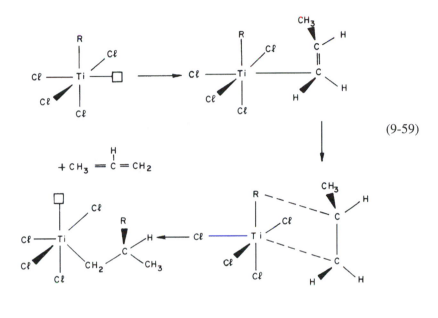

(9-59)

☐ = octahedral vacancy

In the polymerization scheme of reaction (9-59), insertion of a monomer results in an interchange of the polymer substituent and the lattice vacancy. These are not equivalent positions in the crystal lattice of the catalyst. Under normal polymerization conditions the macromolecular alkyl appears to shift back to its original position before the next monomer is added. Isotactic polymers are produced from olefins and catalysts of this type because the monomer always inserts by cis addition with the unsubstituted carbon of the olefin attached to the transition metal which always has the same chirality. If the polymer chain and vacant orbital were to exchange initial positions, however, then the placements of successive monomers would alternate stereochemically, providing syndiotacticity.

Small changes in the catalyst system or in the choice of monomer can result in significant product variations. The stereospecificity and productivity of the polymerization process can often be improved by procedures such as those described in Section 9.5.3.

Termination can be by several different paths, all of which involve transfer reactions:

1. Chain transfer to monomer:

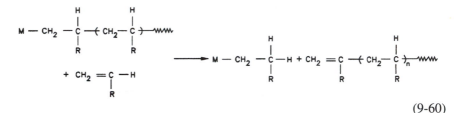

(9-60)

where M is the transition metal;

2. Chain transfer to organometallic compound:

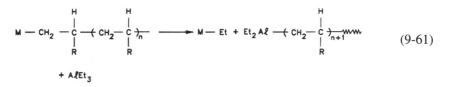

(9-61)

3. Chain transfer to active hydrogen transfer agent:

$$M - CH_2 \overset{\underset{\textstyle H}{|}}{C} \xleftarrow{} CH_2 - \overset{\underset{\textstyle R}{|}}{\underset{|}{C}} \xrightarrow{}_n \text{www} + TH \longrightarrow MT + CH_3 - \overset{\underset{\textstyle R}{|}}{\underset{|}{C}} \xleftarrow{} CH_2 - \overset{\underset{\textstyle R}{|}}{\underset{|}{C}} \xrightarrow{}_n \text{www}$$

(9-62)

where TH is the transfer agent containing a labile H.

4. Internal hydrogen atom transfer:

$$M - CH_2 \overset{\underset{\textstyle H}{|}}{C} \xleftarrow{} CH_2 - \overset{\underset{\textstyle R}{|}}{\underset{|}{C}} \xrightarrow{}_n \text{www} \longrightarrow MH + CH_2 = \overset{\underset{\textstyle R}{|}}{C} \xleftarrow{} CH_2 - \overset{\underset{\textstyle R}{|}}{\underset{|}{C}} \xrightarrow{}_n \text{www}$$ (9-63)

Polymer molecular weight is often regulated in these reactions by using H_2 as a transfer agent. β-hydrogen transfer (reaction 9-63) and transfer to monomer (reaction 9-60) yield unsaturated polymer end-groups. In a process where either of these mechanisms controls polymer growth, $\overline{M}_n$ of the product is readily measured by infrared analysis of terminal C—C double bonds at a wave number of 908 cm^{-1}.

9.5.2 Kinetics of Ziegler–Natta Polymerizations

Generalizations are difficult to perceive in these systems because there is such a wide variety of catalysts and reaction conditions. Industrial processes are usually heterogeneous and the rate of propagation is governed by the coordination of the monomer on the catalyst site. Active sites comprise only a small proportion, usually less than 1%, of the transition metal and are normally not uniformly active.

Conversion-time curves for batch polymerizations are usually S-shaped, with a slow rate initial period followed by a steady reaction rate interval and then a declining rate period. The first period corresponds mainly to the formation of active sites by reaction of the organometallic compound with the transition metal compound. The second, steady polymerization rate interval indicates that the number of active sites and the rate of diffusion of the monomer to the reactive sites are constant. The decreasing reaction rate period corresponds to progressive destruction of active sites and/or to a slowing of the diffusion rate of monomer, as access to the catalyst is hindered by the surrounding polymer.

The rate of monomer consumption is:

$$-\frac{d[M]}{dt} = k_p[C*]\theta_M + K_{tr,M}[C*]\theta_M \tag{9-64}$$

where $[C*]$ is the number of active sites per unit volume of crystalline transition metal compound (in mol/liter), and θ_M is the fraction (unitless) of those sites on which monomer is adsorbed and the two rate constants are first order, with units in sec^{-1}. Usually, a Langmuir adsorption relation is applied to calculate θ_M, assuming that the monomer, organometallic compound and any other species, like H_2, are adsorbed at equilibrium levels on the surface of the transition metal. The number average degree of polymerization is given by:

$$\overline{DP}_n = \frac{k_p[C*]\theta_M}{\sum (\text{rates of transfer reactions})} \tag{9-65}$$

The chain transfer processes are those listed above, and the rate for each would have the form $k_{tr}[C*]\theta_i$, where the θ's are the fractions of catalyst surface occupied by the respective transfer agents of Eqs. (9-60)–(9-63). When the monomer concentration is high, as is usual, then θ_M dominates, the $[C^*]$ terms in the numerator and denominator of the preceding equation cancel, and

$$DP_n = \frac{k_p}{k_{tr,M}} \tag{9-66}$$

Equations (9-64)–(9-66) are deceptively simple, since SEC and TREF studies (Chapter 3) show that the catalytic sites do not have uniform activities (with the exception of single-site metallocene catalysts described in Section 9.5.4). The rate constants quoted in the foregoing equations are thus lumped values, which are not known *a priori* for untested polymerizations.

9.5.3   Practical Features of Ziegler–Natta Polymerizations

The basic combination of transition metal and organometallic compounds has been greatly modified and improved in industrial practice. The activity and stereospecificity of the process was enhanced by the introduction of a third component comprising electron-donors, such as amines, esters, and ethers. Major innovations were achieved with the introduction of supported catalyst systems. The supports are primarily magnesium compounds, such as anhydrous $MgCl_2$, which is isomorphous with $TiCl_3$ [i.e., exchange of the cations does not affect the crystal form]. Catalyst production is a complicated process involving mechanical treatment of mixtures of $TiCl_4$ with $MgCl_2$ and Lewis base electron donors. Electron donors have a plurality of functions, including improvement of the stereoregularity of polymers of 1-olefins. Esters of aromatic acids (e.g., dibutyl phthalate) and alkoxysilanes enjoy the greatest industrial application as electron donors at the time of writing. The functions and mechanisms whereby electron donors act remain to be clarified.

Ti atoms in the catalyst are located on exposed sites in the $MgCl_2$ crystals. About 10% of the Ti atoms are active polymerization sites in these supported catalysts. The fraction of active sites and the propagation rate are both much higher in $MgCl_2$-supported catalysts. These catalysts are used as slurries in hydrocarbon diluents. Productivity exceeds 2000 kg of isotactic polypropylene per gram of Ti. The low residual metal level in the polymer eliminates the need for catalyst removal processes. Catalyst particles are broken up by the growing polymer in the early stages of polymerization, producing an expanding reaction locus in which fragments of the initial catalyst granule are embedded in polymer.

Supported catalysts are also used in gas phase olefin polymerizations. The most widely used of these processes employ fluidized bed reactors, described in more detail in Section 10.4.2(v). Various catalysts are used, including bis(triphenylsilyl) chromate supported on high surface area silica gel and reduced with diethylaluminum ethoxide. This catalyst produces polyethylene with a broad molecular weight distribution. Heterogeneous catalysts with appropriate structures allow control of the granular shape of the product polyolefins (and hence their packing density and transport properties) because the polymer replicates the shape of the catalyst. The monomer is able to reach all the active sites on and in the catalyst particle, when the latter has a porous structure composed of primary crystals of the right size. As a result, the initial granule of catalyst expands during the polymerization reaction and is always enveloped in a skin of polymer.

Important copolymerizations with Ziegler–Natta catalysts are between hydrocarbon monomers. An example is the reaction of ethylene, propylene, and a nonconjugated diene, such as 5-ethylidene-2-norbornene, to produce the so-called EPDM (ethylene-propylene-diene monomer) elastomers. These products have little or no crystallinity, depending on the ethylene content, and can be vulcanized with sulfur systems (Section 1.3.3), because of the residual unsaturation provided by the unconjugated diene comonomer. Copolymerization of ethylene with low levels of 1-olefins, principally butene, hexene and octene, produces polyethylene with controlled short branch contents. These are the LLDPEs, which dominate the packaging film market.

The copolymerization theory presented in Chapter 7 is of limited applicability to processes involving heterogeneous Ziegler–Natta catalysts. The simple copolymer model assumes the existence of only one active site for propagation, whereas the supported catalysts described above have reaction sites that vary in activity and stereoregulating ability. In addition, the catalytic properties of the active sites may vary with polymerization time. The simple copolymer model can be used with caution, however, by employing average or overall reactivity ratios to compare different catalysts and monomers.

Because of the nonuniformity of active sites on supported Ziegler–Natta catalysts, LLDPEs produced by these processes contain a mixture of olefin copolymers, ranging from unbranched linear polyethylene to amorphous copolymers high in 1-olefin content. As a result, the properties of such LLDPEs depend strongly on the details of the distributions of comonomers and molecular weights. Production quality of polyolefins is often controlled on melt flow index (a single-point melt viscosity) and polymer density. These properties are unfortunately not adequate to ensure product uniformity since both the above-mentioned distributions can vary while the quality control parameters remain within control limits. Marked performance differences have been noted between nominally uniform LLDPE batches even when the measured molecular weight distributions were invariant [11]. Such observations have been the impetus for the development of TREF analyses (p. 113).

Impact-modified polypropylenes are produced by combining the homopolymer with an ethylene-propylene copolymer rubber. Ziegler–Natta processes yield such products in cascaded reactors. The first reactor in the sequence produces a rigid polymer with a high propylene content and feeds the second reactor, where the ethylene-propylene elastomer is polymerized in intimate mixture with the first material.

The various regular polymers that can be produced by polymerization of butadiene and isoprene are summarized in reactions (4-3) and (4-4). In addition to the structures shown in these reactions, it should be remembered that 1, 4 polymerization can incorporate the monomer with *cis* or *trans* geometry at the double bond and that the carbon atom that carries the vinyl substituent is chiral in 1,2 and 3,4 polymers. It is therefore possible to have isotactic or syndiotactic polybutadiene or polyisoprene in the latter cases. Further, these various monomer residues can all appear in the same polymer molecule in regular or random sequence. It is remarkable that all these conceivable polymers can be synthesized with the use of suitable catalysts comprising transition metal compounds and appropriate ligands.

In diene copolymerization, also, the monomer sequence can be regulated by the choice of catalyst and polymerization conditions.

9.5.4 Comparisons of *Cis*-1,4-Polydienes

Cis-1,4-polyisoprene (coded in the industry as IR) is available as natural rubber and synthetically by two routes: living anionic polymerization with lithium alkyls in nonpolar solvents or by Ziegler–Natta polymerization with Ti-Al catalysts. Both syntheses are indeed scientific triumphs. It is instructive, however, to note that the differences between the three products, while slight in terms of polymer stereoregularity, are actually serious to rubber compounders. (This illustrates the point that polymers are materials of construction and must meet *all* of a range of performance requirements.) Lithium alkyl living polymerization produces about 92% *cis* content, with linear structure. Ziegler–Natta catalysis produces a polymer with 96–98% *cis* content, broad molecular weight distribution, long chain branching, and some gel. Natural rubber likewise is 98% *cis*, with branching, very broad molecular weight distribution, high mean molecular weight, and considerable gel content, depending on storage time. The viscosities of Li-IR and many grades of natural rubber are too high to allow good mixing with other ingredients of rubber compounds. Their molecular weights therefore must be degraded, usually by a combination of mechanical and chemical means. Ziegler–Natta IR does not require such mastication, because molecular weight can be controlled during polymerization with chain transfer agents. The molecular weight of the Li-IR could be reduced by operating with higher alkyl lithium concentrations, but this would also lower the *cis*-1,4 content (Section 9.2.7).

The effects of stereoregularity are seen in the tensile and tear properties of the three polyisoprene types, which fall in the decreasing order: natural rubber) Ziegler–Natta IR) Li-IR. However, since Li-IR is unbranched and less stereoregular, it flows well during tire building and in injection molding, where the flow of higher cis-content polyisoprenes may be hindered by crystallization under shear.

Synthetic polyisoprenes are superior to natural rubber in terms of consistency of properties. The are also freer of contaminants, and are preferred for applications that require lighter color, for personal care items and for derivatization to chlorinated and cyclized rubber products that are used in the adhesives and coatings industries.

Table 9-2 summarizes differences between polybutadienes produced by different processes. The low *cis*-content polybutadienes are branched. They have lower solution viscosities than their linear counterparts and are preferred for manufacture of high impact polystyrene (HIPS) in which polymerization takes place in a solution of the elastomer in styrene. As the reaction proceeds under agitation, polystyrene becomes the continuous phase, with dispersed droplets of rubber (see Chapter 11). The high *cis*-content, linear polybutadienes are more elastomeric

TABLE 9-2

Catalysts for Polybutadienes

Catalyst system	Polybutadiene structure (%)		
	cis	trans	vinyl
Co octoate/$Et_2AlCl.H_2O$	97	2	1
OrganoNi/$Et_3Al/BF_3.Et_2O$	95	5	2
$TiCl_4/Et_3Al/I_2$	92	4	4
BuLi	38	5	11

(Chapter 4) and perform better on rubber processing equipment. They are favored for tire applications.

9.5.5 Metallocene Catalysts

Homogeneous catalysts capable of producing polyethylene [12] and isotactic polypropylene [13] employ metallocenes. These are organometallic coordination compounds comprising two cyclopentadienyl rings bonded to a metal atom, as in **9-22**.

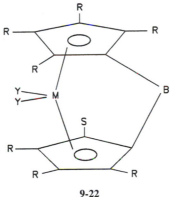

9-22

The cyclopentadienyl ring may be part of an indenyl (**9-16**) structure. Catalysts vary in the number of rings, the nature of their substitutents (R), the transition metal (M), and its substitutents (Y) as well as the type of bridge (B). Their essential feature is that the transition metal is constrained between two rings, providing a sterically hindered catalytic site. They differ also from the heterogeneous Ziegler–Natta catalysts described above in being homogeneous, i.e., soluble in hydrocarbons and, most notably, in producing polyolefins with different characteristics. This is because the metallocene catalysts are single-site systems. They produce more uniform ethylene 1-olefin copolymers than can be obtained with Ziegler–Natta catalysis. In conventional LLDPE the lower molecular weight species are richest in the olefin comonomer, resulting in a fraction of a few weight percent that is highly branched and not crystallizable (for reasons that are elaborated in Chapter 11). This is a disadvantage in certain food packaging applications since the amorphous low-molecular-weight material is extractable by greases and oils. Metallocene-based catalytic systems permit incorporation of the higher olefin comonomer uniformly or preferentially in the higher molecular weight species of the copolymer, thus reducing the extractable content.

The generally even comonomer distribution with respect to molecular weight that is characteristic of metallocene-derived polyolefins is not always advantageous in polyethylene processing operations. However, use of a combination of metallocenes makes it possible to prepare multisite catalysts to produce polymers with multimodal compositions that are designed for particular applications. Appropriate metallocene catalysts yield polypropylenes with different stereochemical structures, including isotactic, syndiotactic, atactic and stereoblock configurations.

Metallocene catalysts are particularly able to polymerize a wide range of monomers, including 1-olefins, various aromatics, cyclic olefins, nitriles, (meth) acrylates, and silanes.

These catalyst systems consist of two components. The catalyst itself is a soluble Group 4B transition metal [Ti, Zr, Hf] attached to two bulky π-carbocyclic ligands, usually cyclopentadienyl (Cp), fluorenyl, indenyl or their substituted structures. At the time of writing the cocatalyst comprises oligomeric alumoxanes formed by hydrolysis of trialkyl aluminum. Methylalumoxane (MAO), the preferred form, contains some 6–20 Al(CH$_3$) repeating units. Its structure is not known in detail, but it is thought to consist of a mixture of linear and cyclic oligomers (**9-23**).

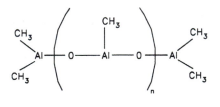

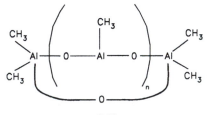

9-23

The MAO hydrolysis reaction can be represented generally as:

$$nA\ell(CH_3)_3 + nH_2O \longrightarrow \left(\underset{O}{\overset{CH_3}{\underset{|}{A\ell}}} \right)_n + 2n\ CH_4 \qquad (9\text{-}67)$$

The active species in MAO cocatalyzed metallocene polymerization is likely a coordinatively unsaturated cationic complex, as in **9-24** (for a zirconium-based metallocene), where substituents on the metal atoms have been omitted for clarity. The Zr—O—Al bond withdraws electron density from the Zr atom. The mechanism for ethylene polymerization is proposed to be as follows, where the monomer forms a π-complex with the transition metal [14]:

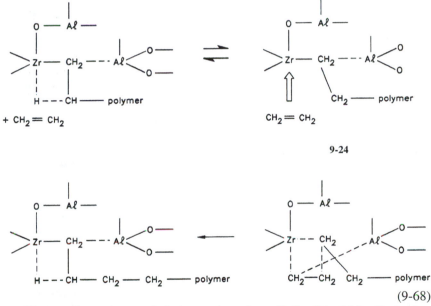

9-24

(9-68)

The growing polymer chain is terminated usually by β-hydride elimination (Eq. 9-63), yielding vinylidene ($=CH_2$) end groups.

When propylene and higher olefins are polymerized the configuration of the polymer is controlled by the catalyst structure. Catalyst **9-25** contains two equivalent Cl atoms, because the molecule is symmetrical about the Zr-indenyl bonds. Recall that in the active catalyst one Cl will be replaced by the growing polymer and the other by the incoming monomer. The polymer chain could occupy either of the two Cl positions, while the monomer could coordinate at the other position. As a result, both positions are equivalent and this metallocene produces isotactic polypropylene. In structure **9-26**, the two Cl sites are not equivalent. At one position, the monomer would find the same environment as in **9-25**, but the second site is less sterically hindered and the configuration of the inserting monomer is not controlled. As a result, this metallocene would produce a mixture of isotactic and atactic polypropylene.

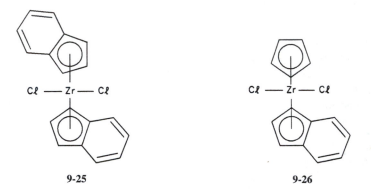

9-25	9-26

Polyethylene is most commonly polymerized with cyclopentadiene metallocenes of Ti, Zr, and Hf, with zirconenes preferred for their stability. Polymerization rates are very high and are a function of the Al/transition metal ratio. Polymer molecular weight decreases with increased reaction temperature, because higher temperatures accelerate β-hydride transfer.

The metallocenes generally used for ethylene polymerization are achiral, and yield atactic polypropylene. Stereoregular polypropylene requires the use of chiral metallocene catalysts, such as [Et(Indenyl)$_2$ZrMe]$^{\oplus}$ [B(C$_6$F$_5$)]$^{\ominus}$.

Supported metallocene catalyst systems are preferred to soluble versions in conventional polyolefin plants, which were designed to use supported Ziegler–Natta or Cr$_2$O$_3$-based catalysts. Metallocenes can be supported on a number of substrates, such as SiO$_2$, MgCl$_2$ or Al$_2$O$_3$. Supported catalysts also provide polypropylene with fewer stereochemical defects.

Production problems with current metallocene-aluminoxane catalysts reflect the high aluminoxane/metal ratio needed for good catalyst productivity and stereochemical control. Aluminoxane production is slow and relatively inefficient and the large amounts used make post-polymerization catalyst removal processes necessary. These deficiencies are expected to be remedied by different cocatalysts, which are being disclosed in the patent literature.

9.6   OLEFIN METATHESIS CATALYSTS

Metathesis means "transposition of parts." It is a disproportionation reaction, known in synthetic organic chemistry as a route to substituted olefins. The reaction leaves the number of C=C bonds unchanged, but shuffles their substitutents.

$$
\begin{array}{ccc}
R_1\!-\!CH_2\!=\!CH\!-\!R_1 & & R_1\!-\!CH\!=\!CH\!-\!R_2 \\
+ & \rightleftharpoons & + \\
R_2\!-\!CH\!=\!CH\!-\!R_2 & & R_1\!-\!CH\!=\!CH\!-\!R_2
\end{array}
\tag{9-69}
$$

It is applied to macromolecules in the ring-opening polymerization of cycloalkenes and bicycloalkenes. Polypentenamer is obtained by the polymerization of cyclopentene catalyzed by transition metal compounds.

$$
n \;\square \longrightarrow -\!(\!CH\!=\!CH\!-\!CH_2\!-\!CH_2\!-\!CH_2\!)\!- \tag{9-70}
$$

It is part of a homologous series of linear unsaturated polymers that are termed *polyalkenamers*. Polypentenamer is an interesting member of this series because cyclopentene is economically available from petrochemical by-products and because the polymer is a readily vulcanizable elastomer (since it contains residual C—C double bonds; cf. p. 10).

The configuration of the polymer about the double bond can be controlled by varying polymerization conditions. A structure with about 85% *trans* configurations has received the most attention. This product retains its elastomeric character down to about $-95°C$. It forms easily melted crystallites on stretching and is readily processed in conventional rubber equipment. Norbornene, **9-27**, also yields a useful polymer:

$$
n \;\bigcirc \xrightarrow[\text{(BuOH)}]{\text{RuC}\ell} +\!\!\bigcirc\!\!-\!CH\!=\!CH\!\rightarrow_{\!n} \tag{9-71}
$$

9-27

Metathesis catalysis has also been used to prepare polyacetylene, which is of interest because of its electrical conductivity.

Metathesis catalysts vary widely. They always contain a transition metal compound which is usually employed in combination with one or more cocatalysts like $AlEt_3$. The most important catalysts for cyclopentene polymerization are derived from W or Mo. Low-molecular-weight alcohols are used in low concentrations to reduce the formation of oligomeric products.

Catalysts comprising a metal–carbon double bond (metallocarbenes, or metallocenes) are efficient. With these initiators, the polymerization mechanism appears to involve coordination of the C=C double bond in the cyclo- or dicycloalkene at a vacant d orbital on the metal. The metallocyclobutane intermediate which is formed decomposes to produce a new metal carbene and a new C=C bond. Propagation consists of repeated insertions of cycloalkenes at the metal carbene.

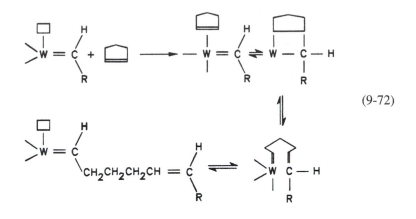

$$(9\text{-}72)$$

PROBLEMS

9-1. Isobutene is polymerized commercially by a cationic mechanism initiated by strong acids like $AlCl_3$. It is not polymerized by free-radicals or anionic initiators. Acrylonitrile is polymerized commercially by free-radical means. It can also be polymerized by anionic initiators like potassium amide but does not respond to cationic initiators. Account for the difference in behavior of isobutene and acrylonitrile in terms of monomer structure.

9-2. Suggest practical ionic initiators and solvents for the ionic polymerization of the following monomers:
   (a) $CH_2=C(CN)_2$
   (b)   isoprene
   (c)   tetrahydrofuran
   (d)   propylene
   (e)   coumarone (**9-17**)

9-3. Outline reaction schemes to produce each of the following block copolymers:
   (a)   (styrene)-$_x$(methyl methacrylate)-$_y$(styrene)-$_z$
   (b)   (α-methylstyrene)-$_x$(isoprene)-$_y$(styrene)-$_z$
   (c)   (styrene)-$_x$(methacrylonitrile)-y

9-4. Show how you would synthesize the following block polymer:

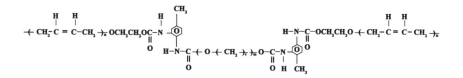

9-5. Show the initiating step in the use of each of the following initiators. Use an appropriate monomer in each case:

(a) n-C_4H_9Li

(b) $AlCl_3 + HCl$

(c) $BF_3 + H_2O$

(d) $Na + naphthalene$

(e) $TiCl_3 + (C_2H_5)_3Al$

9-6. A scrupulously clean and dry solution of styrene (5 g) in 50 ml tetrahydrofuran was held at $-70°C$. Sodium (1.0 g) and naphthalene (6.0 g) were stirred together in 50 ml dry tetrahydrofuran to form a dark green solution of sodium naphthalide (Eq. 9-10). When 1.0 ml of this green solution was injected into the styrene solution the latter turned reddish orange. After a few minutes the reaction was complete. The color was quenched by adding a few milliliters methanol, the reaction mixture was allowed to warm to warm temperature, and the polymer formed was precipitated and washed with methanol. What is $\overline{M}_n$ of the polystyrene formed in the absence of side reactions? What should $\overline{M}_w$ of the product be if the polymerization were carried out so that the growth of all macromolecules was started and ended simultaneously?

REFERENCES

[1]   M. Szwarc, Ed., "Ions and Ion Pairs in Organic Reactions." Wiley, New York, 1974.

[2]   J. P. Kennedy, "Cationic Polymerization of Olefins," Wiley (Interscience), New York, 1975.

[3]   A. Ledwith and D. C. Sherrington, *in* "Comprehensive Chemical Kinetics" (C. H. Bamford and C. F. H. Tipper, Eds.), Vol. 15, Chap. 2. Elsevier, Amsterdam, 1976.

[4]   J. T. Dennison and J. B. Ramsey, *J. Am. Chem. Soc.* **77**, 2615 (1955).

[5]   F. Buckley and A. A. Maryott, *J. Res. Nat. Bur. Std.* **53**, 229 (1954).

[6]   S. Pasynkiewicz, *Polyhedron* **9**, 429 (1990).

[7]   M. Sawamoto, *Prog. Polym. Sci.* **16**, 111 (1971).

[8]   T. Higashimura, S. Aoshima, and M. Sawamoto, *Makromol. Chem. Macromol. Symp.* **13/14**, 457 (1988).

[9]   G. Kaszas, J. Puskas, and J. P. Kennedy, *Makromol. Chem. Macromol. Symp.* **13/14**, 473 (1988).

[10] P. Cossee, *J. Catal.* **3**, 80 (1964).

[11] E. J. Arlman and P. Cossee, *J. Catal.* **3**, 99 (1964).

[12] E. Karbashewski, A. Rudin, L. Kale, W. J. Tchir, and H. P. Schreiber, *Polymer Eng. Sci.* **31**, 1581 (1991).

[13] J. A. Ewen, *J. Am. Chem. Soc.* **106**, 6355 (1984).

[14] W. Kaminsky and R. Steiger, *Polyhedron* **7**, 2375 (1988).

[15] H. Sinn, W. Kaminsky, H. J. Vollmer, and R. Woldt, *Angew. Chem. Int. Ed. Engl. Ed.* **19**, 390 (1980).

Chapter 10

Polymer Reaction Engineering

To define it rudely, but not inaptly, engineering is the art of doing that
well with one dollar which any bungler can do after a fashion.
—Arthur M. Wellington, *The Economic Theory of Railway Location*

10.1 SCOPE

Engineering of polymerization reactions requires a detailed knowledge of the
phenomena that take place in the polymer reactor. This entails a model of the
polymerization kinetics and the heat and mass transfer features of the particular
polymerization and process. Polymerization reactions are usually complex and a
certain degree of mathematical sophistication is required for effective modeling.
Excursions into the details of particular reactions or modeling techniques are beyond the scope of this introductory text and this chapter is therefore limited to a
review of the special considerations that apply in the case of various polymerizations and processes. Most of the following discussions are necessarily qualitative,
for space reasons.

10.2 STEP-GROWTH POLYMERIZATIONS

There are some fundamental differences in the engineering of step-growth and
chain-growth polymerizations because of basic distinctions in the mechanisms of
these reactions. A propagation reaction (Section 6.3.2) in a kinetic chain sequence
must be fast or the series of monomer additions will not be long enough to produce

high-molecular-weight polymers before the intervention of termination or transfer reactions. This is not generally true for step-growth polymerizations where only an addition reaction is involved and the growth of macromolecules can occur in a series of starts and stops.

In step-growth polymerizations, overall costs of monomers, solvent recovery, and preparing the polymer for further processing usually dictate a preference for reactions that are slow at room temperature. (The reasons behind this generalization are summarized in Section 5.3.1.) The ratio of rates of macromolecular growth reactions in typical chain and step-growth polymerizations is often of the order of 10^4.

It is normal, then, to use catalysts and elevated temperatures to accelerate step-growth polymerizations. The activation energies that characterize most such polymerizations are about 84 kJ/mol. This means that a reaction at 200°C is 300 times as fast as at 100°C and it is 2000 times as rapid at 250°C as at 100°C. The exothermic enthalpy of polymerization is about -8.4 to -25 kJ/mol for polyester and polyamide syntheses. This low exotherm makes it possible to carry out such polymerizations in viscous media at high temperatures without danger of a runaway reaction due to limitations of heat transfer rates through the reactor walls. When step-growth reactions are accelerated by application of heat and catalysts, depolymerization reactions also become important. This affects the conditions under which the polymerization can be carried out. To illustrate the basic principles, we consider a step-growth polymerization in which the initial concentrations of reacting groups are equal as in

$$
\text{HOC}\sim\!\sim\!\sim\text{C-OH} + \text{HO}\sim\!\sim\text{OH} \rightleftharpoons \text{HOC}\sim\!\sim\text{CO}\sim\!\sim\text{OH} + \text{H}_2\text{O}
$$

$$\tag{10-1}$$

Here $[\text{COOH}] = [\text{OH}] = C$ at any time, barring side reactions which could consume one or the other functional group. Then, if C_0 is the initial concentration and p is the extent of reaction (p. 170) at any instant during the polymerization,

$$
C = C_0(1 - p) = [\text{COOH}] = [\text{OH}] \tag{10-2}
$$

and

$$
[\sim\!\sim\overset{\text{O}}{\underset{\|}{\text{C}}}-\text{O}\sim\!\sim] = C_0 p \tag{10-3}
$$

since a fraction p of the initial C_0 groups of either kind will now be part of ester groups. The equilibrium constant K for reaction (10-1) is

$$K = \frac{[\sim\underset{\underset{O}{\|}}{C}-O\sim]\,[\,H_2O]}{[\,COOH]\,[\,OH]} \tag{10-4}$$

or with Eqs. (10-2) and (10-3),

$$K = p[H_2O]/C_0(1 - p)^2 \tag{10-5}$$

To obtain a high polymer in equilibrium step-growth reactions, p must be close to unity (Section 5.4.2). Then, with p very close to 1,

$$(1 - p)^2/p \simeq (1 - p)^2 = [H_2O]/C_0K \tag{10-6}$$

and the limiting conversion $p_{\lim}$, which can be attained under these conditions, will be given approximately by

$$p_{\lim} \simeq 1 - ([H_2O]/C_0K)^{1/2} \tag{10-7}$$

Note also that since $f_{av} = 2$ in this example, the Carothers equation (Eq. 5-20) shows the number average degree of polymerization $\overline{X}_n$ to be

$$\overline{X}_n = \frac{1}{1 - p} = \left(\frac{C_0K}{[H_2O]}\right)^{1/2} \tag{10-8}$$

It is evident from Eqs. (10-7) and (10-8) that a high conversion and high molecular weight require low concentrations of the condensation product, water. The lower the value of K, the more essential it is that the water concentration be reduced. [The same considerations apply of course when an equilibrium exists between polymer and any condensation product such as in reaction (b) of Fig. 5-2.]

In step-growth polymerizations with unfavorable values of K, it is therefore standard practice to operate at high temperatures and reduced pressures to remove the condensation products. This is typical of the manufacture of linear polyesters where the final stages of the polymerization are at pressures near 1 mm Hg and temperatures near 280°C. Alkyds (Section 5.4.2) are branched polyesters produced by esterification reactions of mixtures of polyhydric alcohols and acids with varying functionalities. They are used primarily in surface coatings. Alkyd syntheses are completed at temperatures near 240°C. It is not necessary to reduce the pressure to pull residual water out of the reaction mixture, because the final products are relatively low-molecular-weight fluids that are diluted with organic solvents before further use. In one process variation, a small amount of a solvent like xylene is added to the reactants to facilitate water removal by azeotropic distillation. Xylene residues in the final product are of no significance in this instance.

The equilibrium is much more favorable to polymer formation in the production of nylons than polyesters. This can be explained as being due to the greater stability of amide as compared to ester linkages. Resonance structures can be written for both groups as

$$
R - \underset{\underset{\displaystyle \bullet\bullet}{|}}{\overset{\overset{\displaystyle H}{|}}{N}} - \underset{\overset{\displaystyle \|}{O}}{C} - R' \quad\longleftrightarrow\quad R - \underset{\overset{\displaystyle \oplus}{N}}{\overset{\overset{\displaystyle H}{|}}{N}} = \underset{\overset{\displaystyle |}{O^{\ominus}}}{C} - R' \qquad (10\text{-}9)
$$

$$
R - \underset{\bullet\bullet}{\overset{\bullet\bullet}{O}} - \underset{\overset{\displaystyle \|}{O}}{C} - R' \quad\longleftrightarrow\quad R - \underset{\bullet\bullet}{\overset{\oplus}{O}} = \underset{}{\overset{\overset{\displaystyle O^{\ominus}}{|}}{C}} - R' \qquad (10\text{-}10)
$$

Oxygen is more electronegative than nitrogen and is less able to support a positive charge and so the amide linkage is more resistant to hydrolysis. As a result, the equilibrium in reaction (10-11):

$$
H_2N \,\text{\textasciitilde\textasciitilde}\, NH_2 + HO\underset{\overset{\displaystyle \|}{O}}{C} \,\text{\textasciitilde\textasciitilde}\, \underset{\overset{\displaystyle \|}{O}}{C}-OH \quad\rightleftharpoons\quad H_2N \,\text{\textasciitilde\textasciitilde}\, \underset{\overset{\displaystyle \|}{O}}{\overset{\overset{\displaystyle H}{|}}{N}}-C \,\text{\textasciitilde\textasciitilde}\, \underset{\overset{\displaystyle \|}{O}}{C}-OH + H_2O
$$

$$(10\text{-}11)$$

lies much further to the right than that in reaction (10-1).

This difference is reflected in the synthesis of nylon-6,6, where the initial step is the production of a salt (**10-1**) that is recrystallized from methanol to ensure exact equivalence of reactants.

$$
H_2N \xleftarrow{\hspace{2pt}} CH_2 \xrightarrow{\hspace{2pt}}_6 NH_2 + HOOC \xleftarrow{\hspace{2pt}} CH_2 \xrightarrow{\hspace{2pt}}_4 COOH \quad\longrightarrow\quad \overset{\oplus}{H_3N} \xleftarrow{\hspace{2pt}} CH_2 \xrightarrow{\hspace{2pt}}_6 \overset{\oplus}{NH_3}
$$

$$
O = \underset{\overset{\displaystyle |}{O^{\ominus}}}{C} \xleftarrow{\hspace{2pt}} CH_2 \xrightarrow{\hspace{2pt}}_4 \underset{\overset{\displaystyle |}{O^{\ominus}}}{C} = O
$$

10-1

$$(10\text{-}12)$$

An aqueous slurry of the salt is heated with acetic acid end-group stabilizer

(Section 5.4.2) and the reaction is completed at 270-280°C and atmosphere pressure:

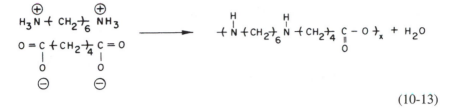

$$(10\text{-}13)$$

The equilibrium in this case allows formation of polyamide linkages even in the presence of high concentrations of water. This is not possible with polyesters, and practical processes for production of both polymer types differ fundamentally for this reason.

The need to drive the polymerizations to completion is common to all step-growth reactions that are carried out under conditions in which polymerization–depolymerization equilibria are significant (Section 5.4.2). This is accomplished in general by removal of a volatile product such as water or an alcohol. The rate of polymerization is often limited by the rate of transfer of such condensation products into the vapor state. A complete kinetic description of the process must then involve both the chemical reaction rate and the rate of mass transfer. The latter depends on the details of reactor design and stirring and therefore so does the rate of polymer production [1].

10.3   CHAIN-GROWTH POLYMERIZATIONS

The major commercial examples of chain-growth polymerizations involve reactions across C=C double bonds to produce polymers with all-carbon backbones. The enthalpies of polymerization of such monomers are of the order of 60–85 kJ/mol. (For example, ΔH_{polym} is 56.5 kJ mol^{-1} for methyl methacrylate and 95.0 kJ mol^{-1} for ethylene polymerization.) This is relatively high compared to the heats of polymerization in the more common step-growth polymerizations. The activation energy for free-radical polymerizations is about the same as for step-growth reactions (Section 6.15) but small increases in temperature give rise to larger increases in heat generation in free-radical reactions because of their greater exothermic heats of polymerization.

Chain-growth polymerizations that are carried out in the absence of a diluent are characterized by high viscosity and poor heat transfer. High-molecular-weight polymer coexists with low-molecular-weight species and monomer, and molecular weight does not increase with conversion. Mixing becomes difficult in such chain-growth polymerizations at intermediate levels of conversion, and the heat transfer efficiency of the reactor decreases as a consequence. The low heat transfer coefficients become more serious as the conversion increases, because the rate of polymerization and the rate of heat generation often accelerate in these stages of the reaction (Section 6.13.2). Since the decomposition rate of thermal initiators is highly dependent on temperature (Section 6.5), faster initiation can further accelerate an increased rate of free radical polymerization. Runaway reactions can thus occur unless special precautions are taken.

10.4   HOMOGENEOUS AND HETEROGENEOUS POLYMERIZATION PROCESSES

Polymerization processes are usefully classified as follows:

   1.   *Homogeneous systems* comprising (a) bulk reactions and (b) solution systems.
   2.   *Heterogeneous systems* comprising (a) heterogeneous bulk polymerizations, (b) heterogeneous solution polymerizations, (c) suspension systems, (d) emulsion systems, (e) dispersion polymerization, (f) gas phase polymerization, and (g) interfacial polymerizations.

10.4.1   Homogeneous Systems

(i)   *Homogeneous Bulk Reactions*

In bulk polymerizations, the initial reaction mixture consists essentially of monomer. If the process is a chain-growth reaction, the mixture will also contain initiator and chain transfer agent, if needed. If the polymer and monomer are miscible, the system remains homogeneous during the polymerization reaction.

Bulk reactions are attractive for step-growth polymerizations. Heat removal is not a serious problem, because such polymerizations are not highly exothermic. Mixing and stirring are also not difficult until the last stages of the reaction, since the product molecular weight and the mixture viscosity remain relatively low until high conversions are reached.

Heat removal and mixing are problems in bulk chain-growth polymerizations for the reasons outlined in the previous section. Thus, homogeneous bulk step-growth reactions are driven to high conversions to achieve high molecular weights but the corresponding chain-growth polymerizations are often limited to lower conversions because of problems in keeping the reaction temperature under control.

Poly(ethylene terephthalate) and nylon-6,6 manufacture are homogeneous bulk step-growth reactions. The molecular weight of the polymer produced is limited by the high viscosity of the reaction mixture at very high conversions. Post polymerization techniques such as that described in connection with reaction (5-39) can be used to increase the polymer molecular weight for some applications.

Polystyrene and poly(methyl methacrylate) polymerizations are typical of homogeneous bulk chain-growth reactions. The molecular weight distributions of the products made in these reactions are broader than predicted from consideration of classical, homogeneous phase free-radical polymerization kinetics because of autoacceleration (Section 6.13.2) and temperature rises at higher conversions.

Crystal polystyrene is produced by thermally initiated (Section 6.5.4) bulk polymerization of styrene at temperature of 120°C or more. (The term *crystal* refers to the optical clarity of products made from this polymer, which is not crystalline.) The rate of polymerization would decrease with increasing conversion and decreasing monomer concentration if the reaction were carried out at constant temperature. For this reason, the polymerization is performed at progressively increasing temperatures as the reaction mixture moves through a series of reactors. The exothermic heat of polymerization is useful here in raising the reaction temperature to about 250°C as the process nears completion.

(ii)   Homogeneous Solution Reactions

Both the monomer and polymer are soluble in the solvent in these reactions. Fairly high polymer concentrations can be obtained by judicious choice of solvent. Solution processes are used in the production of *cis*-polybutadiene with butyl lithium catalyst in hexane solvent (Section 9.2.7). The cationic polymerization of isobutene in methyl chloride (Section 9.4.4) is initiated as a homogeneous reaction, but the polymer precipitates as it is formed. Diluents are necessary in these reactions to control the ionic polymerizations. Their use is avoided where possible in free-radical chain growth or in step-growth polymerizations because of the added costs involved in handling and recovering the solvents.

Advantages of solution reactions include better thermal control and mixing than in bulk polymerizations. Initiator efficiency is also usually better because of the lower viscosity and better agitation. Disadvantages include the costs of solvent removal and recovery. Thermoplastic polymers are recovered from solution polymerizations as fluffy powders or slabs which will not flow readily enough in the hoppers of downstream processing machinery. The powders must be compacted in a separate melting and granulation process. Rubbers like polybutadiene or butyl rubber can be used directly in the slab forms in which they are recovered from solution polymerizations, because natural rubber is handled in that form and the processing equipment in the rubber industry is designed to accommodate it.

10.4.2 Heterogeneous Systems

(i) Heterogeneous Bulk Reactions

If the monomer and polymer are not mutually soluble, the bulk reaction mixture will be heterogeneous. The high pressure free radical process for the manufacture of low density polyethylene is an example of such reactions. This polyethylene is branched because of self-branching processes illustrated in reaction (6-89). Branches longer than methyls cannot fit into the polyethylene crystal lattice, and the solid polymer is therefore less crystalline and rigid than higher density (0.935-0.96 g cm^{-3}) species that are made by coordination polymerization (Section 9.5).

Ethylene is polymerized in high-pressure processes by free-radical reactions at pressures of 1000-3000 atm and temperatures of about 200–280°C. Ethylene is a supercritical fluid with density 0.4–0.5 g cm^{-3} under these conditions. Polyethylene remains dissolved in ethylene at high pressures and temperatures but separates as an ethylene-swollen liquid in the lower ranges. The extent of long-chain branching from chain transfer to polymer (Section 6.8.4) depends on the local reaction temperature and concentrations of monomer and polymer. These factors are determined in turn by the prevailing conversion of monomer to polymer, the efficiency of mixing, and the local ethylene-polyethylene phases. Various reactor designs can be employed to produce polyethylenes with about the same average molecular weight and frequency of short branching. The molecular weight distributions and long-chain branching will depend strongly on reactor geometry and operation, however, and such polyethylenes are often clearly distinguishable in their processing properties and in some physical characteristics.

High-pressure ethylene polymerizations are continuous processes in which ethylene and any comonomers, like vinyl acetate, are fed into tubular or stirred autoclave reactors. The reaction is ignited and sustained by periodic injections of peroxide solutions while the polymerizing mixture travels through the reactor. The ethylene and polyethylene leave the reactor and go into a primary separation vessel which operates at a much lower pressure than the reactor. Most of the ethylene is flashed off in this unit and recycled through compressors to the tube inlet. Conversion per pass is of the order of 30% with ethylene flow rates about 40,000 kg/h. Since the polymerization is not isothermal, polymer properties reflect the reaction history. Polyethylene made in the initial, cooler reactor regions will have higher molecular weight and less branching than material made in subsequent, hotter zones. The final product is a mixture of polyethylenes with different molecular sizes as well as branch types and levels. Reactor operating parameters can be varied to optimize manufacturing costs and polymer properties for different applications, and a complete characterization of the molecular structure of the polymer would require careful SEC and TREF analyses.

Poly(vinyl chloride) (PVC) is produced by mass, suspension, and emulsion processes. Mass polymerization is an example of a heterogeneous bulk system. PVC is virtually insoluble in vinyl chloride because the polymer is about 35% more dense than the monomer under normal polymerization conditions. Vinyl chloride, however, is quite soluble in polymer. The two phases in PVC polymerizations are pure monomer and monomer-swollen polymer. Polymerization proceeds in both phases, but it is very much faster in the polymer-rich phase because the mobility of macro radicals and mutual termination reactions are severely restricted (cf. Section 6.13.2).

Cessation of the growth of PVC radicals is caused almost completely by chain transfer to monomer (Section 6.8.2) rather than by termination by disproportionation or combination. In other words, the relative magnitudes of the various terms in Eq. (6-75) are such that the controlling factor is the $C_M (=k_{tr,M}/k_p)$ term. Since the ratio of these rate constants depends on temperature, the number average molecular weight of the product polymer is controlled simply by the reaction temperature and shows little dependence on initiator concentration or rate of polymerization.

A comparison of the basic elements of the bulk polymerizations of styrene and vinyl chloride is instructive. Styrene and polystyrene are miscible, and the rate of polymerization and the consequent rate of heat generation will decay slowly with conversion in the absence of autoacceleration, which is a relatively weak effect in this system. For this reason polystyrene bulk reactions are operated with rates of heat generation close to the limits of the heat exchange systems in the reactor walls. It is not possible to produce PVC with the same reactors, however. The maximum rate at which heat can be removed from a bulk polymerization always decreases with conversion because of the progressive increase in viscosity of the reaction mixture. The decay in heat transfer efficiency is much greater in bulk PVC production than in polystyrene manufacture, because the suspension of PVC in its monomer becomes very viscous at low conversions and is converted to a poorly conducting monomer-wetted powder at about 20% reaction. In addition, the rate of polymerization and the rate of heat generation rise steadily in PVC syntheses because the physical state of the system suppresses normal radical termination reactions. The PVC reaction would run out of control unless provisions were made to increase the rate of heat removal beyond that needed in homogeneous polymerizations like that of styrene. This is accomplished by using suspension, emulsion, or special bulk reaction systems.

PVC bulk polymerizations are two-stage processes in which a very porous PVC seed particle is produced in a vessel provided with very high speed agitation. The wet polymer powder from this stage is fed to horizontal autoclaves in which the polymerization is finished. Reaction heat is removed by cooling the helical ribbon blender-type agitators as well as the vessel jacket.

Bulk polymerization is the main process for making high-impact polystyrene (HIPS). Polybutadiene is dissolved in styrene at 3–10% (w/w) concentration and the styrene is polymerized with careful agitation. Phase separation occurs with polybutadiene-g-polystyrene separating out. The final product is a dispersion of polybutadiene particles, which themselves contain occluded polystyrene. Polymerization conditions are adjusted to control the size and volume of these particles, which range respectively from 0.1 to 6.0 μm and 0.1 to 0.4 volume fraction of the material.

A variation of the HIPS process uses diblock polybutadiene-polystyrene rubbers to produce core-shell rubber particles with polystyrene cores and thin polybutadiene shells. The small particle size of 0.1 to 0.4 μm is less than optimum for toughening but provides a high-gloss material.

(ii)   Heterogeneous Solution Polymerizations

Solution systems are heterogeneous when the monomer is soluble but the polymer is not. This is typical of many coordination polymerizations of polyolefins (Section 9.5). The process, which is commonly termed a slurry process, consists basically of these steps:

1.   A catalyst preparation step. The catalysts, which are generally solids, are produced with the careful exclusion of water and oxygen.

2.   Polymerization occurs at pressures usually less than 50 atm and at temperatures below 110°C (to avoid dissolving the polymer) to form a slurry of about 20% polymer in an aliphatic liquid diluent. The diluent can be liquid propylene itself in the manufacture of polypropylene.

3.   Polymer recovery is done by stripping of the diluent, washing to remove residual catalyst, and extraction of undesirable polymer components, if necessary.

4.   A compounding step is used to mix various stabilizers and additives into the polymer melt, which is finally chilled and pelletized.

Catalyst removal steps can be eliminated in very efficient processes (Section 9.5) in which the residual catalyst concentration is negligible. Conversion levels are generally higher than in the free-radical, high-pressure polymerization process, and less monomer recycle is therefore required. The reaction temperature in a typical slurry processes is controlled by refluxing the solvent.

(iii)   *Suspension Systems*

Suspension polymerization is also known as pearl or bead polymerization. Kinetically, suspension polymerizations are water-cooled bulk reactions. Monomer droplets with dissolved initiator are dispersed in water. As the polymerization proceeds the droplets become transformed into sticky, viscous monomer-swollen particles. Eventually, they become rigid particles with diameters in the range of about $(50–500)\ 10^{-4}$ cm. The final reaction mixture typically contains 25–50% of polymer dispersed in water. The viscosity of the system remains fairly constant during the reaction and is determined mainly by the continuous water phase.

Note that suspension polymerization is only superficially related to emulsion polymerization, which was outlined in Chapter 8. In suspension processes the coagulation of the dispersion is controlled by agitation plus the action of a water-soluble polymer and/or a fine particle size inorganic powder. The role of water is to act primarily as a heat transfer medium. In vinyl chloride suspension polymerization the specific heat of the monomer and polymer are about equal and are one-quarter that of water, on an equal weight basis. Thus, at the typical 1.5/1 water/vinyl chloride mass ratio the heat capacity of the aqueous phase is about six times that of the organic phase. Another use of water is, of course, to keep the viscosity of the reaction medium at a useful level. Water/monomer ratios of 1.5/1 to 1.75/1 provide a good compromise between suspension concentration and viscosity.

Since emulsion polymerization is initiated in the aqueous phase, the undesirable formation of latex polymer can be minimized in suspension systems by using water-soluble inhibitors, like sodium nitrite.

Prevention of the coalescence of the sticky, partially polymerized particles is a major problem in suspension polymerizations, and proper selection of stabilizing agents is important. Two kinds of additives are used to hinder coalescence of particles in suspension polymerizations. These are platelet-like mineral particles that concentrate at the organic-water interface, like $Ca_3(PO_4)_2$, and/or macromolecular species that are soluble in water and insoluble in the particular organic phase. Poly(vinyl alcohol) and starch products are examples of the latter type.

The normal sequence of operations is as follows:

1.   Premix the initiator and monomer(s).
2.   Add mix from step 1 to water, up to about 35% (v/v) with agitation to produce the desired droplet sizes.
3.   Add inorganic or organic stabilizers, which should concentrate at the droplet/H_2O interfaces.
4.   Reduce stirrer speed so as to minimize coalescence and prevent separation of the droplets and water, because of their density differences.

5. Raise reaction temperature and hold at the desired polymerization level until the softening temperature of the particles reaches or exceeds the reactor temperature (this occurs with depletion of the monomer, which normally plasticizes the polymer).

6. Increase the temperature to complete monomer conversion.

7. Dump the reactor contents, cool the reaction mixture, and separate the polymer.

Formation of a scale of polymer on the reactor walls is normally less than in corresponding bulk or solution polymerizations. Scale formation is troublesome in PVC suspension polymerizations, however, because the polymer is not soluble in its monomer, and a deposit formed on the wall will not be washed off by fresh monomer. This build-up has to be removed in order to maintain satisfactory heat transfer and prevent inclusion of gelled polymer ("fish eyes") in the product. Cleanliness of the reactor walls is very important because the productivity of the equipment is enhanced by longer intervals between shutdowns for cleaning. To this end, some phenolic coatings have been designed that inhibit polymer buildup by terminating free-radical reactions on the walls (cf. Section 6.9).

Modern reactors are made of stainless steel and have capacities up to as much as 180 m^3 (50,000 U.S. gal.). Designs vary, and include top and bottom entry stirrers and multiple impellers. Agitation of the reaction mixture must be given serious attention since many monomers are less dense than water while their polymers are more dense. The mixing systems must then pull monomer down from the surface of the charge and lift polymer off the floor of the reaction vessel.

The major process for poly(vinyl chloride) production is the suspension system. Typical reaction temperatures are 50–65°C. As the reaction proceeds, a conversion (~76%) is reached at which the only monomer left in the system is that absorbed in the polymer particles. This occurs when the monomer concentration is about 30 wt % in the particles. The occurrence of this phenomenon is signaled by a drop in the reactor pressure. Normal pressures in the autoclaves are initially about 150 psig (pounds per square inch, gauge), and it is usual to carry out polymerizations until the pressure drops to about 20–70 psig, depending on the reaction temperature. Water may be injected into the reaction vessel as the polymerization proceeds, to compensate for the volumetric contraction between monomer and polymer. This also helps prevent the reaction mixture from becoming too viscous. As well, the water addition enhances the cooling capacity of the reactor because it increases the heat transfer area on the walls.

Porosity is a desirable characteristic of the particles in many applications of poly(vinyl chloride). If the product is to be used as a "dry-blend" resin, it is required to soak up substantial quantities of liquid plasticizers and still remain free-flowing. The structure and porosity of PVC granules is affected strongly by the choice of organic suspending agents, which are different types of partially hydrolyzed poly(vinyl alcohol). The required porosity is enhanced also by rapidly removing the unreacted monomer which is occluded in the particles. For this reason, such suspension polymerizations of poly(vinyl chloride) are not driven to conversions much greater than about 80%. Polymer intended for extrusion into pipe and other nonplasticized applications is taken to higher conversions to enhance the reactor productivity and the bulk density of the PVC granules. Higher bulk densities result in greater production rates during subsequent extrusion, because the polymer powder is fed more efficiently from the extruder hopper into the conveying screws. Higher polymerization temperatures also result in higher final bulk densities, but these conditions produce lower PVC molecular weights.

Sudden increases in reaction rate, called "heat kicks," are sometimes encountered toward the end of the PVC reaction. This probably results from the deteriorating heat conductivity of the polymer particles as the monomer concentration in these loci decreases. The interior of the particles becomes hot because of poor heat transfer to the surrounding water, and this causes progressive accelerations in the propagation rate and temperature rise. If heat kicks are not subdued, the reactor can go out of control. The effect can be moderated by venting monomer or by adding pentane to absorb thermal energy as it vaporizes. Another alternative is the addition of styrene as a "shortstop." Styryl radicals are too stable to reinitiate vinyl chloride polymerization, and the growth of macroradicals with styrene ends is essentially terminated (Section 7.10.1).

Ideally, a suspension polymerization is run so that the heat of reaction just equals the maximum rate of heat removal through the reactor walls. The total heat transfer from the reaction medium, H_T is:

$$\frac{1}{H_T} = \frac{1}{H_F} + \frac{1}{H_W} + \frac{1}{H_J} \tag{10-14}$$

where H_F = film coefficient on the inside of the autoclave, H_W = thermal conductivity of the jacket wall, and H_J = jacket side film coefficient. Good agitation and freedom from wall fouling are necessary, to keep H_F as high as possible.

Commercial suspension polymerizations are not strictly isothermal, since the reactor contents must be heated to the final reaction temperature. Mixtures of initiators are therefore used in an attempt to maintain a rate of heat generation close to the cooling capacity of the reactor. Particular initiators are useful only over a limited temperature range. Most initiators for suspension polymerizations have half-lives of about 2 h in the 50-70°C range. After 6 h, then, the final initiator concentration will be 10-15% of the amount charged initially to the reactor (from Eq. (6-32). In PVC synthesis, it is fairly common to use one initiator with a $t_{1/2}$ of 1-2 h and another with a longer $t_{1/2}$ of 4-6 h. Other factors that affect the usefulness of initiators include:

1. Storage stability and safety
2. Color development in the polymer (this is a problem with some azo initiators)

3.   Water insolubility and resistance to hydrolysis; water solubility could lead to more reaction in the aqueous phase and wall fouling. Other expedients to reduce aqueous phase reactions include use of a water-soluble free-radical scavenger or a chelating agent to minimize redox reactions in the aqueous phase. (Such water-soluble chelating agents include salts of oxalic acid and ethylene diamine tetraacetic acid.)

The kinetic features of suspension polymerizations are thought to be as described in Chapter 6 for free-radical reactions, in general. However, the particle sizes and structures that are produced are very important polymer properties, and these depend on factors other than the chemistry of the polymerization. In vinyl chloride polymerization, the particle character is related to the agitation level, which depends on the impeller diameter and rotational speed. At low agitation levels large monomer droplets and polymer particles are formed. The droplet and particle sizes decrease at higher agitation levels. At even higher agitation levels, the particles may become larger again, because there is insufficient suspending agent for the many fine droplets that are produced. Increases in reaction temperature are generally accompanied by a modest decrease in interfacial tension and lower droplet and subsequent particle sizes. It is important to control the bead size distribution to avoid very small and very large granules. This is accomplished through selection of the level and nature of suspending agents and other ingredients and an optimized stirring protocol.

Other major products of suspension processes include expandable polystyrene, where a volatile hydrocarbon is diffused into the polymer beads, and spherical divinylbenzene-based beads for chromatographic and ion-exchange applications. PVC is different from most other suspension process polymers in that it is produced by precipitation polymerization, as described earlier.

Suspension polymerization is frequently employed as the second stage following a preliminary bulk polymerization, such as in the manufacture of some HIPS and ABS polymers. Polybutadiene or another elastomer is dissolved in liquid styrene, and this monomer or a mixture of styrene and acrylonitrile is polymerized in a batch kettle. The syrup produced at 30–35% conversion is too viscous for effective mixing and heat transfer. It is therefore dispersed in water, and the polymerization is finished as a suspension reaction.

(iv)   *Emulsion Systems*

Emulsion polymerizations were described in Chapter 8. These reactions yield high-molecular-weight products at fast reaction rates when the corresponding suspension, bulk, or solution free-radical polymerizations are inefficient, because both R_p and $\overline{DP}_n$ can be changed in parallel by altering the soap concentration in emulsion reactions. Thermal control and mixing are relatively easy, in common with suspension polymerizations. Emulsion reactions are more convenient than suspension polymerizations with soft or tacky polymers, because emulsion systems employ higher surfactant concentrations. It is correspondingly more difficult to remove the soaps from the finished product, and this purification is rarely attempted.

The major emulsion processes include the copolymerization of styrene and butadiene to form SBR rubber, polymerization of chloroprene (Fig. 1-4) to produce neoprene rubbers, and the synthesis of latex paints and adhesives based mainly on vinyl acetate and acrylic copolymers. The product is either used directly in emulsion form as a paint or else the surfactants used in the polymerization are left in the final, coagulated rubber product.

Emulsion polymerizations normally produce polymer particles with diameters of 0.1–1 μm (1 μm = 1 micron = 10^{-4} cm), although much larger particles can be made by special techniques mentioned in Chapter 8. The polymer particles made by suspension reactions have diameters in the range of 50–500 μm. Recall that free-radical initiation in suspension reactions is in the monomer phase, whereas the aqueous phase is the initiation site in emulsion polymerizations. The two processes often differ also in the types of stabilizers that are used. Microsuspension polymerization is an alternative technique which can yield particles in the same size range as emulsion processes. This method uses a monomer-soluble initiator and anionic emulsifiers similar in nature and concentration to those used in emulsion polymerizations. A microdispersion of the mixture of the reaction ingredients is first produced mechanically and is then polymerized to provide polymer with essentially the initial fine particle size distribution.

Emulsion polymerization reactors are made of stainless steel and are normally equipped with top-entry stirrers and ports for addition of reactants. Control of the reaction exotherm and particle size distribution of the polymer latex is achieved most readily by semibatch (also called semicontinuous) processes, in which some or all of the reactants are fed into the reactor during the course of the polymerization. Examples are given in Chapter 8. In vinyl acetate copolymerizations, a convenient monomer addition rate is such that keeps the vinyl acetate/water azeotrope refluxing, at about 70°C.

Vinyl acetate and acrylic emulsion copolymers usually contain significant proportions of insoluble "gel" material. This fraction results from chain transfer to polymer (Section 6.8.4). It is not deleterious in products like surface coatings and adhesives, and may even confer some advantages, like faster drying after application to substrates. The molecular weight distributions of such polymers are not of practical interest, since insoluble material has infinite molecular weight, on the scale of the methods summarized in Chapter 3. However, the properties of these latexes are affected by their particle size distributions. Industrial-scale emulsion polymerizations are characterized by variable initial induction periods, as the inhibiting effects of dissolved oxygen in the water feed (Section 6.9) are overcome by decomposing initiator. (It is more economical to waste initiator for this purpose than to eliminate dissolved oxygen in large volumes of water by sparging with an inert gas like nitrogen.) As a consequence, however, it is very difficult to produce polymers with consistent particle size distributions, by starting emulsion polymerizations with a charge of water, monomer, surfactants and the other ingredients listed in Chapter 8. Particle sizes of latex polymers are neatly controlled, however, by including a small quantity of "seed latex" in the initial charge to the reactor. The seed latex has an appropriate small particle size, that has been measured beforehand. The polymer emulsion is grown on the seed latex, controlling the feed rate of other reactants as outlined in Chapter 8, in connection with the production of "core-shell" particles. The important factors here are the particle size distribution of the seed latex and its availability for a large number of seeded polymerizations. The seed polymer need not even have the same chemical composition as the final polymer.

(v)   Gas Phase Polymerizations

Transition metal catalysts that produce high yields of olefin polymers per unit weight of catalyst metal were mentioned in Section 9.5.3. In the gas phase polyethylene processes, ethylene is polymerized or copolymerized in a solvent-free fluidized bed reactor. (Fluidized beds are suspensions of solid particles in fast-moving gas streams. Major applications are in hydrocarbon cracking and other catalytic processes and in drying of solids. Fluidization ensures that the gas contacts the solid particles efficiently. Vigorous agitation of the solid materials prevents clumping and minimizes temperature variations.) In this process, ethylene gas and solid catalysts are fed continuously to a fluidized bed reactor. The fluidized material is polyethylene powder which is produced as a result of the polymerization of the ethylene on the catalyst. The ethylene, which is recycled, supplies monomer for the reaction, fluidizes the solids, and serves as a heat removal medium. The reaction is exothermic and is run normally at temperatures $25–50°C$ below the softening temperature of the polyethylene powder in the bed. This operation requires very good heat transfer to avoid hot spots and means that the gas distribution and fluidization must be very uniform.

The keys to the process are active catalysts. As mentioned in Chapter 6, these are organochromium compounds on particular supports. The catalysts will yield up to about 10^6 kg of polymer per kilogram of metallic chromium. Branching is controlled by use of comonomers like propylene or 1-butene, and hydrogen is used as a chain transfer agent. The catalyst is so efficient that its concentration in the final product is negligible. The absence of a solvent and a catalyst removal step gives the process operating and capital cost advantages over the older slurry processes for low-pressure polyethylene. However, while the granules produced directly by gas phase polymerizations are free-flowing, they are smaller than the pellets produced by the slurry processes. As a result, their bulk density is lower and transportation costs are correspondingly increased. Because of this and because end-users frequently have conveying equipment that is tailored to the properties of polymer pellets, it is not unusual to find that gas phase polyolefins are extruded and pelletized prior to shipment. This illustrates the common wisdom that the costs of production of a polymer are *all* those incurred in its synthesis and finishing operations.

Although the density of the polymer can be varied by copolymerization with higher olefins to match that of polyethylene produced by high-pressure free-radical processes, the two types differ in branch frequency and character and in molecular weight distributions. As a result, they do not have comparable processing and mechanical properties.

Gas phase polymerizations, using other supported catalysts, are also employed to make isotactic polypropylene, with productivities of the same order as those reported for polyethylene manufacture.

(vi)   *Interfacial Polymerizations*

Section 5.5 should be consulted for a general description of this process, which applies only to the production of condensation polymers.

The growth of macromolecules in interfacial reactions is often observed to be into the organic phase, indicating that the active hydrogen compound which is initially in the water phase is the migrating entity. The polymer at the interface serves primarily to control the penetration of active hydrogen compound into the organic phase. The rate of mass transfer of the active hydrogen compound is the rate controlling step, because the basic chemical reactions must be very fast to be useful in interfacial reactions.

Properties of fibers can be altered by carrying out interfacial polymerizations on their surfaces. Thus the shrink resistance of wool can be improved by immersing the fiber first in a solution containing one component of a condensation polymer and then immersing it in another solution containing the other component. Polyamides, polyurethanes, polyureas, and other polymers and copolymers may be grafted on wool in this manner.

10.5   BATCH, SEMIBATCH, AND
CONTINUOUS PROCESSES

Polymerization reactions can be further classified into batch and continuous processes. Continuous operation is feasible for the production of large quantities of polymers with uniform properties. Frequent product changes are not economical, because off-grade polymer is made during start-ups until the reaction conditions are stabilized and because there may be long lags between changes in operating variables and subsequent steady production of uniform product. Continuous processing will sometimes permit production of polymers with narrower property ranges than can be obtained in batch processes.

Semibatch operation involves the continuous or intermittent addition of monomer or other ingredients during polymerization. It is often employed in copolymerizations (Section 7.5) to minimize the drift of copolymer composition when the reactivities differ greatly.

Periodic addition of monomer to an operating batch polymerization assists in controlling the reaction temperature because of its cooling effect.

In emulsion polymerizations semibatch operation provides better control of the particle size of the product. The properties of the product polymers can be modified, also, by continuous or intermittent changes in the composition of the monomer feed in emulsion copolymerizations, where a given monomer can be preferentially concentrated in the interior or on the surface of the final particles, as described in Chapter 8.

10.6 POLYMERIZATION REACTORS

Reactors are conveniently considered in three idealized categories: batch, tubular, and continuous stirred tank reactors (CSTR). The operations of real reactors may be modeled on the basis of one of these types or combination thereof.

The detailed course of a polymerization is determined by the nature of the particular reaction as well as by the characteristics of the reactor which is used. The design and control of the operation are greatly aided by mathematical modeling of the process. Such models may be based on empirical relations between the independent and dependent operating variables. This is not as satisfactory, however, as a model that is derived from accurate knowledge of the polymerization process and reactor operation, because only the latter tool permits extrapolation to reaction conditions that have not yet been tried.

10.6.1 Batch Reactors

Good mixing is important to ensure uniform temperatures and prevent the occurrence of localized inhomogenieties. It is difficult to generalize about mixing in bulk polymerizations in batch reactors, because the viscosity and density of the reaction medium is continuously changing as the reaction proceeds. The corresponding changes in emulsion and suspension systems are, of course, much less pronounced.

The major problem in temperature control in bulk and solution batch chain-growth reactions is the large increase in viscosity of the reaction medium with conversion. The viscosity of styrene mixtures at 150°C will have increased about 1000-fold, for example, when 40 wt % of the monomer has polymerized. The heat transfer to a jacket in a vessel varies approximately inversely with the one-third power of the viscosity. (The exact dependence depends also on the nature of the agitator and the speed of fluid flow.) This suggests that the heat transfer efficiency in a jacketed batch reactor can be expected to decrease by about 40% for every 10% increase in polystyrene conversion between 0 and 40%.

Heat transfer can be increased up to a point by rotating the agitator faster. An increase in speed by a factor of 10^3 will increase the heat transfer rate by 10^2 and the power to the agitator shaft by 10^7. In viscous systems a speed of maximum net heat input is reached beyond which rate of power input into the batch increases faster than the rate of heat removal.

Heat removal is accomplished by transfer to the vessel jacket, use of internal cooling coils, circulation of the reaction fluid through an external cooling loop, or by use of an overhead condenser to remove heat from the monomer or diluent in the vapor phase.

All reactors are jacketed to permit heat removal through the vessel walls. It is frequently necessary to add extra heat removal means as the reaction vessels are scaled up because the heat transfer area of the reactor walls increases with reactor volume to the two-thirds power while the rate of heat generation is proportional to the volume itself.

The energy balance for an isothermal reaction can be written as

$$V(-\Delta H_p)R_p = U A_w(T - T_l) + q_E \tag{10-15}$$

where $V = $ reactor volume, $-\Delta H_p = $ heat of polymerization, which is negative for exothermic reactions, $R_p = $ polymerization rate (cf. Chapter 6), $U = $ overall heat transfer coefficient, $A_w = $ wall area, $T = $ reaction temperature, $T_l = $ temperature of coolant, and $q_E = $ heat removed by the condenser or other auxiliary devices. The two terms on the right-hand side of this equation represent alternative methods for removing the exothermic heat of reaction, with $U A_w(T - T_l)$ corresponding to heat transfer through the reactor walls. Substitution of representative values for the various parameters into Eq. (10-15) shows that q_E increases in relative importance as the reactor size is increased. Limiting operating conditions for nonisothermal polymerizations may be estimated by modifying Eq. (10-15) to take account of the activation energy for rate of energy release, which is set equal to that for rate of polymerization. The modified equation can be solved numerically for different values of R_p, T_l, and so on.

Reflux cooling is the most common method for additional heat removal if an ingredient of the polymerization mixture is volatile at the reaction temperature and extensive foaming does not occur. External heat exchangers can be used in some processes in which a portion of the reaction mixture is continuously removed, pumped through a heat exchanger to cool it, and returned to the reactor. This method is used sometimes in emulsion processes where the mixture viscosity is low and the mechanical stability of the latex is good. It is not practical in suspension systems, however, because continuous agitation is required in these reactions to prevent coalescence of the polymer particles. Bulk and solution polymerizations do not ordinarily rely on external heat exchangers, because the high viscosity and poor agitation in the heat exchanger leads to polymer build-up on the cool walls of this unit. Internal cooling coils can be used only in reactions where the mixture viscosity is low and polymer scale build-up is not a problem. Otherwise, poor mixing around the coils can result in poor product quality and long cleanup times between batches.

10.6.2   Tubular Reactors

Tubular reactors consist in principle of unstirred vessels with very high length/ radius ratios. They are attractive reactors for production of some micromolecular species but are limited in their application to polymer production. This is because the relatively high viscosities that are encountered at intermediate conversions in polymer syntheses lead to difficulties in controlling the reaction temperature. Polymer tends to form a slow moving layer on the cool reactor walls, reducing the flow-through capacity of the tube and the effective heat transfer coefficient.

In general in tubular reactors, the material at the tube center will be at a higher temperature than the reaction mixture at the tube wall. The temperature rise increases with the tube radius, because heat transfer in this reactor type is entirely by convection through the reaction mixture. Thus a larger tube radius increases production rates because of the greater volumetric flow rate, but there is an additional augmentation of production resulting from the higher center line temperatures in the larger bore vessel. The broad temperature distribution is reflected, of course, in a greater polydispersity of polymer molecular weights. As a corollary, thermal runaways are possible with increasingly larger tube diameters.

For a given tube radius there exists a particular wall temperature that gives maximum conversions in free-radical polymerizations. This can be seen qualitatively from the following considerations. If the tube wall is too cool, the initiator will be slowly decomposed and some of it will leave the reactor unconsumed. However, the activation energy for initiator decomposition exceeds that for consumption of monomer (Section 6.16.1), and the initiator can be entirely decomposed at low monomer conversions if the wall temperature is too high for the particular reaction system [2].

Problems from inadequate mixing in tubular vessels can be alleviated to some extent by inserting stationary mixing sections in the reaction train. To visualize a stationary mixer, imagine a long strip of metal that is alternately twisted 180° in clockwise and counterclockwise directions. Fluid flowing through the resulting path is repeatedly divided, subdivided, and recombined, and the result is efficient distributive mixing of volume elements which were originally far apart. The costs of operation of such devices include increased flow resistance and the danger of fouling or plugging.

The only important current application of tubular reactors in polymer syntheses is in the production of high pressure, low density polyethylene. In tubular processes, the newer reactors typically have inside diameters about 2.5 cm and lengths of the order of 1 km. Ethylene, a free-radical initiator, and a chain transfer agent are injected at the tube inlet and sometimes downstream as well. The high heat of polymerization causes nonisothermal conditions with the temperature increasing towards the tube center and away from the inlet. A typical axial temperature profile peaks some distance down the tube where the bulk of the initiator has been consumed. The reactors are operated at 200–300°C and 2000–3000 atm pressure.

The ethylene and polyethylene leave the reactor and pass into a primary separation vessel which operates at a much lower pressure than the reactor itself. Most of the ethylene (and any comonomer) is flashed off in this unit and recycled through compressors to the tube inlet. Conversion per pass is of the order of 30% with ethylene flow rates about 40,000 kg/h.

In many cases the reactor exit valve is opened and partially closed periodically to impose a pressure and flow pulse that helps keep the tube from plugging with polymer. Substantial pressure fluctuations occur in the reactor with this mode of operation.

The other major reactor type used for high-pressure free-radical polymerization of ethylene is a stirred autoclave type. There are very many variations and modifications of this type, as there are of tubular reactors. Stirred autoclaves usually have length-to-diameter ratios of about 20. If they are well agitated with good end-to-end mixing the reactor will approximate a CSTR. In many cases, however, a high degree of directional flow is imposed and mixing is restricted by baffles so that the autoclave operates more like a tubular vessel. Molecular weight and branching distributions are strongly affected by the mode of operation of polyethylene reactors.

10.6.3   Continuous Stirred Tank Reactors

An ideal CSTR is deliberately backmixed, in contrast to a tubular reactor where plug flow and zero backmixing are ideal concepts. The feed is assumed to blend instantly with the tank contents in a perfect CSTR, and the effluent composition and temperature are the same as those of the contents. Neither reactor type is ideal in actual practice, of course, and finite feed blending times, inhomogeneities, and short-circuiting and stagnation are observed in the contents of real CSTRs.

In a free-radical polymerization, the molecular weight distribution produced in a CSTR will be narrower than that made in a comparable batch or tubular reactor. This will be true in any polymerization where molecular weights are controlled by mutual termination reactions of macroradicals and chain transfer to polymer is negligible. If the growth of polymer molecules is halted primarily by chain transfer to monomer or other species the molecular weight distribution will be random regardless of reactor type. The molecular weight distribution obtained in a series of CSTRs can be broadened, if desired, by operating the individual reactors in series or parallel with each unit at a different reaction temperature or mean residence time. The composition of copolymers is more uniform with CSTR reactors than with batch, semibatch, or tubular reactors.

In a single CSTR, monomer and other ingredients of the polymerization recipe are continually fed into the vessel while polymer and the rest of the reaction mixture are removed. The effluent can, of course, serve as the feed to the next CSTR in a series operation. Problems with heat removal are alleviated to some extent because of the beneficial effects of cold monomer feed and the removal of reaction heat with the effluent. CSTR reactors are economically attractive for large-scale production with relatively infrequent changes in product properties.

Consider a mass balance for monomer in a CSTR:

$$v[M]_0 - v[M] = R_p \qquad (10\text{-}16)$$

where $v[M]_0$ is the molar inflow of monomer, $v[M]$ is the corresponding outflow, and R_p is the rate of polymerization (all quantities here are in units of mol/time). Alternatively,

$$(1/\theta)([M]_0 - [M]) = R_p \qquad (10\text{-}17)$$

where θ is the mean residence time equal to the ratio of reactor volume V and volumetric flow rate v and $[M]_0$ and $[M]$ are the molar concentrations of monomer in the influent and effluent, respectively. In terms of conversion $p(=([M]_0 - [M])/[M]_0)$:

$$p[M]_0/\theta = R_p \qquad (10\text{-}18)$$

The solution of Eq. (10-18) depends on the variation of R_p with p. If R_p does not vary with p or decreases with increasing p, the equation has a single solution for a reactor with stipulated residence time. However, if autoacceleration occurs then the same R_p can be observed at different values of [M] in the reactor, providing these changes also correspond to different conversions.

Multiple steady states are theoretically possible in many free radical polymerizations, but they are not usually observed in practice because the reaction is controlled at relatively low conversions (high [M]) where the viscosity of the medium presents less of a problem. This is particularly true of bulk polymerizations such as those in the high-pressure polyethylene processes.

(i)   Residence Time Distributions

The polymerization time in continuous processes depends on the time the reactants spend in the reactor. The contents of a batch reactor will all have the same residence time, since they are introduced and removed from the vessel at the same times. The continuous flow tubular reactor has the next narrowest residence time distribution, if flow in the reactor is truly plug-like (i.e., not laminar). These two reactors are best adapted for achieving high conversions, while a CSTR cannot provide high conversion, by definition of its operation. The residence time distribution of the CSTR contents is broader than those of the former types. A cascade of CSTR's will approach the behavior of a plug flow continuous reactor.

The residence times in a continuous flow reactor have a distribution that can be characterized by any of a trio of functions. One of these is the cumulative probability function $F(t)$, which is the fraction of exiting material that was in the reactor for a time less than t.

Physically, $F(t)$ represents the time dependence of the concentration of a nonreactive tracer that was instantaneously injected as a slug into the flowing reaction medium. A related expression gives $R(t)$, the decay function,

$$R(t) = 1.0 - F(t) \qquad (10\text{-}19)$$

If a nonreactive tracer were being continuously and steadily injected into the stream and then abruptly turned off, $R(t)$ would represent its relative concentration in the reactor effluent. The frequency function $f(t)$ defines the fraction of exiting material that had residence times between t and $t + dt$ in the reactor. It is given by

$$f(t) = dF(t) = -dR(t) \qquad (10\text{-}20)$$

Figure 10-1 illustrates these functions for a general CSTR.

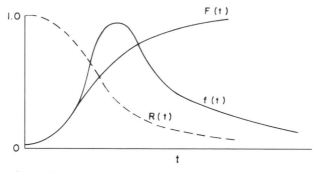

Fig. 10-1.   Curves illustrating the residence time distribution $F(t)$, residence time decay function $R(t)$, and residence time frequency function $f(t)$ for a CSTR.

Now imagine a perfectly mixed CSTR operating with a tracer material with inlet concentration c_{in} and effluent concentration c_{out}. The mass balance for this system is

$$V \, dc_{out}/dt = v(c_{in} - c_{out}) \tag{10-21}$$

If a step change is imposed on c_{in}, so that

$$c_{in} = \begin{cases} 0 & \text{at } t < 0 \\ 1 & \text{at } t > 0 \end{cases}$$

then for $t > 0$ the preceding equation gives

$$\ln(1 - c_{out}) = -vt/V = -t/\theta \tag{10-22}$$

or

$$c_{out} = F(t) = 1 - e^{-t/\theta} \tag{10-23}$$

In this case $R(t)$ has the form

$$R(t) = e^{-t/\theta} \tag{10-24}$$

The exponential distribution of residence times defines a *well-stirred reactor*.

Analogous expressions can be written for a plug flow tubular reactor that ideally has zero mixing in the axial direction and is completely homogeneous radially. A step change imposed on c_{in} in such a system produces an identical step change in c_{out} after a lag of θ sec. Thus, for $t > \theta$ for such a reactor,

$$R(t) = 0 \qquad \text{for } t > 0 \tag{10-25}$$

Real tubular reactors approach axial plug flow if the viscosity of the fluid decreases with increasing rate of shearing and the resulting velocity profile is flat across the tube.

(ii)   State of Mixing in CSTRs

Note that the residence time distribution itself does not completely define the state of mixing of the components of the reaction mixture. It actually defines the state of mixing of volume elements which are small compared to the capacity of the reactor. Any fluid that consists of such volume elements that do not comingle on a molecular scale is called a *macrofluid*, and the corresponding mixing process is termed macromixing or segregated flow mixing [3]. The polymer beads in a continuous suspension polymerization process can have a distribution of residence times with no mass transfer between them. This is then an example of macromixing, with the particles corresponding to an ensemble of batch reactors operating in parallel with a distribution of reaction times given by Eq. (10-24).

The other extreme condition is one in which mixing takes place on a molecular scale. The incoming fluid elements quickly lose their identity as a result of mass transfer between them, and the reaction medium becomes a *microfluid*.

Real flows in continuous polymerization vessels are always intermediate between the completely micromixed and completely segregated conditions. Most studies of CSTR reactions assume one or the other type. Despite this artificiality, there have been some successes in modelling actual polymerizations.

The degree of micromixing has little effect on the conversion of polymer per pass through the reactor, except as it may influence the initiator efficiency in free radical reactions. Other criteria, such as molecular weight and branching distributions of the product polymers, can be strongly affected. The number distribution of molecular weights in a perfectly mixed CSTR will, or course, be the same as the instantaneous molecular weight distribution in a batch reactor at the same temperature, initiation rate, and conversion. The molecular weight distribution in a segregated CSTR is expected to be broader than in a micromixed CSTR. A perfectly mixed CSTR should generally produce a product with a narrower-molecular-weight distribution than a batch reactor if the lifetime of the growing macromolecules is short compared to the mean residence time in the reactor. The perfectly stirred tank will give a broader distribution product if the life of the growing chain is long compared to θ.

The most significant differences between perfectly mixed and segregated flow in a CSTR occur in copolymerizations. In a batch reaction, the copolymer composition varies with conversion, depending on the reactivity ratios and initial monomer feed composition. In a perfectly mixed CSTR, there will be no composition drifts but the distribution of product compositions will broaden as mixing in the reactor approaches segregated flow.

10-1. The equilibrium constant for reaction (10-13) is

$$\frac{[-CONH-][H_2O]}{[-COOH-][-NH_2]} = K_1 \simeq 300 \quad \text{at } 280°C$$

The enthalpy change for this polymerization is $\Delta H_p = -6.5$ kcal mol^{-1}. The polymerization reaction in this problem is finished at a fixed steam pressure (1 atm). The equilibrium concentration of H_2O in the polymer melt varies with temperature and steam pressure in this case. The enthalpy of vaporization of H_2O is about 8 kcal mol^{-1}. Compare the limiting values of number average molecular weight of the polyamide produced at 280 and 250°C final polymerization temperatures. [*Hint:* Recall that the variation of an equilibrium constant K with temperature is given by $d(\ln K)/d(1/T) = -\Delta H/R$, where ΔH is the enthalpy change of the particular process and R is the universal gas constant. Calculate K_1 and the equilibrium concentration of H_2O in the melt at 250°C and use Eq. (10-8).]

10-2. Acrylonitrile has considerable solubility in water, while styrene is negligibly soluble. In the copolymerization of these two monomers in a suspension process, it is not desirable that much polymerization take place in solution or emulsion in the aqueous phase, as this will produce material that differs significantly in molecular weight and structure from that polymerized in the monomer droplets. How could you ensure that polymerization is confined essentially to the suspended monomer droplets?

10-3. Following is a recipe for the suspension polymerization of styrene to give polymer beads with diameters about 0.5 ± 0.25 mm. Such products have surface/mass ratios that are sufficiently high to ensure good dispersion of dry pigments when the colorants and solid polymer are stirred together.

Aqueous phase: 20 kg demineralized H_2O, 0.13 kg $(Ca_3(PO_4)_2$ (insoluble suspending agent), 0.15 kg Na-β-naphthalene sulfonate (suspending agent), and 0.001 kg Na polyacrylate (polymeric suspending agent).

Monomer phase: 17 kg styrene and three initiators, each present in the amount of about 0.02 mol.

The polymerization is not carried out isothermally but with stepwise temperature rises. This is because styrene polymerizes relatively slowly. A mixture of initiators is therefore used, and it has been found useful in this context to select initiators with half-lives similar to the polymerization time in a particular temperature range. The following temperature program has been used:

Hours	Temperature (°C)	Hours	Temperature (°C)
0–1	40–80	8–13	90
1–7	80	13–15	90–110
7–8	80–90	15–27	110

Azodiisobutyronitrile is a suitable initiator for the 80°C stage of this polymerization. Find the most suitable initiators of those listed in Table 6-1 for the 90 and 110°C intervals. Assume an approximate activation energy for peroxy-type initiators of 30 kcal mol^{-1}.

10-4. In industrial polymerization, the monomer is often added slowly to the reaction mixture. Consider the following experiments [J.J. Krackeler and H. Naidus, *J. Polym. Sci. Part C* **27**, 207 (1969)] with polystyrene polymerizations. The simple recipe is styrene: 32 parts, 4.4% solution of $K_2S_2O_8$: 3.4 parts, surfactant solution (containing 7% octyl phenol ethylene oxide adduct and 1% sodium lauryl sulfate): 58 parts, and reaction temperature 70°C. Conversion-time plots are shown below for three modes of operation: (a) In batch (B), polymerization all reactants were added at once. (b) In monomer addition (M ADD), 10% of the monomer was added to the precharge of all the other ingredients. The addition of the remaining 90% of the monomer was started 5 min after the precharge was added. Monomer addition was completed in 90 min. (c) In emulsion addition (E ADD), 10% of the monomer and 10% of the surfactant solution were added in the precharge. The balance of the monomer and emulsifiers was added in a time schedule like that for monomer addition.

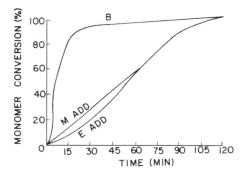

Describe the differences you would expect in the molecular weight of the polymer produced and in the particle size distribution. [*Hint:* From the plot above it can be seen that the rate of monomer addition in cases (b) and (c) was low enough that there was little unconverted monomer in the system at any time. The rate of polymerization was evidently controlled by the rate of addition of monomer rather than by the normal kinetics described in Chapter 7.] Note from this example how changing the operation of the emulsion reaction with a fixed recipe is an effective way to vary properties of polymers and latexes.

10-5. Consider an emulsion polymerization in a CSTR. The feed contains 6 mol monomer per liter of liquid (monomer + water) phase. The initial volumes of monomer and water have volume ratios of 2/1, and the ratio of the densities of polymer and monomers is 1.27. The reactor volume is 50 m³. The rate of polymerization in interval II of the emulsion polymerization is 50%.

(a) What flow rate is required to produce an average conversion of 50% in the reactor?

(b) What fraction of the exiting material is in the reactor for a time less than the mean residence time? Assume a perfectly stirred, isothermal CSTR.

10-6. Propylene can be polymerized in liquid monomer with a $TiCl_3/Al(C_2H_5)_2Cl$ catalyst. Typical pressures are about 400 psi, to keep propylene in the liquid phase at 60°C. Catalyst is continuously fed to the reactor and the slurry of polymer, which is at about 30-50% solids, is continuously removed from the reactor and fed to a separation vessel which is maintained at atmospheric pressure. The reaction temperature is controlled by cooling through the reactor walls and by removing propylene vapor, condensing it, and returning the liquid to the polymerizer. Propylene is polymerized by this method in a cylindrical reactor with diameter 0.5 m and height 2 m. To keep the temperature constant at 70°C in this case, it is found necessary to remove 20% of the exothermic process heat by taking off and condensing propylene vapor. A larger reactor (1-m diameter and 4-m height) is to be run at the same temperature and rate of polymerization (moles monomer/volume/time) as the smaller unit. What fraction of the process heat of the larger reactor must be accounted for by removing propylene, condensing it, and returning the liquid to the vessel?

REFERENCES

[1]   P. J. Hoftyzer, *Appl. Polym. Symp.* **26**, 349 (1975).
[2]   J. P. A. Wallis, R. A. Ritter, and H. Andre, *AIChE J.* **21**, 691 (1975).
[3]   O. Levenspiel, "Chemical Reaction Engineering." Wiley, New York, 1962.

Chapter 11

Mechanical Properties of Polymer Solids and Liquids

"παντα ρει" *(All things flow.)*
—Heraclitos, about 500 B.C.

11.1 INTRODUCTION

Polymers are in general use because they provide good mechanical properties at reasonable cost. The efficient application of macromolecules requires at least a basic understanding of the mechanical behavior of such materials and the factors which influence this behavior.

The mechanical properties of polymers are not single-valued functions of the chemical nature of the macromolecules. They will vary also with molecular weight, branching, cross-linking, crystallinity, plasticizers, fillers and other additives, orientation, and other consequences of processing history and sometimes with the thermal history of the particular sample.

When all these variables are fixed for a particular specimen, it will still be observed that the properties of the material will depend strongly on the temperature and time of testing compared, say, to metals. This dependence is a consequence of the viscoelastic nature of polymers. Viscoelasticity implies that the material has the characteristics both of a viscous liquid which cannot support a stress without flowing and an elastic solid in which removal of the imposed stress results in complete recovery of the imposed deformation.

Although the mechanical response of macromolecular solids is complex, it is possible to gain an understanding of the broad principles that govern this behavior. Polymeric articles can be designed rationally, and polymers can be synthesized for

377

particular applications. This chapter summarizes the salient factors which influence some important properties of solid polymers.

11.2   THERMAL TRANSITIONS

All liquids contract as their temperatures are decreased. Small, simple molecules crystallize quickly when they are cooled to the appropriate temperatures. Larger and more complex molecules must undergo translational and conformational reorganizations to fit into crystal lattices and their crystallization rates may be so reduced that a rigid, amorphous glass is formed before extensive crystallization occurs on cooling. In many cases, also, the structure of polymers is so irregular that crystalline structures cannot be formed. If crystallization does not occur, the viscosity of the liquid will increase on cooling to a level of 10^{14} Ns/m^2 (10^{15} poises) where it becomes an immobile glass. Conformational changes associated with normal volume contraction or crystallization can no longer take place in the glassy state and the thermal coefficient of expansion of the material falls to about one-third of its value in the warmer, liquid condition.

Most micromolecular species can exist in the gas, liquid, or crystalline solid states. Some can also be encountered in the glassy state. The behavior of glass-forming high polymers is more complex, because their condition at temperatures slightly above the glassy condition is more accurately characterized as rubbery than liquid. Unvulcanized elastomers described in Section 4.5 are very viscous liquids that will flow gradually under prolonged stresses. If they are cross-linked in the liquid state, this flow can be eliminated. In any case these materials are transformed into rigid, glassy products if they are cooled sufficiently. Similarly, an ordinarily glassy polymer like polystyrene is transformed into a rubbery liquid on warming to a high enough temperature.

The change between rubbery liquid and glassy behavior is known as the glass transition. It occurs over a temperature range, as shown in Fig. 11-1, where the temperature–volume relations for glass formation are contrasted with that for crystallization. Line $ABCD$ is for a substance which crystallizes completely. Such a material undergoes an abrupt change in volume and coefficient of thermal expansion at its melting point T_m. Line $ABEG$ represents the cooling curve for a glass-former. Over a short temperature range corresponding to the interval EF, the thermal coefficient of expansion of the substance changes but there is no discontinuity in the volume–temperature curve. By extrapolation, as shown, a temperature T_g' can be located that may be regarded as the glass transition temperature for the particular substance at the given cooling rate. If the material is cooled more slowly, the volume–temperature curve is like $ABEG'$ and the glass transition temperature

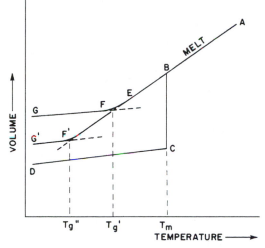

Fig. 11-1.   Volume–temperature relations for a glass-forming polymer and a material that crystallizes completely on cooling. T_m is a melting point, and T_g' and T_g'' are glass transition temperatures of an uncrystallized material that is cooled quickly and slowly, respectively.

T_g'' is lower than in the previous case. The precise value of T_g will depend on the cooling rate in the particular experiment.

Low-molecular-weight molecules melt and crystallize completely over a sharp temperature interval. Crystallizable polymers differ in that they melt over a range of temperatures and do not crystallize completely, especially if they have high molecular weights. Figure 11-2 compares the volume–temperature relation for such a polymer with that for an uncrystallizable analog. Almost all crystallizable polymers are considered to be "semicrystalline" because they contain significant fractions of poorly ordered, amorphous chains. Note that the melting region in this sketch is diffuse, and the melting point is identified with the temperature at B, where the largest and most perfect crystallites would melt. The noncrystalline portion of this material exhibits a glass transition temperature, as shown. It appears that T_g is characteristic generally of amorphous regions in polymers, whether or not other portions of the material are crystalline.

The melting range of a semicrystalline polymer may be very broad. Branched (low-density) polyethylene is an extreme example of this behavior. Softening is first noticeable at about 75°C although the last traces of crystallinity do not disappear until about 115°C. Other polymers, like nylon-6,6, have much narrower melting ranges.

Measurements of T_m and melting range are conveniently made by thermal analysis techniques like differential scanning calorimetry (dsc). The value of T_m is usually taken to be the temperature at which the highest melting crystallites

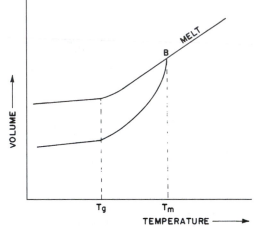

Fig. 11-2.   Volume–temperature relation for an amorphous (upper line) polymer and semicrystalline (lower line) polymer.

disappear. This parameter depends to some extent on the thermal history of the sample since more perfect, higher melting crystallites are produced by slower crystallization processes in which more time is provided for the conformational changes needed to fit macromolecular segments into the appropriate crystal pattern.

The onset of softening is usually measured as the temperature required for a particular polymer to deform a given amount under a specified load. These values are known as *heat deflection temperatures*. Such data do not have any direct connection with results of X-ray, thermal analysis, or other measurements of the melting of crystallites, but they are widely used in designing with plastics.

Both T_m and T_g are practically important. T_g sets an upper temperature limit for the use of amorphous thermoplastics like poly(methyl methacrylate) or polystyrene and a lower temperature limit for rubbery behavior of an elastomer like SBR rubber or 1,4-*cis*-polybutadiene. With semicrystalline thermoplastics, T_m or the onset of the melting range determines the upper service temperature. Between T_m and T_g, semicrystalline polymers tend to be tough and leathery. Brittleness begins to set in below T_g of the amorphous regions although secondary transitions below T_g are also important in this connection. As a general rule, however, semicrystalline plastics are used at temperatures between T_g and a practical softening temperature which lies above T_g and below T_m.

Changes in temperature and polymer molecular weight interact to influence the nature and consequences of thermal transitions in macromolecules. Warming of glassy amorphous materials converts them into rubbery liquids and eventually into viscous liquids. The transition between these latter states is very ill marked,

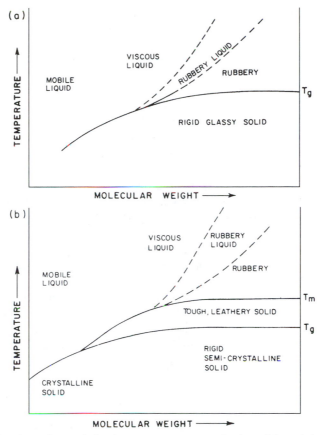

Fig. 11-3.   Approximate relations between temperature, molecular weight, and physical state for (a) an amorphous polymer and (b) a semicrystalline polymer.

however, as shown in Fig. 11-3a. Enhanced molecular weights increase T_g up to a plateau level which is encountered approximately at $\overline{DP}_n = 500$ for vinyl polymers. The rubbery nature of the liquid above T_g becomes increasingly more pronounced with higher molecular weights. Similar relations are shown in Fig. 11-3b for semicrystalline polymers where T_m at first increases and then levels off as the molecular weight of the polymer is made greater. T_m depends on the size and perfection of crystallites. Chain ends ordinarily have different steric requirements from interchain units, and the ends will either produce lattice imperfections in crystallites or will not be incorporated into these regions at all. In either case, T_m is reduced when the polymer contains significant proportions of lower molecular weight species and hence of chain ends.

11.3 CRYSTALLIZATION OF POLYMERS

Order is Heaven's first law.
—Alexander Pope, *Essay on Man*

Sections of polymer chains must be capable of packing together in ordered periodic arrays for crystallization to occur. This requires that the macromolecules be fairly regular in structure. Random copolymerization will prevent crystallization. Thus, polyethylene would be an ideal elastomer except for the fact that its very regular and symmetrical geometry permits the chains to pack together closely and crystallize very quickly. To inhibit crystallization and confer elastomeric properties on this polymer, ethylene is commonly copolymerized with substantial proportions of another olefin or with vinyl acetate.

A melting temperature range is observed in all semicrystalline polymers, because of variations in the sizes and perfection of crystallites. The crystal melting point is the highest melting temperature observed in an experiment like differential scanning calorimetry. It reflects the behavior of the largest, defect-free crystallites. For high-molecular-weight linear polyethylene this temperature, labeled T_m, is about 141°C. Other regular, symmetrical polymers will have lower or higher melting points depending on chain flexibility and interchain forces. At equilibrium at the melting point, the Gibbs free-energy change of the melting process, ΔG_m, is zero and

$$T_m = \Delta H_m / \Delta S_m \qquad (11\text{-}1)$$

The conformations of rigid chains will not be much different in the amorphous state near T_m than they are in the crystal lattice. This means that the melting process confers relatively little additional disorder on the system; ΔS_m is low and T_m is increased correspondingly. For example, ether units in poly(ethylene oxide) (**1-42**) make this structure more flexible than polyethylene, and T_m of high-molecular-weight versions of the former species is only 66°C. By contrast, poly(*p*-xylene) (**11-1**) is composed of stiff chains and its crystal melting point is 375°C.

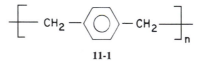

11-1

Stronger intermolecular forces result in greater ΔH_m values and an increase in T_m. Polyamides, which are hydrogen bonded, are higher melting than polyolefins with the same degree of polymerization, and the melting points of polyamides decrease with increasing lengths of hydrocarbon sequences between amide groupings. Thus the T_m of nylon-6 and nylon-11 are 225 and 194°C, respectively.

Bulky side groups in vinyl polymers reduce the rate of crystallization and the ability to crystallize by preventing the close approach of different chain segments. Such polymers require long stereoregular configurations (Section 4.2.2) in order to crystallize.

Crystal perfection and crystallite size are influenced by the rate of crystallization, and T_m is affected by the thermal history of the sample. Crystals grow in size by accretion of segments onto stable nuclei. These nuclei do not exist at temperatures above T_m, and crystallization occurs at measurable rates only at temperatures well below the melting point. As the crystallization temperature is reduced, this rate accelerates because of the effects of increasing concentrations of stable nuclei. The rate passes eventually through a maximum, because the colder conditions reduce the rate of conformational changes needed to place polymer segments into proper register on the crystallite surfaces. When T_g is reached, the crystallization rate becomes negligible. For isotactic polystyrene, for example, the rate of crystallization is a maximum at about 175°C. Crystallization rates are zero at 240°C (T_m) and at 100°C (T_g). If the polystyrene melt is cooled quickly from temperatures above 240° to 100°C or less, there will be insufficient time for crystallization to occur and the solid polymer will be amorphous. The isothermal crystallization rate of crystallizable polymers is generally a maximum at temperatures about halfway between T_g and T_m.

Crystallinity should be distinguished from molecular orientation. Both phenomena are based on alignment of segments of macromolecules but the crystalline state requires a periodic, regular placement of the atoms of the chain relative to each other whereas the oriented molecules need only be aligned without regard to location of atoms in particular positions. Orientation tends to promote crystallization because it brings the long axes of macromolecules parallel and closer together. The effects of orientation can be observed, however, in uncrystallized regions of semicrystalline polymers and in polymers which do not crystallize at all.

11.3.1 Degree of Crystallinity

High-molecular-weight flexible macromolecules do not crystallize completely. When the polymer melt is cooled, crystallites will be nucleated and start to grow independently throughout the volume of the specimen. If polymer chains are long enough, different segments of the same molecule can be incorporated in more than one crystallite. When these segments are anchored in this fashion the intermediate portions of the molecule may not be left with enough freedom of movement to fit into the lattice of a crystallite. It is also likely that regions in which threadlike polymers are entangled will not be able to meet the steric requirements for crystallization.

Several methods are available for determining the average crystallinity of a polymer specimen. One technique relies on the differences between the densities of completely amorphous and entirely crystalline versions of the same polymer and estimates crystallinity from the densities of real specimens, which are intermediate between these extremes. Crystalline density can be calculated from the dimensions of the unit cell in the crystal lattice, as determined by X-ray analysis. The amorphous density is measured with solid samples which have been produced by rapid quenching from melt temperatures, so that there is no experimental evidence of crystallinity. Polyethylene crystallizes too rapidly for this expedient to be effective [the reason for this is suggested in Section 11.3.2(i)], and volume–temperature relations of the melt like that in Fig. 11-1 are extrapolated in order to estimate the amorphous density at the temperature of interest. Crystalline regions have densities on the average about 10% higher than those of amorphous domains, since chain segments are packed more closely and regularly in the former.

The density method is very convenient, because the only measurement required is that of the density of a polymer sample. It suffers from some uncertainties in the assignments of crystalline and amorphous density values. An average crystallinity is estimated as if the polymer consisted of a mixture of perfectly crystalline and completely amorphous regions. The weight fraction of material in the crystalline state w_c is estimated assuming that the volumes of the crystalline and amorphous phases are additive:

$$w_c = \rho_c(\rho - \rho_a)/\rho(\rho_c - \rho_a) \qquad (11\text{-}2)$$

where ρ, ρ_c, and ρ_a are the densities of the particular specimen, perfect crystal, and amorphous polymer, respectively. Alternatively, if additivity of the masses of the crystalline and amorphous regions is assumed, then the volume fraction ϕ_c of polymer in the crystalline state is estimated from the same data:

$$\phi_c = (\rho - \rho_a)/(\rho_c - \rho_a) \qquad (11\text{-}3)$$

X-ray measurements can be used to determine an average degree of crystallinity by integrating the intensities of crystalline reflections and amorphous halos in diffraction photographs. Broadline nuclear magnetic resonance (NMR) spectroscopy is also suitable for measuring the ratio of amorphous to crystalline material in a sample because mobile components of the polymer in amorphous regions produce narrower signals than segments which are immobilized in crystallites. The composite spectrum of the polymer specimen is separated into crystalline and amorphous components to assign an average crystallinity. Infrared absorption spectra of many polymers contain bands which are representative of macromolecules in crystalline and in amorphous regions. The ratio of absorbances at characteristically crystalline and amorphous frequencies can be related to a crystalline/amorphous ratio for the specimen. An average crystallinity can also be inferred from measurements of the enthalpy of fusion per unit weight of polymer when the specific enthalpies of the crystalline and amorphous polymers at the melting temperature can be estimated. This method, which relies on differential scanning calorimetry, is particularly convenient and popular.

Each of the methods cited yields a measure of average crystallinity, which is really only defined operationally and in which the polymer is assumed artificially to consist of a mixture of perfectly ordered and completely disordered segments. In reality, there will be a continuous spectrum of structures with various degrees of order in the solid material. Average crystallinities determined by the different techniques cannot always be expected to agree very closely, because each method measures a different manifestation of the structural regularities in the solid polymer.

A polymer with a regular structure can attain a higher degree of crystallinity than one that incorporates branches, configurational variations, or other features that cannot be fitted into crystallites. Thus linear polyethylene can be induced to crystallize to a greater extent than the branched polymer. However, the degree of crystallinity and the mechanical properties of a particular crystallizable sample depend not only on the polymer structure but also on the conditions under which crystallization has occurred.

Quenching from the amorphous, melt state always produces articles with lower average crystallinities than those made by slow cooling through the range of crystallization temperatures. If quenched specimens are stored at temperatures higher than the glass transition of the polymer, some segments in the disordered regions will be mobile enough to rearrange themselves into lower energy, more ordered structures. This phenomenon, which is known as *secondary crystallization*, will result in a progressive increase in the average crystallinity of the sample.

For the reasons given, a single average crystallinity level cannot be assigned to a particular polymer. Certain ranges of crystallinity are fairly typical of different macromolecular species, however, with variations due to polymer structure, methods for estimating degree of crystallinity, and the histories of particular specimens. Some representative crystallinity levels are listed in Table 11-1. The ranges listed for the olefin polymers in this table reflect variations in average crystallinities which result mainly from different crystallization histories. The range shown for cotton specimens is due entirely to differences in average values measured by X ray, density, and other methods, however, and this lack of good coincidence of different estimates is true to some extent also of polyester fibers.

TABLE 11-1

Representative Degrees of Crystallinity (%)

Low-density polyethylene	45–74
High-density polyethylene	65–95
Polypropylene fiber	55–60
Poly(ethylene terephthalate) fiber	20–60
Cellulose (cotton)	60–80

Crystallization cannot take place at temperatures below T_g, and T_m is therefore always at a higher temperature than T_g. The presence of a crystalline phase in a polymer extends its range of mechanical usefulness compared to strictly amorphous versions of the same species. In general, an increased degree of crystallinity also reduces the solubility of the material and increases its rigidity. The absolute level of crystallinity that a polymer sample can achieve depends on its structure, but the actual degree of crystallinity, which is almost always less than this maximum value, will also reflect the crystallization conditions.

11.3.2   Microstructure of Semicrystalline Polymers

When small molecules crystallize, each granule often has the form of a crystal grown from a single nucleus. Such crystals are relatively free of defects and have well-defined crystal faces and cleavage planes. Their shapes can be related to the geometry of the unit cell of the crystal lattice. Polymers crystallized from the melt are polycrystalline. Their structures are a conglomerate of disordered material and clusters of crystallites that developed more or less simultaneously from the growth of many nuclei. Distinct crystal faces cannot be distinguished, and the ordered regions in semicrystalline polymers are generally much smaller than those in more perfectly crystallized micromolecular species. X-ray maxima are broadened by small crystallite sizes and by defects in larger crystals. In either case such data may be interpreted as indicating that the highly ordered regions in semicrystalline polymers have dimensions of the order of 10^{-5}–10^{-6} cm. These domains are held together by "tie molecules" which traverse more than one crystallite. This is what gives a semicrystalline polymer its mechanical strength. Aggregates of crystals of small molecules are held together only by secondary forces and are easily split apart. Such fragility is not observed in a polymer sample unless the ordered regions are large enough to swallow most macromolecules whole and leave few interregional molecular ties.

The term *crystallite* is used in polymer science to imply a component of an interconnected microcrystalline structure. Metals also belong to the class of microcrystalline solids, since they consist of tiny ordered grains connected by strong boundaries.

(i)   Nucleation of Crystallization

Crystallization begins from a nucleus that may derive from surfaces of adventitious impurities (heterogeneous nucleation) or from the aggregation of polymer segments at temperatures below T_m (homogeneous nucleation). The latter process is reversible up to the point where a critical size is reached, beyond which further growth results in a net decrease of free energy of the system. Another source of nuclei in polymer melts is ordered regions that are not fully destroyed during the prior melting process. Such nuclei can occur if segments in ordered regions find it difficult to diffuse away from each other, because the melt is very viscous or because these segments are pinned between regions of entanglement. The dominant effect in bulk crystallization appears to be the latter type of nucleation, as evidenced by in nuclear magnetic resonance spectroscopy relaxation experiments and other observations that indicate that polyolefins contain regions with different segmental densities at temperatures above their melting temperatures [1–3]. Although segments of macromolecules in the most compact of these regions are not crystalline, as measured by calorimetry or X-ray diffraction, they would remain close together even when the bulk of the polymer is molten and can reform crystallites very readily when the temperature is lowered. The number of such nuclei that are available for crystal growth is a function of the degree of supercooling of the polymer. Incidentally, this explains why polyethylene has never been observed in the completely amorphous state; even when the melt is quenched in liquid N_2 crystallites will form since they are produced simply by the shrinkage of the polymer volume on cooling. An alternative mechanism that is postulated involves heterogeneous nucleation on adventitious impurities. The nature of such adventitious nuclei has not been clearly established. The growing crystal has to be able to wet its nucleus, and it has been suggested that the surfaces of the effective heterogeneities contain crevices in which crystalline polymer is trapped.

The control of nucleation density can be important in many practical applications. A greater number of nucleation sites results in the formation of more ordered regions, each of which has smaller overall dimensions. The average size of such domains can affect many properties. An example is the transparency of packaging films made from semicrystalline polymers. The refractive indexes of amorphous and crystalline polymer domains differ, and light is refracted at their boundaries. Films will appear hazy if the sizes of regions with different refractive indexes approach the wavelength of light. Nucleating agents are sometimes deliberately added to a polymer to increase the number of nuclei and reduce the dimensions of ordered domains without decreasing the average degree of crystallinity. Such agents are generally solids with colloidal dimensions, like silica and various salts. Sometimes a higher melting semicrystalline polymer will nucleate the crystallization of another polymer. Blending with small concentrations of isotactic polypropylene ($T_m \simeq 176°C$) improves the transparency of sheets and films of polyethylene ($T_m \simeq 115–137°C$), for example.

(ii)   Crystal Lamellae

Once nucleated, crystallization proceeds with the growth of folded chain ribbon-like crystallites called lamellae. The arrangement of polymer chains in the lamellae has some resemblance to that in platelike single crystals which can be produced by precipitating crystallizable polymers from their dilute solutions. In such single crystals the molecules are aligned along the thinnest dimension of the plate. The lengths of extended macromolecules are much greater than the thickness of these crystals and it is evident that a polymer chain must fold outside the plate volume and reenter the crystallite at a different point. When polymer single crystals are carefully prepared, it is found that the dimensions are typically a few microns ($1 \ \mu m = 1$ micron $= 10^{-6}$ m) for the length and breadth and about $0.1 \ \mu m$ for the thickness. The thickness is remarkably constant for a given set of crystallization conditions but increases with the crystallization temperature. Perfect crystallinity is not achieved, because the portions of the chains at the surfaces and in the folds are not completely ordered.

There is uncertainty about the regularity and tightness of the folds in solution-grown single crystals. Three models of chain conformations in a single crystal are illustrated in Fig. 11-4.

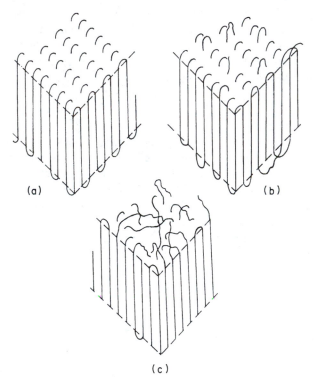

Fig. 11-4.   Possible conformations of polymer chains at the surfaces of chain-folded single crystals. (a) Adjacent reentry model with smooth, regular chain folds, (b) adjacent reentry model with rough fold surface, and (c) random reentry (switchboard) model.

Folded-chain crystals grow by extension of the length and breadth, but not the thickness. The supply of polymer segments is much greater in the melt than in dilute solution, and crystallization in the bulk produces long ribbonlike folded chain structures. These lamellae become twisted and split as a result of local depletion of crystallizable material and growth around defect structures. The regularity of chain folding and reentry is very likely much less under these conditions than in the single crystals produced by slow crystallization from dilute solution.

Another major difference in crystallization from the melt and from dilute solution is that neighboring growing lamellae will generally be close together under the former conditions. Segments of a single molecule are thus likely to be incorporated in different crystallites in bulk crystallized polymer. These "tie molecules" bind the lamellae together and make the resulting structure tough. The number of tie molecules increases with increasing molecular weight and with faster total crystallization rates. The crystallization rate is primarily a function of the extent of supercooling. Cooler crystallization temperatures promote more nuclei but retard the rates of conformational changes required for segmental placement on growing nuclei. (It is observed empirically that the maximum rate of isothermal crystallization occurs at about $0.8T_\mathrm{m}$, where the maximum crystal melting temperature T_m is expressed in K degrees.) The impact resistance and other mechanical characteristics of semicrystalline polymers are dependent on crystallization conditions. The influence of fabrication conditions on the quality of articles is much more pronounced with semicrystalline polymers than with metals or other materials of construction, as a consequence.

(iii)   Morphology of Semicrystalline Polymers

The morphology of a crystallizable polymer is a description of the forms that result from crystallization and the aggregation of crystallites. The various morphological features which occur in bulk crystallized polymers are reviewed in this section.

Crystalline lamellae are the basic units in the microstructures of solid semicrystalline polymers. The lamellae are observed to be organized into two types of larger structural features depending on the conditions of the bulk solidification process.

The major feature of polymers that have been bulk crystallized under quiescent conditions are polycrystalline structures called *spherulites*. These are roughly spherical supercrystalline structures which exhibit Maltese cross-extinction patterns when examined under polarized light in an optical microscope. Spherulites are characteristic of semicrystalline polymers and are also observed in low-molecular-weight materials that have been crystallized from viscous media. Spherulites are aggregates of lamellar crystallites. They are not single crystals and include some

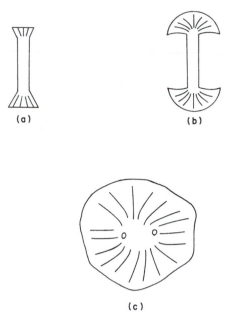

(a)

(b)

(c)

Fig. 11-5.   Successive stages in the development of a spherulite by fanning growth from a nucleus.

disordered material within their boundaries. The sizes of spherulites may vary from somewhat greater than a crystallite to dimensions visible to the naked eye.

A spherulite is built up of lamellar subunits that grow outward from a common nucleus. As this growth advances into the uncrystallized polymer, local inhomogeneities in concentrations of crystallizable segments will be encountered. The folded chain fibril will inevitably twist and branch. At some early stage in its development the spherulite will resemble a sheaf of wheat, as shown schematically in Fig. 11-5a. Branching and fanning out of the growing lamellae tend to create a spherical shape, but neighboring spherulites will impinge on each other in bulk crystallized polymers and prevent the development of true spherical symmetry. The main structural units involved in a spherulite include branched, twisted lamellae with polymer chain directions largely perpendicular to their long axes and interfibrillar material, which is essentially uncrystallized. This is sketched in Fig. 11-6.

The growth of polymer spherulites involves the segregation of noncrystallizable material into the regions between the lamellar ribbons. The components that are not incorporated into the crystallites include additives like oxidation stabilizers, catalyst residues, and so on, as well as comonomer units or branches. The spherulite structures and interspherulitic boundaries are held together primarily by polymer molecules which run between the twisted lamellar subunits and the spherulites themselves. Slow crystallization at low degrees of supercooling

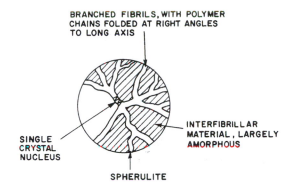

Fig. 11-6.   Basic structure of a polymer spherulite.

produces fewer nuclei and larger spherulites. The polymeric structures produced under such conditions are more likely to be brittle than if they were produced by faster cooling from the melt. This is because there will be fewer interspherulitic tie molecules and because low-molecular-weight uncrystallizable matter will have had more opportunity to diffuse together and produce weak boundaries between spherulites.

The supermolecular structures developed on fast-cooling of crystallizable polymers change with time because of secondary crystallization. A parallel phenomenon is the progressive segregation of mobile uncrystallizable low-molecular-weight material at storage temperatures between T_g and T_m. This will also result in a gradual embrittlement of the matrix polymer. A useful way to estimate whether an additive at a given loading can potentially cause such problems over the lifetime of a finished article is to accelerate the segregation process by deliberately producing some test specimens under conditions that facilitate slow and extensive crystallization.

The type of nucleation that produces spherulitic supercrystalline structures from quiescent melts is not the same as that which occurs more typically in the industrial fabrication of semicrystalline polymer structures. The polymer molecules are under stress as they crystallize in such processes as extrusion, fiber spinning, and injection molding. The orientation of chain segments in flow under stress results in the formation of elongated crystals that are aligned in the flow direction. These are not folded chain crystallites. The overall orientation of the macromolecules in these structures is along the long crystal axis rather than transverse to it as in lamellae produced during static crystallization. Such elongated chain fibrils are probably small in volume, but they serve as a nuclei for the growth of a plurality of folded chain lamellae, which develop with their molecular axes parallel to the parent fibril and their long axes initially at right angles to the long direction of the nucleus. These features are called *row structures*, or *row-nucleated structures*, as distinguished from spherulites. Row-nucleated microstructures are as complex as spherulites and include tie molecules, amorphous regions, and imperfect crystallites. The relative amounts and detailed natures of row-nucleated and spherulitic supercrystalline structures in a particular sample of polymer are determined by the processing conditions used to form the sample. The type or organization that is produced influences many physical properties.

Other supercrystalline structures can be produced under certain conditions. A fibrillar morphology is developed when a crystallizable polymer is stretched at temperatures between T_g and T_m. (This is the orientation operation mentioned on page 15.) Similar fibrillar regions are produced when a spherulitically crystallized specimen is stretched. In both cases, lamellae are broken up into folded-chain blocks that are connected together in microfibrils whose widths are usually between 60 and 200×10^{-8} cm. In each microfibril, folded-chain blocks alternate with amorphous sections that contain chain ends, chain folds, and tie molecules. The tie molecules connecting crystalline blocks along the fiber axis direction are principally responsible for the strength of the structure. Microfibrils of this type make up the structure of oriented, semicrystalline, synthetic fibers. Figure 11-7 is

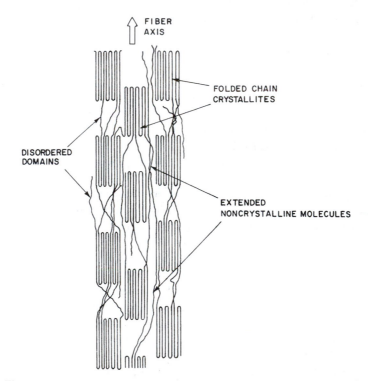

Fig. 11-7.   Schematic representation of structure of a microfibril in an oriented fiber.

a simplified model of such a structure. The gross fiber is made up of interwoven microfibrils that may branch, end and fuse together.

The mechanical properties of polymer crystallites are anisotropic. Strengths and stiffnesses along the molecular axis are those of the covalent bonds in the polymer backbone, but intermolecular cohesive forces in the transverse directions are much weaker. For example, in the chain direction the modulus of polyethylene is theoretically ~ 200 GPa (i.e., 200×10^9 Pa), while the moduli of the crystallites in the two transverse directions are ~ 2 GPa. Oriented extended chain structures are produced by very high orientations. In conventional spinning of semicrystalline fibers or monofilaments (the distinction is primarily in terms of the diameters of these products) the polymer melt is extruded and cooled, so that stretching of the solid polymer results in permanent orientation. The degree to which a high molecular weight polymer can be stretched in such a process is limited by the "natural draw ratio" of the polymer, which occurs because entanglements in the material prevent its extension beyond a certain extent without breaking. These limitations are overcome industrially by so-called "gel spinning." In this operation a mixture of the polymer and diluent is extruded and stretched, the diluent is removed and the product is given a final stretch. Use of a diluent, such as a low-molecular-weight hydrocarbon in the case of polyethylene, facilitates slippage of entanglements and high elongations. Full extension of all the macromolecules in a sample requires that the ratio of the stretched to unstretched fiber lengths (draw ratio) exceed the ratio of contour length to random coil end-to-end distance [Section 4.4.2(i)]. For a polyethylene of molecular weight 10^5 and degree of polymerization about 3600, this ratio would be 60 is the molecules behaved like fully oriented chains. When allowance is made for the effects of fixed bond angles and restricted rotational freedom on the random coil dimensions and the contour length, this ratio is calculated to be about 27. This corresponds more or less to the degrees of stretch that are achieved in the production of "superdrawn fibers" of thermoplastics, although not all the macromolecules need be fully extended to achieve optimum properties in such materials. These products have stiffnesses and tensile strengths that approach those of glass or steel fibers. The crystal superstructures of fibers of the rodlike macromolecules mentioned in Section 4.6 are similar to those of superdrawn thermoplastics. The former do not require high draw ratios to be strong, however, because their molecules are already in a liquid crystalline order even in solution.

During high-speed extrusion processes such as those in fiber and film manufacturing processes, crystallization occurs under high gradients of pressure or temperature. The molecules in the polymer melts become elongated and oriented under these conditions, and this reduces their entropy and hence the entropy change ΔS_m when these molecules crystallize. Since ΔH_m is not affected, the equilibrium crystallization temperature is increased (Eq. 11-1) and nucleation and crystallization start at higher temperatures and proceed faster in such processes than in melts that are cooled under low stress or quiescent conditions.

In addition to the various morphological features listed, intermediate super-molecular structures and mixtures of these entities will be observed. The me-chanical properties of finished articles will depend on the structural state of a semicrystalline polymer, and this in turn is a function of the molecular structure of the polymer and to a significant extent also of the process whereby the object was fabricated.

11.4   THE GLASS TRANSITION

The mechanical properties of amorphous polymers change profoundly as the temperature is decreased through the glass transition region. The corresponding changes in the behavior of semicrystalline polymers are less pronounced in general, although they are also evident.

11.4.1   Modulus–Temperature Relations

At sufficiently low temperatures a polymer will be a hard, brittle material with a modulus greater than 10^9 N m^{-2} (10^{10} dyn/cm^2). This is the glassy region. The tensile modulus is a function of the polymer temperature and is a useful guide to mechanical behavior. Figure 11-8 shows a typical modulus–temperature curve for an amorphous polymer.

In the glassy region the available thermal energy (RT energy units/mol) is insufficient to allow rotation about single bonds in the polymer backbone, and movements of large-scale (about 50 consecutive chain atoms) segments of macro-molecules cannot take place. When a material is stressed, it can respond by de-forming in a nonrecoverable or in an elastic manner. In the former case there must be rearrangements of the positions of whole molecules or segments of molecules that result in the dissipation of the applied work as internal heat. The mechanism whereby the imposed work is absorbed irreversibly involves the flow of sections of macromolecules in the solid specimen. The alternative, elastic response is char-acteristic of glasses, in which the components cannot flow past each other. Such materials usually fracture in a brittle manner at small deformations, because the creation of new surfaces is the only means available for release of the strain energy stored in the solid (window glass is an example). The glass transition region is a temperature range in which the onset of motion on the scale of molecular displace-ments can be detected in a polymer specimen. An experiment will detect evidence of such motion (Section 11.4.4) when the rate of molecular movement is appro-priate to the time scale of the experiment. Since the rate of flow always increases

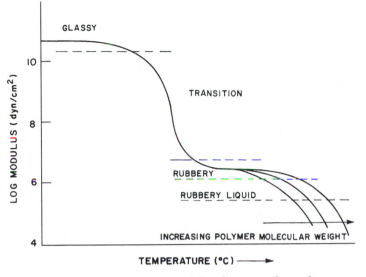

Fig. 11-8.   Modulus–temperature relations for an amorphous polymer.

with temperature, it is not surprising that techniques that stress the specimen more quickly will register higher transition temperatures. For a typical polymer, changing the time scale of loading by a factor of 10 shifts the apparent T_g by about 7°C. In terms of more common experience, a plastic specimen which can be deformed in a ductile manner in a slow bend test may be glassy and brittle if it is struck rapidly at the same temperature.

As the temperature is raised the thermal agitation becomes sufficient for segmental movement and the brittle glass begins to behave in a leathery fashion. The modulus decreases by a factor of about 10^3 over a temperature range of about 10–20°C in the glass-to-rubber transition region.

Let us imagine that measurement of the modulus involves application of a tensile load to the specimen and measurement of the resulting deformation a few seconds after the sample is stressed. In such an experiment a second plateau region will be observed at temperatures greater than T_g. This is the rubbery plateau. In the temperature interval of the rubbery plateau, the segmental displacements that give rise to the glass transition are much faster than the time scale of the modulus measurement, but flow of whole macromolecules is still greatly restricted. Such restrictions can arise from primary chemical bonds as in cross-linked elastomers (Section 4.5.1) or by entanglements with other polymer chains in uncross-linked polymers. Since the number of such entanglements will be greater the higher the molecular weight of the polymer, it can be expected that the temperature range corresponding to the rubbery plateau in uncross-linked polymers will be extended to higher values of T with increasing M. This is shown schematically in Fig. 11-8. A cross-plot of the molecular weight–temperature relation is given in Fig. 11-3a.

The rubbery region is characterized by a short-term elastic response to the application and removal of a stress. This is an entropy-driven elasticity phenomenon of the type described in Chapter 4. Polymer molecules respond to the gross deformation of the specimen by changing to more extended conformations. They do not flow past each other to a significant extent, because their rate of translation is restricted by mutual entanglements. A single, entangled molecule has to drag along its attached neighbors or slip out of its entanglement if it is to flow. The amount of slippage will increase with the duration of the applied stress, and it is observed that the temperature interval of the rubbery plateau is shortened as time between the load application and strain measurement is lengthened. Also, molecular flexibility and mobility increase with temperature, and continued warming of the sample causes the scale of molecular motions to increase in the time scale of the experiment. Whole molecules will begin to slip their entanglements and flow during the several seconds required for this modulus experiment. The sample will flow in a rubbery manner. When the stress is released, the specimen will not contract completely back to its initial dimensions. With higher testing temperatures, the flow rate and the amount of permanent deformation observed will continue to increase.

If the macromolecules in a sample are cross-linked, rather than just entangled, the intermolecular linkages do not slip and the rubbery plateau region persists until the temperature is warm enough to cause chemical degradation of the macromolecules. The effects of cross-linking are illustrated in Fig. 11-9. A lightly cross-linked specimen would correspond to the vulcanized rubber in an automobile tire. The modulus of the material in the rubbery region is shown as increasing with temperature because the rubber is an entropy spring (cf. Fig. 1-3a and Section 4.5.2). The modulus also rises with increased density of cross-linking in accordance with Eq. (4-48). At high cross-link densities, the intermolecular linkages will be spaced so closely as to eliminate the mobility of segments of the size ($\sim$50 main chain bonds) involved in motions that are unlocked in the glass–rubber transition region. Then the material remains glassy at all usage temperatures. Such behavior is typical of tight network structures such as in cured phenolics (Fig. 5-1).

In a solid semicrystalline polymer, large-scale segmental motion occurs only at temperatures between T_g and T_m and only in amorphous regions. At low degrees of crystallinity the crystallites act as virtual cross-links, and the amorphous regions exhibit rubbery or glassy behavior, depending on the temperature and time scale of the experiment. Increasing levels of crystallinity have similar effects to those shown in Fig. 11-9 for variations in cross-link density. Schematic modulus–temperature relations for a semicrystalline polymer are shown in Fig. 11-10. As with moderate cross-linking, the glass transition is essentially unaffected by the

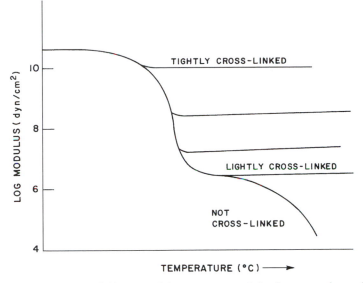

Fig. 11-9.   Effect of cross-linking on modulus–temperature relation for an amorphous polymer.

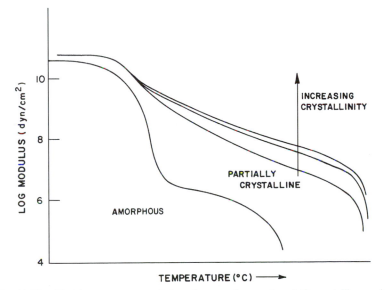

Fig. 11-10.   Modulus–temperature relations for amorphous and partially crystalline versions of the same polymer.

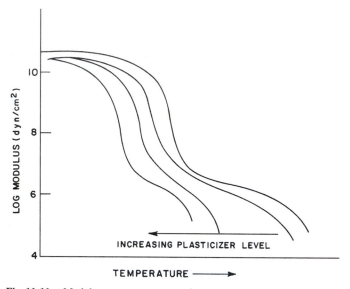

Fig. 11-11. Modulus versus temperature for a plasticized amorphous polymer.

presence of crystallites. At very high crystallinity levels, however, the polymer is very rigid and little segmental motion is possible. In this case the glass transition has little practical significance. It is almost a philosophical question whether a T_g exists in materials like the superdrawn thermoplastic fibers noted on page 393 or the rodlike structures mentioned in Section 4.6.

Modulus–temperature behavior of amorphous polymers is also affected by admixture with plasticizers. These are the soluble diluents described briefly in Section 12.3.2. As shown in Fig. 11-11, the incorporation of a plasticizer reduces T_g and makes the polymer more flexible at any temperature above T_g. In poly(vinyl chloride), for example, T_g can be lowered from about 85°C for unplasticized material to -30°C for blends of the polymer with 50 wt % of dioctyl phthalate plasticizer. A very wide range of mechanical properties can be achieved with this one polymer by variations in the type and concentration of plasticizers.

11.4.2   Effect of Polymer Structure on T_g

Observed T_g's vary from -123°C for poly(dimethyl siloxane) (**1-43**) to 273°C for polyhydantoin (**11-2**) polymers used as wire enamels and to even higher temperatures for other polymers in which the main chain consists largely of aromatic structures. This range of behavior can be rationalized, and the effects of polymer structure on T_g can be predicted qualitatively. Since the glass-to-rubber transition

TABLE 11-2

Glass Transition and Crystal Melting Temperatures of Polymers (°C)

	T_g	$T_m{}^a$		T_g	$T_m{}^a$
Poly(dimethyl siloxane)	−127	—	Poly(ethyl methacrylate)	65	—
Polyethylene	−120[b]	140	Poly(propyl methacrylate)	35	—
Polypropylene(isotactic)	−8	176	Poly(n-butyl methacrylate)	21	—
Poly(1-butene) (isotactic)	−24	132	Poly(n-hexyl methacrylate)	−5	—
Polyisobutene	−73	—	Poly(phenyl methacrylate)	110	—
Poly(4-methyl 1-penetene) (isotactic)	29	250	Poly(acrylic acid)	106	—
cis-1,4-Polybutadiene	−102	—	Polyacrylonitrile	97	—
trans-1,4-Polybutadiene	−58[b]	96[b]	Poly(vinyl chloride) (conventional)	87	—
cis-1,4-Polyisoprene	−73	—	Poly(vinyl fluoride)	41	200
Polyformaldehyde	−82[b]	175	Poly(vinylidene chloride)	−18	200
Polystyrene (atactic)	100	—	Poly(vinyl acetate)	32	—
Poly(alpha-methyl styrene)	168	—	Poly(vinyl alcohol)	85	—
Poly(methyl acrylate)	10	—	Polycarbonate of bisphenol A	157	—
Poly(ethyl acrylate)	−24	—	Poly(ethylene terephthalate) (unoriented)	69	267
Poly(propyl acrylate)	−37	—	Nylon-6,6 (unoriented)	50[b]	265
Poly(phenyl acrylate)	57	—	Poly(p-xylene)	—	375
Poly(methyl methacrylate) (atactic)	105	—			

^{a} T_m is not listed for vinyl polymers in which the most common forms are atactic nor for elastomers, which are not crystalline in the unstretched state.

^{b} Conflicting data are reported.

reflects the onset of movements of sizable segments of the polymer backbone, it is reasonable to expect that T_g will be affected by the flexibility of the macro-molecules and by the intensities of intermolecular forces. Table 11-2 lists T_g and T_m values for a number of polymers. The relations between intra- and interchain features of the macromolecular structure and T_g are summarized in the following paragraphs.

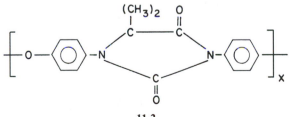

11-2

The kinetic flexibility of a macromolecule is directly related to the ease with which conformational changes between *trans* and *gauche* states can take place. The lower the energy barrier ΔE in Fig. 4-2, the greater the ease of rotation about main chain bonds. Polymers with low chain stiffnesses will have low T_g's in the absence of complications from interchain forces. Chain backbones with

$$-\overset{|}{\underset{|}{Si}}-O- \text{ or } -\overset{|}{\underset{|}{C}}-O-$$

bonds tend to be flexible and have low glass transitions. Insertion of an aromatic ring in the main chain causes an increase in T_g, and this is of importance in the application of amorphous glassy polymers like poly(phenylene oxide) (**1-14**) and polycarbonate (**1-52**).

Bulky, inflexible substituents on chain carbons impede rotations about single bonds in the main chain and raise T_g. Thus the T_g of polypropylene and poly(methyl methacrylate) are respectively higher than those of polyethylene and poly(methyl acrylate). However, the size of the substituent is not directly related to T_g; a flexible side group like an alkyl chain lowers T_g because a segment containing the substituent can move through a smaller unoccupied volume in the solid than one in which the pendant group has more rigid steric requirements. Larger substituents prevent efficient packing of macromolecules in the absence of crystallization, but motion of the polymer chain is freed only if the substituent itself can change its conformation readily. The interplay of these two influences is shown in Table 11-2 for the methacrylate polymers.

Stronger intermolecular attractive forces pull the chains together and hinder relative motions of segments of different macromolecules. Polar polymers and those in which hydrogen bonding or other specific interactions are important therefore have high T_g. Glass transition temperatures are in this order: polyacrylonitrile > poly(vinyl alcohol) > poly(vinyl acetate) > polypropylene.

Polymers of vinylidene monomers (1,1-disubstituted ethylenes) have lower T_g's than the corresponding vinyl polymers. Polyisobutene and polypropylene comprise such a pair and so do poly(vinylidene chloride) and poly(vinyl chloride). Symmetrical disubstituted polymers have lower T_g's than the monosubstituted macromolecules because no conformation is an appreciably lower energy form than any other (cf. the discussion of polyisobutene in Section 4.3).

For a given polymer type, T_g increases with number average molecular weight according to

$$T_g = T_g^\infty - u/\overline{M}_n \tag{11-4}$$

where T_g^∞ is the glass-to-rubber transition temperature of an infinitely long polymer chain and u is a constant that depends on the polymer. Observed T_g's level off within experimental uncertainty at a degree of polymerization between 500 and 1000, for vinyl polymers. Thus T_g is 88°C for polystyrene with $\overline{M}_n = 10{,}000$ and 100°C for the same polymer with $\overline{M}_n > 50{,}000$.

Cross-linking increases the glass transition temperature of a polymer when the average size of the segments between cross-links is the same or less than the lengths of the main chain that can start to move at temperatures near T_g. The glass transition temperature changes little with degree of cross-linking when the cross-links are widely spaced, as they are in normal vulcanized rubber. Large shifts of T_g with increased cross-linking are observed, however, in polymers that are already highly cross-linked, as in the "cure" of epoxy [p. 11] and phenolic (Fig. 5-1) thermosetting resins.

The glass transition temperature of miscible polymer mixtures can be calculated from

$$\frac{1}{T_g} \simeq \frac{w_A}{T_{g_A}} + \frac{w_B}{T_{g_B}} \tag{11-5}$$

where T_{g_i} and w_i are the glass temperature (in K) and weight fraction of component i of the compound. This equation is useful with plasticizers (Section 12.3.2) which are materials that enhance the flexibility of the polymer with which they are mixed. The T_g values of plasticizers themselves are most effectively estimated by using Eq. (11-5) with two plasticized mixtures of known compositions and measured T_g's. The foregoing equation cannot be applied to polymer blends in which the components are not mutually soluble, because each ingredient will exhibit its characteristic T_g in such mixtures. The existence of a single glass temperature is in fact a widely used criterion for miscibility in such materials (Section 12.1).

Equation (11-5) is also a useful guide to the glass transition temperatures of statistical copolymers. In that case T_{g_A} and T_{g_B} refer to the glass temperatures of the corresponding homopolymers. It will not apply, however, to block and graft copolymers in which a separate T_g will be observed for each component polymer if the blocks or branches are long enough to permit each homopolymer type to segregate into its own region. This separation into different domains is necessary for the use of styrene–butadiene block polymers as thermoplastic rubbers.

11.4.3   Correlations between T_m and T_g

A rough correlation exists between T_g and T_m for crystallizable polymers, although the molecular mechanisms that underly both transitions differ. Any structural feature that enhances chain stiffness will raise T_g, since this is the temperature needed for the onset of large-scale segmental motion. Stronger intermolecular forces will also produce higher T_g's. These same factors increase T_m, as described on page 382, in connection with Eq. (11-1).

Statistical copolymers of the types described in Chapter 8 tend to have broader glass transition regions than homopolymers. The two comonomers usually do not fit into a common crystal lattice and the melting points of copolymers will be lower and their melting ranges will be broader, if they crystallize at all. Branched and linear polyethylene provide a case in point since the branched polymer can be regarded as a copolymer of ethylene and higher 1-olefins.

11.4.4   Measurement of T_g

Glass transition temperatures can be measured by many techniques. Not all methods will yield the same value because this transition is rate dependent. Polymer segments will respond to an applied stress by flowing past each other if the sample is deformed slowly enough to allow such movements to take place at the experimental temperature. Such deformation will not be recovered when the stress is released if the experiment has been performed above T_g. If the rate at which the specimen is deformed in a particular experiment is too fast to allow the macromolecular segments to respond by flowing, the polymer will be observed to be glassy. It will either break before the test is completed or recover its original dimensions when the stress is removed. In either event, the experimental temperature will have been indicated to be below T_g. As a consequence, observed glass transition temperatures vary directly with the rates of the experiments in which they are measured.

The T_g values quoted in Table 11-2 are either measured by very slow rate methods or are obtained by extrapolating the data from faster, nonequilibrium techniques to zero rates. This is a fairly common practice, in order that the glass transition temperature can be considered as characteristic only of the polymer and not of measuring method.

Many relatively slow or static methods have been used to measure T_g. These include techniques for determining the density or specific volume of the polymer as a function of temperature (cf. Fig. 11-1) as well as measurements of refractive index, elastic modulus, and other properties. Differential thermal analysis and differential scanning calorimetry are widely used for this purpose at present, with simple extrapolative corrections for the effects of heating or cooling rates on the observed values of T_g. These two methods reflect the changes in specific heat of the polymer at the glass-to-rubber transition. Dynamic mechanical measurements, which are described in Section 11.5, are also widely employed for locating T_g.

In addition, there are many related industrial measurements based on softening point, hardness, stiffness, or deflection under load while the temperature is being varied at a stipulated rate. No attempt is usually made to compensate for heating rate in these methods, which yield transition temperatures about 10–20° higher than those from the other procedures mentioned. Some technical literature that is used for design with plastics quotes brittleness temperatures rather than T_g. The former is usually that temperature at which half the specimens tested break in a specified impact test. It depends on the polymer and also on the nature of the impact, sample thickness, presence or absence of notches, and so on. Since the measured brittleness temperature is influenced very strongly by experimental conditions, it cannot be expected to correlate closely with T_g or even with the impact behavior of polymeric articles under service conditions which may differ widely from those of the brittleness test method.

Heat distortion temperatures (HDTs) are widely used as design criteria for polymeric articles. These are temperatures at which specimens with particular dimensions distort a given amount under specified loads loads and deformations. Various test methods, such as ASTM D648, are described in standards compilations. Because of the stress applied during the test, the HDT of a polymer is invariably higher than its T_g.

11.5   POLYMER VISCOELASTICITY

An ideal elastic material is one which exhibits no time effects. When a stress σ is applied the body deforms immediately to a strain ϵ. (These terms were defined broadly in Section 1.8). The sample recovers its original dimensions completely and instantanously when the stress is removed. Further, the strain is always proportional to the stress and is independent of the rate at which the body is deformed:

$$\sigma = Y\epsilon \tag{11-6}$$

where Y is Young's modulus if the deformation mode is a tensile stretch and Eq. (11-6) is an expression of the familiar Hooke's law. The changes in the shape of an isotropic, perfectly elastic material will always be proportional to the magnitude of the applied stress if the body is twisted, sheared, or compressed instead of extended, but the particular stress/strain (modulus) will differ from Y. Figure 11-12 summarizes the concepts and symbols for the elastic constants in tensile, shear, and bulk deformations.

An experiment such as that in Fig. 11-12a can produce changes in the volume as well as the shape of the test specimen. The elastic moduli listed in this figure are related by Poisson's ratio β, which is a measure of the lateral contraction accompanying a longitudinal extension:

$$\beta = \frac{1}{2}[1 - (1/V)\partial V/\partial \epsilon] \tag{11-7}$$

where V is the volume of the sample. When there is no significant volume change, $\partial V/\partial \epsilon = 0$ and $\beta = 0.5$. This behavior is characteristic of ideal rubbers. Real solids dilate when extended, and values of β down to about 0.2 are observed for rigid,

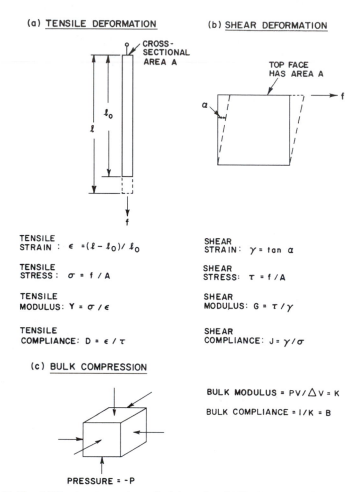

Fig. 11-12.   (a) Elastic constants in tensile deformation. (b) Elastic constants in shear deformation.
(c) Elastic constants in volume deformation.

brittle materials. The moduli in the elastic behavior of isotropic solids are related
by

$$Y = 2G(1 + \beta) = 3K(1 - 2\beta) \qquad (11\text{-}8)$$

At very low extensions when there is no significant amount of permanent deformation Y/G is between about 2.5 for rigid solids and 3 for elastomeric materials.

An ideal, Newtonian fluid was described on page 433. Such a material has no
elastic character; it cannot support a strain and the instantaneous response to a

shearing stress τ is viscous flow:

$$\tau = \eta\dot{\gamma} \tag{11-9}$$

Here $\dot{\gamma}$ is the shear rate or velocity gradient ($=d\gamma/dt$) and η is the viscosity which was first defined in connection with Eq. (3-32).

Polymeric (and other) solids and liquids are intermediate in behavior between Hookean, elastic solids, and Newtonian, purely viscous fluids. They often exhibit elements of both types of response, depending on the time scale of the experiment. Application of stresses for relatively long times may cause some flow and permanent deformation in solid polymers while rapid shearing will induce elastic behavior in some macromolecular liquids. It is also frequently observed that the value of a measured modulus or viscosity is time dependent and reflects the manner in which the measuring experiment was performed. These phenomena are examples of *viscoelastic* behavior.

Three types of experiments are used in the study of viscoelasticity. These involve creep, stress relaxation, and dynamic techniques. In creep studies a body is subjected to a constant stress and the sample dimensions are monitored as a function of time. When the polymer is first loaded an immediate deformation occurs, followed by progressively slower dimensional changes as the sample creeps towards a limiting shape. Figure 1-3 shows examples of the different behaviors observed in such experiments.

Stress relaxation is an alternative procedure. Here an instantaneous, fixed deformation is imposed on a sample, and the stress decay is followed with time. A very useful modification of these two basic techniques involves the use of a periodically varying stress or deformation instead of a constant load or strain. The dynamic responses of the body are measured under such conditions.

11.5.1 Phenomenological Viscoelasticity

Consider the tensile experiment of Fig. 11-12a as a creep study in which a steady stress τ_0 is suddenly applied to the polymer specimen. In general, the resulting strain $\epsilon(t)$ will be a function of time starting from the imposition of the load. The results of creep experiments are often expressed in terms of compliances rather than moduli. The tensile creep compliance $D(t)$ is

$$D(t) = \epsilon(t)/\sigma_0 \tag{11-10}$$

The shear creep compliance $J(t)$ (see Fig. 11-12b) is similarly defined as

$$J(t) = \gamma(t)/\tau_0 \tag{11-11}$$

where τ_0 is the constant shear stress and $\gamma(t)$ is the resulting time-dependent strain.

Stress relaxation experiments correspond to the situations in which the deformations sketched in Fig. 11-12 are imposed suddenly and held fixed while the resulting stresses are followed with time. The tensile relaxation modulus $Y(t)$ is then obtained as

$$Y(t) = \sigma(t)/\epsilon_0 \tag{11-12}$$

with ϵ_0 being the constant strain. Similarly, a shear relaxation experiment measures the shear relaxation modulus $G(t)$:

$$G(t) = \tau(t)/\gamma_0 \tag{11-13}$$

where γ_0 is the constant strain.

Although a compliance is the inverse of a modulus for an ideal elastic body this is not true for viscoelastic materials. That is,

$$Y(t) = \sigma(t)/\epsilon_0 \neq \epsilon(t)/\sigma_0 = D(t) \tag{11-14}$$

and

$$G(t) = \tau(t)/\gamma_0 \neq \gamma(t)/\tau_0 = J(t) \tag{11-15}$$

Consider two experiments carried out with identical samples of a viscoelastic material. In experiment (a) the sample is subjected to a stress σ_1 for a time t. The resulting strain at t is ϵ_1, and the creep compliance measured at that time is $D_1(t) = \epsilon_1/\sigma_1$. In experiment (b) a sample is stressed to a level σ_2 such that strain ϵ_1 is achieved immediately. The stress is then gradually decreased so that the strain remains at ϵ_1 for time t (i.e., the sample does not creep further). The stress on the material at time t will be σ_3, and the corresponding relaxation modulus will be $Y_2(t) = \sigma_3/\epsilon_1$. In measurements of this type, it can be expected that $\sigma_2 > \sigma_1 > \sigma_3$ and $Y(t) \neq (D(t))^{-1}$, as indicated in Eq. (11-14). $G(t)$ and $Y(t)$ are obtained directly only from stress relaxation measurements, while $D(t)$ and $J(t)$ require creep experiments for their direct observation. These various parameters can be related in the linear viscoelastic region described in Section 11.5.2.

(i)   Terminology of Dynamic Mechanical Experiments

A complete description of the viscoelastic properties of a material requires information over very long times. Creep and stress relaxation measurements are limited by inertial and experimental limitations at short times and by the patience of the investigator and structural changes in the test material at very long times. To supplement these methods, the stress or the strain can be varied sinusoidally in a dynamic mechanical experiment. The frequency of this alternation is ν cycles/s or $\omega(= 2\pi\nu)$ rad/s. An alternating experiment at frequency ω is qualitatively equivalent to a creep or stress relaxation measurement at a time $t = (1/\omega)$ sec.

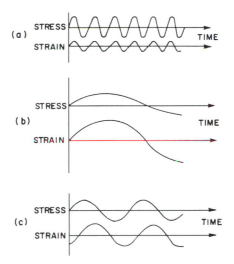

Fig. 11-13. Effect of frequency on dynamic response of an amorphous, lightly cross-linked polymer: (a) elastic behavior at high frequency—stress and strain are in phase, (b) liquid-like behavior at low frequency—stress and strain are 90° out of phase, and (c) general case—stress and strain are out of phase.

In a dynamic experiment, the stress will be directly proportional to the strain if the magnitude of the strain is small enough. Then, if the stress is applied sinusoidally the resulting strain will also vary sinusoidally. In special cases the stress and the strain will be in phase. A cross-linked, amorphous polymer, for example, will behave elastically at sufficiently high frequencies. This is the situation depicted in Fig. 11-13a where the stress and strain are in phase and the strain is small. At sufficiently low frequencies, the strain will be 90° out of phase with the stress as shown in Fig. 11-13c. In the general case, however, stress and strain will be out of phase (Fig. 11-13b).

In the last instance, the stress can be factored into two components one of which is in phase with the strain and the other which leads the strain by $\pi/2$ rad. (Alternatively, the strain could be decomposed into a component in phase with the stress and one which lagged behind the stress by 90°.) This is accomplished by use of a rotating vector scheme, as shown in Fig. 11-14. The magnitude of the stress at any time is represented by the projection OC of the vector **OA** on the vertical axis. Vector **OA** rotates counterclockwise in this representation with a frequency ω equal to that of the sinusoidally varying stress. The length of **OA** is the stress amplitude (maximum stress) involved in the experiment. The strain is represented by the projection OD of vector **OB** on the vertical axis. The strain vector **OB** also rotates counterclockwise with frequency ω but it lags **OA** by an angle δ. The *loss tangent* is defined as tan δ.

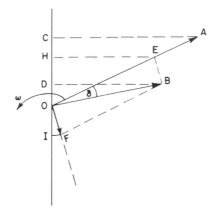

Fig. 11-14. Decomposition of strain vector into two components in a dynamic experiment.

The strain vector **OB** can be resolved into vector **OE** along the direction of **OA** and **OF** perpendicular to **OA**. Then the projection OH of **OE** on the vertical axis is the magnitude of the strain which is in phase with the stress at any time. Similarly, projection OI of vector **OF** is the magnitude of the strain which is 90° (one-quarter cycle) out of phase with the stress. The stress can be similarly resolved into two components with one along the direction of **OB** and one leading the strain vector by $\pi/2$ rad.

When the stress is decomposed into two components the ratio of the in-phase stress to the strain amplitude (γ_a, maximum strain) is called the *storage modulus*. This quantity is labeled $G'(\omega)$ in a shear deformation experiment. The ratio of the out-of-phase stress to the strain amplitude is the loss modulus $G''(\omega)$. Alternatively, if the strain vector is resolved into its components, the ratio of the in-phase strain to the stress amplitude τ_a is the storage compliance $J'(\omega)$, and the ratio of the out-of-phase strain to the stress amplitude is the loss compliance $J''(\omega)$. $G'(\omega)$ and $J'(\omega)$ are associated with the periodic storage and complete release of energy in the sinusoidal deformation process. The loss parameters $G''(\omega)$ and $J''(\omega)$ on the other hand reflect the nonrecoverable use of applied mechanical energy to cause flow in the specimen. At a specified frequency and temperature, the dynamic response of a polymer can be summarized by any one of the following pairs of parameters: $G'(\omega)$ and $G''(\omega)$, $J'(\omega)$ and $J''(\omega)$, or τ_a/γ_a (the absolute modulus $|G|$) and $\tan\delta$.

An alternative set of terms is best introduced by noting that a complex number can be represented as in Fig. 11-15 by a point P (with coordinates x and y) or by a vector **OP** in a plane. Since dynamic mechanical behavior can be represented by a rotating vector in Fig. 11-13, this vector and hence the dynamic mechanical response is equivalent to a single complex quantity such as G^* (complex modulus)

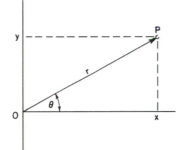

Fig. 11-15.   Representation of a complex number $z = x + iy$ by a vector on the xy plane ($i = \sqrt{-1}$). $z = x + iy = re^{i\theta} = r(\cos\theta + i\sin\theta)$; $r = |z| = |x + y| = (x^2 + y^2)^{1/2}$; $1/z = e^{i\theta}/r = (1/r) \cdot (\cos\theta - i\sin\theta)$.

or J^* (complex compliance). Thus, in shear deformation,

$$G^*(\omega) = G'(\omega) + iG''(\omega) \qquad (11\text{-}16)$$

$$J^*(\omega) = \frac{1}{G^*(\omega)} = \frac{1}{G'(\omega) + G''(\omega)} = J'(\omega) - iJ''(\omega) \qquad (11\text{-}17)$$

[Equation (11-17) can be derived from Eq. (11-16) by comparing the expressions for z and z^{-1} in Fig. 11-15.] It will also be apparent that

$$|G^*| = [(G')^2 + (G'')^2]^{1/2} \qquad (11\text{-}18)$$

and

$$\tan\delta = G''(\omega)/G'(\omega) = J''(\omega)/J'(\omega) \qquad (11\text{-}19)$$

The real and imaginary parts of the complex numbers used here have no physical significance. This is simply a convenient way to represent the component vectors of stress and strain in a dynamic mechanical experiment.

Tan δ measures the ratio of the work dissipated as heat to the maximum energy stored in the specimen during one cycle of a periodic deformation. The conversion of applied work to thermal energy in the sample is called *damping*. It occurs because of flow of macromolecular segments past each other in the sample. The energy dissipated per cycle due to such viscoelastic losses is $\pi \gamma_a^2 G''$.

For low strains and damping the dynamic modulus G' will have the same magnitude as that obtained from other methods like stress relaxation or tensile tests, provided the time scales are similar in these experiments.

Viscosity is the ratio of a stress to a strain rate [Eq. (11-9)]. Since the complex modulus G^* has the units of stress, it is possible to define a complex viscosity η^* as the ratio of G^* to a complex rate of strain:

$$\eta^*(\omega) = \frac{G^*(\omega)}{i\omega} = \frac{G'(\omega) + iG''(\omega)}{i\omega} = \eta'(\omega) - i\eta''(\omega) \qquad (11\text{-}20)$$

Then it follows that

$$\eta'(\omega) = G''(\omega)/\omega \qquad (11\text{-}21)$$

and

$$\eta''(\omega) = G'(\omega)/\omega \qquad (11\text{-}22)$$

The $\eta'(\omega)$ term is often called the *dynamic viscosity*. It is an energetic dissipation term related to $G''(\omega)$ and has a value approaching that of the steady flow viscosity η in very low frequency measurements on polymers which are not cross-linked.

11.5.2   Linear Viscoelasticity

In linear viscoelastic behavior the stress and strain both vary sinusoidally, although they may not be in phase with each other. Also, the stress amplitude is linearly proportional to the strain amplitude at given temperature and frequency. Then mechanical responses observed under different test conditions can be interrelated readily. The behavior of a material in one condition can be predicted from measurement made under different circumstances.

Linear viscoelastic behavior is actually observed with polymers only in very restricted circumstances involving homogeneous, isotropic, amorphous specimens subjected to small strains at temperatures near or above T_g and under test conditions that are far removed from those in which the sample may be broken. Linear viscoelasticity theory is of limited use in predicting service behavior of polymeric articles, because such applications often involve large strains, anisotropic objects, fracture phenomena, and other effects which result in nonlinear behavior. The theory is nevertheless valuable as a reference frame for a wide range of applications, just as the thermodynamic equations for ideal solutions help organize the observed behavior of real solutions.

The major features of linear viscoelastic behavior that will be reviewed here are the superposition principle and time–temperature equivalence. Where they are valid, both make it possible to calculate the mechanical response of a material under a wide range of conditions from a limited store of experimental information.

(i)   Boltzmann Superposition Principle

The Boltzmann principle states that the effects of mechanical history of a sample are linearly additive. This applies when the stress depends on the strain or rate of strain or alternatively, where the strain is considered a function of the stress or rate of change of stress.

In a tensile test, for example, Eq. (11-10) relates the strain and stress in a creep experiment when the stress τ_0 is applied instantaneously at time zero. If this loading were followed by application of a stress σ_1 at time u_1, then the time-dependent strain resulting from this event alone would be

$$\epsilon(t) = \sigma_1 D(t - u_1) \tag{11-23}$$

The total strain from the imposition of stress σ_0 at $t = 0$ and σ_1 at $t = u_1$ is

$$\epsilon(t) = \sigma_0 D(t) + \sigma_1 D(t - u_1) \tag{11-24}$$

In general, for an experiment in which stresses $\sigma_0, \sigma_1, \sigma_2, \ldots, \sigma_n$ were applied at times $t = 0, u_1, u_2, \ldots, u_n$,

$$\epsilon(t) = \sigma_0 D(t) + \sum_{i=1}^{n} \sigma_n D(t - u_n) \tag{11-25}$$

If the loaded specimen is allowed to elongate for some time and the stress is then removed, creep recovery will be observed. An uncross-linked amorphous polymer approximates a highly viscous fluid in such a mechanical test. Hence the elongation-time curve of Fig. 1-3c is fitted by an equation of the form

$$\epsilon(t) = \sigma_0 [D(t) - D(t - u_1)] \tag{11-26}$$

Here a stress σ_0 is applied at $t = 0$ and removed at $t = u_1$. (This is equivalent to the application of an additional stress equal to $-\sigma_0$.)

(ii)   Use of Mechanical Models

Equation (11-6) summarized purely elastic response in tension. The analogous expression for shear deformation (Fig. 11-12) is

$$\tau = G\gamma \tag{11-27}$$

This equation can be combined conceptually with the viscous behavior of Eq. (11-9) in either of two ways. If the stresses causing elastic extension and viscous flow are considered to be additive, then

$$\tau = \tau_{\text{elastic}} + \tau_{\text{viscous}} = G\gamma + \eta \, d\gamma/dt \tag{11-28}$$

A mechanical model for such response would include a parallel arrangement of a spring for elastic behavior and a dashpot for the viscous component. (A dashpot is a piston inside a container filled with a viscous liquid.) This model, shown in Fig. 11-16a, is called a Kelvin or a Voigt element. When a force is applied across such a model, the stress is divided between the two components and the elongation of each is equal.

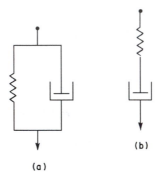

(b)

(a)

Fig. 11-16. Simple mechanical models of viscoelastic behavior. (a) Voigt or Kelvin element and (b) Maxwell element.

Another way to combine the responses of Eqs. (11-9) and (11-27) is to add the strains. Then

$$\gamma = \gamma_{\text{elastic}} + \gamma_{\text{viscous}}$$

$$\frac{d\gamma}{dt} = \frac{d\gamma_{\text{elastic}}}{dt} + \frac{d\gamma_{\text{viscous}}}{dt} = \frac{1}{G}\frac{d\tau}{dt} + \frac{\tau}{\eta} \qquad (11\text{-}29)$$

The mechanical analog for this behavior is a spring and dashpot in series. This body, called a Maxwell element, is shown in Fig. 11-16b.

Mechanical models are useful tools for selecting appropriate mathematical functions to describe particular phenomena. The models have no physical relation to real materials, and it should be realized that an infinite number of different models can be used to represent a given phenomenon. Two models are mentioned here to introduce the reader to such concepts, which are widely used in studies of viscoelastic behavior.

The Maxwell body is appropriate for the description of stress relaxation, while the Voigt element is more suitable for creep deformation. In a stress relaxation experiment, a strain γ_0 is imposed at $t = 0$ and held constant thereafter $(d\gamma/dt = 0)$ while τ is monitored as a function of t. Under these conditions, Eq. (11-29) for a Maxwell body behavior becomes

$$0 = \frac{1}{G}\frac{d\tau}{dt} + \frac{\tau}{\eta} \qquad (11\text{-}30)$$

This is a first-order homogeneous differential equation and its solution is

$$\tau = \tau_0 \exp(-Gt/\eta) \qquad (11\text{-}31)$$

where τ_0 is the initial value of stress at $\gamma = \gamma_0$.

Another way of writing Eq. (11-29) is

$$\frac{d\gamma}{dt} = \frac{1}{G}\frac{d\tau}{dt} + \frac{\tau}{\zeta G} \qquad (11\text{-}32)$$

where ζ (zeta) is a relaxation time defined as

$$\zeta \equiv \eta/G \tag{11-33}$$

An alternative form of Eq. (11-31) is then

$$\tau = \tau_0 \exp(-t/\zeta) \tag{11-34}$$

The relaxation time is the time needed for the initial stress to decay to $1/e$ of its initial value.

If a constant stress τ_0 were applied to a Maxwell element, the strain would be

$$\gamma = \tau_0/G + \tau_0 t/\eta \tag{11-35}$$

This equation is derived by integrating Eq. (11-29) with boundary condition $\gamma = 0$, $\tau = \tau_0$ at $t = 0$. Although the model has some elastic character the viscous response dominates at all but short times. For this reason, the element is known as a *Maxwell fluid*.

A simple creep experiment involves application of a stress τ_0 at time $t = 0$ and measurement of the strain while the stress is held constant. The Voigt model (Eq. 11-28) is then

$$\tau_0 = G\gamma + \eta \, d\gamma/dt \tag{11-36}$$

or

$$\frac{\tau_0}{\eta} = \frac{G\gamma}{\eta} + \frac{d\gamma}{dt} = \frac{\gamma}{\zeta} + \frac{d\gamma}{dt} \tag{11-37}$$

where $\zeta = G/\eta$ is called a *retardation time* in a creep experiment. Equation (11-37) can be made exact by using the multiplying factor $e^{t/\zeta}$. Integration from $\tau = \tau_0$, $\gamma = 0$ at $t = 0$ gives

$$G\gamma/\tau_0 = 1 - \exp(-Gt/\eta) = 1 - \exp(-t/\zeta) \tag{11-38}$$

If the creep experiment is extended to infinite times, the strain in this element does not grow indefinitely but approaches an asymptotic value equal to τ_0/G. This is almost the behavior of an ideal elastic solid as described in Eq. (11-6) or (11-27). The difference is that the strain does not assume its final value immediately on imposition of the stress but approaches its limiting value gradually. This mechanical model exhibits delayed elasticity and is sometimes known as a *Kelvin solid*. Similarly, in creep recovery the Maxwell body will retract instantaneously, but not completely, whereas the Voigt model recovery is gradual but complete.

Neither simple mechanical model approximates the behavior of real polymeric materials very well. The Kelvin element does not display stress relaxation under constant strain conditions and the Maxwell model does not exhibit full recovery of strain when the stress is removed. A combination of the two mechanical models can be used, however, to represent both the creep and stress relaxation behaviors of polymers. This is the standard linear solid, or Zener model, comprising either a spring in series with a Kelvin element or a spring in parallel with a Maxwell model. Details of this construction are outside the scope of this introductory text.

Limitations to the effectiveness of mechanical models occur because actual polymers are characterized by many relaxation times instead of single values and because use of the models mentioned assumes linear viscoelastic behavior which is observed only at small levels of stress and strain. The linear elements are nevertheless useful in constructing appropriate mathematical expressions for viscoelastic behavior and for understanding such phenomena.

(iii) Time–Temperature Correspondence

The left-hand panel of Fig. 11-17 contains sketches of typical stress relaxation curves for an amorphous polymer at a fixed initial strain and a series of temperatures. Such data can be obtained much more conveniently than those in the experiment summarized in Fig. 11-8, where the modulus was measured at a given time and a series of temperatures. It is found that the stress relaxation curves can be caused to coincide by shifting them along the time axis. This is shown in the right-hand panel of Fig. 11-17 where all the curves except that for temperature T_8 have been shifted horizontally to form a continuous "master curve" at temperature T_8. The glass transition temperature is shown here to be T_5 at a time of 10^{-2} min. The polymer behaves in a glassy manner at this temperature when a strain is imposed within 10^{-2} min or less.

Similar curves can be constructed for creep or dynamic mechanical test data of amorphous polymers. Because of the equivalence of time and temperature,

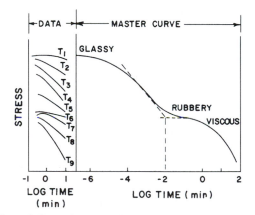

Fig. 11-17. Left panel: Stress decay at various temperatures $T_1 < T_2 < \cdots < T_9$. Right panel: Master curve for stress decay at temperature T_8.

the temperature scale in dynamic mechanical experiments can be replaced by an inverse log frequency scale.

Master curves permit the evaluation of mechanical responses at very long times by increasing the test temperature instead of prolonging the experiment. A complete picture of the behavior of the material is obtained in principle by operating in experimentally accessible time scales and varying temperatures.

Time–temperature superposition can be expressed mathematically as

$$G(T_1, t) = G(T_2, t/a_T) \tag{11-39}$$

for a shear stress relaxation experiment. The effect of changing the temperature is the same as multiplying the time scale by shift factor a_T. A minor correction is required to the formulation of Eq. (11-39) to make the procedure complete. The elastic modulus of a rubber is proportional to the absolute temperature T and to the density ρ of the material, as summarized in Eq. (4-48). It is therefore proper to divide through by T and ρ to compensate for these effects of changing the test temperature. The final expression is then

$$\frac{G(T_1, t)}{\rho(T_1)T_1} = \frac{G(T_2, t/a_T)}{\rho(T_2)T_2} \tag{11-40}$$

If a compliance were being measured at a series of temperatures T, the data could be reduced to a reference temperature T by

$$J(T_1, t) = \frac{\rho(T)T}{\rho(T_1)T_1} J\left(\frac{T, t}{a_T}\right) \tag{11-41}$$

where $\rho(T_1)$ is the material density at temperature T_1.

It is common practice now to use the glass transition temperature measured by a very slow rate method as the reference temperature for master curve construction. Then the shift factor for most amorphous polymers is given fairly well by

$$\log_{10} a_T = -\frac{C_1(T - T_g)}{C_2 + T - T_g} \tag{11-42}$$

where the temperatures are in Kelvin degrees. Equation (11-42) is known as the WLF equation, after the initials of the researchers who proposed it [4]. The constants C_1 and C_2 depend on the material. "Universal" values are $C_1 = 17.4$ and $C_2 = 51.6°C$. The expression given holds between T_g and $T_g + 100°C$. If a different reference temperature is chosen an equation with the same form as (11-42) can be used, but the constants on the right-hand side must be reevaluated.

Accumulation of long-term data for design with plastics can be very inconvenient and expensive. The equivalence of time and temperature allows information about mechanical behavior at one temperature to be extended to longer times by using data from shorter time studies at higher temperature. It should be used with caution, however, because the increase of temperature may promote changes in the material, such as crystallization or relaxation of fabrication stresses, that affect mechanical behavior in an irreversible and unexpected manner. Note also that the master curve in the previous figure is a semilog representation. Data such as those in the left-hand panel are usually readily shifted into a common relation but it is not always easy to recover accurate stress level values from the master curve when the time scale is so compressed.

The following simple calculation illustrates the very significant temperature and time dependence of viscoelastic properties of polymers. It serves as a convenient, but less accurate, substitute for the accumulation of the large amount of data needed for generation of master curves. Suppose that a value is needed for the compliance (or modulus) of a plastic article for 10 years service at 25°C. What measurement time at 80°C will produce an equivalent figure? We rely here on the use of a shift factor, a_T, and Eq. (11-39). Assume that the temperature dependence of the shift factor can be approximated by an Arrhenius expression of the form

$$a_T = \exp \frac{\Delta H}{R} \left[\frac{1}{T} - \frac{1}{T_0} \right]$$

where the activation energy, ΔH, may be taken as 0.12 MJ/mol, which is a typical value for relaxations in semicrystalline polymers and in glassy polymers at temperatures below T_g. (The shift factor could also have been calculated from the WLF relation if the temperatures had been around T_g of the polymer.) In the present case:

$$a_T = \exp \left[\frac{0.12 \times 10^6}{8.31} \right] \left[\frac{1}{353} - \frac{1}{298} \right] = 0.53 \times 10^{-3}$$

The measurement time required at 80°C is $[0.53 \times 10^{-3}]$ [10 years] [365 days/year] [24 hours/day] = 45.3 hours, to approximate 10 years service at 25°C.

11.6   DYNAMIC MECHANICAL BEHAVIOR AT THERMAL TRANSITIONS

The storage modulus $G'(\omega)$ behaves like a modulus measured in a static test and decreases in the glass transition region (cf. Fig. 11-8). The loss modulus $G''(\omega)$ and tan δ go through a maximum under the same conditions, however. Figure 11-18 shows some typical experimental data. T_g can be identified as the peak in the tan δ or the loss modulus trace. These maxima do not coincide exactly. The maximum in tan δ is at a higher temperature than that in $G''(\omega)$, because tan δ is the ratio of $G'(\omega)$ and $G''(\omega)$ (Eq. 11-19) and both these moduli are changing in the transition region. At low frequencies (about 1 Hz) the peak in tan δ is about 5°C warmer than

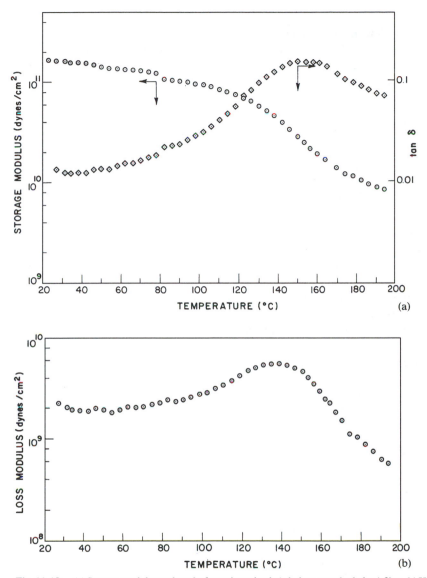

Fig. 11-18.   (a) Storage modulus and tan δ of an oriented poly(ethylene terephathalate) fiber. 11 Hz frequency. (b) Loss modulus of the same fiber.

T_g from static measurements or the maximum in the loss modulus–temperature curve.

The development of a maximum in tan δ or the loss modulus at the glass-to-rubber transition is explained as follows. At temperatures below T_g the polymer behaves elastically, and there is little or no flow to convert the applied energy into internal work in the material. Now h, the energy dissipated as heat per unit volume of material per unit time because of flow in shear deformation, is

$$h = \tau \, d\gamma/dt = \eta(d\gamma/dt)^2 \qquad (11\text{-}43)$$

[To check this equation by dimensional analysis in terms of the fundamental units mass (m), length (l), and time (t):

$$\tau = ml^{-1}t^{-2}, \qquad d\gamma/dt = t^{-1}, \qquad \eta = ml^{-1}t^{-1}, \qquad \text{force} = mlt^{-2},$$

$$\text{work} = ml^2t^{-2}, \qquad \text{work/volume/time} = ml^{-1}t^{-3} = \text{Eq. (11-43).}]$$

Thus the work dissipated is proportional to the viscosity of the material at fixed straining rate $d\gamma/dt$. At low temperatures, η is very high but γ and $d\gamma/dt$ are vanishingly small and h is negligible. As the structure is loosened in the transition region, n decreases but $d\gamma/dt$ becomes much more significant so that h (and the loss modulus and tan δ) increases. The effective straining rate of polymer segments continues to increase somewhat with temperature above T_g but η, which measures the resistance to flow, decreases at the same time. The net result is a diminution of damping and a fall-off of the magnitudes of the storage modulus and tan δ.

An interesting application of dynamic mechanical data is in blends of rubbers for tire treads. Rolling is at low frequencies, while skidding is at high frequencies. (Compare the hum of tires on the pavement with the screech of a skid.) Therefore, low rolling friction requires low damping (i.e., little dissipation of mechanical energy) at low frequencies while skid resistance implies high damping at higher frequencies. One way to achieve the desired property balance is to blend elastomers with the respective properties.

11.6.1   Relaxations at Temperatures below T_g

In the glass transition region, the storage modulus of an amorphous polymer drops by a factor of $\sim$1000, and tan δ is generally one or more. (The tan δ in Fig. 11-18a is less than this because the polymer is oriented and partially crystalline.) In addition to T_g, minor transitions are often observed at lower temperatures, where the modulus may decrease by a factor of $\sim$2 and tan δ has maxima of 0.1 or less. These so-called secondary transitions arise from the motions of side groups or segments of the main chain that are smaller than those involved in the displacements associated with T_g. Secondary transitions increase in temperature

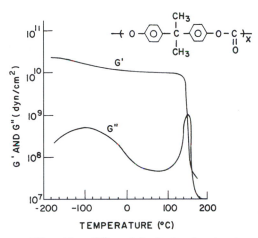

Fig. 11-19. Storage (G') and loss (G'') moduli of polycarbonate polymer [5]. The broad low-temperature peak is probably composed of several overlapping maxima.

with increasing frequency in a manner similar to the main glass transition. They can be detected by dynamic mechanical and also by dielectric loss factor and nuclear magnetic resonance measurements.

Some amorphous polymers are not brittle at temperatures below T_g. Nearly all these tough glasses have pronounced secondary transitions. Figure 11-19 is a sketch of the temperature dependence of the shear storage and loss moduli for polycarbonate, which is one such polymer. The molecular motions that are responsible for the ductile behavior of some glassy polymers are probably associated with limited range motions of main chain segments. Polymers like poly(methyl methacrylate) that exhibit secondary transitions due to side group motions are not particularly tough.

11.7 STRESS–STRAIN TESTS

Stress-strain tests were mentioned on page 24 and in Fig. 11-12. In such a tensile test a parallel-sided strip is held in two clamps that are separated at a constant speed, and the force needed to effect this is recorded as a function of clamp separation. The test specimens are usually dogbone shaped to promote deformation between the clamps and deter flow in the clamped portions of the material. The load-elongation data are converted to a stress-strain curve using the relations mentioned on p. 24. These are probably the most widely used of all mechanical tests on polymers. They provide useful information on the behavior of isotropic specimens, but their relation to the use of articles fabricated from the same polymer as the test specimens is generally not straightforward. This is because such articles are anisotropic, their properties may depend strongly on the fabrication history, and the use conditions may vary from those in the tensile test. Stress–strain tests are discussed here, with the above cautions, because workers in the field often develop an intuitive feeling for the value of such data with particular polymers and because they provide useful general examples of the effects of testing rate and temperature in mechanical testing.

Dynamic mechanical measurements are performed at very small strains in order to ensure that linear viscoelasticity relations can be applied to the data. Stress–strain data involve large strain behavior and are accumulated in the nonlinear region. In other words, the tensile test itself alters the structure of the test specimen, which usually cannot be cycled back to its initial state. (Similarly, dynamic deformations at large strains test the fatigue resistance of the material.)

Figure 1-2 records some typical stress–strain curves for different polymer types. Some polymers exhibit a yield maximum in the nominal stress, as shown in part (c) of this figure. At stresses lower than the yield value, the sample deforms homogeneously. It begins to neck down at the yield stress, however, as sketched in Fig. 11-20. The necked region in some polymers stabilizes at a particular reduced diameter, and deformation continues at a more or less constant nominal stress until the neck has propagated across the whole gauge length. The cross-section of the necking portion of the specimen decreases with increasing extension, so the true stress may be increasing while the total force and the nominal stress

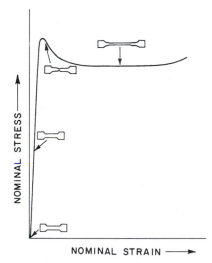

Fig. 11-20.   Tensile stress–strain curve and test specimen appearance for a polymer which yields and cold draws.

(Section 1.8) are constant or even decreasing. The process described is variously called yielding, necking, cold flow, and cold-drawing. It is involved in the orientation processes used to confer high strengths on thermoplastic fibers. Tough plastics always exhibit significant amounts of yielding when they are deformed. This process absorbs impact energy without causing fracture of the article. Brittle plastics have stress–strain curves like that in Fig. 1-2b and do not cold flow to any noticeable extent under impact conditions. Many partially crystalline plastics yield in tensile tests at room temperature but this behavior is not confined to such materials.

The yield stress of amorphous polymers is found to decrease linearly with temperature until it becomes almost zero near T_g. Similarly, the yield stress of partially crystalline materials becomes vanishingly small near T_m, as the crystallites that hold the macromolecules in position are melted out. Yield stresses are rate dependent and increase at faster deformation rates.

The shear component of the applied stress appears to be the major factor in causing yielding. The uniaxial tensile stress in a conventional stress–strain experiment can be resolved into a shear stress and a dilational (negative compressive) stress normal to the parallel sides of test specimens of the type shown in Fig. 11-20. Yielding occurs when the shear strain energy reaches a critical value that depends on the material, according to the von Mises yield criterion, which applies fairly well to polymers.

Yield and necking phenomena can be envisioned usefully with the Considère construction shown in Fig. 11-21. Here the initial conditions are initial gauge length and cross-sectional area l_i and A_i, respectively and the conditions at any instant in the tensile deformation are length l and cross-sectional area A, when the force applied is F. The true stress, σ_t, defined as the force divided by the corresponding

Consideres Construction

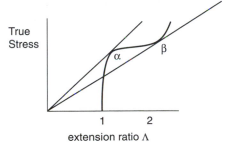

Fig. 11-21.   Sketch of conditions for yielding in tensile deformation.

instantaneous cross-sectional area A_i, is plotted against the extension ratio, $\Lambda (\Lambda = l/l_0 = \epsilon - 1$, as defined in Fig. 11-12). If the deformation takes place at constant volume then:

$$A_i l_i = Al \tag{11-44}$$

The nominal stress (engineering stress) $= \sigma = F/A_i$, and therefore:

$$\sigma = \sigma_t / \Lambda \tag{11-45}$$

$$\frac{d\sigma}{d\Lambda} = \frac{1}{\Lambda} \frac{d\sigma_t}{d\Lambda} - \frac{\sigma_t}{\Lambda^2} \tag{11-46}$$

Since $d\epsilon = d\Lambda$, then at yield

$$\frac{d\sigma}{d\epsilon} = 0 = \frac{d\sigma}{d\Lambda} \tag{11-47}$$

and the yield condition is characterized by:

$$\frac{d\sigma_t}{d\Lambda} = \frac{\sigma_t}{\Lambda} \tag{11-48}$$

A maximum in the plot of *engineering* stress against strain occurs only if a tangent can be drawn from $\lambda = 0$ to touch the curve of *true* stress against extension ratio at a point, labeled α in Fig. 11-21. In this figure a second tangent through the origin touches the curve at point β. This defines a minimum in the usual plot of nominal stress against extension ratio where the orientation induced by the deformation stiffens the polymer in the necked region. This phenomenon is called *strain hardening*. The neck stabilizes and travels through the specimen by incorporating more material from the neighboring tapered regions. As the tensile deformation proceeds the whole parallel gauge length of the specimen will yield. If the true stress-extension ratio relation is such that a second tangent cannot be drawn the material will continue to thin until it breaks. Molten glass exhibits this behavior.

The phenomenon of strain hardening in polymers is a consequence of orientation of molecular chains in the stretch direction. If the necked material is a semicrystalline polymer, like polyethylene or a crystallizable polyester or nylon, the crystallite structure will change during yielding. Initial spherulitic or row nucleated structures will be disrupted by sliding of crystallites and lamellae, to yield morphologies like that shown in Fig. 11-7.

Yielding and strain hardening are characteristic of some metals as well as polymers. Polymer behavior differs, however, in two features. One is the temperature rise that can occur in the necked region as a result of the viscous dissipation of mechanical energy and orientation-induced crystallization. The other feature is an increase of the yield stress at higher strain rates. These opposing effects can be quite significant, especially at the high strain rates characteristic of industrial orientation processes for fibers and films.

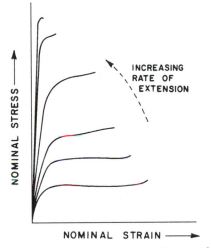

Fig. 11-22.   Effect of strain rate on the tensile stress–strain curve of a polymer which yields at low straining rates.

11.7.1   Rate and Temperature Effects

Most polymers tend to become more rigid and brittle with increasing straining rates. In tensile tests, the modulus (initial slope of the stress–strain curve) and yield stress rise and the elongation at fracture drops as the rate of elongation is increased. Figure 11-22 shows typical curves for a polymer which yields. The work to rupture, which is the area under the stress–strain curve, is a measure of the toughness of the specimen under the testing conditions. This parameter decreases at faster extension rates.

   The influence of temperature on the stress–strain behavior of polymers is generally opposite to that of straining rates. This is not surprising in view of the correspondence of time and temperature in the linear viscoelastic region (Section 11.5.2.iii). The curves in Fig. 11-23 are representative of the behavior of a partially crystalline plastic.

11.8   CRAZING IN GLASSY POLYMERS

When a polymer sample is deformed, some of the applied energy can be dissipated by movement of sections of polymer molecules past each other. This yielding process uses energy that might otherwise be available to enlarge preexisting microcracks into new fracture surfaces. The two major mechanisms for energy dissipation in glassy polymers are crazing and shear yielding.

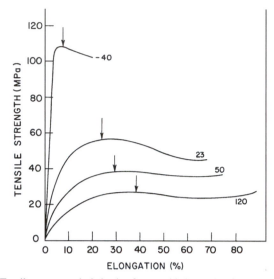

Fig. 11-23.   Tensile stress–strain behavior for a molded sample of a nylon-6,6 at the indicated temperatures (°C). The arrows indicate the yield points which become more diffuse at higher temperatures.

Crazes are pseudocracks that form at right angles to the applied load and that are traversed by many microfibrils of polymer that has been oriented in the stress direction. This orientation itself is due to shear flow. Energy is absorbed during the crazing process by the creation of new surfaces and by viscous flow of polymer segments. Although crazes appear to be a fine network of cracks, the surfaces of each craze are connected by oriented polymeric structures and a completely crazed specimen can continue to sustain appreciable stresses without failure. Crazing detracts from clarity, as in poly(methyl methacrylate) signage or windows, and enhances permeability in products such as plastic pipe. Mainly, however, it functions as an energy sink to inhibit or retard fracture.

The term *crazing* is apparently derived from an Anglo-Saxon verb *krasen*, meaning "to break." In this process polymer segments are drawn out of the adjoining bulk material to form cavitated regions in which the uncrazed surfaces are joined by oriented polymer fibrils, as depicted in Fig. 11-24. Material cohesiveness in amorphous glassy polymers, like polystyrene, arises mainly through entanglements between macromolecules and entanglements are indeed essential for craze formation and craze fibril strength in such polymers [6].

Glassy polymers with higher cohesiveness, like polycarbonate and cross-linked epoxies, preferentially exhibit shear yielding [7], and some materials, such as rubber-modified polypropylene, can either craze or shear yield, depending on the deformation conditions [8]. Application of a stress imparts energy to a body which

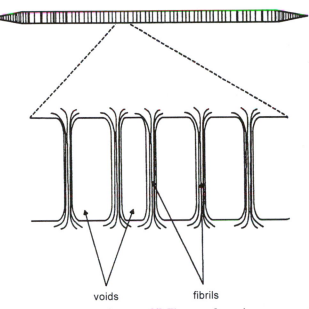

voids   fibrils

Fig. 11-24.   Sketch of a craze in polystyrene [6]. The upper figure shows a craze, with connecting fibrils between the two surfaces. The lower figure is magnification of a section of the craze showing voids and fibrils. Actual crazes in this polymer are about 0.1–2 μm thick; this figure is not to scale.

can be dissipated either by complete recovery (on removal of the load), by catastrophic rupture, or by polymer flow in the stress application region. The latter process, called shear yielding, or shear banding, is a useful mechanism for absorbing impact forces.

A few comments on the distinction between crazing and shear yielding may be appropriate here. A material which undergoes shear yielding is essentially elastic at stresses up to the yield point. Then it suffers a permanent deformation. There is effectively no change in the volume of the material in this process. In crazing, the first craze initiates at a local stress less than the shear stress of the bulk material. The stress required to initiate a craze depends primarily on the presence of stress-raising imperfections, such as crack tips or inclusions, in the stressed substance, whereas the yield stress in shear is not sensitive to such influences. Permanent deformation in crazing results from fibrillation of the polymer in the stress direction.

Craze formation is a dominant mechanism in the toughening of glassy polymers by elastomers in "polyblends." Examples are high-impact polystyrene (HIPS), impact poly(vinyl chloride), and ABS (acrylonitrile-butadiene-styrene) polymers. Polystyrene and styrene-acrylonitrile (SAN) copolymers fracture at strains of ~10^{-2}, whereas rubber-modified grades of these polymers (e.g., HIPS and ABS) form many crazes before breaking at strains around 0.5. Rubbery particles in polyblends act as stress concentrators to produce many craze cracks and to induce orientation of the adjacent rigid polymer matrix. Good adhesion between the glassy polymer and rubbery inclusion is important so that cracks do not form and run between the rubber particles. Crazes and yielding are usually initiated at the equators of rubber particles, which are the loci of maximum stress concentration in stressed specimens, because of their modulus difference from the matrix polymer. Crazes grow outward from rubber particles until they terminate on reaching other particles. The rubber particles and crazes will be able to hold the matrix polymer together, preventing formation of a crack, as long as the applied stress are not catastrophic. The main factors that promote craze formation are a high rubber particle phase volume, good rubber-matrix polymer adhesion and an appropriate rubber particle diameter [10, 12]. The latter factor varies with the matrix polymer, being about 2 μm for polystyrene and about one-tenth of that value for unplasticized PVC. It is intuitively obvious that good adhesion between rubber and matrix polymer is required for transmission of stresses across phase boundaries. Another interesting result stems from the differences in thermal expansion coefficients of the rubber and glassy matrix polymer. For the latter polymers the coefficient of linear expansion (as defined by ASTM method D696) is $\sim 10^{-6}$ K^{-1}, while the corresponding value for elastomers is $\sim 10^{-2}$ K^{-1}. When molded samples of rubber-modified polymers are cooled from the melt state the elastomer phase will undergo volume dilation. This increases the openess of the rubber structure and causes a shift of the T_g of the rubber to lower values than that of unattached elastomer [13].

11.9   FRACTURE MECHANICS

This discipline is based on the premise that all materials contain flaws, and that fracture occurs by stress-induced extension of these defects. The theory derives from the work of A. E. Griffith [12], who attempted to explain the observation that the tensile strengths (defined as breaking force $\div$ initial cross-sectional area) of fine glass filaments were inversely proportional to the sample diameter. He assumed that every object contained flaws, that failure is more likely the larger the defect, and that larger bodies would break at lower tensile stresses because they contained larger cracks. The basic concept is that a crack will grow only if the total energy of the body is lowered thereby. That is to say, the elastic strain energy which is relieved by crack growth must exceed the energy of the newly created surfaces. It is important to note also that the presence of a crack or inclusion changes the stress distribution around it, and the stress may be amplified greatly around the tips of sharp cracks. The relation which was derived between crack size and failure stress is known as the Griffith criterion:

$$\sigma_f = \left[\frac{2\gamma Y}{\pi a}\right]^{1/2},\tag{11-49}$$

where: σ_f = failure stress, based on the initial cross-section, a = crack depth, Y = tensile (Young's) modulus, and γ = surface energy of the solid material (the factor 2 is inserted because fracture generates two new surfaces). This equation applies to completely elastic fractures; all the applied energy is consumed in generating the fracture surfaces. Real materials are very seldom completely elastic, however, and a more general application of this concept allows for additional energy dissipation in a small plastic deformation region near the crack tip. With this amendment, Eq. (11-49) is applicable with the 2γ term replaced by G, the *strain energy release rate*, which includes both plastic and elastic surface work done in extending a preexisting crack [13]:

$$\sigma_f = \left[\frac{YG}{\pi a}\right]^{1/2}\tag{11-50}$$

The general equation to describe the applied stress field around a crack tip is [15]:

$$\sigma = \frac{K}{[\pi a]^{1/2}}\tag{11-51}$$

where K is the stress intensity factor and σ is the local stress. Equation (11-51) applies at all stresses, but the stress intensity reaches a critical value, K_c, at the stress level where the crack begins to grow. K_c is a material property, called the *fracture toughness*, and the corresponding strain energy release rate becomes the *critical strain energy release rate*, G_c.

The equations cited above are for an ideal semi-infinite plate, with no boundary effects. Application to real specimens requires calibration factors, so that the fracture toughness of Eq. (11-51), at the critical point is given by:

$$K_c = \sigma_f \Gamma [\pi a]^{1/2}\tag{11-52}$$

where Γ (gamma) is a calibration factor which is itself a function of specimen geometry and crack size. The Γ values have been tabulated (mainly for metals) for a variety of shapes [16]. The independent variable in Eq. (11-52) is the crack depth, a. To measure K_c, sharp cracks of various known depths are made in specimens with fixed geometry and plots of $\sqrt{a}$ versus $\sigma_f \Gamma$ are linear with slope K_c. Instrumented impact tests yield values for the specimen fracture energy, U_f. With such data, the critical strain energy release rate can be calculated according to [17]:

$$G_c = \frac{U_f}{BD\Phi}\tag{11-53}$$

where B and D are the specimen depth and width, respectively, and the calibration factor ϕ is a function of the specimen geometry and the ratio of the crack depth and specimen width. Here again, the independent variable is the crack depth, a, as manifested in parameter ϕ.

The total work of crack formation equals $G_c \times$ the crack area. Catastrophic failure is predicted to occur when $\sigma [\pi a]^{1/2} = [YG_c]^{1/2}$, or when $K_c = [YG_c]^{1/2}$. K_c and G_c are the parameters used in *linear elastic fracture mechanics* (LEFM). Both factors are implicitly defined to this point for plane stress conditions. To understand the term *plane stress*, imagine that the applied stress is resolved into three components along Cartesian coordinates; plane stress occurs when one component is $= 0$. Such conditions are most likely to occur when the specimen is thin.

This reference to specimen thickness leads to a consideration of the question of why a polymer that is able to yield will be less brittle in thin than in thicker sections. Polycarbonate is an example of such behavior. Recall that yielding occurs at constant volume (tensile specimens neck down on extension). In thin objects the surfaces are load-free and can be drawn inward as a yield zone grows ahead of a crack tip. In a thick specimen the material surrounding the yield zone is at a lower stress than that in the crack region. It is not free to be drawn into the yield zone and acts as a restraint on plastic flow of the region near the crack tip. As a consequence, fracture occurs with a lower level of energy absorption in a thick specimen. The crack tip in a thin specimen will be in a state of plane stress while the corresponding condition in a thick specimen will be plane strain. Plane strain is the more dangerous condition.

The parameters which apply to plane strain fracture are G_{Ic} and K_{Ic}, where the subscript I indicates that the crack opening is due to tensile forces. K_{Ic} is measured by applying Eq. (11-52) to data obtained with thick specimens. To illustrate the differences between plane stress and plane strain fracture modes, thin polycarbonate specimens, with thicknesses ≤ 3 mm reported to have G_c values of 10 kJ/m^2, while the G_{Ic} of thick specimens is 1.5 kJ/m^2.

It will be useful to consider a practical application of LEFM here. Consider a study of the effect of preexisting flaws on the ability of PVC pipe to hold pressure [18]. The critical stress intensity factor, K_{Ic}, is reported to be 1.08 MPa·m$^{1/2}$ for PVC under static load at 20°C. The stress (σ) in a pipe $=$ the internal pressure, P, $+$ the hoop stress at the inner surface:

$$\sigma = P + \frac{P[D - t]}{2t} \tag{11-54}$$

where D is the outside diameter and t is the pipe wall thickness. In this case, $D = 250$ mm and $t = 17$ mm. For pipe of this type Γ (in Eq. 11-52) is about 1.12 [19]. Assuming that the presence of 1 mm flaws will give a conservative estimate of pipe service life, we take $a = 1$ mm in Eq. (11-52). The critical stress for failure is

$$\sigma_f = \frac{K_{Ic}}{\Gamma[\pi a]^{1/2}} = \frac{1.08\,\text{MPa m}^{1/2}}{1.12[10^{-3}\pi\,\text{m}]^{1/2}} = 17.2\,\text{MPa}$$

$$\sigma_f = P + \frac{P[D - t]}{2t} = 7.85 P_c$$

where P_c is the critical pressure for brittle fracture of the pipe. Hence,

$$P_c = \frac{17.2\,\text{MPa}}{7.85} = 2.19\,\text{MPa} = 318\,\text{psi}$$

An otherwise well-made pipe will sustain steady pressures up to 2.19 MPa (318 psi) without failing by brittle fracture if it contains initial flaws as large as 1 mm. Similar calculations show that initial flaws or inclusions smaller than 4.5 mm permit steady operation at 150 psi (1.03 MPa).

LEFM discussed to this point refers to the resistance of bodies to crack growth under static loads. Crack growth under cyclic loading is faster than under static loads at the same stress amplitudes, because the rate of loading and the damage both increase with higher frequencies.

Polymers which yield extensively under stress exhibit nonlinear stress–strain behavior. This invalidates the application of linear elastic fracture mechanics. It is usually assumed that the LEFM approach can be used if the size of the plastic zone is small compared to the dimensions of the object. Alternative concepts have been proposed for rating the fracture resistance of tougher polymers, like polyolefins, but empirical pendulum impact or dart drop tests are deeply entrenched for judging such behavior.

11.10   TOUGHNESS AND BRITTLENESS

Many polymers that yield and exhibit tough, ductile behavior under the conditions of normal tensile tests prove to be brittle when impacted. This is particularly true when the sample contains notches or other stress concentrators. Fracture behavior is characterized by a variety of empirical tests. None of these can be expected to correlate very closely with service performance, because it is very difficult to analyze stress and deformation behavior of complex real articles under the variety of loads that may be encountered in practice. Impact tests aim to rate the fracture resistance of materials by measuring the energy required to break specimens with standard dimensions. The values obtained relate to the experimental conditions and the geometry and history of the specimen. A single figure for impact strength is of limited value in itself but such data can be useful for predicting serviceability of materials for different applications if they are obtained, say, from a series of impact tests at various temperatures and sample shapes and are combined with experience of the performance of similar materials and part shapes under related service conditions.

Impact tests are often used to locate the brittle-ductile temperature or brittleness temperature. This parameter is generally defined as the temperature at which half the specimens tested fail in a given test. Because of the nature of most of these impact tests, this approximates the temperature range in which yielding processes begin to absorb substantial portions of the applied energy. As the test temperature is increased through the brittle-to-tough transition region, the measured impact energies increase substantially and the specimens exhibit more evidence of having flowed before fracturing.

Ductile–brittle transitions are more accurately located by variable temperature tests than by altering impact speed in an experiment at a fixed temperature. This is because a linear fall in temperature is equivalent to a logarithmic increase in straining rate. The ductile–brittle transition concept can be clarified by sketches such as that in Fig. 11-25. In the brittle region, the impact resistance of a material is related to its LEFM properties, as described above. In the mixed mode failure

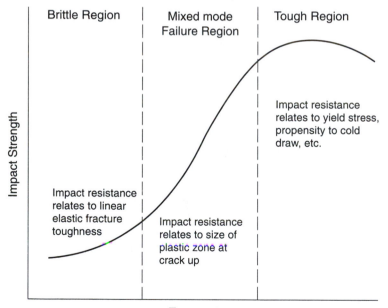

Fig. 11-25. Ductile-brittle behavior in impact resistance [20]. The transition between the zones varies with rate of impact and type of test.

region, fracture resistance is proportional to the size of the yield zone that deveops at a crack tip during impact. If the yield zone (also sometimes called a plastic zone) is small, fracture tends to be brittle and can be described by LEFM concepts. If yielding takes place on a large scale, then the material will absorb considerable energy before fracturing and its behavior will be described as tough.

The relative importance of the yield zone can be estimated, for a given product, by comparing its yield stress and;'fracture toughness. The parameter proposed for this purpose is $[K_{Ic}/\sigma_y]^2$, where σ_y is the yield stress [21]. This ratio has units of length and has been suggested to be proportional to the size of the yield zone. Higher values indicate tougher materials. In the third region of Fig. 11-25, impact resistance is determined by the capacity of the product to absorb energy by localized necking and related mechanisms, after yielding.

Notches act as stress raisers and redistribute the applied stress so as to favor brittle fracture over plastic flow. Some polymers are much more notch sensitive than others, but the brittleness temperature depends in general on the test specimen width and notch radius. Polymers with low Poisson ratios tend to be notch sensitive. Comparisons of impact strengths of unnotched and notched specimens are often used as indicators of the relative danger of service failures with complicated articles made from notch sensitive materials.

Weld lines (also known as knit lines) are a potential source of weakness in molded and extruded plastic products. These occur when separate polymer melt flows meet and weld more or less into each other. Knit lines arise from flows around barriers, as in double or multigating and use of inserts in injection molding. The primary source of weld lines in extrusion is flow around spiders (multiarmed devices that hold the extrusion die). The melt temperature and melt elasticity (which is mentioned in the next section of this chapter) have major influences on the mechanical properties of weld lines. The tensile and impact strength of plastics that fail without appreciable yielding may be reduced considerably by in double-gated moldings, compared to that of samples without weld lines. Polystryrene and SAN copolymers are typical of such materials. The effects of weld lines is relatively minor with ductile amorphous plastics like ABS and polycarbonate and with semicrystalline polymers such as polyoxymethylene. This is because these materials can reduce stress concentrations by yielding [22].

Semicrystalline polymers are impact resistant if their glass transition temperatures are much lower than the test temperature. The impact strength of such materials decreases with increasing degree of crystallinity and particularly with increased size of supercrystalline structures like spherulites. This is because these changes are tantamount to the progressive decrease in the numbers of tie molecules between such structures.

The impact strength of highly cross-linked thermoset polymers is little affected by temperature since their behavior is generally glassy in any case.

11.11   RHEOLOGY

Rheology is the study of the deformation and flow of matter. The processing of polymers involves rheological phenomena. They cannot be evaded. It is important, therefore, that practitioners have at least some basic knowledge of this esoteric subject. This section is a very brief review, with the aim of guiding the perplexed to at least ask the right questions when confronted with a rheological problem. Following is a summary of the major points, which are elaborated below.

    1.   The rheological behavior of materials is generally very complex, and polymers are usually more complex than alternative materials of construction.

    2.   A complete rheological characterization of a material is very time consuming and expensive and much of the data will be irrelevant to any particular process or problem.

    3.   Rheological measurements must be tailored to the particular process and problem of interest. This is the key to successful solution of rheological and processing problems. Relevant rheological experiments are best made at the same temperatures, flow rates, and deformation modes that prevail in the process of interest.

    4.   Following are some important questions that should be asked in the initial stages of enquiry:

    a.   Is the process isothermal? Most standard rheological measurements are isothermal; many processes are not.

    b.   Is the material behavior entirely viscous or does it also comprise elastic components?

    c.   Does processing itself change the rheological properties of the material?

    d.   Do the material purchase specifications ensure rheological behavior?

    e.   Do steady-state rheological measurements characterize the material *in a particular process*?

    f.   Is it best to look to rheological measurements or to process simulations for answers to a particular problem?

The coefficient of viscosity concept, η, was introduced in connection with Eq. (3-32) as the quotient of the shearing force per unit area divided by the velocity gradient. The numerator here is the shearing stress, τ, and the denominator is termed the shear rate, $\dot{\gamma}$ (γ is the strain and $\dot{\gamma} \equiv d\gamma/dt$). With these changes, Eq. (3-32) reads:

$$\eta = \frac{(F/A)}{(dv/dr)} = \frac{\tau}{\dot{\gamma}} \tag{11-55}$$

The viscosity of water at room temperaure is 10^{-3} Pa $\cdot$ sec [$=1$ centipoise (cP)], while that of molten thermoplastics at their processing temperataures is in the

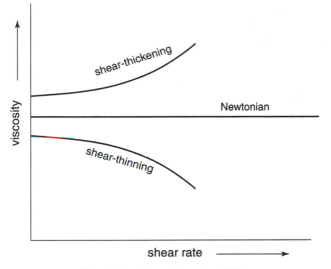

Fig. 11-26.   Time-independent fluids.

neighborhood of 10^3–10^4 Pa · sec. Lubricating oils are characterized by η values up to about 1 Pa · sec. In SI units the dimensions of viscosity are N · sec/m^2 = Pa·sec.

If η is independent of shear history, the material is said to be time independent. Such liquids can exhibit different behavior patterns, however, if, as is frequently the case with polymers, η varies with shear rate. A material whose viscosity is independent of shear rate, e.g., water, is a Newtonian fluid. Figure 11-26 illustrates shear-thickening, Newtonian and shear-thinning η–$\dot{\gamma}$ relations. Most polymer melts and solutions are shear-thinning. (Low-molecular-weight polymers and dilute solutions often exhibit Newtonian characteristics.) Wet sand is a familiar example of a shear-thickening substance. It feels hard if you run on it, but you can sink down while standing still.

A single figure for η is not appropriate for non-Newtonian substances, and it common practice to plot "flow curves" of such materials in terms of η_a (apparent viscosity) against corresponding values of $\dot{\gamma}$. Many equations have been proposed to describe non-Newtonian behavior. Generally, however, the mathematics involved is not worth the effort except for the simplest problems. It is most efficient to read the required viscosity values from experimental η_a–$\dot{\gamma}$ plots. These relations can usually be described over limited shear rate ranges by power law expressions of the form:

$$\tau = C(\dot{\gamma})^n \qquad (11\text{-}56)$$

where n is the power law index. If $n < 1$ the material is shear-thinning; if $n > 1$, it is shear-thickening. The constant C has no real physical significance because its units will vary with n. Equation (11-56) indicates that a log–log plot of τ vs. γ is linear over the shear rate range of applicability. An alternative expression is

$$\eta_a = \mu(\dot{\gamma})^{n-1} \tag{11-57}$$

where μ, sometimes called the consistency, also has limited significance since its units also are dependent on n. This problem can be circumvented by referencing the apparent viscosity to that at a specified shear rate, $\dot{\gamma}_{aref}$, which is conveniently taken as 1 sec^{-1}. Then

$$\eta_a = \eta_{aref}|\dot{\gamma}_a/\dot{\gamma}_{aref}|^{n-1} \tag{11-58}$$

Substances that do not flow at shear stresses less than a certain level exhibit yield properties. Then

$$\dot{\gamma} = \frac{\tau - \tau_y}{\eta} \tag{11-59}$$

where τ_y is the yield stress. The yield stress may be of no significance, as in high-speed extrusion of plastics, or it could be an important property of materials as in the application of architectural paints and in rotational molding.

Most polymeric substances are time dependent to some extent and $\eta = \eta(\dot{\gamma}, t)$, where t here refers to the time under shear. If shearing causes a decrease in viscosity the material is said to *thixotropic*; the opposite behavior character-izes a *rheopectic* substance. These patterns are sketched in Fig. 11-27. After

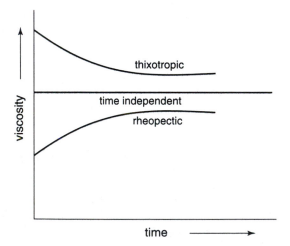

Fig. 11-27. Viscosity–time relations for time-dependent fluids.

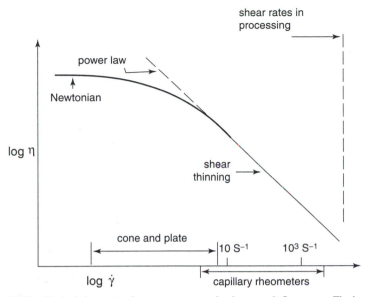

Fig. 11-28.   Typical rheometer shear rate ranges and polymer melt flow curve. The lower shear rate region of the flow curve exhibits viscosities that appear to be independent of $\dot{\gamma}$. This is the lower Newtonian region.

shearing has been stopped, time-dependent fluids recover their original condition in due course. PVC plastisols [mixtures of poly(vinyl chloride) emulsion polymers and plasticizers] and some mineral suspensions often exhibit rheopexy. Thixotropy is a necessary feature of house paints, which must be reasonably fluid when they are applied by brushing or rolling, but have to be viscous in the can and shortly after application, in order to minimize pigment settling and sagging, respectively.

A variety of laboratory instruments have been used to measure the viscosity of polymer melts and solutions. The most common types are the coaxial cylinder, cone-and-plate, and capillary viscometers. Figure 11-28 shows a typical flow curve for a thermoplastic melt of a moderate molecular weight polymer, along with representative shear rate ranges for cone-and-plate and capillary rheometers. The last viscometer type, which bears a superficial resemblance to the orifice in an extruder or injection molder, is the most widely used and will be the only type considered in this nonspecialized text.

Equation (11-60) [cf. Eq. (3-60)] gives the relation between flow rate and viscosity for a fluid under pressure P in a tube with radius r and length l. In such a device the apparent shear stress, $\tau_a = Pr/2l$; and the apparent shear rate, $\dot{\gamma}_a = 4Q/\pi r^3$, where Q, the volumetric flow rate, is simply the Q/t term of

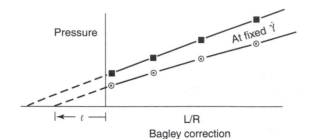

Fig. 11-29. Bagley end correction plot.

Eq. (3-60). That is,

$$\eta = \frac{\pi P r^4 t}{8Ql} = \frac{Pr/2l}{4Q/\pi r^3} = \frac{\tau_a}{\dot{\gamma}_a} \qquad (11\text{-}60)$$

The shear stress and shear rate here are termed apparent, as distinguished from the respective true values at the capillary wall, τ_w and $\dot{\gamma}_w$. The Bagley correction to the shear stress allows for pressure losses incurred primarily by accelerating the polymer from the wider rheometer barrel into the narrower capillary entrance [23]. It is measured by using a minimum of two dies, with identical radii and different lengths. The pressure drop, at a given apparent shear rate, is plotted against the l/r ratio of the dies, as shown in Fig. 11-29. The absolute values of the negative intercepts on the l/r axis are the Bagley end-corrections, e. The true shear stress at each shear rate is given by

$$\tau_w = \frac{P}{\left[2\frac{l}{r} + e\right]} \qquad (11\text{-}61)$$

Alternatively, τ_w can be measured directly by using a single long capillary with l/r about 40. The velocity gradient in Fig. 3-5 is assumed to be parabolic, but this is true strictly only for Newtonian fluids. The Rabinowitsch equation [24] corrects for this discrepancy in non-Newtonian flow, such as that of most polymer melts:

$$\dot{\gamma}_w = \left[\frac{3n+1}{4n}\right]\frac{4Q}{\pi r^3} \qquad (11\text{-}62)$$

where n is the power law index mentioned earlier. Application of the Bagley and Rabinowitsch corrections (in that order) converts the apparent flow curve from capillary rheometer measurements to a true viscous flow curve such as would be obtained from a cone-and-plate rheometer. However, this manipulation has sacrificed all information on elastic properties of the polymer fluid and is not useful for prediction of the onset of many processing phenomena, which we will now consider.

Measurements made under standardized temperature and pressure conditions from a simple capillary rheometer and orifice of stipulated dimensions provide *melt*

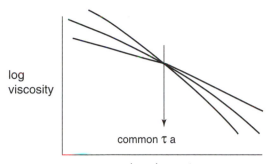

log
viscosity

common τ a

log shear stress

Fig. 11-30.   Apparent flow curves of different polymers with the same MFI (at the intersection point).

flow index (MFI) or *melt index* characteristics of many thermoplastics. The units of MFI are grams output/10 min extrusion time. The procedure, which amounts to a measurement of flow rate at a standardized value of τ_a, is very widely used for quality and production control of polyolefins, styrenics and other commodity polymers. A lower MFI shows that the polymer is more viscous, *under the conditions of the measurement.* This parameter can be shown to be inversely related to a power of an average molecular weight of the material [25] [(MFI)$^{-1}$ $\propto$ $M_w^{3.4-3.7}$]. MFI, which is easy to measure, is often taken to be an inverse token of polymer molecular size. The problem with this assumption is that MFI, or η_a for that matter, scales with average molecular weight only so long as the molecular weight distribution shape is invariant. This assumption is useful then for consideration of the effects of variations in a particular polymerization process, but may be prone to error when comparing products from different sources.

A more serious deficiency resides in reliance on MFI to characterize different polymers. No single rheological property can be expected to provide a complete prediction of the properties of a complex material like a thermoplastic polymer. Figure 11-27 shows log η_a – log τ_a flow curves for polymers having the same melt index, at the intersection of the curves, but very different viscosities at higher shear stress where the materials are extruded or molded. This is the main reason why MFI is repeatedly condemned by purer practitioners of our profession. The parameter is locked into industrial practice, however, and is unlikely to be displaced.

Viscoelasticity was introduced in Section 11.5. A polymer example may be useful by way of recapitulation. Imagine a polymer melt or solution confined in the aperture between two parallel plates to which it adheres. One plate is rotated at a constant rate, while the other is held stationary. Figure 11-31a shows the time dependence of the shear stress after the rotation has been stopped. τ decays immediately to zero for an inelastic fluid but the decrease in stress is much more gradual if the material is viscoelastic. In some cases, the residual stresses may

Viscoelasticity

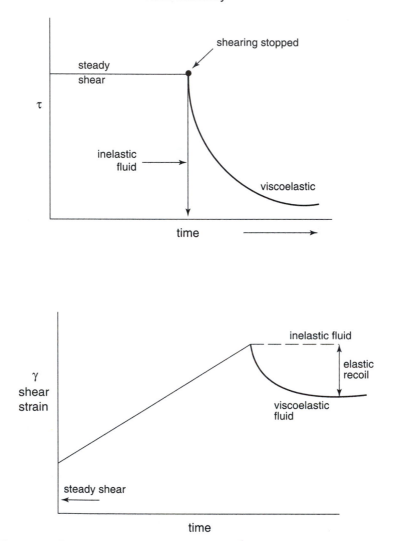

Fig. 11-31.   Comparisons of inelastic and viscoelastic behavior on the cessation of steady shearing.

not reach zero, as in molded or extruded thermoplastics that have been quenched from the molten state. Such articles contain molded-in stresses that are relieved by gradual decay over time, resulting in warpage of the part. Figure 11-31b sketches the dependence of the deformation once the steady shearing has stopped. An inelastic fluid maintains the final strain level, while a viscoelastic substance will undergo some elastic recoil.

Whether a polymer exhibits elastic as well as viscous behavior depends in part on the time scale of the imposition of a load or deformation compared to the characteristic response time of the material. This concept is expressed in the dimensionless Deborah number:

$$N_{\text{Deb}} = \frac{\text{reponse time of material}}{\text{time scale of process}}$$

(This parameter was named after the prophetess Deborah to whom Psalm 114 has been ascribed. This song states correctly that even mountains flow during the infinite observation time of the Lord, viz., *"The mountains skipped like rams, The hills like lambs"*.) The process is primarily elastic if $N_{\text{Deb}} > 1$ and essentially viscous if $N_{\text{Deb}} < 1$. The reponse time of the polymer, and its tendency to behave elastically, will increase with higher molecular weight and skewing of the molecular weight distribution toward larger species.

A number of polymer melt phenomena reveal elastic performance [26]. A common example is extrudate swell, in which the cross section of an extruded profile is observed to be larger than that of the orifice from which it was produced. Melt elasticity is required during extrusion coating operations where the molten polymer sheet is pulled out of the sheet die of the extruder by a moving substrate of paper or metal foil. Since the final laminate must be edge-trimmed it is highly desireable that the edges of the polymer match those of the substrate without excessive edge waviness or tearing. This requires a good degree of melt cohesion provided by intermolecular entanglements which also promote elasticity. Other elasticity-related phenomena include undesireable extrudate surface defects, called melt fracture and sharkskin, which appear with some polymers as extrusion speeds are increased.

A number of modern devices provide accurate measurements of both viscous and elastic properties (although some have limited shear rate ranges). An inexact but very convenient indicator of relative elasticity is extrudate swell (or die swell) which is inferred from the ratio of the diameter of the leading edge of a circular extrudate to that of the corresponding orifice. Since MFI is routinely measured, its limited value can be augmented by concurrent die swell data. As an example, polyethylenes intended for extrusion coating should be monitored for minimum die swell, at given melt index, while polymers for high speed wire covering require maximum die swell values which can be set by experience.

The emphasis to this point has been on viscous behavior in shearing modes of deformation. However, any operation which reduces the thickness of a polymeric liquid must do so through deformations that are partly extensional and partly shear. In many cases polymers respond very differently to shear and to extension. A prime industrial example involves low density and linear low density polyethylenes, i.e., LDPE and LLDPE, respectively (Section 9.5.3). LDPE grades intended for extrusion into packaging film have relatively low shear viscosities and high elongational viscosities. As a result, extrusion of tubular film involves reasonable power requirements and stable inflated film "bubbles" between the die and film take-off. LLDPE's of comparable MFI's require much more power to extrude. Their melts can, however, be drawn down to much thinner gauges (an advantage), but the tubular film bubbles are more prone to wobble and tear (a disadvantage). The best of both worlds can be realized by blending minor proportions of selected LDPE's into LLDPE's.

When problems occur during polymer processing it is necessary to perform at least a preliminary analysis of the particular fabrication process. Experiments on production equipment are time-consuming, difficult to control and expensive. Therefore, equivalent laboratory experiments are very desireable. Ideally, one would be able to analyze the production process in terms of fundamental physical quantities and measure these with rheological equipment. It is necessary to make sure that the laboratory measurements correspond to the actual production process and to select the rheological characteristics that bear on the particular problem. That is to say, do not measure viscosity to try to get information about a phenomenon that is affected mainly by the elastic character of the material. Note in this connection that most laboratory data are obtained from steady-state measurements, while the polymers in some processes never reach equilibrium condition. (The ink in a high-speed printing operation is a good example.) If this rheological analysis is not feasible the production process can sometimes be simulated on a small, simplified scale, while paying attention to the features that are critical in the simulation.

There a number of fine recent rheological references [27–29] that should be consulted for more details than can be considered in an introductory text.

11.12   EFFECTS OF FABRICATION PROCESSES

An important difference between thermoplastics and other materials of construction lies in the strong influence of fabrication details on the mechanical properties of plastic articles. This is exhibited in the pattern of frozen-in orientation and fabrication stresses. The manufacturing process can also have marked effects on crystalline texture and qualities of products made from semicrystalline polymers. Orientation generally produces enhanced stiffness and strength in the stretch direction and weakness in the transverse direction. In semi-crystalline polymers, the final structure is sensitive to the temperature–time sequence of the forming and subsequent cooling operations and to the presence or absence of orientation during cooling. For example, different properties are produced by stretching a crystallized sample at temperatures between T_g and T_m, or by orienting the molten polymer before crystallizing the product.

In summary, the final properties of thermoplastic articles depend both on the molecular structure of the polymer and on the details of the fabrication operations. This is a disadvantage, in one sense, since it makes product design more complicated than with other materials that are less history-dependent. On the other hand, this feature confers an important advantage on plastics because fabrication particulars are additional parameters that can be exploited to vary the costs or balance of properties of the products.

PROBLEMS

11-1. Nylon-6,6 can be made into articles with tensile strengths around 12,000 psi or into other articles with tensile strengths around 120,000 psi. What is the basic difference in the processes used to form these two different articles? Why do polyisobutene properties not respond in the same manner to different forming operations?

11-2. Suggest a chemical change and/or a process to raise the softening temperature of articles.

11-3. Which of the following polymers would you expect to have a lower glass transition temperature? Which would have a higher melting point? Explain why. (Assume equal degrees of polymerization.)

(a) $-(OCH_2CH_2-O-\underset{\underset{O}{\|}}{C}+CH_2)_4\underset{\underset{O}{\|}}{C})_x$

(b) $+OCH_2-CH_2-O\underset{\underset{O}{\|}}{C}-\bigcirc-\underset{\underset{O}{\|}}{C})_n$

11-4. A rubber has a shear modulus of 10^7 dyn/cm^2 and a Poisson's ratio of 0.49 at room temperature. A load of 5 kg is applied to a strip of this material which is 10 cm long, 0.5 cm wide, and 0.25 cm thick. How much will the specimen elongate?

11-5. Estimate the glass transition temperature of a copolymer of vinyl chloride and vinyl acetate containing 10 wt% vinyl acetate. The T_g's of the homopolymers are listed in Table 11-2.

11-6. The raw data from a tensile test (Fig. 11-20) are obtained in terms of force and corresponding elongation for a test specimen of given dimensions. The area under such a force–elongation curve can be equated to the impact strength of an isotropic polymer specimen if the tensile test is performed at impact speeds. Show that this area is proportional to the work necessary to rupture the sample.

11-7. The stress relaxation modulus for a polyisobutene sample at 0°C is 2.5 × 10^5 N/m². The stress here is measured 10 min after imposition of a fixed deformation. Use the WLF equation (Eq. 11-42) to estimate the temperature at which the relaxation modulus is 2.5 × 10^5 N/m² for a measuring time of 1 min.

11-8. A Maxwell model (Eq. 11-29, Fig. 11-16b) is being deformed at a constant rate $d\gamma/dt = C$. What is the stress on the element $[\sigma(t)]$ at a time t after the imposition of the fixed straining rate? Express your answer in terms of the constants G and η of the Maxwell element and the strain $\gamma(t)$ which corresponds to $\sigma(t)$.

11-9. Commercial polymer films are usually produced with some orientation. The orientation is generally different in the longitudinal (machine) direction than in the transverse direction. How could you tell which is the machine direction from a stress-strain test? (*Hint:* Refer back to the effects of orientation mentioned in Section 1.8.)

11-10. (a) Calculate the fraction of crystallinity of polythylene samples with densities at 20°C of 926, 940, and 955 kg/m³. Take the specific volume of crystalline polyethylene as 0.989 × 10^{-3} m³/kg and that of amorphous polyethylene as 1.160 × 10^{-3} m³/kg. (b) What assumption did you make in this calculation?

11-11. The Clausius–Clapeyron equation for the effects of pressure on an equilibrium temperature is

$$dP/dT = \Delta H/T \Delta V$$

where ΔH is the enthalpy change and ΔV is the volume change associated with a phase change. Calculate the melting temperature for polyethylene in an injection molding operation under a hydrostatic pressure of 80 MPa. Take $\Delta H = 7.79$ kJ/mol of ethylene repeat units and $T_m = 143.5$°C at 1 atm.

11-12. The hoop stress, σ_h, at the outer surface of a pipe with internal pressure P, outside diameter d_1, and inside diameter d_2 is

$$\sigma_h = \frac{2P d_2^2}{d_1^2 - d_1^2}$$

A well-made PVC pipe has outside diameter 150 mm and wall thickness 15 mm. Measurements on this pipe give $K_{Ic} = 2.4$ MPa $\cdot$ m$^{1/2}$ and yield stress, $\sigma_y = 50$ MN $\cdot$ m^{-2}. Assume that the largest flaws are 100-μm cracks and that the calibration factor for the specimens used to measure fracture toughness is 1.12. The pipe is subjected to a test in which the internal pressure is gradually raised until the pipe fails. Will the pipe fail by yielding or brittle fracture in this burst test? At what pressure will the pipe rupture?

REFERENCES

[1] T. Bremner and A. Rudin, *J. Polym. Sci., Phys. Ed.* **30**, 1247 (1992).
[2] T. A. Kavassalis and P. R. Sundararajan, *Macromolecules* **26**, 4146 (1993).
[3] V. P. Privalko and Y. S. Lipatov, *Makromol. Chem.* **175**, 641 (1974).
[4] M. L. Williams, R. F. Landel, and J. D. Ferry, *J. Am. Chem. Soc.* **77**, 3701 (1955).
[5] K. H. Illers and H. Breuer, *Kolloid. Z.* **17b**, 110 (1961).
[6] A. M. Donald and E. J. Kramer, *J. Polym. Sci., Polym. Phys. Ed.* **20**, 899 (1982).
[7] S. Hashemi and J. G. Williams, *Polym., Eng. Sci.* **16**, 760 (1986).
[8] A. C. Yang and E. J. Kramer, *Macromolecules* **19**, 2010 (1986).
[9] D. G. Cook, Ph. D. Thesis, University of Waterloo, 1992.
[10] B. Z. Jang, D. R. Uhlmann, and J. B. Vander Sande, *J. Appl. Polym. Sci.* **29**, 3409 (1984).
[11] C. B. Bucknall, C. A. Correa, V. P. Soares, and X. C. Zang, *8th Int. Conf. on Deformation, Yield and Fracture of Polymers*, The Institute of Materials, London, Conference Preprints, 9/1–9/3 (1994).
[12] K. C. E. Lee, Ph. D. Thesis, University of Waterloo, 1995.
[13] L. Morbitzer, *J. Appl. Polym. Sci.* **20**, 2691 (1976).
[14] A. E. Griffith, *Phil. Trans. Roy. Soc.* **A221**, 163 (1921).
[15] G. R. Irwin, in "Fracturing of Metals," p. 147. *American Society of Metals*, Cleveland, 1948.
[16] ASTM Standard E399. American Society for Testing Materials, Philadelphia, D. P. Rooke and D. J. Cartwright, "Compendium of Stress Intensity Factors," HMSO, London, (1976).
[17] E. Plati and J. Williams, *Polym. Eng. Sci.* **15**, 470 (1975).
[18] R. W. Truss, *Plast. Rubber Process. Appl.* **10**, 1 (1988).
[19] N. G. McCrum, C. P. Buckley, and C. B. Bucknall, "Principles of Polymer Engineering," p. 204. Oxford University Press, Oxford, 1990.
[20] S. Turner and G. Dean, *Plast. Rubber Process. Appl.* **14**, 137 (1990).
[21] D. R. Moore, *Polymer Testing* **5**, 255 (1985).
[22] R. M. Criens and H. G. Mosle, *Soc. Plast. Eng. Antec.* **40**, 22 (1982).
[23] E. B. Bagley, *J. Appl. Phys.* **28**, 624 (1957).
[24] B. Rabinowitsch, *Z. Physik. Chem.* **A145**, 1 (1929).
[25] T. Bremner, D. G. Cook, and A. Rudin, *J. Appl. Poly. Sci.* **41**, 1617 (1990); **43**, 1773 (1991).
[26] E. B. Bagley and H. P. Schreiber, in "Rheology," Vol. 5 (F. R. Eirich, Ed.). Academic Press, New York, 1969.
[27] F. N. Cogswell, "Polymer Melt Rheology." Woodhead Publishing, Cambridge, England, 1997.
[28] J. M. Dealy, "Rheometers for Molten Plastics." Van Nostrand Reinhold, New York, 1982.
[29] J. M. Dealy and K. F. Wissbrun, "Melt Rheology and Its Role in Plastics Proicessing." Van Nostrand Reinhold, New York, 1990.

Chapter 12

Polymer Mixtures

*Nature is a rag merchant who works up every shred and ort and end
into new creations; like a good chemist whom I found the other day, in
his laboratory, converting his old shirts into pure white sugar.*
—Ralph Waldo Emerson, *Conduct of Life:
Considerations by the Way*, Chapter 12

12.1 COMPATIBILITY

The term *compatibility* is often assumed to mean the miscibility of polymers with
other polymers, plasticizers, or diluents. Decisions as to whether a mixture is
compatible are not always clear-cut, however, and may depend in part on the
particular method of examination and the intended use of the mixture.

A common criterion for compatibility requires the formation of transparent
films even when the refractive indices of the components differ. This means that
the polymer molecules must be dispersed so well that the dimensions of any
segregated regions are smaller than the wavelength of light. Such a fine scale of
segregation can be achieved most readily if the components are miscible. It is
possible, however, that mixtures that are otherwise compatible may appear not to
be, by this standard, if it is difficult to produce an intimate mixture. This may
happen, for example, when two high-molecular-weight polymers are blended or
when a small quantity of a very viscous liquid is being dispersed in a more fluid
medium.

Another criterion is based on the observation that miscible polymer mixtures
exhibit a single glass transition temperature. When a polymer is mixed with
compatible diluents the glass–rubber transition range is broader and the glass
transition temperature is shifted to lower temperatures. A homogeneous blend
exhibits one T_g intermediate between those of the components. Measurements
of this property sometimes also show some dependence on mixing history or on
solvent choice when test films are formed by casting from solution.

Heterogeneous blends with very fine scales of segregation may have very broad glass transition regions and good optical clarity. It is a moot point, then, whether such mixtures are compatible. If the components are not truly miscible, the blend is not at equilibrium but the user may not be able to distinguish between a persistent metastable state and true miscibility.

Many investigators have opted to study polymer compatibility in solution in mutual solvents, because of uncertainty as to whether a bulk mixture is actually in an equilibrium state. Compatible components form a single, transparent phase in mutual solution, while incompatible polymers exhibit phase separation if the solution is not extremely dilute.

Equilibrium is relatively easily achieved in dilute solutions and studies of such systems form the foundation of modern theories of compatibility. Application of such theories to practical problems involves the assumption that useful polymer mixtures require the selection of miscible ingredients and that compatibility can therefore ultimately be explained in terms of thermodynamic stability of the mixture.

This assumption is not necessarily useful technologically. A more practical definition would consider components of a mixture compatible if the blend exhibits an initially desirable balance of properties that does not deteriorate over a time equal to the useful life that is expected of articles made from the mixture. Miscible mixtures are evidently compatible by this criterion. Compatibility is not restricted to such behavior since a blend of immiscible materials can be very useful so long as no significant desegregation occurs while the mixture is being mixed.

12.2   THERMODYNAMIC THEORIES

The terminology in this area is sometimes a little obscure, and Table 12-1 is provided to summarize the classification of solution types.

Thermodynamic theories assume that a necessary requirement for solution and compatibility is a negative Gibbs free energy change when the blend components are mixed. That is to say,

$$\Delta G_m = \Delta H_m - T\Delta S_m < 0 \qquad (12\text{-}1)$$

where the subscript m refers to the change of state corresponding to formation of the mixture and the other symbols have their usual significance. There will be no volume change ($\Delta V_m = 0$) or enthalpy change ($\Delta H_m = 0$) when an ideal solution is formed from its components. The properties of ideal solutions thus depend entirely on entropy effects and

$$\Delta G_m = -T\Delta S_m \qquad \text{(ideal solution)} \qquad (12\text{-}2)$$

TABLE 12-1

Solution Behavior[a]

Solution type	ΔH_m	ΔS_m
Ideal	Zero	Ideal
Regular	Nonzero	Ideal
Athermal	Zero	Nonideal
Irregular	Nonzero	Nonideal

[a] The ideal entropy of mixing ΔS_m is

$$\Delta S_m^{ideal} = -R\sum N_i \ln x_i \qquad (12\text{-}3)$$

where x_i is the mole fraction of component i in the mixture and N_i is the number of moles of species i. Equation (12-3) represents the entropy change in a completely random mixing of all species. The components of the mixture must have similar sizes and shapes for this equation to be true.

12.2.1   Regular Solutions; Solubility Parameter

If ΔH_m is not required to be zero, a so-called "regular solution" is obtained. All deviations from ideality are ascribed to enthalpic effects. The heat of mixing ΔH_m can be formulated in terms of relative numbers of intermolecular contacts between like and unlike molecules. Nonzero ΔH_m values are assumed to be caused by the net results of breaking solvent (1–1) contacts and polymer (2–2) contacts and making polymer–solvent (1–2) contacts [1, 2].

Consider a mixture containing N_1 molecules of species 1, each of which has molecular volume v_1 and can make c_1 contacts with other molecules. The corresponding values for species 2 are N_2, v_2, and c_2, respectively. Each (1–1) contact contributes an interaction energy w_{11}, and the corresponding energies for (2–2) and (1–2) contacts are w_{22} and w_{12}. Assume that only first-neighbor contacts need to be taken into consideration and that the mixing is random. If a molecule of species i is selected at random, one assumes further that the probability that it makes contact with a molecule of species j is proportional to the volume fraction of that species (where i may equal j). If this randomly selected molecule were of species 1 its energy of interaction with its neighbors would be $c_1 w_{11} N_1 v_1 / V + c_1 w_{12} N_2 v_2 / V$, where the total volume of the system V is equal to $N_1 v_1 + N_2 v_2$. The energy of interaction of N_1 molecules of species 1 with the rest of the system is $N_1/2$ times the first term in the previous sum and N_1 times the second term [it takes two species 1 molecules to make a (1–1) contact]; i.e., $c_1 w_{11} N_1^2 v_1 / 2V + c_1 w_{12} N_1 N_2 v_2 / V$. Similarly, the interaction energy of N_2 species 2 molecules with the rest of the system is $c_2 w_{22} N_2^2 v_2 / 2V + c_2 w_{12} N_1 N_2 v_1 / V$. The total contact energy of the system E is the sum of the expressions for (1–1) and (2–2) contacts plus half the sum of the expressions for (1–2) contacts (because we have counted the latter once in connection with N_1 species 1 molecules and again with reference to the N_2 species 2 molecules):

$$E = \frac{c_1 w_{11} N_1^2 v_1 + w_{12} N_1 N_2 (c_1 v_2 + c_2 v_1) + c_2 w_{22} N_2^2 v_2}{2(N_1 V_1 + N_2 V_2)} \tag{12-4}$$

Equation (12-4) can be manipulated to

$$E = N_1 \left(\frac{1}{2} c_1 w_{11} \right) + N_2 \left(\frac{1}{2} c_2 w_{22} \right) + \frac{1}{2} \frac{N_1 N_2}{N_1 v_1 + N_2 v_2}$$
$$\times [w_{12}(c_1 v_2 + c_2 v_1) - w_{11} c_1 v_2 - w_{22} c_2 v_1] \tag{12-5}$$

To eliminate w_{12} it is assumed that

$$\frac{1}{2} w_{12} \left(\frac{c_1}{v_1} + \frac{c_2}{v_2} \right) = \left[\frac{c_1 w_{11}}{v_1} \frac{c_2 w_{22}}{v_2} \right]^{1/2} \tag{12-6}$$

In effect, this takes w_{12} to be equal to the geometric mean of w_{11} and w_{22}. Then

$$E = N_1 \frac{c_1 w_{11}}{2} + N_2 \frac{c_2 w_{22}}{2} - \frac{N_1 N_2 v_1 v_2}{N_1 v_1 + N_2 v_2} \left[\left(\frac{c_1 w_{11}}{2v_1} \right)^{1/2} - \left(\frac{c_2 w_{22}}{2v_2} \right)^{1/2} \right]^2 \tag{12-7}$$

The first two terms on the right-hand side of Eq. (12-7) represent the interaction energies of the isolated components, and the last term is the change in internal energy ΔU_m of the system when the species are mixed. If the contact energies can be assumed to be independent of temperature, the enthalpy change on mixing, ΔH_m, is then

$$\Delta H_m = \Delta U_m = \frac{N_1 N_2 v_1 v_2}{N_1 v_1 + N_2 v_2} \left[\left(\frac{c_1 w_{11}}{2v_1} \right)^{1/2} - \left(\frac{c_2 w_{22}}{2v_2} \right)^{1/2} \right]^2 \tag{12-8}$$

The terms in $(c_i w_{ii}/2v_i)^{1/2}$ are solubility parameters and are given the symbol δ_i. It is convenient to recast Eq. (12-8) in the form

$$H_m = [N_1 N_2 v_1 v_2 / (N_1 v_1 + N_2 v_2)][\delta_1 - \delta_2]^2$$
$$= (N_1 v_1 / V)(N_2 v_2 / V)[\delta_1 - \delta_2]^2 V = V \phi_1 \phi_2 [\delta_1 - \delta_2]^2 \tag{12-9}$$

where the ϕ_i are volume fractions. Hence the heat of mixing per unit volume of mixture is

$$\Delta H_m / V = \phi_1 \phi_2 [\delta_1 - \delta_2]^2 \tag{12-10}$$

where V is the total volume of the mixture. For solutions, subscript 1 refers to the solvent and subscript 2 to the polymeric solute.

Miscibility occurs only if $\Delta G_m \leq 0$ Eq. (12-1). Since ΔS_m in Eq. (12-3) is always positive (the ln of a fraction is negative), the components of a mixture are assumed to be compatible only if $\Delta H_m \leq T \Delta S_m$. Thus solution depends in this analysis on the existence of a zero or small value of ΔH_m. Note that this theory allows only positive (endothermic) heats of mixing, as in Eq. (12-10). In general, then, miscibility is predicted if the absolute value of the $(\delta_1 - \delta_2)$ difference is zero or small.

The convenience of the solubility parameter approach lies in the feasibility of assigning δ values a priori to individual components of the mixture. This is accomplished as follows.

Operationally, the cohesion of a volatile liquid can be estimated from the work required to vaporize unit amount of the material. In this process the molecules are transported from their equilibrium distances in the liquid to an infinite separation in the vapor. The cohesive energy density (sum of the intermolecular energies per unit volume) is at its equilibrium value in the liquid state and is zero in the vapor. By this reasoning, the cohesive energy density in the liquid state is $\Delta U_v / V^0$, in which ΔU_v is the molar energy of vaporization and V^0 is the molar volume of the liquid.

From inspection of Eq. (12-8), it is clear that the solubility parameter δ is the square root of the cohesive energy density. That is,

$$\delta = (\Delta U_v / V^0)^{1/2} \tag{12-11}$$

If the vapor behaves approximately like an ideal gas

$$\delta^2 = (\Delta H_v - RT)/V^0 = (\Delta H_v - RT)\rho/M \tag{12-12}$$

where ρ is the density of liquid with molecular weight M. Thus the heat of vaporization ΔH_v can serve as an experimental measure of δ.

Cohesive energy densities and solubility parameters of low-molecular-weight species can be determined in a straightforward manner by direct measurement of ΔH_v or by various computational methods that are based on other thermodynamic properties of the substance. A polymer is ordinarily not vaporizable, however, and its δ is therefore assessed by equating it to the solubility parameter of a solvent in which is dissolves readily. If solution occurs it is assumed that $\Delta H_m = 0$ and $\delta_1 = \delta_2$ (Eq. 12-10). Experimentally, δ is usually taken as equal to that of a solvent that will produce the greatest swelling of a lightly cross-linked version of the polymer or the highest intrinsic viscosity of a soluble polymer sample. These two experimental methods may, however, give somewhat different results for the same polymer, depending on the polarity and hydrogen-bonding character of the solvent. Such solvent effects are mentioned in more detail in Section 12.2.3.

A more convenient procedure relies on calculations of δ values rather than experimental assessments. Solubility parameters of solvents can be correlated

TABLE 12-2

Group Molar Attraction Constants [4]

Group	Molar attraction F_i $(cal/cc)^{1/2}/mol$	Group	Molar attraction F_i $(cal/cc)^{1/2}$ mol
—CH₃	148.3	—H acidic dimer	−50.47
—CH₂—	131.5	OH aromatic	170.99
>CH—	85.99	NH₂	226.56
—C— with no H	32.03	>NH	180.03
CH₂=olefin	126.54	>N—	61.08
—CH=olefin	121.53	C≡N	354.56
>CH=olefin	84.51	NCO	358.66
—CH=aromatic	117.12	—S—	209.42
—C=aromatic	98.12	Cl₂	342.67
—O—(ether, acetal)	114.98	Cl primary	205.06
—O—(epoxide)	176.20	Cl secondary	208.27
—COO—	326.58	Cl aromatic	161.0
>C=O	262.96	Br	257.88
—CHO	292.64	Br aromatic	205.60
(CO)₂O	567.29		
—OH—	225.84	F	41.33

Structure feature		Structure feature	
Conjugation	23.26	6-Membered ring	−23.44
Cis	−7.13	Ortho substitution	9.69
Trans	−13.50	Meta substitution	6.6
4-Membered ring	77.76	Para substitution	40.33
5-Membered ring	20.99		

with the structure, molecular weight, and density of the solvent molecule [3]. The same procedure is applied to polymers, where

$$\delta = \rho \sum F_i / M_0 \tag{12-13}$$

In Eq. (12-13), ρ is the density of the amorphous polymer at the solution temperature, M_0 is the formula weight of the repeating unit, and $\sum F_i$ is the sum of all the molar attraction constants. A modified version of a compilation [4] of molar attraction constants is reproduced in Table 12-2.

Examples of the use of the tabulated molar constants are given in Fig. 12-1. Such group contribution methods are often used in engineering estimations of other thermodynamic porperties.

The solubility parameter of random copolymers δ_c may be calculated from

$$\delta_c = \sum \delta_i w_i \tag{12-14}$$

where δ_i is the solubility parameter of the homopolymer that corresponds to monomer i in the copolymer and w_i is the weight fraction of repeating unit i

(a) $-\!\!\!+\!CH_2-\overset{\overset{\displaystyle H}{|}}{C}\!\!+\!\!\!-$

with phenyl ring attached

Group	F_i	No. groups	$\sum F_i$	Description
				Density $= 1.05\,\mathrm{g\,cm^{-3}}$
—CH$_2$—	131.5	1	131.5	$M_r = 104\,\mathrm{g\,mol^{-1}}$
>CH—	85.99	1	85.99	$\delta = 1.05(896.77)/104$
—C=(aromatic)	117.12	6	702.72	$= 9.0\,(\mathrm{cal\,cm^{-3}})^{1/2}$
6-membered ring	−23.44	1	−23.44	
			$896.77\left(\frac{\mathrm{cal\,cm^{-3}}}{\mathrm{mol}}\right)^{1/2}$	

(b) $-\!\!\!+\!CH_2-\overset{\overset{\displaystyle H}{|}}{\underset{\underset{\displaystyle CN}{|}}{C}}\!\!+\!\!\!-$

Group	F_i	Description
—CH$_2$—	131.5	Density $= 1.18\,\mathrm{g\,cm^{-3}}$
>CH—	85.99	$M_r = 53$
CN	354.56	$\delta = (1.18)(572.05)/53$
	572.05	$= 12.7\,(\mathrm{cal\,cm^{-3}})^{1/2}$

(c)

Group	F_i	No. groups	$\sum F_i$	Description
—CH$_3$	148.3	2	296.40	Density $= 1.15\,\mathrm{g\,cm^{-3}}$
—CH$_2$—	131.5	2	263.0	$M_r = 284$
>CH—	85.99	1	85.99	$\delta = (1.15)(2572.44)/284$
$-\overset{|}{\underset{|}{C}}-$	32.03	1	32.03	$= 10.4$
6-membered ring	−23.44	2	−46.88	
Para substitution	40.33	2	80.66	
—OH	225.84	1	225.84	
—O—(ether)	114.98	2	229.96	
—C=(aromatic)	117.12	12	1405.44	
			2572.44	

[Conversion factors:
$1\,\mathrm{MPa}^{1/2} = 1\,(\mathrm{J\,cm^{-3}})^{1/2}$
$= 0.49\,(\mathrm{cal\,cm^{-3}})^{1/2};$
$1\,\mathrm{cm^3\,mol^{-1}} = 10^{-6}\,\mathrm{m^3\,mol^{-1}}.]$

Fig. 12-1. Calculation of solubility parameters from molar attraction constants.

is the copolymer [5]. Alternating copolymers can be treated by taking the copolymer repeating unit as that of a homopolymer (see Fig. 12-1c for example). No satisfactory method exists for assigning values to block or graft copolymers.

Mixtures of solvents are often used, especially in formulating surface coatings. It is not unusual to find that a mixture of two nonsolvents will be a solvent for a given polymer. This occurs if one nonsolvent δ value is higher and the other is lower than the solubility parameter of the solute. The solubility parameter of the mixture δ_m is usually approximated from

$$\delta_m = \delta_A \phi_A + \delta_B \phi_B \tag{12-15}$$

where the ϕ's are volume fractions.

It is believed that the temperature dependence of δ can be neglected over the range normally encountered in industrial practice. Most tabulated solubility parameters refer to 25°C.

Solubility can be expected if $\delta_1 - \delta_2$ is less than about $2(\text{cal cm}^{-3})^{1/2}$ $[4(\text{Mpa})^{1/2}]$ and there are no strong polar or hydrogen-bonding interactions in either the polymer or solvent. Crystalline polymers, however, will be swollen or softened by solvents with matching solubility parameters but will generally not dissolve at temperatures much below their crystal melting points.

Table 12-3 lists solubility parameters for some common polymers and solvents. The units of δ are in $(\text{energy/volume})^{1/2}$ and those tabulated, in $\text{cal}^{1/2} \text{ cm}^{-3/2}$, are called *hildebrands*.

The use of the geometric mean expedient to calculate w_{12} in Eq. (12-6) in effect assumes that the cohesion of molecules of both species of the mixture is entirely due to dispersion forces. To allow for the influence of hydrogen-bonding interactions, it has been found useful to characterize solvents qualitatively as poorly, moderately, or strongly hydrogen-bonded. The solvents listed in Table 12-3 are grouped according to this scheme. Mutual solubility may not be achieved even if $\delta_1 \simeq \delta_2$ when the two ingredients of the mixture have different tendencies for hydrogen bond formation.

The practice of matching solubility parameters and hydrogen-bonding tendency involves some serious theoretical problems, but it is useful if used with caution. For example, polystyrene, which is classed as poorly hydrogen-bonded and has a δ value of 18.4 $(\text{Mpa})^{1/2}$ is highly soluble in the poorly hydrogen-bonded solvents benzene and chloroform, both of which have matching solubility parameters. The polymer can be dissolved in methyl ethyl ketone ($\delta = 19.0$ medium hydrogen bonding), but the latter is not nearly as good as solvent as either of the first pair. (The intrinsic viscosity of a polystyrene of given molecular weight is higher in chloroform or benzene than in methyl ethyl ketone.) On the other hand, poly(methyl methacrylate) has practically the same δ as polystyrene but is classed as medium hydrogen-bonded. The two polymers are regarded as incompatible when both have high molecular weights, but benzene and chloroform do not seem to be weaker solvents than methyl ethyl ketone for poly(methyl methacrylate).

Some of the problems noted here probably reflect the use of an oversimplified view of hydrogen bonding, in general. However, any attempt to correct this deficiency will most likely complicate the predictive method without a commensurate gain in practical utility.

Improvements on the simple solubility parameter approach are summarized in Section 12.2.3.

TABLE 12-3

Solubility Parameters

(a) Solubility Parameters for Some Common Solvents $(MPa)^{1/2a}$

(i) Poorly hydrogen-bonded (generally hydrocarbons and derivatives containing halogen, nitrate, and cyano groups)		(ii) Moderately hydrogen-bonded (generally esters, ethers, ketones)	
Solvent	δ	Solvent	δ
n-Hexane	14.9	Diisodecyl phthalate	14.7
Carbon tetrachloride	17.6	Diethyl ether	15.1
Toluene	18.2	Isoamyl acetate	16.0
Benzene	18.8	Dioctyl phthalate	16.2
Chloroform	19.0	Isobutyl chloride	16.6
Tetrahydronaphthalene	19.4	Methyl isobutyl ketone	17.2
Methylene chloride	19.8	Dioctyl adipate	17.8
Carbon disulfide	20.5	Tetrahydrofuran	18.6
Nitrobenzene	20.5	Methyl ethyl ketone	19.0
Nitroethane	22.7	Acetone	20.3
Acetonitrile	24.4	1,4-Dioxane	20.5
Nitromethane	26.0	Diethylene glycol monomethyl ether	20.9
		Furfural	22.9
		Dimethyl sulfoxide	24.6

(iii) Strongly hydrogen-bonded (generally alcohols, amides, amines, acids)

Solvent	δ	Solvent	δ
Lauryl alcohol	16.6	1-Butanol	23.3
Piperidene	17.8	Diethylene glycol	24.8
Tetraethylene glycol	20.3	Propylene glycol	25.8
Acetic acid	20.7	Methanol	29.7
Meta-cresol	20.9	Ethylene glycol	29.9
t-Butanol	21.7	Glycerol	33.8
Neopentyl glycol	22.5	Water	47.9

TABLE 12-3

Continued

(b) Solubility parameters of Polymers $(MPa)^{1/2}$

Polymer[b]	δ	H-bonding group[c]
Polytetrafluoroethylene	12.7	Poor
Polyethylene	16.4	Poor
Polyisobutene	17.0	Poor
Polypropylene	17.0	Poor
Polybutadiene	17.2	Poor
Polyisoprene	17.4	Poor
Poly(butadiene-*co*-styrene) 75/25	17.4	Poor
Poly(tetramethylene oxide)	17.6	Medium
Poly(butyl methacrylate)	18.0	Poor?
Polystyrene	18.4	Poor
Poly(methyl methacrylate)	19.0	Medium
Poly(butadiene-*co*-acrylonitrile) 75/25	19.2	Poor
Poly(ethyl acrylate)	19.2	Medium
Poly(vinyl acetate)	19.7	Medium
Poly(vinyl chloride)	19.9	Medium
Poly(methyl acrylate)	20.7	Medium
Polyformaldehyde	20.9	Medium
Ethyl cellulose	21.1	Strong
Poly(vinyl chloride-*co*-vinyl acetate) 87/13	21.7	Medium
Cellulose diacetate	23.3	Strong
Poly(vinyl alcohol)	26.0	Strong
Polyacrylonitrile	26.0	Poor
Nylon-6,6	28.0	Strong

[a] Selected data from Ref. [6].

[b] Compositions of copolymers are in parts by weight.

[c] The hydrogen-bonding group of each polymer has been taken as equivalent to that of the parent monomer. (The hydrogen-bonding tendency can be assigned qualitatively in the order alcohols > ethers > ketones > aldehydes > esters > hydrocarbons or semiquantitatively from infrared absorption shifts of CH_3OD in a reference solvent and in the liquid of interest [7].)

12.2.2   Flory–Huggins Theory

Nonideal thermodynamic behavior has been observed with polymer solutions in which ΔH_m is practically zero. Such deviations must be due to the occurrence of a nonideal entropy, and the first attempts to calculate the entropy change when a long chain molecule is mixed with small molecules were due to Flory [8] and Huggins [9]. Modifications and improvements have been made to the original theory, but none of these variations has made enough impact on practical problems of polymer compatibility to occupy us here.

The Flory–Huggins model uses a simple lattice representation for the polymer solution and calculates the total number of ways the lattice can be occupied by small molecules and by connected polymer segments. Each lattice site accounts for a solvent molecule or a polymer segment with the same volume as a solvent molecule. This analysis yields the following expression for ΔS_m, the entropy of mixing N_1 moles of solvent with N_2 moles of polymer.

$$\Delta S_m = -R(N_1 \ln \phi_1 + N_2 \ln \phi_2) \qquad (12\text{-}16)$$

where the ϕ_i are volume fractions and subscripts 1 and 2 refer to solvent and polymer, respectively. The polymer consists of r_2 segments, each of which can displace a single solvent molecule from a lattice site. Thus r_2 is defined as

$$r_2 = M/\rho V_1^0 \qquad (12\text{-}17)$$

where M is the molecular weight of the polymer that would have density ρ in the corresponding amorphous state at the solution temperature and V_1^0 is the molar volume of the solvent. The number of lattice sites needed to accommodate this mixture is $(N_1 + N_2 r_2)L$, where L is Avogadro's constant.

Equation (12-16) is similar to Eq. (12-1), except that volume fractions have replaced mole fractions. This difference reflects the fact that the entropy of mixing of polymers is small compared to that of micromolecules because there are fewer possible arrangements of solvent molecules and polymer segments than there would be if the segments were not connected to each other.

Equation (12-16) applies also if two polymers are being mixed. In this case the number of segments r_i in the ith component of the mixture is calculated from

$$r_i = M_i/\rho_i V_r \qquad (12\text{-}16a)$$

where V_r is now a reference volume equal to the molar volume of the smallest polymer repeating unit in the mixture. The corresponding volume fraction ϕ_i is

$$\phi_i = N_i r_i \Big/ \sum N_i r_i \qquad (12\text{-}18)$$

The entropy gain per unit volume of mixture is much less if two polymers are mixed than if one of the components is a low-molecular-weight solvent, because N_1 is much smaller in the former case.

To calculate ΔH_m (the enthalpy of mixing) the polymer solution is approximated by a mixture of solvent molecules and polymer segments, and ΔH_m is estimated from the number of 1, 2 contacts, as in Section 12.2.1. The terminology is somewhat different in the Flory–Huggins theory, however. A site in the liquid lattice is assumed to have z nearest neighbors and a line of reasoning similar to that developed above for the solubility parameter model leads to the expression

$$\Delta H_m = zw(N_1 + N_2 r_2)\phi_1 \phi_2 L \qquad (12\text{-}19)$$

for the enthalpy of mixing of N_1 moles of solvent with N_2 moles of polymer. Here w is the increase in energy when a solvent-polymer bond is formed from molecules that were originally in contact only with species of like kind.

Now the Flory–Huggins interaction parameter χ (chi) is defined as

$$\chi = zwL/RT \tag{12-20}$$

This dimensionless quantity is the polymer–solvent interaction energy per mole of solvent, divided by RT, which itself has the dimensions of energy. Since $\phi_1 = N_1(N_1 + N_2 r_2)$, Eq. (12-19) can be recast to give the enthalpy of forming a mixture with volume fraction ϕ_2 of polymer in N_1 moles of solvent as

$$\Delta H_m = RT \chi N_1 \phi_2 \tag{12-21}$$

The total volume V of this solution is $(N_1 + N_2 r_2) V_1^0$, where V_1^0 is the molar volume of the solvent. Then the enthalpy of mixing per unit volume of mixture is

$$\frac{\Delta H_m}{V} = \frac{RT \chi N_1 \phi_2}{(N_1 + N_2 r_2) V_1^0} = \frac{RT \chi \phi_1 \phi_2}{V_1^0} \tag{12-22}$$

If the Flory–Huggins value in Eq. (12-22) is now equated to the solubility parameter expression of Eq. (12-10), it can be seen that

$$\chi = V_1^0 (\delta_1 - \delta_2)^2 / RT \tag{12-23}$$

Equation (12-23) suffers from the same limitations as the simple solubilty parameter model, because the expression for H_m is derived by assuming that intermolecular forces are only nondirectional van der Waals interactions. Specific interactions like ionic or hydrogen bonds are implicitly eliminated from the model. The solubility parameter treatment described to this point cannot take such interactions into account because each species is assigned a solubility parameter that is independent of the nature of the other ingredients in the mixture. The χ parameter, on the other hand, refers to a pair of components and can include specific interactions even if they are not explicitly mentioned in the basic Flory–Huggins theory. Solubility parameters are more convenient to use because they can be assigned *a priori* to the components of a mixture. χ values are more realistic, but have less predictive use because they must be determined by experiments with the actual mixture.

From Eqs. (12-16) and (12-21) the Gibbs free energy change on mixing at temperature T is

$$\Delta G_m = \Delta H_m - T \Delta S_m = RT(\chi N_1 \phi_2 + N_1 \ln \phi_1 + N_2 \ln \phi_2) \tag{12-24}$$

Now, since

$$(\partial G_m / \partial N_1)_{T,P,N_2} = (\partial G_{\text{solution}} / \partial N_1)_{N_2,T,P} - \left(\partial G_1^0 / \partial N_1\right)_{N_2,T,P}$$

then

$$(\partial G_m / \partial N_1)_{T,P,N_2} = \mu_1 - G_1^0 \tag{12-25}$$

Thus, the difference in chemical potential of the solvent in the solution (μ_1) and in the pure state at the same temperature (G_1^0) can be expressed in terms of χ by differentiating Eq. (12-24) with respect to N_1. Experimentally, $\mu_1 - G_1^0$ can be obtained from measurements of any of several thermodynamic properties of the polymer solution, as explained in Section 2.10.2. It can be shown then that the second virial coefficient (Eq. 2-76) is given by

$$A_2 = (0.5 - \chi)/\rho^2 V_1^0 \tag{12-26}$$

where ρ is the polymer density at the particular temperature. Since $A_2 = 0$ in theta mixtures (Section 2.11.1) where the polymer is insoluble, the condition for compatibility is $\chi < 0.5$. When the mixture is produced from two polymers A and B, Eq. (12-24) can be recast in the form

$$\Delta G_m = RTV[\chi_{AB}\phi_A(1 - \phi_A) + (\phi_A/V_A)\ln\phi_A$$
$$+ (\phi_B/V_B)\ln(1 - \phi_A)] \tag{12-27}$$

where V is the total volume of the mixture, V_i is the molar volume of species i, and χ_{AB} is the interaction parameter for the two polymeric species. Since $V_i = M_i/\rho_i$ this is also equivalent to

$$\Delta G_m = RTV[\chi_{AB}\phi_A(1 - \phi_A) + (\phi_A\rho_A/M_A)\ln\phi_A$$
$$+ (\phi_B\rho_B/M_B)\ln(1 - \phi_B)] \tag{12-28}$$

The logarithmic terms are negative because the ϕ_i are less than one. Therefore ΔG_m is less negative and the mixture is less stable the higher the molecular weights of the components. In fact, mixtures of high polymers are indicated to be always incompatible unless $\chi_{AB} \leq 0$. This situation will occur only when the enthalpy of mixing is less than or equal to zero, i.e., when there are some specific interactions (not of the van der Waals type) between the components of the mixture.

The Flory–Huggins theory predicts that the solubility of polymers will be inversely related to their molecular sizes. Compatibility of polymers with other materials is certainly affected by the molecular weight of the macromolecules. Higher molecular weight materials are generally less soluble in solvents. The influence of molecular weight on the stability of other mixtures is more complex. Higher molecular weight species are generally more difficult to disperse, especially if they are minor components of mixtures in which the major species are lower molecular weight, less viscous substances. If they can be dispersed adequately, however, their diffusion rates and consequent rates of segregation will be correspondingly less and the dispersion may appear to be stable as a result.

The Flory–Huggins model differs from the regular solution model in the inclusion of a nonideal entropy term and replacement of the enthalpy term in solubility parameters by one in an interaction parameter χ. This parameter characterizes a *pair* of components whereas each δ can be deduced from the properties of a single component.

In the initial theory, χ was taken to be a function only of the nature of the components in a binary mixture. It became apparent, however, that it depends on concentration and to some extent on molecular weight. It is now considered to be a free energy of interaction and thus consists of enthalpic and entropic components with the latter accounting for its temperature dependence.

The Flory–Huggins theory has been modified and improved and other models for polymer solution behavior have been presented. Many of these theories are more satisfying intellectually than the solubility parameter model but the latter is still the simplest model for predictive uses. The following discussion will therefore focus mainly on solubility parameter concepts.

12.2.3   Modified Solubility Parameter Models

The great advantage of the solubility parameter model is in its simplicity, convenience, and predictive ability. The stability of polymer mixtures can be predicted from a knowledge of the solubility parameters and hydrogen-bonding tendencies of the components. The predictions are not always very accurate, however, because the model is so oversimplified. Some examples have been given in the preceding section. More sophisticated solution theories are not predictive. They contain parameters that can only be determined by analysis of particular mixtures, and it is not possible to characterize individual components *a priori*. The solubility parameter scheme is therefore the model that is most often applied in practice.

Numerous attempts have been made to improve the predictive ability of the solubility parameter method without making its use very much more cumbersome. These generally proceed on the recognition that intermolecular forces can involve dispersion, dipole–dipole, dipole–induced dipole, or acid–base interactions, and a simple δ value is too crude an overall measurement of these specific interactions.

The most comprehensive approach has been that of Hansen [10,11], in which the cohesive energy δ^2 is divided into three parts:

$$\delta^2 = \delta_d^2 + \delta_p^2 + \delta_H^2 \tag{12-29}$$

where the subscripts d, p, and H refer, respectively, to the contributions due to dispersion forces, polar forces, and hydrogen-bonding. A method was developed for the determination of these three parameters for a large number of solvents. The value of δ_d was taken to be equal to that of a nonpolar substance with nearly the same chemical structure as a particular solvent. Each solvent was assigned a point in δ_d, δ_p, δ_H space in which these three parameters were plotted on mutually perpendicular axes. The solubility of a number of polymers was measured in a series of solvents, and the δ_p and δ_H values for the various solvents which all dissolved a given polymer were shifted until the points for these solvents were close. This is a very tedious and inexact technique. More efficient methods include molecular dynamics calculations [12] and inverse gas chromatographic analyses [13].

TABLE 12-4

Three-Dimensional [Hansen] Solubility Parameters $(Mpa)^{1/2}$

	δ_d	δ_p	δ_H
N-Hexane	14.9	0	0
Benzene	18.4	0	0
Chloroform	17.8	3.1	5.7
Nitrobenzene	18.8	12.4	4.1
Diethyl ether	14.7	2.9	5.2
Iso-amyl acetate	15.5	3.1	7.0
Dioctyl phthalate	16.7	7.0	3.1
Methyl isobutyl ketone	15.5	6.2	4.1
Tetrahydrofuran	16.9	5.8	8.1
Methyl ethyl ketone	16.1	9.1	5.2
Acetone	15.7	10.5	7.0
Diethylene glycol monomethyl ether	16.3	7.8	12.8
Dimethyl sulfoxide	18.6	16.5	10.3
Acetic acid	14.7	8.0	13.6
m-Cresol	18.2	5.2	13.0
1-Butanol	16.1	5.8	15.9
Methylene glycol	16.3	14.9	20.7
Methanol	15.3	12.4	22.5

The three-dimensional solubility parameter concept defines the limits of compatibility as a sphere. Values of these parameters for some of the solvents listed earlier in Table 12-3 are given in Table 12-4. More complete lists are available in handbooks and technological encyclopedias. The recommended procedure in conducting a solubility parameter study is to try to dissolve the polymeric solute in a limited number of solvents that are chosen to encompass the range of subsolubility parameters. A three-dimensional plot of solubility then reveals a "solubility volume" for the particular polymer in δ_d, δ_p, δ_H space.

Three dimensional presentations are cumbersome and it is more convenient to transform the Hansen parameters into fractional parameters as defined by [14]

$$f_d = \delta_d/\delta \tag{12-29a}$$

$$f_p = \delta_p/\delta \tag{12-29b}$$

$$f_H = \delta_H/\delta \tag{12-29c}$$

The data can now be represented more conveniently in a triangular diagram, as in Fig. 12-2. This plot shows the approximate limiting solubility boundaries for poly(methyl methacrylate). The boundary region separates efficient from poor solvents. The probable solubility parameters of the solute polymer will be at the heart of the solubility region. The boundaries are often of greater interest than the central region of such loops because considerations of evaporation rates, costs, and other properties may also influence the choice of solvents.

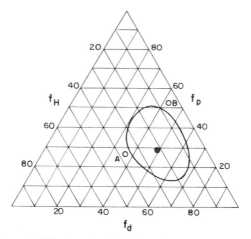

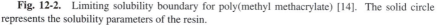

Fig. 12-2. Limiting solubility boundary for poly(methyl methacrylate) [14]. The solid circle represents the solubility parameters of the resin.

The design of blended solvents is facilitated by use of these subparameters, along with graphical analyses. Thus, referring again to Fig. 12-2, the polymer will be insoluble in solvents A and B but a mixture of the two should be a solvent. It has been suggested also that a plot of δ_p versus δ_H should be sufficient for most practical purposes, since δ_d values do not vary greatly, at least among common solvents.

The procedures outlined have a practical use, but it should be realized that the subparameter models have some empirical elements. Assumptions such as the geometric mean rule (Eq. 12-6) for estimating interaction energies between unlike molecules may have some validity for dispersion forces but are almost certainly incorrect for dipolar interactions and hydrogen bonds. Experimental uncertainties are also involved since solubility "loops" only indicate the limits of compatibility and always include doubtful observations. Some of the successes and limitations of various versions of the solubility parameter model are mentioned in passing in the following sections which deal briefly with several important polymer mixtures.

12.3 SOLVENTS AND PLASTICIZERS

12.3.1 Solvents for Coating Resins

The most widespread use of the solubility parameter has been in the formulation of surface coatings. Single solvents are rarely used because the requirements for evaporation rates, safety, solvency, and so on generally mean that a solvent blend is more effective and less costly. Further, use of nonsolvents is often effective for cost reductions. The cheapest organic solvents are hydrocarbons, whereas most solvents for film-forming polymers are moderately hydrogen bonded and have δ values in the range 16–20 (Mpa)$^{1/2}$. The simple example that was given in connection with Fig. 12-2 illustrates how such blends can be formulated. The procedure can be used to blend solvents with nonsolvents or even to make a solvent mixture from nonsolvents. The latter procedure must be used with caution for surface coatings, however, since the effective solubility parameter of the system will drift toward those of the higher boiling components as the solvents evaporate. If these residual liquids are nonsolvents the final coalesced polymer film may have poor clarity and adhesion to the substrate. The slowest evaporating component of the solvent blend should be a good solvent for the polymer in its own right, since the last solvent to leave the film has a strong influence on the quality of the film.

12.3.2 Plasticization of Polymers

A plasticizer is a material that enhances the processibility or flexibility of the polymer with which it is mixed. The plasticizer may be a liquid or solid or another polymer. For example, rigid poly(vinyl chloride) is a hard solid material used to make credit cards, pipe, house siding, and other articles. Mixing with about 50–100 parts by weight of phthalate ester plasticizers converts the polymer into leathery products useful for the manufacture of upholstery, electrical insulation, and other items. Plasticizers in surface coatings enhance the flow and leveling properties of the material during application and reduce the brittleness of the dried film.

Some degree of solvency of the plasticizer for the host polymer is essential for plasticization. Not surprisingly, a match of solubility parameters of the plasticizer and polymer is often a necessary but not a sufficient condition for compatibility. In the case of PVC, the dielectric constants of the plasticizer should also be near that of the polymer.

It is often useful to employ so-called "secondary plasticizers," which have limited compatibility with the host polymer. Thus, aliphatic diesters are poorly compatible with PVC, but they can be combined with the highly compatible phthalate ester plasticizers to improve low temperature properties of the blend.

Continued addition of a plasticizer to a polymer results in a progressive reduction in the glass transition temperature of the mixture. This suggests that the plasticizer acts to facilitate relative movement of macromolecules. This can happen if the plasticizer molecules are inserted between polymer segments to space these segments further apart and thus reduce the intensity of polymer-polymer interactions. Such a mode of action is probably characteristic of low temperature

plasticizers for PVC, like dioctyl adipate. Plasticizers with more specific interactions with the polymer will reduce the effective number of polymer–polymer
contacts by selectively solvating the polymer at these contact points. PVC plasticizers like diisoctyl phthalate seem to act in the latter fashion.

Rubbers are plasticized with petroleum oils, before vulcanization, to improve
processability and adhesion of rubber layers to each other and to reduce the cost and
increase the softness of the final product. Large quantities of these "oil-extended"
rubbers are used in tire compounds and related products. The oil content is frequently about 50 wt% of the styrene–butadiene rubber. The chemical composition
of the extender oil is important. Saturated hydrocarbons have limited compatibility
with most rubbers and may "sweat-out." Aromatic oils are more compatible and
unsaturated straight chain and cyclic compounds are intermediate in solvent power.

12.4   FRACTIONATION

The properties of a polymer sample of given composition, structure, and average
molecular weight are not uniquely determined unless the distribution of molecular weight about the mean is also known. Methods to determine this distribution
include gel permeation (size exclusion) chromatography and various fractionation
techniques. Fractionation is a process for the separation of a chemically homogeneous polymer specimen into components (called "fractions") which differ in
molecular size and have narrower-molecular-weight distributions than the parent
material. Ideally, each fraction would be monodisperse in molecular weight but
such a separation has not been approached in practice and the various fractions
that are collected always overlap to some extent.

It should be noted that all the fractionation process does is provide narrower-
molecular-weight distribution materials. The molecular weight distribution of the
original material cannot be reconstructed until the average molecular weight of
each fraction is obtained by other independent measurements.

Fractionation depends on the differential solubility of macromolecules with
different sizes. It has been displaced in many cases by size exclusion chromatography as a means for measuring molecular weight distributions, but it is till often
the only practical way of obtaining narrow fractions in sufficient quantities for the
study of physical properties of well-characterized specimens. It is also part of
the original procedure for the calibration of solution viscosity measurements for
estimation of molecular weights.

The Flory–Huggins theory leads to some useful rules for fractionation operations. Only the results will be summarized here. Details of the theory and
experimental methods are available in Refs. [15], [16], and other sources.

Consider a polymeric species with degree of polymerization i in solution. The homogeneous solution can be caused to separate into two phases by decreasing the affinity of the solvent for the polymer by lowering the temperature or adding some poorer solvent, for example. If this is done carefully, a small quantity of polymer-rich phase will separate and will be in equilibrium with a larger volume of a solvent-rich phase. The chemical potential of the i-mer will be the same in both phases at equilibrium, and the relevant Flory–Huggins expression is

$$\ln\left(\phi'_i/\phi_i\right) = \sigma_i \tag{12-30}$$

where ϕ'_i and ϕ_i are the volume fraction of polymer of degree of polymerization i in the polymer-rich and solvent-rich phases, respectively. Sigma (σ) is a function of the volume fractions mentioned and the number average molecular weights of all the polymers in each phase, as well as the dimensionless parameter χ (Eq. 12-23). Sigma cannot be calculated exactly, but it can be shown to be always positive [15]. It follows then from Eq. (12-30) that $\phi'_i > \phi_i$, regardless of i. This means that all polymer species tend to concentrate preferentially in the polymer-rich phase. However, since ϕ'_i/ϕ_i increases exponentially with i, the latter phase will be relatively richer in the larger than in the smaller macromolecules.

Fractionation involves the adjustment of the solution conditions so that two liquid phases are in equilibrium, removal of one phase and then adjusting solution conditions to obtain a second separated phase, and so on. Polymer is removed from each separated phase and its average molecular weight is determined by some direct measurement such as osmometry or light scattering.

It is evident that both phases will contain polymer molecules of all sizes. The successive fractions will differ in average molecular weights but their distributions will overlap. Various mathematical techniques have been used to allow for such overlapping in the reconstruction of the molecular weight distribution of the parent polymer from the average molecular weights measured with fractions.

Although refractionation may narrow the molecular weight distributions of "primary" fractions such operations are subject to a law of diminishing returns because of the complications of Eq. (12-30) which have just been mentioned.

If the volumes of the polymer rich and solvent-rich phases are V' and V, respectively, then the fraction f_i of i-mer that remains in the solvent-rich phases is given by

$$f_i = \frac{\phi_i V}{\phi_i V + \phi'_i V'} = \frac{1}{1 + R e^{\sigma_i}} \tag{12-31}$$

where $R = V'/V$. Similarly, the fraction of i-mer in the polymer-rich phase is

$$f'_i = R e^{\sigma_i}/(1 + R e^{\sigma_i}) \tag{12-32}$$

and

$$f_i/f_i' = 1/\mathrm{Re}^{\sigma_i} \tag{12-33}$$

If the volume of the solvent-rich phase is much greater than that of the polymer-rich phase ($R \ll 1$), then most of the smaller macromolecules will remain in the former phase (Eq. 12-33). Also, as i increases, the proportion of i-mer in the polymer-rich phase will increase.

Dilute solutions are needed for efficient fractionation. When fractionation is effected by gradual precipitation of polymer from solution, good practice requires that the initial polymer concentration decrease with increasing molecular weight of the whole polymer. A 10-g sample of a low-molecular-weight polymer should be dissolved in about 1 liter of solvent while a high-molecular-weight polymer might easily require 10 liters.

Temperature rising elution fractionation (TREF) is a useful technique for characterizing the distribution of branches and other uncrystallizable entities in semicrystalline polymers. Recall that regularity of polymer structure is necessary for crystallizability (Section 4.1.2) and branches and comonomer residues cannot usually fit into crystal lattices. This method is particularly valuable with polyolefins like polyethylene, whose properties are affected by the distributions of both molecular weight and branching [17]. The procedure involves dissolution of the sample in a solvent, followed by slow cooling to deposit successive layers of less and less crystallizable species onto an inert substrate, like silanized silica. The material here consists of onion-skin layers of polymer, with the least regular (i.e., most branched) species on the outside. The foregoing procedure is then reversed, as the precipitated polymer is eluted by flowing solvent at progressively increasing temperatures. The concentration of eluting dissolved polymer and the corresponding branch concentration can be monitored by infra-red detection at different wavelengths [18].

12.5 PRACTICAL ASPECTS OF POLYMER BLENDING [19]

Polymer blends have become a very important subject for scientific investigation in recent years because of their growing commercial acceptance. Copolymerization and blending are alternative routes for modifications of properties of polymers. Blending is the less expensive method. It does not always provide a satisfactory alternative to copolymerization, of course, but polymer blends have been successfully used in an increasing number of applications in recent years. Such successes encourage more attempts to apply this technique to a wider range of problems in polymer-related industries.

12.5.1   Objectives in Making Blends

It is usually, but not always, desired to make a blend whose properties will not change significantly during its normal usage period. Unstable blends are sometimes required, however. An example is the use of slip agents (surface lubricants) in polyolefin films. The additives must be sufficiently compatible with the host resin not to exude from the polymer melt onto the extruder barrel walls during film extrusion. If the slip agent migrated to this boundary the resin would turn with the screw and would not be extrudable. The slip agent must exude from the solid polymer, however, since its lubrication function is exercised only on the surface of the final film. Amides of long chain fatty acids have the right balance of controlled immiscibility for such applications in polyolefin plastics. Lubricants and antistatic agents are other examples of components of blends which are not designed to be stable.

This section will concentrate on stable blends since these are of greater general interest. It should be noted that stability in this context does not necessarily imply miscibility or even that the mixture attains a state of thermodynamic equilibrium during its useful lifetime. More generally, all that is required is that the components of the mixture adhere to each other well enough to maintain an adequate mechanical integrity for the particular application and that this capacity be maintained for the expected reasonable lifetime of the particular article.

12.5.2   Blending Operations

The manufacture of useful, stable blends involves two major steps: (1) The components are mixed to a degree of dispersion that is appropriate for the particular purpose for which the blend is intended; (2) additional procedures are followed, if necessary, to ensure that the dispersion produced in step 1 will not demix during its use period.

Note that it is useful to consider step 2 as a problem involving retardation of a kinetic process (demixing). The viewpoint that focuses on blending as a problem in thermodynamic stability is included here as a special case but should not exclude other routes to stabilization which may be practical under some circumstances.

To illustrate this point consider the production of lacquers for PVC films and sheeting. Such lacquers contain a PVC homopolymer or low-acetate vinyl chloride-vinyl acetate copolymer, poly(methyl methacrylate), a plasticizer and perhaps some stabilizers, dulling agents (such as silica), pigments, and so on. Methyl ethyl ketone (MEK) is the solvent of choice because it gives the best balance of low toxicity, volatility, and low cost. Any other solvent is effectively excluded for a variety of reasons such as cost, inadequate volatility for coating machines designed to dry MEK, unfamiliar odor, toxicity, and so on. Unfortunately MEK is really a poor solvent for this mixture. The solids concentrations required for effective coatings produce a mucuslike consistency if the lacquer is produced by conventional slow speed stirring and heating. The mixture is very thixotropic and tends to form uneven coatings and streaks when applied by the usual roller coating methods. For reasons listed above, addition of better solvents is not an acceptable route to improvement of the quality of the coating mixture.

A practical procedure is readily apparent, however, if one proceeds by steps 1 and 2 above. A good dispersion is first made by intensive mechanical shearing. High energy mixers are available which can boil the solvent in a few minutes just from the input of mechanical work. The solid ingredients are added slowly to the initially cold solvent while it is being sheared in such an apparatus. This produces a finely dispersed, hot mixture. It is not a true solution, however, and will revert eventually to a mucuslike state. To retard this demixing process one can add a small concentration of an inexpensive nonsolvent like toluene. This makes the liquid environment less hospitable for the solvated polymer coils which shrink and are thus less likely to overlap and segregate. The final mixture is still not stable indefinitely, but it can be easily redispersed by whipping with an air mixer at the coating machine.

Although the scientific principles behind this simple example of practical technology are easily understood, it illustrates the benefits that can be realized by considering the blending process as a dispersion operation that may be followed, if necessary, by an operation to retard the rate at which the ingredients of the blend demix. In special cases, of course, the latter operation may be rendered unnecessary by the selection of blend ingredients that are miscible in the first instance.

The basic requirements for achieving good dispersions of polymeric mixtures have been reviewed elsewhere [20–22] and will not be repeated here in any detail. Extruders and intensive mixers produce mainly laminar mixing in which the interfacial area between components of the mixture is increased in proportion to the total amount of shear strain which is imparted to the fluid substrate. Better laminar mixing is realized if the viscosities of the components of the blend are reasonably well matched. Such mixers operate by moving their inner metal surfaces relative to each other. Shear strain is imparted to the polymer mixture if it adheres to the moving walls of the mixer. When the ingredients of the mixture have different viscosities, the more fluid component will take up most or all of the imparted strain, particularly if it is the major ingredient of the mixture. Thus it is easy to melt blend a minor fluid component with a major, more viscous ingredient, but a minor viscous component may swim in a more fluid sea of the major component without being dispersed. Similar considerations apply if there are serious mismatches in the melt elasticities of the components of a mixture.

It is also well known in compounding technology that the quality of a dispersion may be sensitive to the condition of the blend that is fed to the mixing machine and in some cases also to the order in which the ingredients are added to the mixer.

Some mixers provide dispersive as well as laminar mixing. In dispersive mixing, the volume elements of the compound are separated and shuffled. Dispersive mixing processes can be added to laminar mixing operations by introducing mixing sections into extruder screws or installing stationary mixers in the extruder discharge sections.

The correlation between quality of a laminar mixture and the total shear strain which the material has undergone applies particularly to blends of polymers. When hard or agglomerated components are being mixed, however, it is necessary to subject such materials to a high shear stress gradient, and special equipment and processes have been developed for such purposes. The rubber and coatings industries in particular abound with examples of such techniques.

Special note should be taken of the difficulty of forming intimate mixtures of some semicrystalline polymers, and particularly of polyethylenes. Experimental and theoretical studies have shown that local structure persists in such polymers even at temperatures above the T_m measured by differential scanning calorimetry (DSC). These structures consist of folded chain domains in polyethylenes and of helical entities in polypropylene. That is to say, in these polymers, at least, the lowest energy states of the uncrystallized material are characterized by a minima in free energy, rather maxima in entropy. Molecular dynamics simulations of mixtures of linear polyethylene and isotactic polypropylene indicate that the two species will segregate into distict domains in the melt even when the initial state was highly interpenetrating [23]. The formation of locally ordered regions is expected to be more significant for longer polymer chains. From a theoretical point of view, such observations imply that the mixing of polymers cannot be described adequately by a purely statistical model as in the original Flory–Huggins formulation. This theory has been generalized by some researchers to decompose the interaction parameter, χ, into enthaplic and entropic terms, where the latter may be construed as reflecting "local structures" [24].

Practically, the foregoing phenomena indicate the difficulty, or perhaps the impossibility, of forming molecular level mixtures of polyethylenes with other polymers, or with other polyethylenes, by conventional techniques which operate on polymers in which some local order has already been established during polymerization. In a sense, then, "polyethylene is not compatible with polyethylene," as can be seen in the persistence of separate DSC melting patterns after intensive melt mixing of relatively branched and unbranched versions of this polymer.

It is assumed in what follows that a satisfactory dispersion can be obtained despite the problems that may be encountered in special cases, described above. We consider the various procedures that may retard or eliminate the demixing of such dispersions in the following section.

TABLE 12-5

Procedures to Retard or Eliminate Demixing

1. Use of miscible components (i.e., $\Delta G_m = \Delta H_m - T\Delta S_m \leq 0$) Low-molecular-weight polymers (b) Specific interactions to produce negative ΔH_m (c) Generally match solubility parameters	2. Rely on slow diffusion rates (a) Mix high-molecular-weight polymers (b) Cocrystallization
3. Prevent segregation (a) Cross-linking (b) Forming interpenetrating networks (c) Mechanical interlocking of components	4. Use "compatibilizing agents" (a) Statistical copolymers (b) Graft copolymers (c) Block copolymers

12.5.3   Procedures to Retard or Eliminate Demixing of Polymer Mixtures

The various procedures that will be discussed are listed in Table 12-5 in order to present an overview of the basic ideas. Each heading in this table is considered briefly in this section.

(i)   Use of Miscible Components

Thermodynamically stable mixtures will of course form stable blends. This implies miscibility on a molecular level. It is desirable for some applications but not for others, like rubber modification of glassy polymers.

1.   A particular polymer mixture can be made more miscible by reducing the molecular weights of the components. From Eq. (12-1) any measure that increases the entropy of mixing ΔS_m will favor a more negative ΔG_m. The Flory–Huggins theory shows that the entropy gain on mixing a polymer is inversely related to its number average size. This is observed in practice. Low-molecular-weight polystyrenes and poly(methyl methacrylate) polymers are miscible but the same species with molecular weights around those of commercial molding grades ($\sim$100,000) are not.

Advantage can be taken of this enhanced stability of blends of low molecular polymers by chain-extending or cross-linking the macromolecules in such mixtures after they have been formed or applied to a substrate. This procedure is the basis of many formulations in the coatings industry.

2.   Many synthetic polymers are essentially nonpolar and do not participate in specific interactions like acid-base reactions, hydrogen-bonding, or dipole-dipole interactions. In that case, intermolecular interactions are of the van der Waals type

and ΔH_m in Eq. (12-1) is positive. The only contribution to a negative ΔG_m then comes from the small ΔS_m term.

If specific interactions do occur between the components of a polymer blend, the mixing process will be exothermic (ΔH_m negative) and miscibility can be realized. Water-soluble polymers are often miscible with each other, for example, because they participate in hydrogen bonding.

3.   The most widely used method for predicting mixture stability relies on the selection of ingredients with matching solubility parameters and hydrogen-bonding tendencies, as outlined earlier. A small or negative Flory–Huggins interaction parameter value is also characteristic of a stable mixture. Predictions of blend stability can be made quickly from tabulations of solubility and Flory–Huggins parameters. Although such calculations are very useful, they cannot be expected to be universally accurate because the solubility parameter model does not take account of polymer molecular weight and the Flory–Huggins parameter may be concentration dependent.

(ii)   Reliance on Slow Diffusion Rates

High-Molecular-Weight Polymers.   A given blend of two or more polymers can be made more stable by decreasing the molecular weights of the components to the level of oligomers, as mentioned above in connection with polymer miscibility. When a particular blend is not sufficiently stable, it can also paradoxically be improved in this regard by increasing the molecular weights of the ingredients.

Since demixing is a diffusional process, it can be reduced to an acceptable level by using higher molecular weight, more viscous polymers. The difficulty of dispersing such materials to a fine level of dispersion is correspondingly increased, of course, but if this can be achieved the rate of segregation will also be retarded.

Cocrystallization.   An additional factor that is operable in some cases involves the ability of the ingredients of a mixture to cocrystallize. These components cannot then demix since portions of each are anchored in the ordered regions in which they both participate. This may be particularly useful for hydrocarbon polymers where favorable enthalpies of mixing do not exist.

Copolymers of ethylene, propylene, and unconjugated diene (EPDM) polymers vary in their usefulness as blending agents for polyethylene. It has been shown that EPDMs with relatively high levels of ethylene can cocrystallize with branched polyethylene or high-ethylene- content copolymers of ethylene with vinyl acetate or methyl methacrylate [25]. Such blends are stable and may have particularly good mechanical properties. Ethylene/propylene copolymers can serve as compatibilizing agents for blends of polypropylene and low density polyethylene.

Those copolymers which have residual crystallinity because of longer ethylene sequences are preferable to purely amorphous materials for this application [26]. Cocrystallization is probably also a factor in this case.

(iii)   *Prevention of Segregation*

Once a satisfactory initial dispersion has been produced various operations can be conducted to reduce or eliminate the rate of demixing. These are considered separately below.

Cross-Linking.   A thermoset system is produced when a polymer is cross-linked under static conditions, as in a compression mold. This is the basis of the production of vulcanized articles or cross-linked polyethylene pipe and wire insulation. If the same polymer is lightly cross-linked while it is being sheared in the molten state, however, it will remain thermoplastic. If it is more heavily cross-linked during this process, the final product may contain significant quantities of gel particles, but the whole mass will still be tractable.

This technique provides a method for incorporating fillers or reinforcing agents into some polymers which ordinarily do not tolerate such additions. A high loading of carbon black cannot normally be put into polyethylene, for example, without serious deterioration of the mechanical properties of that polymer. Various hydrocarbon elastomers will accept high carbon black contents. Such black-loaded rubbers do not normally form stable mixtures with polyethylene, but strong, permanent blends can be made by carrying out simultaneous blending and crosslinking operations in an internal mixer. If conductive carbon black is mixed carefully with a peroxide or other free radical source, rubber, and the polyolefin, this technique can yield semiconductive compositions in hydrocarbon matrices. When the peroxide decomposes it produces radicals that can abstract atoms from the main chains of polymers. When the resulting macroradicals combine, the parent polymers are linked by primary valence bonds.

The potential exists for chemical bonding of the two polymeric species in such operations but it is not certain that this is always what happens. It is possible in some instances that the stability of the mixture derives mainly from the entanglement of one polymer in a loose, cross-linked network of the other.

The "dynamic cross-linking" process is used to produce thermoplastic elastomers from mixtures of crystallizable polyolefins and various rubbers. Variations of basically the same method are employed to produce novel, stable polymer alloys by performing chemical reactions during extrusion of such mixtures. In that case, the current industrial term is *reactive extrusion*. Such processes are used, for example, to improve processability of LLDPE's into tubular film (by introducing long chain branches during extrusion with low levels of peroxides) or to modify the molecular weight distribution of polypropylenes (again by extrusion with radical-generating peroxides).

Interpenetrating Networks.   Interpenetrating networks (IPNs) and related materials are formed by swelling a cross-linked polymer with a monomer and polymerizing and cross-linking the latter to produce interlocked networks. In semi-interpenetrating systems, only the first polymer is cross-linked. Most of these materials reveal phase separation but the phases vary in size, shape, and sharpness of boundaries depending on the basic miscibility of the component polymers, the cross-link density in the two polymers, and the polymerization method. Some affinity of the components is needed for ordinary interpenetrating networks because they must form solutions or swollen networks during synthesis. This may not be required for IPNs based on latex polymers, where the second stage monomer is often soluble in the first, cross-linked latex polymer.

Blends of elastomers are routinely used to improve processability of unvulcanized rubbers and mechanical properties of vulcanizates like automobile tires. Thus, *cis*-1,4-polybutdiene improves the wear resistance of natural rubber or SBR tire treads. Such blends consist of micron-sized domains. Blending is facilitated if the elastomers have similar solubility parameters and viscosities. If the vulcanizing formulation cures all components at about the same rate the cross-linked networks will be interpenetrated. Many phenolic-based adhesives are blends with other polymers. The phenolic resins grow in molecular weight and cross-link, and may react with the other polymers if these have the appropriate functionalities. As a result, the cured adhesive is likely to contain interpenetrating networks.

Mechanical Interlocking of Components.   In some instances the polymers in a blend may be prevented from demixing because of numerous mutual entanglements produced by mechanical processing or the polymerization history of the blend.

If the melt viscosities of polypropylene and poly(ethylene terephthalate) polymers are reasonably matched under extrusion conditions, a finely dispersed blend may be produced in fiber form. Orientation of such fibers yields strong filaments in which microfibrils of the two partially crystallized polymers are intertwined and unable to separate. Similar fibers with a sheath of one polymer surrounding a core of the other have no mechanical integrity [27].

Enhanced hydrophilicity or dyeability can be conferred on some acrylonitrile-based polymers by polymerizing them in aqueous media containing polyacrylamide. In this case, also, two separate phases exist but the zones of each component are too highly interpenetrating to permit macroseparation and loss of mechanical strength.

Thermoplastic polyolefins (TPOs) are based on blends of polypropylene with ethylene-propylene rubbers. Many perform well as hose, exterior automotive trim and bumpers without chemical linking of the main polymeric components.

(iv)   Use of "Compatibilizing Agents"

Mixtures of immiscible polymers can be made more stable by addition of another material that adheres strongly to the original components of the blend.

Plasticizers perform this function if a single plasticizer solvates the dissimilar major components of a blend. Phthalate esters help to stabilize mixtures of poly(vinyl chloride) and poly(methyl methacrylate) for example. These materials are also plasticizers for polystyrene, and stable blends of this polymer with poly(vinyl chloride) can be made by adding dioctyl phthalate to a blend of polystyrene and rigid PVC.

The most generally useful compatibilizing agents are copolymers in which each different monomer or segment adheres better to one or other of the blend ingredients. Applications of copolymers are classified here according to structure as statistical, block, or graft copolymers. This seems to be as useful a framework as any within which to organize the review, but it has no fundamental bearing on the properties of blends, and different copolymer types may very well be used in similar applications in polymer mixtures.

It is interesting that a mixture of poly-A and poly-B can sometimes be stabilized by addition of a copolymer of C and D, where A, B, C, and D are different monomers. This occurs if the intermolecular repulsion of C and D units is strong enough that each of these monomer residues is more compatible with one or other of the homopolymer ingredients than with the comonomer to which it is linked chemically.

Statistical Copolymers.   The term *statistical* is used here to refer to copolymers in which the sequence distribution of comonomers can be inferred statistically from the simple copolymer model (Chapter 7) or alternative theory. In the present context "statistical copolymers" excludes block and graft structures and incorporates all other copolymers. It is useful first of all in this section to point out that statistical copolymers are not mutually miscible if the mixture involves abrupt changes in copolymer composition. Coatings chemists observe this phase separation as haze (internal reflections) in films.

Note also that although a conventional high conversion vinyl copolymer may exhibit a wide range of compositions (depending on the reactivity ratios of the comonomers and the monomer feed composition), there are generally so many mutually miscible intermediate compositions that the extremes can be expected to blend well with the rest of the mixture.

Use of statistical copolymers in blends is usually predicated on the existence of a specific interaction between one of the comonomers in the copolymer and other ingredients in the mixture. Thus PVC is miscible with the ethylene/ethyl acrylate/carbon monoxide copolymers [28]. The homogenizing effect here is a weak acid–base interaction between the carbonyl of the copolymer and the weakly acidic hydrogen atoms attached to the chlorine carrying carbons of the PVC. Ethylene/vinyl acetate/carbon monoxide copolymers are more miscible with PVC, and ethylene/vinyl acetate/sulfur dioxide copolymers are miscible with the same polymer over a very wide composition range.

The morphology and stability of mixtures of PVC with copolymers depends on the composition and mixing history of the blend as well as on the nature of the copolymer. Ethylene/vinyl acetate copolymer is reported to behave essentially as a lubricant between PVC particles at low copolymer concentrations and to begin to form single-phase compositions with PVC with increasing copolymer content in the blend. This situation changes with increasing vinyl acetate content in the copolymer and increasing mixing temperatures, both of which increase the solubility of the copolymer in PVC.

In the coextrusion and lamination of polymers the individual layers are sometimes inherently nonadhering. An expedient to improve the strength of such multilayer structures involves the use of intermediate "glue" layers between surfaces that do not adhere well. Copolymers are often useful in such glue layers, particularly when the copolymer contains a comonomer that adheres well to one of the surfaces and a comonomer that interacts or is miscible with the other polymer to be bonded. Acid-containing copolymers are often prepared for this purpose. *Ionomers* consisting of partially neutralized ethylene/methyl methacrylate copolymers have been employed to bond polyethylene with nylons and poly(butylene terephthalate). In this case the acid component of the copolymer is capable of hydrogen-bonding interactions with the nylon or polyester. There is also the potential for some interchange between functional groups in the two polymers during melt processing.

Graft Copolymers.   Graft copolymers themselves may exhibit a two-phase morphology and this influences their behavior in blends. The morphological structure that is observed depends on the relative volume fractions of the backbone and graft polymers and their mutual affinity. If separation occurs it will be on a microphase scale because of the chemical linkages between the two polymer types. Amorphous graft copolymers often have good transparency (if there is no crystalline component) because of the small scale of segregation.

The structure of graft copolymers is generally more complex than that of block polymers in that the trunk polymer may be joined to more than one grafted branch and the nature of the production of such copolymers is such that cross-linking also may occur. For this reason the microphase separation that is observed in graft copolymers alone is less distinct and regular than that seen with block copolymers of the same species.

The component of the graft copolymer that is present in the larger concentration will normally form the continuous phase and exert a strong influence on the physical properties of the unblended material. If both phases are present in nearly equal volume fractions, fabrication conditions will determine which component forms the continuous phase.

Graft copolymers decrease the particle size of the dispersed phase in a binary homopolymer mixture and improve the adhesion of the dispersed and continuous phases. The copolymers accumulate at interfaces because parts of the graft are repelled by the unlike component of the blend. They do not necessarily form optically homogeneous mixtures with homopolymers for this reason.

The major application of graft copolymers is in high-impact polystyrene (HIPS), ABS, and other rubber-toughened glassy polymers. The morphology of such blends depends on their synthesis conditions. They are normally made by polymerizing monomers in which the elastomer is dispersed. The elastomer–monomer mixture will tend to form the continuous phase initially but stirring in the early stages of the polymerization of the glassy polymer produces a phase inversion with a resulting dispersion of monomer-swollen rubber in a polymer/monomer continuous phase. When polymerization is completed, the result is a dispersion of rubbery particles in the rigid matrix.

Requirements for rubber toughening of glassy polymers include (1) good adhesion between the elastomer and matrix, (2) cross-linking of the elastomer, and (3) proper size of the rubber inclusions. These topics are reviewed briefly in the order listed:

1. *Rubber-matrix adhesion.* If adhesion between the glassy polymer and elastomer is not good, voids can form at their interfaces and can grow into a crack. The required adhesion is provided by grafting. Affinity between the matrix and rubber is not needed in such cases. Thus, polybutadiene, which has less affinity for polystyrene than styrene–butadiene copolymer, is a better rubbery additive for polystyrene. The butadiene homopolymer has a lower glass transition temperature and remains rubbery at faster crack propagation speeds than the styrene–butadiene copolymer. The inherently poorer adhesion of the polybutadiene and the matrix is masked by the effectiveness of polystyrene–polybutadiene grafts.

2. *Cross-linking of rubber.* A moderate degree of cross-linking in the rubbery phase of the graft copolymer is required to optimize the contribution of the rubbery phase in blends with glassy polymers. Inadequate cross-linking can result in smearing out of the rubbery inclusions during mechanical working of the blend, while excessive cross-linking increases the modulus of the inclusions and reduces their ability to initiate and terminate the growth of crazes.

3. *Particle size.* In general a critical particle size exists for toughening different plastics. The impact strength of the blend decreases markedly if the average particle size is reduced below this critical size. The decrease in impact strength is not as drastic when the particle size increases beyond the optimum value, but larger particles produce poor surfaces on molded and extruded articles and are of no practical use.

Block Copolymers. Block and graft copolymers have generally similar effects of collecting at interfaces and stabilizing dispersions of one homopolymer in another. Most graft copolymers are made at present by free radical methods whereas most commercial block copolymers are synthesized by ionic or stepgrowth processes. As a result, the detailed architecture of block copolymers is more accurately known and controlled.

Many block copolymers segregate into two phases in the solid state if the sequence lengths of the blocks are long enough. Segregation is also influenced by the chemical dissimilarity of the components and the crystallizability of either or both components. This two-phase morphology is generally on a microscale with domain diameters of the order of 10^{-6}–10^{-5} cm.

The critical block sizes needed for domain formation are greater than those needed for phase separation in physical mixtures of the corresponding homopolymers. This is because the conformational entropy of parts of molecules in the block domains is not as high as in mixtures, since placement of segments is restricted by the unlike components to which they are linked. Thus the minimum molecular weights of polystyrene and *cis*-polybutadiene for domain formation in AB block copolymers of these species are about 5,000 and 40,000, respectively [29].

The properties of block copolymers that are most affected by molecular architecture are elastomeric behavior, melt processability, and toughness in the solid state. The effects of such copolymers in polymer blends can obviously also be strongly influenced by the same factors.

When one component of the block polymer is elastomeric, a thermoplastic rubber can be obtained. This occurs only when the block macromolecules include at least two hard (T_g > usage temperature) blocks. A diblock structure pins only one end of the rubbery segment, and true network structures can therefore not be produced in such AB species. The volume fraction of the hard block must be sufficiently high ($\geq 20\%$) to provide an adequate level of thermally labile crosslinking for good recovery properties. If the volume fraction of hard material is too high, however, the rigid domains may change from spherical, particular regions to an extended form in which elastic recovery is restricted.

A block copolymer is expected to be superior to a graft copolymer in stabilizing dispersions of one polymer in another because there will be fewer conformational restraints to the penetration of each segment type into the homopolymer with which it is compatible. Similarly, diblock copolymers might be more effective than triblock copolymers, for the same reason, although tri- and multiblock copolymers may confer other advantages on the blend because of the different mechanical properties of these copolymers.

Block copolymers serve as blending agents with simple homopolymers as well as stabilizing agents for mixtures of homopolymers. Blends of a homopolymer with an AB-type block copolymer will be weak if the elastomeric segment of the block polymer forms the sole continuous phase or one of the continuous phases.

This problem can be circumvented by cross-linking the rubber after the blend is made or by using an ABA block copolymer in which the central segment (B) is rubbery and the terminal, glassy (A) segments serve to pin both ends of the center portions.

When block copolymers are used in rubber mixes there is no particular advantage to a triblock or multiblock species because the final mixture will be vulcanized in any event.

Linear ABA and $(AB)_n$ block copolymers can form physical networks that persist at temperatures above the glassy regions of the hard segments. Very high melt elasticities and viscosities are therefore sometimes encountered. The accompanying processing problems can often be alleviated by blending with small proportions of appropriate homopolymers. For example, when styrene–butadiene–styrene triblock rubbers are used as thermoplastic elastomers it is common practice to extend the rubbery phase with paraffinic or naphthenic oils to decrease the cost and viscosity of the compound. (Aromatic oils are to be avoided as they will lower T_g of the polystyrene zones.) The accompanying decrease in modulus is offset by addition of polystyrene homopolymer which also reduces elasticity during processing.

While copolymers are generally used in blends to modify the properties of homopolymers or mixtures of homopolymers, the reverse situation also occurs. This is illustrated by the foregoing example and also by mixtures of poly(phenylene oxide) (**1-14**) polymers and styrene–butadiene–styrene triblock thermoplastic elastomers. Minor proportions of the block copolymers can be usefully added to the phenylene oxide polymer to improve the impact strength and processability of the latter. This is analogus to the use of polystyrene or HIPS in such applications. It is interesting also that the incorporation of poly(phenylene oxide) elevates the usage temperature of the thermoplastic elastomer by raising the softening point of the hard zones [30].

A number of studies have been conducted to determine the conditions for production of transparent films when an AB block copolymer is mixed with a hompolymer that is chemically similar to one of the blocks and the blend is cast from a common solvent. All agree that the homopolymer is solubilized into the corresponding domain of the block copolymer when the molecular weight of the homopolymer does not exceed that of the same segment in the block polymer. If the molecular weight of the homopolymer is greater than the molecular weights of the appropriate segments in the block polymer, the system will separate into two phases. When high-molecular-weight polystyrene is added to a styrene–butadiene block copolymer with styrene blocks that are shorter than those of the homopolymer, separate loss modulus transitions can be detected for the polystyrene homopolymer zones and the polystyrene domains in the block copolymer [31].

The behavior observed depends also on the morphology of the block polymer. Thus when the block polymer texture consists of inclusions of poly-B in a continuous matrix of poly-A, addition of homopolymer A will result only in its inclusion in the matrix regardless of the molecular weight of the homopolymer. However, addition of increasing amounts of poly-B can lead to a whole series of morphologies that eventually include separate zones of poly-B.

When a block copolymer is blended with a homopolymer that differs in composition from either block the usual result is a three-phase structure. Miscibility of the various components is not necessarily desirable. Thus styrene–butadiene–styrene block copolymers are recommended for blending with high density polyethylene to produce mixtures that combine the relative high melting behavior of the polyolefin with the good low temperature properties of the elastomeric midsections of the block polymers.

It is claimed that the toughening of polystyrene by styrene–butadiene diblock copolymers is augmented by melt blending the components in the presence of peroxides. The grafting and cross-linking that occur are an instance of dynamic cross-linking processes described earlier. Rubbery triblock styrene–butadiene–styrene copolymers toughen polystyrene without the need for cross-linking, for reasons mentioned above.

The compounding of styrene–butadiene–styrene triblock polymers with graft polymer high impact polystyrene is also interesting. Blends of polystyrene and the thermoplastic rubber show worthwhile impact strength increases only when the elastomer is present at a volume fraction $\gtrsim 25\%$. But when the thermoplastic rubber is added to high-impact polystyrene, which already has about 25 vol % rubber, the result is a product with super high impact strength. Also, the thermoplastic rubber can be used to carry fire retardants into the mixture without loss of impact strength.

The major current applications of block copolymers in blends involve styrene–diene polymers, but other block polymers are also useful. Siloxane–alkylene ether block copolymers are widely used as surfactants in the manufacture of polyurethane foams, for example.

12.6   REINFORCED ELASTOMERS

The service performance of rubber products can be improved by the addition of fine particle size carbon blacks or silicas. The most important effects are improvements in wear resistance of tire treads and in sidewall resistance to tearing and fatigue cracking. This reinforcement varies with the particle size, surface nature, state of agglomeration and amount of the reinforcing agent and the nature of the elastomer. Carbon blacks normally are effective only with hydrocarbon rubbers. It seems likely that the reinforcement phenomenon relies on the physical adsorption of polymer chains on the solid surface and the ability of the elastomer molecules to slip over the filler surface without actual desorption or creation of voids.

12.7 REINFORCED PLASTICS

Particulate fillers are used in thermosets and thermoplastics to enhance rigidity and, mainly, to reduce costs. Examples are calcium carbonate in poly(vinyl chloride) and clays in rubber compounds. Fiber reinforcement is more important technically, however, and the main elements of this technology are reviewed briefly here. Fibers are added to plastics materials to increase rigidity, strength and usage temperatures. Fiber reinforced plastics are attractive construction materials because they are stiff, strong and light. The specific stiffness (modulus/density) and specific strength (tensile strength/density) of glass reinforced epoxy polymers approximate those of aluminum, for example.

Many reinforced thermoplastic articles are fabricated by injection molding. (This is a process in which the polymeric material is softened in a heated cylinder and then injected into a cool mold where the plastiuc hardens into the shape of the mold. The final part is ejected by opening the mold.) Thermosetting resins which are frequently reinforced are epoxies (p. 11) and unsaturated polyesters, of which more is said below. Glass fibers are the most widely used reinforcing agents, although other fibrous materials, like aromatic polyamides (1-23) confer advantages in special applications.

The improved mechanical properties of reinforced plastics require that the fiber length exceed a certain minimum value. The aspect ratio (length/diameter) of the fibers should be at least about 100 for the full benefits of reinforcement. This is why particulates like carbon black are reinforcements only for hydrocarbon elastomers, but not for plastics generally.

To estimate the degree of reinforcement from parallel continuous fibers assume that the deformations of the fibers and matrix polymer will be identical and equal to that of the specimen when the composite material is stretched in a direction parallel to that of the fibers. The applied stress is shared by the fibers and polymer according to:

$$\sigma_C A_C = \sigma_F A_F + \sigma_P A_P \tag{12-34}$$

where σ is the stress (force/cross-sectional area), A is the area normal to the fiber axis and the subscripts, C, F, and P refer to the composite material, fiber and polymer, respectively. Since σ is given by

$$\sigma = Y\epsilon \tag{12-35}$$

(this is Eq. (11-6)) where Y is the tensile modulus (with the same units as stress) and ϵ is the nominal strain (increase in length/original length), then Eq. (12-34) is equivalent to:

$$Y_C A_C = Y_F A_F + Y_P A_P \tag{12-36}$$

(because $\epsilon_C = \epsilon_F = \epsilon_P$ in this case). The weight fraction of fiber, w_F in the composite is:

$$w_F = \frac{A_F \rho_F}{A_F \rho_F + A_P \rho_P} \tag{12-37}$$

(since the lengths of the specimen, fibers and polymer component are all equal). Here ρ_F and ρ_P are the respective densities of the fiber and polymer. The ratio of the load carried by the polymer to that carried by the fiber is:

$$\frac{\sigma_P A_P}{\sigma_F A_F} = \frac{Y_P A_P}{Y_F A_F} = \frac{Y_P}{Y_F} \cdot \frac{\rho_F}{\rho_P} \left[\frac{1}{w_F} - 1 \right] \tag{12-38}$$

To take a specific example, consider a glass-reinforced polyester laminate, where the chemical reactions involved in polyester technology are sketched in Fig. (12-3). A mixture of saturated and unsaturated acids is mixed with polyhydric alcohols (here shown as a diol) to form an unsaturated polyester. The unsaturated acid (maleic anhydride) provides sites for cross-linkages during subsequent styrene polymerization [shown in reaction (ii)]. Some of the diacid needed to provide sufficient polyester molecular weight (~2000) is a saturated species (isophthalic acid in this example) because the cross-linked polymer would be excessively brittle if the cross-links were too close together. The unsaturated polyester produced in step (i) is mixed with a reactive monomer, usually styrene. Glass reinforcement in the proper form is impregnated with the styrene–polyester mixture and "cured" by free-radical polymerization of the styrene across the unsaturated linkages in the polyester. Boats and car bodies are among the products made by this process.

In glass-polyester products typical values of the parameters mentioned above are $Y_F = 70$ GNm^{-2}, $Y_P = 3.5$ GNm^{-2}, $\rho_F = 2.6 \times 10^3$ kg m^{-3}, $\rho_P = 1.15 \times 10$ kg m^{-3}, and $w_F = 0.6$. Then, from Eq. (12-38), the polymer will bear about 8% of the load taken by the glass fibers.

Eq. (12-38) is equivalent to:

$$Y_C = \frac{Y_F A_F + Y_P A_P}{A_C} = Y_F \phi_F + YP[1 - \phi_F] \tag{12-39}$$

where ϕ_F is the volume fraction of fiber in the composite (since $\phi_F = A_F/A_C$ for continuous fibers). This is the "law of mixtures" rule for composite proiperties. With the cited values, ϕ_F in the present example is 0.40 and the modulus of the composite is about 43% of the fiber modulus.

The fiber alignment is also a significant factor in composite properties. If the fibers in the foregoing example were randomly oriented their reinforcing effect would be less than 0.2 of the figure calculated above.

Discontinuous fibers are used when the manufacturing process prohibits the application of continuous fibers, for example, in injection molding. In composites of discontinuous fibers, stress cannot be transmitted from the matrix polymer to the fibers across the fiber ends. Under load, the polymer is subjected to a shear stress

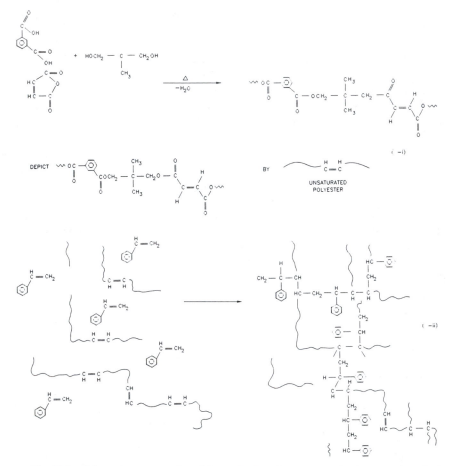

Fig. 12-3. Schematic representation of the production of an unsaturated polyester resin and subsequent cross-linking by polymerizing the styrene in a mixture of this monomer with the polyester.

because the stress along each fiber will be zero at its ends and a maximum, σ_m, at its center. The shear stress at the fiber–polymer interface transmits the applied force between the components of the composite. The shear strength of this interface is typically low and reliance must therefore be placed on having sufficient interfacial area to transmit the load from the polymer to the fiber. This means that the discontinuous fibers must be longer than a certain critical minimum length, l_c, which depends on the interfacial shear stress, τ, fiber diameter and applied load. Experience shows that this minimum length is not difficult to exceed in dough or sheet molding compounds, where unsaturated polyesters are mixed with chopped fiber mats, with fiber lengths about 5–14 mm. These composites are usually compression molded and cured hot in the mold. The process does not damage the fiber to any significant extent. In injection molding, on the other hand, the initial fibers are likely to be shortened by the mechanical action of the compounding process and the shearing action of the reciprocating screw in the injection molder. They are thus less likely to be effective than in sheet molding formulations.

The properties of fiber-polymer composites are influenced by the strength of the bond between the phases, since stresses must be transmitted across their boundaries. Some problems have been encountered in providing strong interfacial bonds because it is difficult to wet hydrophilic glass surfaces with generally hydrophobic viscous polymers. Coupling agents have therefore been developed to bind the matrix and reinforcing fibers together.

These agents often contain silane or chromium groupings for attachment to glass surfaces, along with organic groups that can react chemically with the polymer. Thus vinyltriethoxysilane [$H_2C{=}CHSi(OC_2H_5)_3$] is used for glass-unsaturated polyester systems and γ-aminopropyltriethoxysilane [$H_2NCH_2CH_2\,CH_2Si\,(OC_2H_5)_3$] is a coupling agent for glass-reinforced epoxies and nylons. The silanes which seem to couple effectively to glass are those in which some groups can be hydrolyzed to silanols. Si–O bonds are presumably formed across the interface between the glass and coupling agent.

Coupling agents for more inert polymers like polyolefins are often acid-modified versions of the matrix polymer, with maleic acid grafted polypropylene as a prime example.

PROBLEMS

12-1. Toluene (molecular weight $= 92$, density $= 0.87$ g/cm^3) boils at $110.6°$C at 1 atm pressure. Calculate its solubility parameter at $25°$C. [The enthalpy of vaporization can be approximated from the normal boiling point T_b (K) of a solvent from $\Delta H_{(25°C)} = 23.7T_b + 0.020T_b^2 - 2950$ cal/mol (J. Hildebrand and R. Scott, "The Solubility of Nonelectrolytes," 3rd ed. Van Nostrand Reinhold, New York, 1949).]

12-2. Calculate the solubility parameter for a methyl methacrylate–butadiene copolymer containing 25 mol % methyl methacrylate.

12-3. Calculate the solubility parameter for poly(vinyl butyl ether). Take the polymer density as 1.0 g/cm^3.

12-4. (a)   A vinyl acetate/ethylene copolymer is reported to be soluble only in poorly hydrogen bonded solvents with solubility parameters between 8.5 and 9.5 (cal/cm^3)$^{1/2}$. A manufacturer wishes to make solutions of this copolymer in Varsol No. 2 (a nonaromatic hydrocarbon distillate, $\delta = 7.6$, poorly hydrogen bonded). Suggest another relatively low cost solvent that could be added to the Varsol to increase its solvent power for the copolymer and calculate the composition of this mixed solvent.

(b)   Would you expect this copolymer to form stable mixtures with poly-ethylene? Why or why not?

12-5.   Calculate the composition by volume of a blend of n-hexane, t-butanol, and dioctyl phthalate that would have the same solvent properties as tetrahy-drofuran. (Use Table 12-4 and match δ_P and δ_H values.)

12-6.   The introduction of a minor proportion of an immiscible second polymer reduces the viscosity and elastic character of a polymer melt at the pro-cessing rates used in normal commercial fabrication operations [19]. It is necessary, also, that there be good adhesion between the dissimilar zones in the solid blend to obtain finished articles with good mechanical strength. Suggest polymeric additives that could be used in this connection to modify the processing behavior of

(a)   polyethylene melts.

(b)   styrene–butadiene rubber (SBR).

12-7.   *Note:* (This problem is for **illustrative purposes only**. Methyl isobutyl ketone fumes have been reported to be **hazardous**.) A common solvent mixture for commercial nitrocellulose consists of the following:

Diluent (toluene)	50 parts by volume
"Latent" solvent (1-butanol)	13 parts by volume
"Active" solvent (mixture)	37 parts by volume

The "active" solvent mixture includes:

Methyl ethyl ketone (fast evaporation rate)	32%
Methyl isobutyl ketone (medium evaporation rate)	54%
Diethylene glycol monomethyl ether (slow evaporation rate)	14

From these data estimate whether tetrahydrofuran would be a solvent for this nitrocellulose polymer. Use Table 12-3.

12-8.   A composite consists of 45% by volume of continuous, aligned carbon fibers and an epoxy resin. The tensile strength and modulus of the fibers is 3000 Mpa and 200 GPa, respectively, while the corresponding parameters of the cured epoxy are 70 MPa and 2.5 GPa, respectively. Determine: (a) which component of the composite will fail first when the material is deformed in the fiber direction, and (b) the failure stress of the composite.

REFERENCES

[1]   G. Scatchard, *Chem. Rev.* **8**, 321 (1931).
[2]   J. H. Hildebrand, J. M. Prausnitz, and R. L. Scott, "Regular and Related Solutions." Van Nostrand Reinhold, New York, 1970.

[3]   P. A. Small, *J. Appl. Chem.* **3**, 71 (1953).

[4]   K. L. Hoy, *J. Paint Technol.* **42** (541), 76 (1970).

[5]   S. Krause, *J. Macromol. Sci. Macromol. Rev.* **C7**, 251 (1972).

[6]   E. A. Grulke, *in* "Polymer Handbook" (J. Brandrup and E. Immergut, Eds.), 3rd ed., p. VII/519. Wiley, New York, 1989.

[7]   A. Beerbrower, L. A. Kaye, and D. A. Pattison, *Chem. Eng.*, p. 118 (December 18, 1967).

[8]   P. J. Flory, *J. Chem. Phys.* **9**, 660 (1941); **10**, 51 (1942).

[9]   M. L. Huggins, *J. Chem. Phys.* **9**, 440 (1941); *Ann. N.Y. Acad. Sci.* **43**, 1 (1942).

[10] C. M. Hansen, *J. Paint Technol.* **39**, 104, 511 (1967).

[11] C. M. Hansen, *Ind. Eng. Chem. Prod. Res. Dev.* **8**, 2 (1969).

[12] P. Choi, T. A. Kavassalis and A. Rudin, *IEC Res.* **33**, 3154 (1994).

[13] P. Choi, T. A. Kavassalis and A. Rudin, *J. Coll. Interf. Sci.* **180**, 1 (1996).

[14] J. P. Teas, *J. Paint Technol.* **40**, 519 (1968).

[15] P. J. Flory, "Principles of Polymer Chemistry." Cornell University Press, Ithaca, NY, 1953.

[16] L. H. Tung, Ed., "Fractionation of Synthetic Polymers." Dekker, New York, 1977.

[17] E. Karbashewski, L. Kale, A. Rudin, H. P. Schreiber, and W. J. Tchir, *Polym. Eng. Sci.* **31**, 1581 (1991).

[18] M. G. Pigeon and S. Rudin, *J. Appl. Polym. Sci.* **51**, 303 (1994).

[19] A. Rudin, *J. Macromol. Sci. Rev.* **C19**, 267 (1980).

[20] J. T. Bergen, *in* "Rheology" (F. R. Eirich, Ed.), Vol. 4. Academic Press, New York, 1967.

[21] W. D. Mohr, *in* "Processing of Thermoplastic Materials" (E. C. Bernhardt, Ed.). Van Nostrand Reinhold, New York, 1959.

[22] Z. Tadmor and C. G. Gogos, "Principles of Polymer Processing." Wiley (Interscience), New York, 1979.

[23] P. Choi, H. P. Blom, T. A. Kavassalis and A. Rudin, *Macromolecules* **28**, 8247 (1995).

[24] G. H. Fredrickson, A., J. Liu, and F. S. Bates, *Macromolecules* **27**, 2503 (1994).

[25] H. W. Starkweather, Jr., *J. Appl. Polym. Sci.* **20**, 364 (1980).

[26] E. Nolley, J. W. Barlow, and D. R. Paul, *Polym. Eng. Sci.* **25**, 139 (1980).

[27] A. Rudin, D. A. Loucks, and J. M. Goldwasser, *Polym. Eng. Sci.* **74**, 741 (1980).

[28] C. M. Robeson and J. E. McGrath, *Polym. Eng. Sci.* **17**, 300 (1977).

[29] M. Morton, *Am. Chem. Soc. Polym. Div. Preprints* **10** (2), 512 (1969).

[30] A. R. Shultz and B. M. Beach, *J. Appl. Polym. Sci.* **21**, 2305 (1977).

[31] G. Kraus and K. W. Rollman, *J. Polym. Sci. Phys. Ed.* **14**, 1133 (1976).

Appendix A

Conversion of Units

To convert from	To	Multiply by
newtons/m^2 (N/m^2)	pascals (PA)	1
N/m^2	dyn/cm^2	10.00
N/m^2	pounds per square inch (psi)	1.450×10^{-4}
dyn/cm^2	kg/mm^2	1.02×10^{-6}
dynes	newtons (N)	10^{-5}
psi	kg/mm^2	7.03×10^{-4}
psi	atmospheres (atm)	6.81×10^{-2}
psi	N m^{-2} or Pascal (Pa)	6.897×10^{3}
dynes	newtons	10^{-5}
calories	joules	4.187
ergs	joules	10^{-7}
kilopond, k$_p$	newtons	9.807
g/denier	dyn/cm^2	$8.83 \times 10^{8} \rho^{a}$
g/denier	psi	$1.28 \times 10^{4} \rho^{a}$
poise	N sec/m^2 = Pa sec	10^{-1}
stokes	m^2/sec	10^{-4}

$R = 8.3143\,\text{J}\,\text{K}^{-1}\,\text{mol}^{-1} = 1.9872\,\text{cal}\,\text{K}^{-1}\,\text{mol}^{-1} = 0.08206\,\text{liter atm}\,\text{K}^{-1}\text{mol}^{-1}$
$= 8.3144 \times 10^{7}\,\text{ergs}\,\text{mol}^{-1}\,\text{K}^{-1}$

[a] ρ = density

Appendix B

List of Symbols

A	anionic portion of cationic initiator (Chapter 9)
A	monomer residues in graft and block polymers
A	arithmetic mean (Section 2.3)
A	area
A_w	heat transfer area (Chapter 10)
A	preexponential factor in Arrhenius equations (6-116) and (9-53).
A	Helmholtz free energy (Eq. 4-26)
A	symbol for active species in chain-growth polymerizations (Chapter 6)
A	quantity defined in Eq. (6-110)
A_i	virial coefficients (Eq. 2-76)
B	virial coefficient (Eq. 2-78)
B	bulk compliance (Chapter 11)
B	specimen depth (Section 11.9)
B	monomer residues in block and graft copolymers
B	symbol for nucleophilic portion of anionic initiator (Eq. 9-2)
B_i	general reactivity factor for macroradical ending in monomer M_i (Section 7.11)
C	virial coefficient (Eq. 2-78)
C_1, C_2	Mooney–Rivlin constants (Eq. 4-49)
C_1, C_2	WLF equation constants (Eq. 11-42)
$[C^*]$	concentration of active sites in Ziegler–Natta polymerization (Eq. 9-64)
D	outside diameter of pipe (Eq. 11-54)

D	contour length of polymer molecule (Chapter 4)
D	tensile creep compliance (Chapter 11)
D	specimen width (Section 11.9)
DP	degree of polymerization (Section 1.2)
$(\overline{DP}_n)_0$	number average degree of polymerization of vinyl polymers made in absence of solvent or chain transfer agents (Chapter 6)
E	energy (Section 4.3)
E	Arrhenius activation energy
E	contact energy of a mixture of molecules (Chapter 12)
Et	CH_3CH_2-
F	fraction of vinyl polymer formed by chain disproportionation and/or chain transfer (Chapter 6)
F	quantity (Eq. 2-15)
F	force (Section 3.3)
F	fraction of polymer formed by chain transfer or disproportionation (Eq. 6-109)
F_i	mole fraction of comonomer i in copolymer (Chapter 7)
$F(t)$	cumulative probability function of residence times in a continuous reactor (Chapter 10)
G	Gibbs free energy
G_1^0	molar Gibbs free energy of solvent (Section 2.10.1)
G	function of comonomer concentrations (Section 7.6)
G	shear modulus (Chapter 11)
$G(t)$	shear relaxation modulus
G'	storage modulus
G''	loss modulus
G^*	complex modulus
G_c	critical strain energy release rate (Section 11.9)
G_s	surface free energy (Chapter 8)
H	enthalpy
H	optical constant in light scattering (Eq. 3-25)
H	function of comonomer concentrations (Section 7.6)
H_p	enthalpy of polymerization (Chapter 10)
I	light intensity
I	moment of inertia (Section 4.4.1)
I	symbol for polymerization initiator (Chapters 6 and 9)
I_0	intensity of incident light (Eq. 3-10)
I_θ'	intensity of light scattered at angle θ to incident beam (Eq. 3-10)
J	product of intrinsic viscosity and molecular weight (Eq. 3-73)
J	shear compliance (Chapter 11)
J'	storage compliance
J''	loss compliance

J^* complex compliance

$J(t)$ shear creep compliance

K   degrees Kelvin ($=°C + 273$)

K constant in Mark–Houwink–Sakurada equation (Eq. 3-43)

K optical constant in light scattering measurements (Eq. 3-19)

K' proportionality constant (Eq. 2-46)

K'' proportionality constant (Eq. 2-50)

K equilibrium constant

K bulk modulus (Chapter 11)

K stress intensity factor (Section 11.9)

K_c fracture toughness (Section 11.9)

K_{1c} fracture toughness in plane strain (Section 11.9)

L Avogadro's constant (6.023×10^{23} per mole)

M molecular weight

M concentration (moles/liter)

M   symbol for a monomer unit (Chapter 6)

Me CH_3-

M_0 formula weight for repeating unit in a polymer

$\overline{M}_n$ number average molecular weight

$\overline{M}_v$ viscosity average molecular weight

$\overline{M}_w$ weight average molecular weight

$\overline{M}_z$ z average molecular weight

$\overline{M}_{z+i}$ $z + i$ average molecular weight

N number of polymer molecules (or moles) per unit volume

N number of light scatterers per volume V (Section 3.2)

N newton (SI unit)

N number of latex particles per unit volume of aqueous phase in emulsion polymerization (Chapter 8)

N_{Deb} Deborah number (dimensionless) (Chapter 11)

N_i number average length of sequences of monomer i in a copolymer (Chapter 8)

P pressure

Pa   Pascal, SI unit of pressure or stress ($= N/m^2$)

P_1^0 vapor pressure of pure solvent at given temperature

P' magnitude of a colligative property of a polymer solution (Section 2.9)

P'' intensity of light scattered from a polymer solution (Section 2.9)

P_i general reactivity factor for radical ending in monomer M_i (Section 7.11)

P_y probability of occurrence of a molecule containing y monomers (Chapter 7)

P_{ij} probability that monomer j follows monomer i in a copolymer
        (Chapter 7)
Q   volumetric flow rate (Chapter 11)
Q volume of fluid flow in time t (Eq. 3-60)
Q symbol for inhibitor or retarder molecule (Chapter 6)
Q_i general reactivity factor for monomer M_i (Section 7.11)
R   alkyl group
R universal gas constant
R symbol for rate of a reaction (Chapter 6)
R function of reactivity ratios (Eq. 7-47c)
R ratio of volumes of solvent-rich and polymer-rich phases in
        fractionation (Chapter 12)
R_i rate of initiation (Chapter 6)
$R_{M\cdot}$ rate of formation of monomer-ended radicals (Chapter 6)
R_p rate of propagation (Chapter 6)
R_t rate of termination (Chapter 6)
R_{tc} rate of termination by combination (Chapter 6)
R_{td} rate of termination by disproportionation (Chapter 6)
R_θ reduced scattering intensity (Eq. 3-18)
R_{90} Rayleigh ratio (Eq. 3-21)
$R(t)$ decay function of residence times in a continuous reactor
        (Chapter 10)
S entropy
S function of reactivity ratios (Eq. 7-47a)
T absolute temperature
T_b boiling point
ΔT_b boiling point difference (Eq. 2-70)
T_c ceiling temperature (Chapter 6)
T_f freezing point
T_g glass transition temperature
T_m melting temperature
TH   symbol for chain transfer agent (Chapter 6)
U moment (Section 2.4.1)
U internal energy function (Table 4-1)
U heat transfer coefficient (Chapter 10)
U_f fracture energy (Section 11.9)
V volume
V equivalent volume of a spherical, solvated polymer molecule (Eq. 3-35)
V_r reference volume (Chapter 12)
V_1^0 molar volume of solvent
V_0 Kinetic parameter defined by Eq. (6-55)
$W(M)$ cumulative weight fraction (Section 2.3.2)

X	degree of polymerization
X	symbol for reducing agent (Chapter 6)
$X(M)$	cumulative mole fraction (Section 2.3.1)
Y	Young's (tensile) modulus
Z	function of reactivity ratios (Eq. 7-47)
Z	symbol for counterion in anionic polymerizations (Eq. 9-2)
Z	metal component of a Lewis acid (Eq. 9-34)
a	weight of polymer in end group measurements (Eq. 3-9)
a	number of ways an event can happen (Chapter 5)
a	van der Waals radius (Chapter 9)
a	crack depth (Eq. 11-49)
a_T	shift factor [Section 11.5.2(iii)]
a	(superscript) Mark–Houwink–Sakurada constant (Eq. 3-43)
b	symbol for block in polymer nomenclature (Section 1.10)
b	arbitrary scale factor in plots of light scattering data (Section 3.2.3)
b	number of ways an event can fail to happen (Chapter 5)
c	weight concentration
c	number of intermolecular contacts (Chapter 12)
c_2	concentration of polymer (weight/volume of solution)
d	pipe diameter (Problem 11-12)
d	distance
d	end-to-end distance of a polymer chain (Section 4.4.1)
dq	energy added to a system as heat (Table 4-1)
dw	work (Table 4-1)
e	base of natural logarithms
e	polarity factor (Section 7.11)
e	equivalent weight of reagent used in end group determinations (Eq. 3-9)
e	(subscript) equivalent random polymer chain [Section 4.4.2(iii)]
f	functionality (Chapter 5)
f	initiator efficiency (Chapter 6)
f	number of monomer units in j turns of the polymer helix in a crystallite (Chapter 4)
f	tensile force (Section 4.5)
f	mole fraction of comonomer i in monomer feed (Chapter 7)
f_i	proportion of total sample with molecular weight M_i (Section 3.3)
f_e	internal energy contribution to tensile force [Section 4.5.2(ii)]
$f(t)$	frequency of residence times in a continuous reactor (Chapter 10)
g	symbol for graft in polymer nomenclature (Section 1.10)
g	gravitational acceleration constant
h	height of column of liquid
h	Planck's constant

h	energy dissipated/unit volume/time in shear deformation (Chapter 11)
i	number of repeating units in a polymer molecule
i	$(-1)^{1/2}$
j	dummy index
j	number of turns of polymer helix which accommodate f monomer units in a polymer crystal (Chapter 4)
k	specific reaction rate constant
k	calibration constant in vapor phase osmometry (Eq. 3-8)
k_s	instrument constant in vapor phase osmometry (Eq. 3-7)
k_H	Huggins constant (Eq. 3-62)
k_I	constant in Kraemer equation (Eq. 3-63)
k_p	propagation rate constant
$k_{tr,M}$	rate constant for chain transfer to monomer
l	length
l	bond length (Section 4.4.2)
m	number of repeating units in polymer molecule
m	mass (Section 4.4.1)
m	meter (0.001 km)
n	number of repeating units in polymer molecule
n	refractive index of solution (Eq. 3-13)
n	number of equivalents (Chapter 5)
n	number of radicals per polymer particle in emulsion polymerization (Chapter 7)
n	power law index (Chapter 11)
n	(subscript) number distribution (Section 3.4.3)
n_j	number of molecules (or moles) per unit volume with molecular weight M_j
n_0	refractive index of suspending medium (Eq. 3-13)
p	amount of reagent used in end group measurements (Eq. 3-9)
p	extent of reaction
q	fraction of all monomers in a sample which form part of cross-links (Chapter 7)
q_E	heat removed by condenser and/or other devices (Chapter 10)
q_i	quantity of polymer in unit volume of sample with molecular weight M_i
r	radius (Eq. 3-60)
r	radial position (units of length) (Section 3.3)
r	number of reactive groups per molecule in end group determinations (Eq. 3-9)
r	distance between light scattering source and viewer (Eq. 3-10)
r	number of segments in a polymer (Chapter 12)
r_e	radius of equivalent sphere (Section 3.3)

r_g	radius of gyration (Section 3.2.6)
r_2	number of segments [Section 12.2(ii)]
s	probability that a monomer-ended radical will grow by monomer addition in a free radical polymerization (Chapter 6)
s_n	standard deviation of number distribution (Eq. 2-29)
s_w	standard deviation of weight distribution (Eq. 2-30)
t	time
t	pipe wall thickness (Eq. 11-54)
t_0, t_1	flow times for solvent and solution, respectively, in capillary viscometry of polymer solutions
$t_{1/2}$	half-life (Chapter 6)
u	dummy symbol for time (Chapter 11)
u	probability function (Chapter 5)
u	constant in equation linking, T_g and $\overline{M}_n$ (Chapter 11)
v	flow rate of liquid (Section 3.3)
v	probability function (Chapter 5)
v	volume of a molecule (Chapter 12)
v_2	specific volume of polymer (Eq. 3-3)
w	weight fraction (Section 3.3.1)
w	interaction energy (Chapter 12)
w	(subscript) weight distribution (Section 3.4.3)
w_c	weight fraction of polymer in crystalline state (Eq. 11-2)
x	number of repeating units in polymer molecule
x	mole fraction (Section 3.3.1)
x	Cartesian coordinate
x	ratio of comonomer concentrations in monomer feed (Chapter 7)
y	Cartesian coordinate
$\overline{y}_w$	weight average degree of polymerization (Chapter 7)
z	Cartesian coordinate
z	ratio of comonomers in copolymer (Chapter 7)
z	charge on ion (Chapter 9)
z	number of nearest neighbors in liquid lattice (Chapter 12)
Γ	(gamma) virial coefficients (Eq. 2-77)
Γ	calibration factor, dimensionless (Section 11.19)
Δ	(delta) difference
ΔE	energy difference between trans and skew conformations (Fig. 4-2)
ΔH	activation energy [Section 11.5.2(iii)]
ΔH_f	latent heat of freezing (Eq. 2-71)
ΔH_m	enthalpy change on melting (Eq. 11-1)
ΔH_v	latent heat of vaporization (Eq. 2-70)
ΔG_m	Gibbs free energy change in melting process

ΔS_m entropy change on melting (Eq. 11-1)

$\Delta\epsilon$ difference between energy minima in *trans* and *gauche* staggered conformations (Fig. 4-2)

$\Delta\Omega$ bridge imbalance in vapor phase osmometer (Eq.3-8)

Δ (superscript) exponent in relation between hydrodynamic volume and polymer molecular weight (Section 3.3)

Λ extension ratio (Section 11.7)

$\prod$ product of

$\sum$ sum of

Φ calibration factor (Eq. 11-53)

Ω (omega) drainage factor (unitless) (Section 3.3)

α (alpha) excess polarizability (Section 3.2)

α expansion coefficient [Section 4.4.2(ii)]

α function of comonomer concentrations (Section 7.6)

α deformation angle in shear

α degree of dissociation of ion pairs in cationic polymerization (Chapter 9)

α_3 measure of skewness of molecular weight distribution (Eq. 2-39)

α_η ratio between intrinsic viscosities (Eq. 3-41)

β (beta) Poisson's ratio (Chapter 11)

γ shear strain (Chapter 11)

γ surface energy (Eq. 11-49)

$\dot{\gamma}$ shear rate = velocity gradient = $d\gamma/dt$

γ_a strain amplitude

$\gamma^{\cdot}$ shear rate (units are time^{-1})

γ interfacial tension (Chapter 8)

δ (delta) loss angle (Chapter 11)

δ solubility parameter (Chapter 12)

ϵ (epsilon) energy (Section 4.3)

ϵ number of additional cross-links in a molecule which already contains one cross-link (Chapter 7)

ϵ nominal tensile strain (Chapter 4)

ζ (zeta) relaxation time, retardation time

η (eta) coefficient of viscosity

η viscosity of polymer solution (Section 3.3)

η function of comonomer concentrations and reactivity ratios (Section 7.6)

η_a apparent viscosity

η_0 viscosity of solvent (Section 3.3)

$[\eta]$ intrinsic viscosity, limiting viscosity number (Eq. 3-37)

$[\eta]_\theta$ intrinsic viscosity under theta conditions (Section 3.3)

η^* complex viscosity

η' dynamic viscosity $=$ "real" part of η^*

η'' "imaginary" part of η^*

η_r relative viscosity (unitless) (Table 3-2)

η_{sp} specific viscosity (unitless) (Table 3-2)

η_{inh} inherent viscosity [units of (concentration)$^{-1}$] (Table 3-2)

θ (theta) angle

θ mean residence time in a continuous reactor (Chapter 10)

θ fraction of active Ziegler–Natta catalyst sites (Eq. 9-65)

θ_M fraction of active sites occupied by monomer in Ziegler–Natta polymerization

κ (kappa) Boltzmann constant $= R/L$

λ (lambda) wavelength of light

λ proportionality constant in relation between polymer molecular weight and hydrodynamic volume (Section 3.3)

λ stretched length of rubber specimen [Section 4.5.2(iii)]

λ_0 unstretched length of rubber specimen [Section 4.5.2(iii)]

μ (mu) chemical potential

μ number of moles of monomer reacted per mole of initiator in ionic polymerizations (Chapter 9)

μ micron $= 10^{-6}$ m

μ consistency (Chapter 11)

μ constant in Eqs. 3-70 and 3-71

ν (nu) kinetic chain length (Chapter 6)

ν frequency (cycles/sec)

ξ (xi) function of comonomer concentration (Section 7.6)

π (pi) osmotic pressure

ρ (rho) density

ρ_a density of amorphous polymer (Eq. 11-2)

ρ_c density of perfectly crystalline polymer (Eq. 11-2)

ρ_0 fraction of reactive groups in a copolymerization which are part of a multifunctional cross-linking agent (Chapter 8)

σ (sigma) number of chain elements in a macromolecule

σ tensile stress

σ function defined in Eq. (12-30)

σ_h hoop stress (Eq. 11-54)

σ_0 stress amplitude

σ_y yield stress

τ (tau) turbidity

τ shear stress [Section 4.5.1(iii)]

τ time (Chapter 6)

ϕ (phi) volume fraction (Section 3.3, Section 12.2.2)

ϕ_c volume fraction of polymer in the crystalline state (Eq. 11-3)

χ (chi) Flory–Huggins interaction parameter (Chapter 12)

ω (omega) frequency (rad/sec)

ω shape factor (Eq. 3-33)

Index

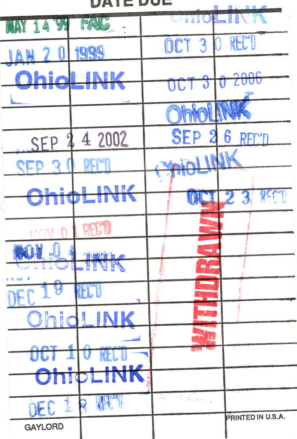